T0145252

Advances in Intelligent Systems and Computing

Volume 877

Series editor

Janusz Kacprzyk, Polish Academy of Sciences, Warsaw, Poland
e-mail: kacprzyk@ibspan.waw.pl

The series "Advances in Intelligent Systems and Computing" contains publications on theory, applications, and design methods of Intelligent Systems and Intelligent Computing. Virtually all disciplines such as engineering, natural sciences, computer and information science, ICT, economics, business, e-commerce, environment, healthcare, life science are covered. The list of topics spans all the areas of modern intelligent systems and computing such as: computational intelligence, soft computing including neural networks, fuzzy systems, evolutionary computing and the fusion of these paradigms, social intelligence, ambient intelligence, computational neuroscience, artificial life, virtual worlds and society, cognitive science and systems, Perception and Vision, DNA and immune based systems, self-organizing and adaptive systems, e-Learning and teaching, human-centered and human-centric computing, recommender systems, intelligent control, robotics and mechatronics including human-machine teaming, knowledge-based paradigms, learning paradigms, machine ethics, intelligent data analysis, knowledge management, intelligent agents, intelligent decision making and support, intelligent network security, trust management, interactive entertainment, Web intelligence and multimedia.

The publications within "Advances in Intelligent Systems and Computing" are primarily proceedings of important conferences, symposia and congresses. They cover significant recent developments in the field, both of a foundational and applicable character. An important characteristic feature of the series is the short publication time and world-wide distribution. This permits a rapid and broad dissemination of research results.

More information about this series at http://www.springer.com/series/11156

Ning Xiong · Zheng Xiao
Zhao Tong · Jiayi Du · Lipo Wang
Maozhen Li
Editors

Advances in Computational Science and Computing

Editors
Ning Xiong
Division of Intelligent Future Technologies
Mälardalen University
Västerås, Sweden

Zheng Xiao
School of Information Science
and Technology
Hunan University
Changsha, Hunan, China

Zhao Tong
School of Computer Science
Hunan Normal University
Changsha, Hunan, China

Jiayi Du
Hunan University of Chinese Medicine
School of Computer Science
Changsha, Hunan, China

Lipo Wang
School of Electrical and Electronic
Engineering
Nanyang Technological University
Singapore, Singapore

Maozhen Li
Design and Physical Sciences Brunel
Electronic and Computer Engineering,
College of Engineering
Uxbridge, Middlesex, UK

ISSN 2194-5357 ISSN 2194-5365 (electronic)
Advances in Intelligent Systems and Computing
ISBN 978-3-030-02115-3 ISBN 978-3-030-02116-0 (eBook)
https://doi.org/10.1007/978-3-030-02116-0

Library of Congress Control Number: 2018957133

This Springer imprint is published by the registered company Springer Nature Switzerland AG
The registered company address is: Gewerbestrasse 11, 6330 Cham, Switzerland

Contents

Crowding or Surround Suppression with a Hybrid Stimulus-Task Combination?

Mingliang Gong[1,2(✉)] and Lynn A. Olzak[2]

[1] School of Psychology, Jiangxi Normal University, Nanchang, China
gongmliang@gmail.com
[2] Department of Psychology, Miami University, Oxford, OH, USA

Abstract. Both crowding and surround suppression result in fuzzy visual patterns, yet they have been treated as different phenomena because they vary in some properties. For example, an inward-outward asymmetry of masking is a hallmark of crowding but does not appear in surround suppression. Studies on surround suppression and crowding usually employ different stimulus patterns and tasks, and it is unclear whether the discrepancies between them derive from the different stimulus patterns, the different tasks, or whether they are just different phenomena. A hybrid of stimulus pattern (large surround annuli, usually used in surround suppression) and task (discrimination task, usually used in crowding studies) was used in a fine orientation discrimination task, with the inward-outward asymmetry as a "critical test" indicator. The results showed that whereas some observers showed no inward-outward asymmetry, others showed such an asymmetry. Intriguingly, this asymmetry was in the direction opposite to that commonly reported. Together, the results cannot be wholly predicted by either surround suppression or crowding, and might reveal a new mechanism because of our particular combination of task and stimulus pattern.

Keywords: Orientation discrimination · Surround suppression
Crowding · Inward-outward asymmetry · Peripheral vision

1 Introduction

Human vision can be divided into central vision and peripheral vision. Visual acuity decreases from fovea towards peripheral vision [1] since the number of neurons devoted to process a visual stimulus of a given size decreases from fovea to periphery in the visual cortex [2]. As a result, patterns in visual periphery are usually fuzzy.

Among the studies of peripheral vision, visual crowding and surround suppression have attracted a lot of attention. Both phenomena reduce performance in psychophysical tasks. Surround suppression is an inhibition of the pattern discrimination when it is surrounded by an annular mask [3, 4]. Visual crowding is the deterioration of object recognition in the presence of other objects. The two phenomena share a lot of common properties such as orientation tuning [4, 5] and spatial frequency tuning [4, 6]. However, they have been treated as different phenomena because they have a few different properties as well as distinct interpretation models.

N. Xiong et al. (Eds.): ISCSC 2018, AISC 877, pp. 1–10, 2019.
https://doi.org/10.1007/978-3-030-02116-0_1

Surround suppression is usually interpreted in the context of an inhibition model: neurons have a specific excitatory region, with activations outside this region giving rise to inhibitory influences on the central region [7]. By contrast, crowding is thought to be caused by the limitation of spatial resolution in periphery [8, 9], and often interpreted by pooling models [10–12], or a probabilistic substitution model [13, 14]. Specifically, pooling models posit that visual systems tend to show compulsory pooling (e.g., averaging) of features from a larger field which includes features from both target and flankers [10, 11]. As a result, the target becomes unrecognizable. However, the results explained by pooling models could also be accommodated by a substitution model, which posits that people sometimes mistakenly report features of the flankers instead of the features of the target. It is likely that both models are playing roles in crowding [10]. Therefore, researchers have started to adopt comprehensive models to interpret visual crowding [15, 16].

In addition to the varying interpretation models, different properties of surround suppression and crowding are also observed. In the seminal work on crowding by Bouma [17], it was found that an outward-facing mask had a more effective inhibition effect than an inward mask. This result was confirmed by other studies [18, 19], suggesting that the inward-outward asymmetry is a robust effect in crowding. However, this asymmetry is not observed in surround suppression [19, 20]. Therefore, the presence of an inward-outward asymmetry, as a unique attribute of crowding, can be regarded as a hallmark of crowding [19, 21]. The examination of inward-outward asymmetry can be taken as a "litmus test" to discriminate between the two phenomena [19].

To sum up, though surround suppression and crowding are very similar, they have their own unique characteristics. The goal of the current study was to explore what factors make them different. Previous studies examining the two phenomena employed different stimulus patterns and different tasks. Crowding experiments usually use small localized nearby "masks" (i.e., flankers) and has been shown to mainly affect discrimination or identification [16], with a negligible influence on detection of target [5, 12]. Thus crowding experiments use discrimination or identification tasks (e.g., "what is the orientation of the central grating?"). By contrast, surround suppression studies most often use large surround annuli and detection tasks (e.g., "Is there a sinusoidal grating?") (see Table 1 and Fig. 1). It is unclear whether the discrepancies between crowding and surround suppression derive from stimuli or task differences, or if they are in reality different phenomena.

Table 1. Stimulus patterns and tasks used in surround suppression and crowding

	Surround suppression	Crowding
Stimuli	Annulus/half-annulus	Flankers
Tasks	Detection	Discrimination/identification

Fig. 1. An example of typical stimulus patterns used in surround suppression (left) and crowding (right).

In the current study we attempted to examine which factor, stimulus pattern or task, differentiates the two frequently entwined phenomena, surround suppression and crowding, in a fine orientation discrimination task. The stimulus was a large surround annulus which is commonly used in surround suppression studies whereas the task was discrimination which is usually used in crowding tasks. Since the hybrid has components from both surround suppression and crowding, it makes sense that characteristics from both phenomena would display. But which factor is more critical? If task is the critical factor, an inward-outward asymmetry will be observed, as is found in crowding; if stimulus is the critical factor, an inward-outward asymmetry will not be observed, as is found in surround suppression.

2 Method

2.1 Observers

Six undergraduates (four males: CJL, MTC, TBC and THO; two females: KAY and JMM) from Miami University participated in the study. All had normal or corrected-to-normal vision and were naive to the purpose of the experiment. None had prior experience in crowding and surround suppression experiments. Before running the experiment, all received massive training with simple grating stimuli and considerable training with masked stimuli. The study followed the tenets of the Declaration of Helsinki and was approved by the IRB at Miami University.

2.2 Apparatus

Grating stimuli, generated on a Dell Dimension XPS R450, were presented on a Viewsonic Professional Series PS775 17-inch monitor in a dim room. The mean luminance was set to 19.8 cd/m2. Gamma correction was applied to linearize the screen display via software. The display resolution was 2 pixels per minute of visual arc, with a viewing distance of about 2.74 m to compensate for a larger pixel size. Responses were recorded by the same computer when observers pressed the keyboard.

2.3 Stimuli

The target stimuli of sinusoidal grating patches tilting slightly to the left or right of vertical were discriminated in the absence (control condition) or presence (test conditions) of the mask. Specifically, the control stimuli were 40-minute diameter circular patches of 4 cycles per degree (cpd), sinusoidal grating, tilted slightly to the left or right of vertical. The tilts were always symmetric around vertical. They were presented at a contrast of 0.1. In test conditions, a hemi-circular annulus with a width of 80 min abutted the target, and was presented at a contrast of 0.5. Mask and test patch were always in phase relative to each other. Relative to the fixation point, the annulus was either facing inward or outward. All stimuli were luminance modulated. See Fig. 2 for example stimulus patterns. To balance any left-right biases, the fixation point was either to the left or right of the target by 9 degrees. The amount of tilt was determined individually for each observer prior to the beginning of the experiment.

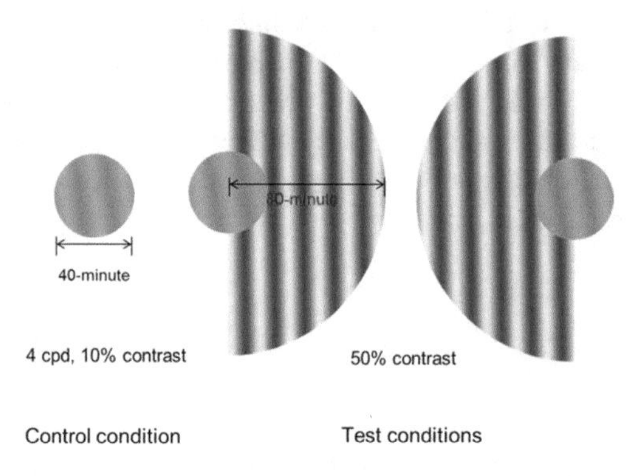

Fig. 2. Stimuli used in control and test conditions.

2.4 Procedure

Before running the actual experiment, all observers were trained with grating stimuli. The training condition was the same as the control condition except for the magnitude of tilting orientations. Training began with an easy orientation discrimination task, with the grating either tilting to the left or to the right. Then the orientation magnitude of tilting was gradually and systematically made smaller with the progress of training, until each individual reached our criterion, i.e., his/her discrimination threshold (d′ stabilized at around 1.5–2.0 in the study). On a given trial, one of the two gratings to be discriminated was presented. Confidence ratings about the orientation of the grating were measured on a 6-point scale. Values between 1 and 6 indicated varying degrees of certainty that the leftward-tilted or rightward-titled grating had been presented. A rating of 1 or 6 indicated very high confidence of left- and right-tilted orientation were presented, respectively.

The experiment consisted of three conditions: a control condition and two test conditions. In the control condition, an isolated grating stimulus was presented; in inward and outward mask conditions, the target was surrounded by half-annulus on the inner and outer sides respectively. In the experiment, the fixation point, displayed 9 degrees either on the left or right side of the stimuli, was present throughout the whole session. Observers were required to focus on the fixation during each trial. They were highly trained and were able to meet this requirement. Also they were asked to hold their response if they did shift their attention to the target, and the trial was intermingled with remaining trials and repeated. On each trial, a single stimulus was presented for 500 ms, accompanied by a middle-pitch tone. Observers needed to perform an orientation discrimination task on target stimuli (single patch under control condition and center-surround gratings under test conditions) presented in the periphery with monocular viewing (right eye). Their responses were collected using the 6-point scale. Immediately following the response, or after a 5-second response interval elapsed without a valid response, a low or high tone sounded, indicating which stimulus actually had been presented. If the subject did not respond within 5 s, or responded with an invalid key, the trial was intermingled with remaining trials and repeated. Each observer performed 18 sessions/replications (JMM performed 12 sessions), with each session consisted of three conditions of 80 trials each. Every observer completed 4320 trials in total (2880 trials for JMM). To minimize the influence of fatigue, observers performed the task no more than one hour per day and were encouraged to take breaks during the experiment.

3 Results

Performance was measured in d'. To compare the suppression strength of inward versus outward masking, we used a dual-axis configuration which included all data from the 18 sessions of each individual (12 for JMM) (see Fig. 3). The graph shows that most data are distributed in the upper left area, suggesting larger inward over outward masking effect in most sessions.

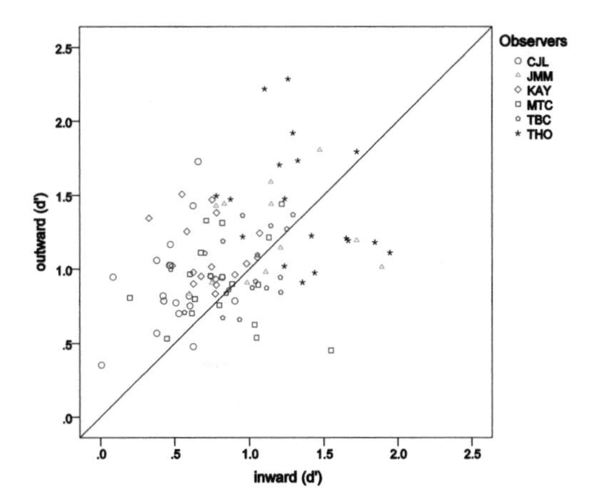

Fig. 3. Performance in inward versus outward masking condition. The solid line indicates equality. Markers in the upper left area indicate a stronger inward masking effect; markers in the lower right area indicate a stronger outward masking effect. Different shape markers represent different observers.

Then we carried out a one-way repeated measures ANOVA on d′ for each individual (the first six columns in Fig. 4) and the data averaged over the six observers (the last column in Fig. 4). The results showed no consistent pattern across all observers. Planned comparisons showed that four observers (i.e., JMM, MTC, TBC and THO) did not show significant differences between outward and inward masking; the other two (i.e., CJL and KAY), showed larger inward masking effect, $t(17) = 5.08$, $p < .001$ and $t(17) = 4.90$, $p < .001$, respectively. This masking effect was in the opposite direction of that in crowding. To evaluate the overall effect of condition, 18 sessions from each individual (12 from JMM) were averaged, followed by an ANOVA. The result revealed a significant effect for condition, $F(2, 10) = 11.34$, $p < 0.01$. Further analysis showed that overall d′ under inward masking condition was significantly smaller than that under outward masking condition, $t(5) = 2.85$, $p < 0.05$. For all observers, the performance was significantly better in the control condition than in the test conditions (inward and outward), indicating strong masking effects for all individuals (all $ps < 0.05$).

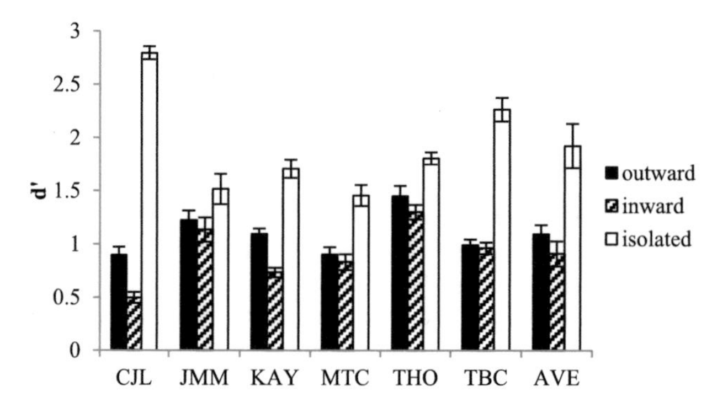

Fig. 4. Performance in d′. Note that lower performance in d′ signifies a stronger masking effect. Error bars indicate ±1 SEM.

4 Discussion

In the current study, we employed a "hybrid stimulus-task combination" and performed a "critical test" [19] to explore whether stimulus or task is the critical factor that differentiates surround suppression and crowding, or whether the particular combination of task and surround involves a distinct mechanism. Results showed that the inward-outward asymmetry effects varied from one observer to another. Some observers showed no significant asymmetry, whereas others showed a significant inward-outward asymmetry. Remarkably, the asymmetry was in the direction opposite to that commonly reported: the inward, rather than the outward, flanker produced a larger masking effect, i.e., a "reversed inward-outward asymmetry". When averaged across observers, the reversed asymmetry effect still existed. This finding was not in line with results from either surround suppression or crowding studies.

According to Pelli [22], crowding is a result of a cortical constraint, and the representation of the space on cortical surface of V1 is a function of the logarithm of that in visual field (i.e., cortical magnification). It follows that the outward mask is closer to the target than the inward mask in the cortex, even though their visual distances in the visual field are equivalent [23]. This theory could perfectly explain the inward-outward asymmetry commonly reported (i.e., outward masking induces a stronger suppression effect than inward masking does), but not our findings in the present study. In fact, the reversed inward-outward asymmetries that some of our observers showed is completely in violation of the cortical magnification explanation.

There are, however, several potential reasons for the results we observed. First, it is worth noting that the larger outward inhibition showed in previous research does not mean that the inward mask dominates every aspect of crowding. When participants are asked to identify the central target, they sometimes mistakenly report a flanker instead of the target ("substitute model"). Studies have shown that participants are more likely to report the inward than the outward flanker [24, 25]. Therefore, although the inward mask disrupts target discrimination less than the outward mask, it has a higher chance to be confused with the target. Together, inward and outward masks may contribute to the crowding effect in different ways [25].

Another possible reason for the reversed asymmetry is the locus of spatial attention. It is well documented that attention plays an important role in crowding [9, 21, 26–30]. Consistent with the current study, Petrov and Meleshkevich [28] also found large individual differences in terms of crowding strength, with half of the observers showed no asymmetry. While their overall inward–outward asymmetry effect across observers was significant, it was much weaker than those reported in other studies (e.g., [18]). They found that the weak asymmetry effect was due to presenting the stimulus at a single location, to which observers could fully allocate their attention. In their control experiment, stimuli were simultaneously presented to the left and right of the fixation point. In this case, the inward-outward asymmetry effect became very strong. The authors proposed that crowding asymmetry can be strongly influenced by attention allocation [28]. This finding was further supported by another of their studies [21]. In this study they found that spatial attention was usually mislocalized to the outward side of the target, so outward masking was stronger than inward masking. When attention was diffused, the asymmetry effect disappeared. Therefore, the locus of spatial attention may in part determine inward-outward asymmetry. In our study, observers who showed a reversed asymmetry effect might somehow allocate their attention to the inward side of the target, which resulted in a larger suppression effect of inward flankers.

With respect to the absence of inward-outward asymmetry that some observers showed, the stimulus that the present study employed might be playing a crucial role. Nurminen et al. [20] found that for the specific half-annulus inward or outward surround (like the one we used here), it was the size in the visual field rather than the size of the corresponding cortical representations that determined the suppression strength. That is, the type of stimulus plays an important role in determining whether or not inward-outward asymmetry occurs. This finding can partially explain our results in terms of the absence of inward-outward asymmetry that some observers showed.

Some research has employed fine spatial frequency or orientation discrimination tasks in conjunction with a surround mask ("center-surround suprathreshold discrimination"). Similar to the current study, observers in these studies are highly trained but naïve to the purpose of the experiments and are asked to discriminate whether the stimulus presented is A or B by rating their certainty. Stimuli are isolated or surrounded, higher or lower contrast, higher or lower luminance, in or out of phase, in the same or different depth plane, or first- or second-order [31–33]. Unlike the current study, these studies were conducted in foveal vision instead of peripheral vision. Overall, results of these studies share some features of surround suppression and some features of crowding; however, they do not fit neatly into either one. For instance, like surround suppression [4, 34], but not crowding [35], center-surround suprathreshold discrimination shows contrast loss [32]. On the other hand, like crowding [36], center-surround suprathreshold discrimination also shows polarity or phase dependence [32, 33] which is not observed in surround suppression in detection [34]. Similarly, the results of the current study cannot fit neatly into either surround suppression or crowding. Therefore, the present study together with others indicates either that current conceptualizations are not complete, or that they are not appropriate under our conditions. Further study on suprathreshold orientation discrimination might focus on the neural activity (e.g., single-unit recording) to examine whether a mechanism that differs from crowding and surround suppression is involved.

In summary, our findings suggested that the asymmetry effect was more complicated than we predicted. The special "hybrid of stimulus and task" cannot be simply classified as either surround suppression or crowding. In addition to the effects of stimulus and task, attention might be playing a role in generating the current results. Alternatively, our particular combination of task and surround may reveal a new phenomenon that differs from either surround suppression or crowding.

5 Conclusions

In this study we examined whether stimulus or task is the critical factor that differentiates surround suppression and crowding, using the inward-outward asymmetry as an indicator. Individual differences were shown: whereas some observers showed no inward-outward asymmetry, others showed an asymmetry. Pooled data across observers also showed an inward-outward asymmetry. Interestingly, this asymmetry was in opposite direction to that commonly reported. This striking finding of the reversed inward-outward asymmetry in suprathreshold orientation discrimination cannot be wholly predicted by either surround suppression or crowding.

References

1. Anton-Erxleben, K., Carrasco, M.: Attentional enhancement of spatial resolution: linking behavioural and neurophysiological evidence. Nat. Rev. Neurosci. **14**(3), 188–200 (2013)
2. Daniel, P.M., Whitteridge, D.: The representation of the visual field on the cerebral cortex in monkeys. J. Physiol. **159**, 203–221 (1961)
3. Cai, Y.C., Lu, S.N., Li, C.Y.: Interactions between surround suppression and interocular suppression in human vision. PLoS One **7**(5), e38093 (2012)

4. Petrov, Y., Carandini, M., McKee, S.: Two distinct mechanisms of suppression in human vision. J. Neurosci. **25**(38), 8704–8707 (2005)
5. Levi, D.M., Hariharan, S., Klein, S.A.: Suppressive and facilitatory spatial interactions in peripheral vision: peripheral crowding is neither size invariant nor simple contrast masking. J. Vis. **2**, 167–177 (2002)
6. Chung, S.T., Levi, D.M., Legge, G.E.: Spatial-frequency and contrast properties of crowding. Vis. Res. **41**(14), 1833–1850 (2001)
7. Jones, H.E., Grieve, K.L., Wang, W., Sillito, A.M.: Surround suppression in primate V1. J. Neurophysiol. **86**(4), 2011–2028 (2001)
8. He, S., Cavanagh, P., Intriligator, J.: Attentional resolution and the locus of visual awareness. Nature **383**(6598), 334–337 (1996)
9. Strasburger, H.: Unfocussed spatial attention underlies the crowding effect in indirect form vision. J. Vis. **5**(11), 8 (2005)
10. Freeman, J., Chakravarthi, R., Pelli, D.G.: Substitution and pooling in crowding. Atten. Percept. Psychophys. **74**(2), 379–396 (2012)
11. Greenwood, J.A., Bex, P.J., Dakin, S.C.: Positional averaging explains crowding with letter-like stimuli. Proc. Natl. Acad. Sci. U.S.A. **106**(31), 13130–13135 (2009)
12. Pelli, D.G., Palomares, M., Majaj, N.J.: Crowding is unlike ordinary masking: distinguishing feature integration from detection. J. Vis. **4**(12), 1136–1169 (2004)
13. Ester, E.F., Klee, D., Awh, E.: Visual crowding cannot be wholly explained by feature pooling. J. Exp. Psychol. Hum. Percept. Perform. **40**(3), 1022–1033 (2014)
14. Ester, E.F., Zilber, E., Serences, J.T.: Substitution and pooling in visual crowding induced by similar and dissimilar distractors. J. Vis. **15**(1), 4.1–4.12 (2015)
15. Harrison, W.J., Bex, P.J.: Visual crowding is a combination of an increase of positional uncertainty, source confusion, and featural averaging. Sci. Rep. **7**, 45551 (2017)
16. Agaoglu, M.N., Chung, S.T.: Can (should) theories of crowding be unified? J. Vis. **16**(15), 10 (2016)
17. Bouma, H.: Visual interference in the parafoveal recognition of initial and final letters of words. Vis. Res. **13**(4), 767–782 (1973)
18. Bex, P.J., Dakin, S.C., Simmers, A.J.: The shape and size of crowding for moving targets. Vis. Res. **43**(27), 2895–2904 (2003)
19. Petrov, Y., Popple, A.V., McKee, S.P.: Crowding and surround suppression: not to be confused. J. Vis. **7**(2), 12.1–12.9 (2007)
20. Nurminen, L., Kilpelainen, M., Vanni, S.: Fovea-periphery axis symmetry of surround modulation in the human visual system. PLoS One **8**(2), e57906 (2013)
21. Petrov, Y., Meleshkevich, O.: Locus of spatial attention determines inward-outward anisotropy in crowding. J. Vis. **11**(4), 1.1–1.11 (2011)
22. Pelli, D.G.: Crowding: a cortical constraint on object recognition. Curr. Opin. Neurobiol. **18**(4), 445–451 (2008)
23. Motter, B.C., Simoni, D.A.: The roles of cortical image separation and size in active visual search performance. J. Vis. **7**(2), 6.1–6.15 (2007)
24. Huckauf, A., Heller, D.: Spatial selection in peripheral letter recognition: in search of boundary conditions. Acta Psychol. (Amst) **111**(1), 101–123 (2002)
25. Strasburger, H., Malania, M.: Source confusion is a major cause of crowding. J. Vis. **13**(1), 24 (2013)
26. Chen, J., He, Y., Zhu, Z., Zhou, T., Peng, Y., Zhang, X., et al.: Attention-dependent early cortical suppression contributes to crowding. J. Neurosci. **34**(32), 10465–10474 (2014)
27. Mareschal, I., Morgan, M.J., Solomon, J.A.: Attentional modulation of crowding. Vis. Res. **50**(8), 805–809 (2010)

28. Petrov, Y., Meleshkevich, O.: Asymmetries and idiosyncratic hot spots in crowding. Vis. Res. **51**(10), 1117–1123 (2011)
29. Scolari, M., Kohnen, A., Barton, B., Awh, E.: Spatial attention, preview, and popout: which factors influence critical spacing in crowded displays? J. Vis. **7**(2), 7.1–7.23 (2007)
30. Yeshurun, Y., Rashal, E.: Precueing attention to the target location diminishes crowding and reduces the critical distance. J. Vis. **10**(10), 16.1–16.12 (2010)
31. Hibbeler, P.J., Olzak, L.A.: No masking between test and mask components in perceptually different depth planes. Perception **40**(9), 1034–1046 (2011)
32. Olzak, L.A., Laurinen, P.I.: Contextual effects in fine spatial discriminations. J. Opt. Soc. Am. A. Opt. Image. Sci. Vis. **22**(10), 2230–2238 (2005)
33. Saylor, S.A., Olzak, L.A.: Contextual effects on fine orientation discrimination tasks. Vis. Res. **46**(18), 2988–2997 (2006)
34. Petrov, Y., McKee, S.P.: The effect of spatial configuration on surround suppression of contrast sensitivity. J. Vis. **6**(3), 224–238 (2006)
35. Levi, D.M.: Crowding–an essential bottleneck for object recognition: a mini-review. Vis. Res. **48**(5), 635–654 (2008)
36. Chung, S.T., Mansfield, J.S.: Contrast polarity differences reduce crowding but do not benefit reading performance in peripheral vision. Vis. Res. **49**(23), 2782–2789 (2009)

Knowledge Discovery of Regional Academic Schools of TCM Based on Bibliometrics

Luan Gao[1,2](✉), Chenling Zhao[1], Zhenguo Wang[2], Fengcong Zhang[2], and Fenggang Li[3]

[1] Clinical College of TCM, Anhui University of TCM, Hefei, China
happygaoluan@163.com
[2] Institute of TCM Literature, Shandong University of TCM, Jinan, China
[3] School of Management, Hefei University of Technology, Hefei, China

Abstract. In this article, bibliometrics will be introduced into the study of regional academic schools of TCM and the typical Chinese medicine academic schools of Xin'an Medicine school will be taken for example. Based on the bibliometric analysis in the past 58 years of Xin'an Medicine science, we will know what achievements have been got in Xin'an Medicine areas in recent years. And the researches will be taken in five parts– age distribution, periodical distribution, author distribution, research topics and research methods, and the article will point out the research methods or feasibility of bibliometrics on regional academic schools of TCM.

Keywords: Bibliometrics · Regional academic schools of TCM
Xin'an Medicine School

The concept of "bibliometrics" was proposed by A. Pritchard in 1969. It is a subject composed of mathematics, statistics and philology, mainly studies the distribution structure, quantity relation, variation and quantitative management, and mainly researches the structure, characteristics or regulations of science and technology in literature, authors, institutions, vocabulary and so on [1–6].

The regional medical school is a general generalization of the characteristics of certain regional physicians and a focus on the tendentiousness of pathogenesis and particularity of treatment [7]. Since the 1970s, the trend of organizing, excavating, and researching the regional academic schools of TCM has been gradually approaching a new climax. Overviewed the study of various regional academic schools, there are a series of books and articles published in recent years. Therefore, the bibliometric method can be introduced into the study to evaluate and analyze the published literature and the relevant literatures of the published regional academic schools of TCM.

1 The Non-equilibrium in the Published Literature of Regional Academic Schools of TCM

At present, the researches on the regional academic schools are mainly related to Xin'an Medicine School, Lingnan Medicine School, Shanghai School of TCM, Qiantang Medicine School, Menghe Medicine School, Yongjia Medicine School,

N. Xiong et al. (Eds.): ISCSC 2018, AISC 877, pp. 11–20, 2019.
https://doi.org/10.1007/978-3-030-02116-0_2

Wuzhong Medicine School and Xujiang Medicine School. In recent years, with the government's attention to the work of TCM, researchers began to carry out the researches on the relevant schools, and published many literatures.

Firstly, we searched the Chinese Journal Full-text Database (CNKI) and conducted preliminary searches on the regional academic schools of TCM, found that there are imbalance between the different regional schools. The results are as follows (Table 1):

Table 1. The number of the published literature of regional academic schools of TCM

Search words	Literature quantity
Xin'an Medicine	945
Lingnan Medicine	616
Menghe Medical School	294
Xujiang Medicine	181
Shanghai style of traditional TCM	160
Wuzhong Medicine	68
Qiantang Medicine	56
Yongjia Medicine	25

From the above results, we can get to know that in the study of regional academic schools of TCM Xin'an Medicine literature achievements get the maximum among the whole schools, it can be retrieved 945 results if we put The school of Xin'an Medicine as the key words; The second is the Lingnan Medicine school, the amount reaches into 616; There are 294 about Menghe. While using the retrieval words only, we cannot search out all the research results of the schools, but at least, it can suggest a general trend of the development and tell us the differences in the number of academic achievements between different regional academic schools of TCM.

The bibliometric, as a kind of mature quantitative method, has been widely used in the study of TCM, so the study can use the bibliometrics method to evaluate and analyse the published articles among the regional academic schools of TCM.

2 Research on Regional Academic Schools of TCM Based on Bibliometrics

With the expectation to explore the research ideas of regional academic schools of TCM based on bibliometrics. The school of Xin'an Medicine will be preferred as the research object in the study, and the method of bibliometrics will be used to carry out quantitative research on regional schools of TCM.

Xin'an Medicine school is recognized as a mature regional medicine school with profound cultural background and clear color of TCM school, has got outstanding academic achievements and historical far-reaching importance, it was formed in Song and Yuan dynasty, flourished in Ming and Qing dynasty, and has survived more than 800 years. In the mid - 1970s, as the rising of "Huixue" research, one of the three important learning in our country, the Chinese Medicine profession, even the historians and cultural circles began to pay great attention in Xin'an Medicine. The academic has always been attaching great importance on sorting and researching medical literature of Xin'an, and articles has published a lot as well. In recent years, the researchers have conducted a comprehensive study of Xin'an Medicine from the literature, the experiment and clinical studies, they published thousands of papers, which gave the significance to Xin'an Medicine research. Professor Yu Yingao once said, "Xin'an Medicine books named by the region of Chinese Medicine can be described as the richest man", and in the contemporary, Xin'an Medicine research results in the regional medical schools are also remarkable.

2.1 Data Sources and Methods

In this study, the Chinese Biomedical Literature Database (CBM) and the Chinese Journal Full-text Database (CNKI) were selected as the database source.

Using field retrieval methods, such as title, keywords, abstract fields, to "xinan medicine" or "huizhou medicine" or "xin" as the search term. At the same time, considering the particularity of Xin'an Medicine research, in order to prevent literature omissions, we also put "the Xin'an famous doctor" as a search term, such as "Sun Yikui, Wang Ji, Cheng Guopeng, Ye Tianshi" and so on; As well as "the Xin'an Medicine classics" for the research, such as "Medical Insight", "Medical Formulas Gathered and Explained", "The Essential Herbal Foundation" etc. Among them, the content has nothing to do with the Xin'an Medicine research literature, repeatedly published literature or published in abstract form of literature, both web searching and manual retrieval are unable to get the full text of literature, are excluded. A total of 4,008 documents were retrieved.

The literature is summarized, and then entered into the database, which is based on "title", "author", "age", "periodical", "research topic", "research method" and so on as the field name, at last, the "Xin'an Medicine research database" is preliminary established. the records amounts to 4008.

2.2 Research Results

By using the method of literature metrology, respectively analyzed from the age distribution, periodical distribution, author distribution, research topics, research methods, and the results are as follows.

Age Distribution. It can be seen from Fig. 1, the number of Xin'an Medicine research papers show a relatively stable growth trend, while there are more obvious differences in different stage, a total of 4008 articles were published, the average annual number of published papers reaches to 69.10.

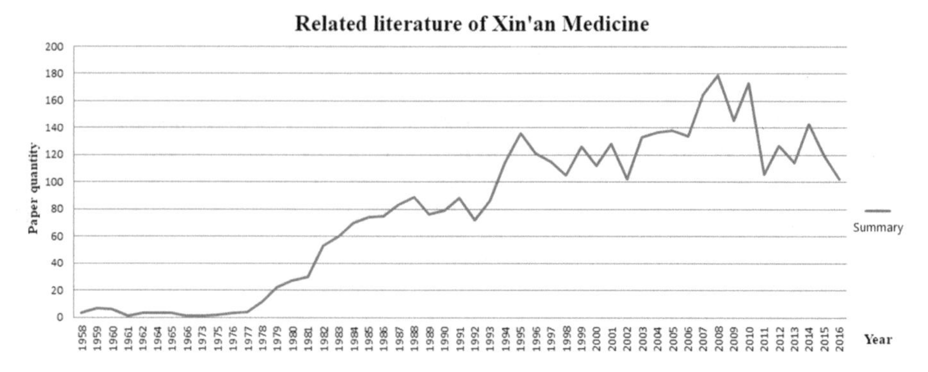

Fig. 1. The line figure of published literature in Xin'an Medicine yearly distribution (1958–2016)

Periodical Distribution. From 1958 to 2016, Xin'an Medicine research literature is distributed in 172 kinds of journals, while each of 100 kinds of journals contained in the volume of more than 100 papers, each contained 1 article in the 49 kinds of journals. Among them, the top one is "Journal of Anhui College of TCM" with 213 articles. The second is "New Chinese Medicine", with 170 articles; Ranked No.3 is "Clinical journal of TCM", which number is 119. See Table 2 for details.

Table 2. The top 20 journals about the published Xin'an Medicine

Rank	Journal Title	Quantity	Percentage
1	Journal of Anhui College of TCM	213	5.94
2	New Chinese Medicine	170	4.49
3	Clinical Journal of TCM	119	4.12
4	TCM in Jilin	115	3.14
5	Shandong Journal of TCM	114	3.01
6	Hunan Journal of TCM	108	2.85
7	Jiangsu Journal of TCM	105	2.77
8	Liaoning Journal of TCM	105	2.77
9	Nei Mongol Journal of TCM	104	2.77
10	Zhejiang Journal of TCM	101	2.69
11	Journal of TCM Literature	96	2.56
12	Journal of Sichuan of TCM	90	2.38
13	Shaanxi Journal of TCM	89	2.35
14	Shanghai Journal of TCM	87	2.30
15	Journal of Zhejiang College of TCM	86	2.27
16	Jiangxi Journal of TCM	82	2.22
17	Journal of Shandong University of Chinese Medicine	78	2.06
18	Modern TCM	73	1.93
19	Yunnan Journal of TCM and Materia Medica	69	1.82
20	TCM in Xinjiang	62	1.64

Author Distribution. After sorting out the literature, removing the literature which have no authors, a total of 4472 authors have published literature in this field, authors who published more than 10 articles are shown in Table 3. Among them, Professor Wang Jian is the leader in the field of Xin'an Medicine research.

Table 3. The top 24 authors of Xin'an Medicine literature

Rank	Author	Signed literature number
1	Wang Jian	67
2	Huang Hui	39
3	Guo Jinchen	30
4	Liu Jian	22
5	Niu Shuping	22
6	Li Yan	17
7	Wan Simei	15
8	Zheng Rixin	15
9	Wang Wei	14
10	Zhang Yucai	14
11	Gao Luan	13
12	Hu Jianpeng	12
13	Tong Guangdong	12
14	Huang Xiaozhou	11
15	Mao Xiao	11
16	Wang Xuguang	11
17	He Shixi	11
18	Yang Qinjun	11
19	Liu Lanlin	11
20	Yang Jin	11
21	Cai Zhongyuan	10
22	Wang Letao	10
23	Tang Wei	10
24	Hu Ling	10

Research Topics. The professional will retrieve the documents according to the literature content keywords, there are a total of 4518 different keywords and 13903 times, and each document remains 3.67 keywords on average. The top 20 keywords accounted for 29.51%. Table 4 and Fig. 2 are for details, what can be seen is that the keywords in top five were Ye Tianshi, Taohong Siwu Tang, Long Dan Xiegan Tang, The Golden Mirror of Medicine, Xin'an Medicine. Therefore, if only the keyword of "Xin'an Medicine" is selected, a lot of literature must be missed.

Table 4. The table of literature research topic in the top 20

Rank	Keyword	Number of times	Percentage
1	Ye Tianshi	928	6.67
2	Taohong Siwu Tang	424	3.05
3	Longdan Xiegan Tang	416	2.99
4	The Golden mirror of medicine	307	2.21
5	Xin'an Medicine	259	1.86
6	TCM therapy	193	1.39
7	A guide to clinical practice with medical record	192	1.38
8	Treat	150	1.08
9	Wuwei Xiaodu Yin	139	1.00
10	Xin'an doctors	127	0.91
11	Zhisou San	127	0.91
12	Consilia	125	0.90
13	Warm disease	116	0.83
14	Treatment application	105	0.76
15	Cough	91	0.65
16	Wang Ji	90	0.65
17	Sun Yikui	90	0.65
18	Medical heart enlightenment	87	0.63
19	Treatment based on syndrome differentiation	70	0.50
20	Banxia Baizhu Tianma Tang	67	0.48

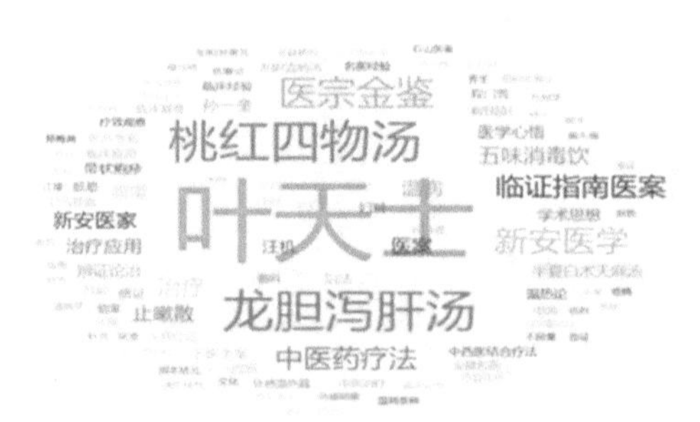

Fig. 2. The cloud atlas of the top 100 research subjects

Research Methods. Throughout Xin'an Medicine research method, there are 16 kinds of research methods, but the existing research is still mainly based on the traditional literature research method, focusing on digging the literature value of Xin'an Medicine, the amount of this kind of article comes to 2388, accounting for 59.58% of all the literature (Table 5).

Table 5. The table of research methods distribution

Rank	Method	Number of papers
1	Literature	2388
2	Clinical	1481
3	Experiment	60
4	Theory	22
5	Papers	14
6	Culture	12
7	News	7
8	Else	6
9	Picture	6
10	Education	3
11	Consilia	2
12	Teaching	2
13	Investigation	2
14	Database	1
15	Language	1
16	Nourishing of life	1

3 Analysis of Research Results and Existing Problems

3.1 Development Trend of Xin'an Medicine Literature

According to the development about the amount of papers, we can roughly divide it into 3 stages: the first stage is from 1958 to 1977, the overall trend is flat, but the growth is limited, the average annual published only 3.08 papers; the second Stage is from 1978 to 1993, it raised steadily, the number of published papers rose from 12 to 86, an average of 62.25 per year; the third stage is from 1994 to 2016 which can be regarded as the prosperity of Xin'an Medicine research, and it reached a peak of 179 in 2008. Although it has been fallen slightly in recent years, but the average annual publication is 129.34, about 41.99 times to the first phase.

3.2 The Main Team of Xin'an Medicine Research

From the point of view on the existing research, the majority of the researchers are focused on the team of Xin'an Medicine research which is led by Professor Wang Jian in Anhui university of Chinese medicine. Anhui is the birth place of Xin'an Medicine. In 1978, Under the advocacy of Wang Renzhi, the deputy director of the provincial health department, Health Bureau in Shexian of anhui province has set up the "Xin'an medical history research group", officially opened the prelude of Xin'an Medicine research. In the next 50 years, Professor Wang Jian will lead the team to search in various perspectives from the formation factors of Xin'an Medicine, representative books, representative physicians, clinical observation and mechanism, which gives the research great significance of The Times.

3.3 Research Topics Are Extensive

In this study, the literature will be retrieved with the keywords according to the contents, there are a total of 4518 different keywords appeared 13903 times. Among them, the keyword of "Ye Tianshi" ranked first. Ye Tianshi is a famous medical scientist in the Qing Dynasty, lived in Shexian of Anhui province. His ancestors took on medicine from generation to generation. Later, his grandfather moved to Wuxian of Jiangsu province (now Suzhou). Due to his ancestral home belongs to Xin'an area, so the study still classified him as Xin'an doctors. Of course, because Ye Tianshi has the great achievements in medicine, both theoretical and clinical experience would make him become the main research object from all the researchers, certainly he will be turned into a research hot spots.

In addition, due to the current entry process of keywords are in accordance with the original literature, therefore, the study now involving 4518 different keywords, leading to the big dispersion of measurement. Considering this, a new theme System for simple measurement should be reseted.

3.4 The Research Methods Are to Be Enriched and Developed

Throughout the decades of literature, Xin'an Medicine research has experienced a process from the beginning to the prosperity, from the traditional methods with original literature and clinical research to the trial of new ideas, and has achieved brilliant achievements, all of those has promoted the development of Xin'an Medicine. But we should also see that although the current research methods of Xin'an Medicine have 16 cases, the existing research are mainly based on the traditional literature research methods, accounting for 59.58% of all, and the vast majority of the researches were done in a single way, without a combination of the various methods. What needs to be seen is that the current research methods for TCM, In addition to literature and clinical, there are still many ways to introduce. Of course, the literature search, for some interdisciplinary, as well as the Xin'an literature extraction of the prescription mechanism research, due to the limitations of the search terms, failed to include comprehensively, and caused a part of the lack of literature either. Therefore, for the involved literature, we have to retrieve comprehensively as far as possible. Only in this way can we likely to provide the results of bibliometrics, and this is a key of research.

4 The Feasibility of Bibliometrics About the Research of Regional Academic Schools of TCM

In this study, we used the method of bibliometrics to study the regional academic schools of TCM, chose Xin'an Medicine for the study subject and analyzed it from five aspects of age distribution, periodical distribution, author distribution, research topic and research method. Although there are still some problems in the study, the application of bibliometrics can reflect the important achievements of Xin'an Medicine research, which embodies the academic value of Xin'an Medicine.

The key to bibliometrics is quantify, and it must be studied from the quantitative point of view. Through the statistical analysis of the characteristics of the literature, we will find out the regularity. At present, bibliometrics has been widely used in various fields of expertise both at home and abroad. The application of bibliometrics in the field of medicine has also been increasing with time. In particularly after 1994, there has been a significant increase in the amount of literature in this aspect. After decades of research and development, bibliometrics has produced many research methods, such as frequency analysis, co-occurrence analysis, cross-correlation analysis and so on.

Therefore, the bibliometrics method can be totally introduced into the academic schools research, the next step of the study is that each of all influencing factors can be carried out with intensive study, so that the formation of quantitative indicators and the construction of "bibliometrics based on regional schools evaluation of TCM Learning system "will be achieved, and then it can be able to evaluate the original academic value of TCM and academic influence. The bibliometric approach is feasible in the study of regional academic schools of TCM. And the following technical road map is preliminarily protocoled and it is expected to be extended to other regional academic schools of TCM (Fig. 3).

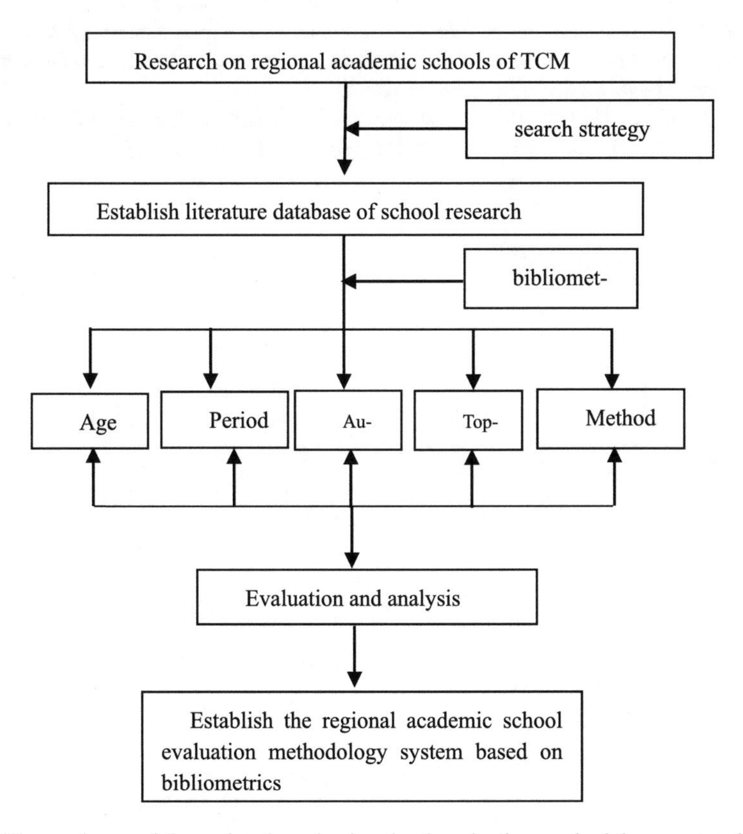

Fig. 3. The road map of the regional academic school evaluation methodology system based on bibliometrics

Acknowledgment. This research has been supported by China post-doctoral science foundation project (2016M592240); and project of Anhui university of Chinese medicine (2016ts001).

References

1. Gasparyan, A.Y., Yessirkepov, M., Duisenova, A., Trukhachev, V.I., Kostyukova, E.I., Kitas, G.D.: Researcher and author impact metrics: variety, value, and context. J. Korean Med. Sci. **33**(18), e139 (2018)
2. Gal, D., Thijs, B., Glänzel, W., Sipido, K.R.: A changing landscape in cardiovascular research publication output: bridging the translational gap. J. Am. Coll. Cardiol. **71**(14), 1584–1589 (2018)
3. Qu, Y., Zhang, C., Hu, Z. Li, S., Kong, C., Ning, Y., Shang, Y., Bai, C.: The 100 most influential publications in asthma from 1960 to 2017: a bibliometric analysis. Respir. Med. **137**, 206–212 (2018)
4. Varzgalis, M., Bowden, D.J., Mc Donald, C.K., Kerin, M.J.: A bibliometric analysis of the citation classics of acute appendicitis. Ann. Med. Surg. **19**, 45–50 (2017)
5. Held, M., Bruins, M.F., Castelein, S., Laubscher, M., Dunn, R., Hoppe, S.: A neglected infection in literature: childhood musculoskeletal tuberculosis - a bibliometric analysis of the most influential papers. Int. J. Mycobacteriology **6**(3), 229–238 (2017)
6. Han, Y., Fu, X., Zhang, F., Li, X., Wang, Z.: The status and exploration of traditional Chinese medicine literature informatization in bibliometric view. World Sci. Technol./Mod. Tradit. Chin. Med. Mater. Med. **17**(3), 427–433 (2015)
7. Hong, J., Wu, H.: Thinking on key issue of inheritance and development of TCM academic schools. China J. Tradit. Chin. Med. Pharm. **28**(6), 1641–1643 (2013)

The Minimum Wiener Index of Unicyclic Graph with Given Pendant Vertices

Yan Wu(✉)

School of Mathematics and Big Data,
Anhui University of Science and Tehnology, Huainan, China
1297659089@qq.com

Abstract. The Wiener index W is the sum of distances between all pairs of a (connected) graph. This paper characterizations of extremal graphs of unicyclic with given pendant vertices.

Keywords: Wiener index · Unicyclic graph · Pendant vertices

1 Introduction

Let G be a connected graph with vertices labeled as $v_1, v_2, ..., v_n$. The distance between vertices v_i and v_j, denoted by $d(v_i, v_j)$, is the length of shortest path between them. The famous Wiener index $W(G)$ is the sum of distances between all pairs of vertices, that is,

$$W(G) = \sum_{i<j} d(v_i, v_j). \quad (1)$$

The name Wiener index or wiener number for the quantity defined in Eq. (1) is usual in chemical literature, sine Harold wiener [11] in 1947 seems to be the first who considered it. Wiener himself conceived W only for acyclic molecules and defined it in a slightly different–yet equivalent–manner; the definition of the Wiener index in terms of distances between vertices of a graph, such as in Eq. (1), was first given by Hosoya [6]. In the mathematical literature W seems to be first studied only in 1976 [2]; for a long time mathematicians were unaware of the (earlier) work on W done in chemistry (cf. the book [1]). The names distance of a graph [2] and transmission [7] were also used to denoted W. The vast majority of chemical applications of the Wiener index deal with acyclic organic molecules; for reviews see [4,5,8–10]. The molecular graphs of these are trees [3]. In view of this, it is not surprising that in the chemical literature there are numerous studies of properties of the Wiener indices.

In this paper, we concentrate on unicyclic graphs. A graph G is called a unicyclic graph if it contains exactly one cycle. We may use the following notation to represent a unicyclic graph:

$$G = U(C_n; T_1, T_2, ..., T_l),$$

N. Xiong et al. (Eds.): ISCSC 2018, AISC 877, pp. 21–27, 2019.
https://doi.org/10.1007/978-3-030-02116-0_3

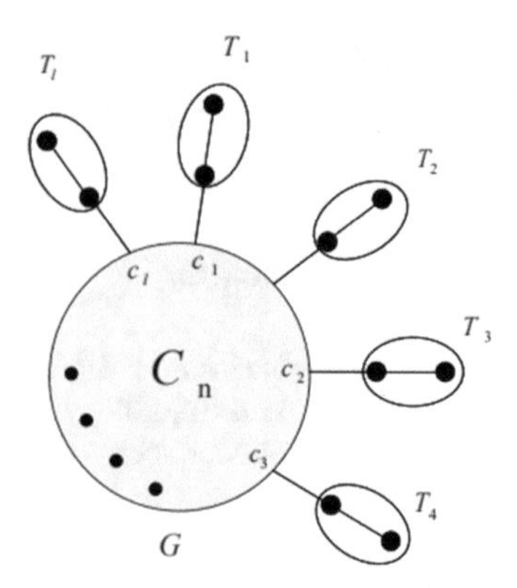

Fig. 1. Unicyclic graph

where C_n is the unique cycle in G with $V(C_n) = \{c_1, c_2, ..., c_n\}$ such that c_i is adjacent to c_{i+1} for $1 \leq i \leq l$. For each i, let T_i be a tree adjacent to C_n at $c_i(i < n)$ (Fig. 1). k_i is a number of pendant in T_i, and $k_1 + k_2 + ... + k_l = k$. The number of vertices in T_i denoted by $\mid T_i \mid$ and $n + \sum_{i=1}^{l} \mid T_i \mid = m$.

If there are no special instructions, the graphs discussed in this paper are connected graphs.

2 Notations and Preliminaries

In this section we shall give some lemmas, which lead to our main results. Before that, we give some notation for easy description.

The sum of distances between all pairs of vertices which one vertex in T_1, another in T_2 denoted by

$$W_G(T_1, T_2) = \sum_{v_i \in T_1 v_j \in T_2} d(v_i, v_j),$$

If two vertices are both in one tree T, we denoted wiener index by

$$W_G(T) = \sum_{v_i, v_j \in T} d(v_i, v_j),$$

the path or cycle are also denoted like that.

Let G be a unicyclic, there are two trees T_i, T_j adjacent to C_n at c_i, c_j. Let G' be the graph obtained from G by moving T_i from c_i to c_j. We say that G' is obtained from G by α-transformation (Fig. 2).

Lemma 2.1 Let G and G' be the graphs defined as above. Than $W(G) > W(G')$.

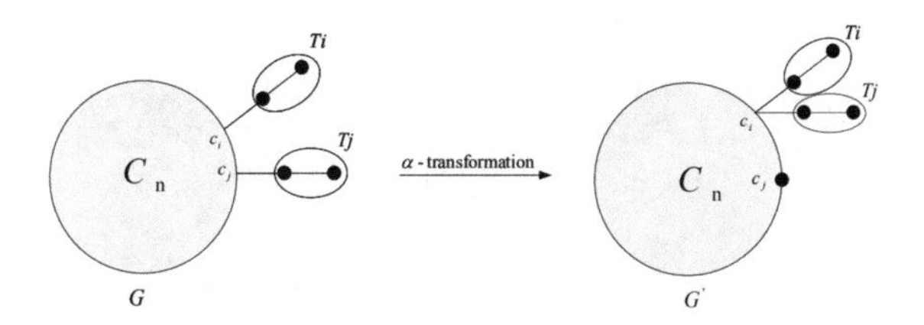

Fig. 2. α-transformation

Proof. For the α-transformation from G to G', we know that

$$W_G(T_i, T_j) > W_{G'}(T_i, T_j),$$
$$W_G(C_n) = W_{G'}(C_n),$$
$$W_G(T_i) = W_{G'}(T_i), W_G(T_j) = W_{G'}(T_j),$$
$$W_G(C_n, T_i) = W_{G'}(C_n, T_i), W_G(C_n, T_j) = W_{G'}(C_n, T_j),$$

so we have,

$$W(G) - W(G') = W_G(T_i, T_j) - W_{G'}(T_i, T_j) > 0,$$

as desired.

Corollary 2.2. Let G be a unicyclic, there are l trees $(0 < l \leq n)$ T_1, T_2,..., T_l adjacent to C_n at c_1, c_2,..., c_l. Let G' be the graph obtained from G by do l times α-transformation until there's only one tree attached to the C_n. Than $W(G) > W(G')$.

Proof. We get G_1 by do α-transformation for G, according to Lemma 2.1, we have $W(G) > W(G_1)$. We do α-transformation to G_1 and get G_2, than $W(G_1) > W(G_2)$, repeat α-transformation until we get G' obtained from G_{l-2}, than we have $W(G) > W(G_1) > W(G_2) > ... > W(G_{l-2}) > W(G')$. Thus the result follows.

Let G be a unicyclic, there are two trees T_i, T_j adjacent to C_n at same vertex c_0, vertex a in T_i and b in T_j are both adjacent to c_0. Let G' be the graph obtained from G by turning a into T_j such that b adjacent to a and a adjacent to c_0. We say that G' is obtained from G by $\beta(T_i, T_j)$-transformation. Similarly, if there are paths, we denoted by $\beta(P_i, P_j)$-transformation (Fig. 3).

Lemma 2.3. Let G and G' be the graphs defined as above. If $|T_i| \geq |T_j| + 1$, Than $W(G) \geq W(G')$. If and only if $|T_i| \doteq |T_j| + 1$, $W(G) \doteq W(G')$.

Proof. For the β-transformation from G to G', we know that

$$W_G(C_n) = W_{G'}(C_n),$$
$$W_G(T_i, T_j) + W_G(T_i) + W_G(T_j) = W_{G'}(T_i, T_j)$$
$$+ W_{G'}(T_i) + W_{G'}(T_j),$$
$$W_G(C_n, T_i) > W_{G'}(C_n, T_i),$$
$$W_G(C_n, T_j) < W_{G'}(C_n, T_j),$$

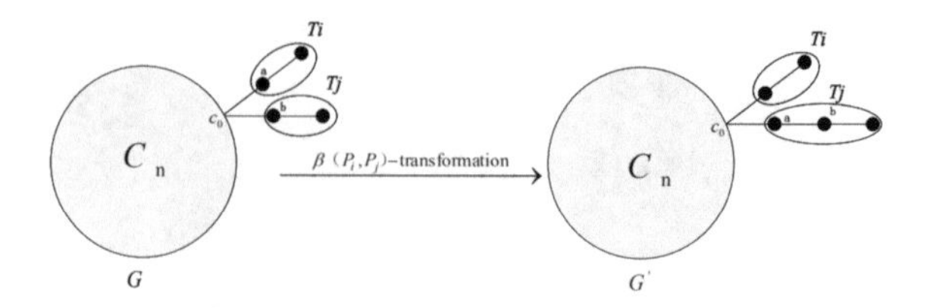

Fig. 3. $\beta(T_i, T_j)$-transformation

so we have

$$\begin{aligned}
&W(G) - W(G') \\
&= W_G(C_n, T_i) + W_G(C_n, T_j) - [W_{G'}(C_n, T_i) \\
&\quad + W_{G'}(C_n, T_j)] \\
&= [W_G(C_n, T_i) - W_{G'}(C_n, T_i)] + [W_G(C_n, T_j) \\
&\quad - W_{G'}(C_n, T_j)] \\
&= \mid C_n \mid\mid T_i - 1 \mid - \mid C_n \mid\mid T_j \mid \\
&= \mid C_n \mid\mid T_j - T_i - 1 \mid \geq 0
\end{aligned}$$

as desired.

Corollary 2.4. Let G be a unicyclic, there is a path adjacent to C_n at c_0, and k paths $P_1, P_2, ..., P_k$ both adjacent to this path at p_0. Let G' be the graph obtained from G by do many times $\beta(P_i, P_j)$-transformation $(1 \leq i, j \leq k)$ until $\mid P_1 \mid = \mid P_2 \mid = \mid p_3 \mid = ... = \mid P_m \mid = \mid P_{m+1} \mid +1 = ... = \mid P_k \mid +1 (1 \leq m \leq k)$. Than $W(G) > W(G')$.

Proof. Without losing generality, assume that $\mid P_i \mid \dot{=} \mid P_j \mid +1$ $(1 \leq i \leq j \leq k)$, we do a $\beta(P_i, P_j)$-transformation on G, according to the proof of Lemma 2.3, other paths without P_i and P_j have the same effect as $\mid C_n \mid$ in Lemma 2.3 on the change of Wiener index after transformation, so wiener index has decreased. And Lemma 2.3 implies that Wiener index keep decreasing by doing many times $\beta(P_i, P_j)$-transformation on G until $\mid P_1 \mid = \mid P_2 \mid = \mid p_3 \mid = ... = \mid P_m \mid = \mid P_{m+1} \mid +1 = ... = \mid P_l \mid +1 (1 \leq m \leq k)$. This follows that $W(G) > W(G')$.

Corollary 2.5. Let G be a unicyclic, there are k paths $P_1, P_2, ..., P_k$ both adjacent to C_n at c_0. Let G' be the graph obtained from G by do many times $\beta(P_i, P_j)$-transformation $(1 \leq i, j \leq k)$ until $\mid P_1 \mid = \mid P_2 \mid = \mid p_3 \mid = ... = \mid P_m \mid = \mid P_{m+1} \mid +1 = ... = \mid P_l \mid +1 (1 \leq m \leq k)$. Than $W(G) > W(G')$.

Proof. By the proof of Corollary 2.4 has shown that Corollary 2.5 is a special case of Corollary 2.4. That is $d(c_0, p_0) = 0$, Thus the result follows.

Let G be a unicyclic, there are l trees $(0 < l \leq n)$ $T_1, T_2, ..., T_l$ adjacent to C_n at same vertex c_0. We find the path from c_0 to p_1, and p_1 is any of pendant vertices in $T_i (1 \leq i \leq l)$, denoted this path by P_1. If $C_n < \mid P_1 \mid$, we definitely can find a vertex a such that $\mid P(p_1, a) \mid = \lceil \frac{1}{2}(\mid C_n \mid + \mid P_1 \mid) \rceil$. There are many trees adjacent to P_1, denote them in $T_1^{'}, T_2^{'}, ..., T_{|p_1|}^{'}$. Let G' be the graph obtained from G by moving $T_1^{'}, T_2^{'}, ..., T_{|p_1|}^{'}$ to a. If $C_n \geq | P_1 |$, let G' be the graph obtained from G by moving $T_1^{'}, T_2^{'}, ..., T_{|p_1|}^{'}$ to c_0. We say that G' is obtained from G by $\gamma(P_1)$-transformation (Figs. 4 and 5).

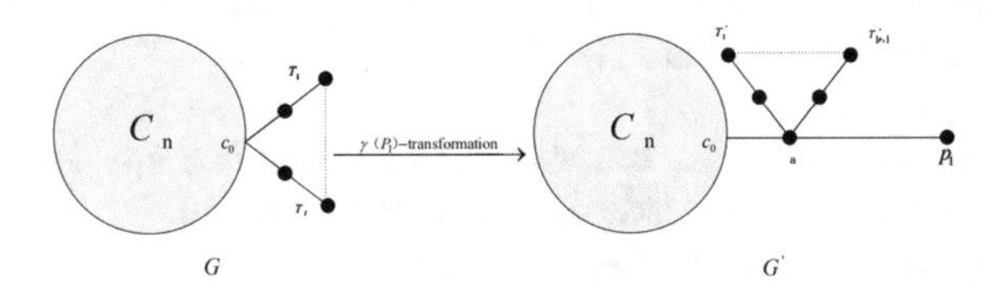

Fig. 4. $\gamma(P_1)$-transformation-case 1

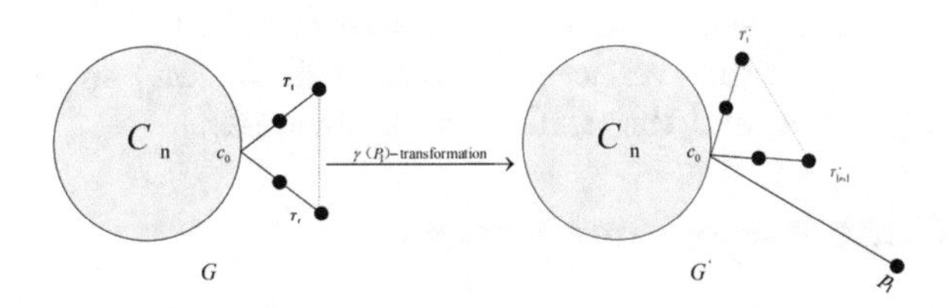

Fig. 5. $\gamma(P_1)$-transformation-case 2

Lemma 2.6. Let G and G' be the graphs defined as above. Than $W(G) \geq W(G')$.

Proof. Without losing generality, we consider about the effect of T_1 on the change of Wiener index after transformation, we denote T_1 adjacent to P_1 at t_1, obviously, $W_G(T_1, T_i) \geq W_{G'}(T_1, T_i)$, $(2 \leq i \leq | p_1 |)$, if and only if T_1 and T_i adjacent to P_1 at same vertex the equality holds.

Case 1 $C_n \geq | P_1 |$.

If T_1 on the left side of a, the distance of vertices in the left side of t_1 to vertices in T_1 is increased, the distance of vertices in the right side of a to vertices in T_1 is decreased, so we have

$$\begin{aligned} & W_G(T_1, P_1) + W_G(T_1, C_n) - [W_G'(T_1, P_1) + W_{G'}(T_1, C_n)] \\ &= |P(a, p_1)||P(t_1, a)| - |P(c_0, t_1)||P(t_1, a)| - |C_n||P(t_1, a)| \\ &= [|P(a, p_1)| - [|C_n| + |P(c_0, t_1)|]]|P(t_1, a)| \geq 0 \end{aligned}$$

If T_1 on the right side of a, the distance of vertices in the right side of t_1 to vertices in T_1 is increased, the distance of vertices in the left side of a to vertices in T_1 is decreased, so we have

$$\begin{aligned} &W_G(T_1, P_1) + W_G(T_1, C_n) - [W'_G(T_1, P_1) + W_{G'}(T_1, C_n)] \\ &= |C_n||P(t_1, a)| + |P(c_0, a)||P(t_1, a)| - |P(t_1, p_1)||P(t_1, a) \\ &= [|C_n| + |P(c_0, a)| - |P(t_1, p_1)|]|P(t_1, a)| \geq 0 \end{aligned}$$

then by the discussions above we conclude that the wiener index has decreased after $\gamma(p_1)$-transformation. Thus, $W(G) \geq W(G')$.

Case 2 $C_n \geq | P_1 |$

By the discussions of case1 has shown that case2 is a special case of case1. That is T_1 always in the right side of a and $|P(c_0, a)| = 0$.

This completes the proof of Lemma 2.6.

Corollary 2.7. Let G be a unicyclic, there is only one path P_0 adjacent to C_n at c_0, and one path P_1 adjacent to P_0 at a,P_2 adjacent to P_1 at c, the end vertex of the P_0 denoted by p_0, the end vertex of the P_0 denoted by p_1, let G' be the graph obtained from G by moving P_2 to a. If $|P_1| \leq d(a, b)$, than $W(G) \geq W(G')$.

Proof. We can easily see the distance from vertices in P_2 to vertices in $P_{(c,p_1)}$ are decreased, and to other vertices are increased. So, we can insure $W(G) \geq W(G')$ if $|P_1| \leq d(a, b)$, and the result follows obviously.

3 Main Conclusions and Proofs

In this section, based on the lemmas introduced in Sect. 2, we determine the graphs with the minimum wiener index among the unicyclic graphs with $|V(G)| = n$ and k pendant vertices.

Let G_1 be a unicyclic, there is a path adjacent to C_n at c_0, and k paths $P_1, P_2, ..., P_k$ both adjacent to this path at p_0, and $|C_n| + d(c_0, p, 0)| =| P_1 |=| P_2 |=| p_3 |= ... =| P_m |=| P_{m+1} | +1 = ... =| P_k | +1 (1 \leq m \leq\ k)$.

Let G_2 be a unicyclic, there are k paths $P_1, P_2, ..., P_k$ both adjacent to C_n at c_0. and $| P_1 |=| P_2 |=| p_3 |= ... =| P_m |=| P_{m+1} | +1 = ... =| P_k | +1 (1 \leq m \leq\ k)$.

Theorem 3.1. The minimum wiener index among the unicyclic graphs with $|V(G)| = n$ and k pendant vertices is G_1 if $|C_n| \leq \lceil \frac{n}{k+1} \rceil$.

Proof. Let G be a unicyclic, according to Corollary 2.2 we can decreasing the wiener index by doing many times α-transformation until there's only one tree attached to the C_n. And we choose the path P_1 which is the longest from c_0 to all of pendant vertices, than we doing $\gamma(p_1)$-transformation on it, if P_1 is not longest path after $\gamma(p_1)$-transformation, choose the longest path P_2 and doing $\gamma(p_2)$-transformation, repeat γ-transformation until the path we selected is still the longest after γ-transformation, the Lemma 2.6 implies that the wiener index is decreased. Than according to Corollary 2.7, we can moving all the pendant paths into one vertex, we doing β-transformation and γ-transformation for these paths, obviously, G_1 is the final form after transformation. Thus the result follows.

Theorem 3.2. The minimum wiener index among the unicyclic graphs with $|V(G)| = n$ and k pendant vertices is G_2 if $|C_n| > \lceil \frac{n}{k+1} \rceil$.

Proof. By the proof of Theorem 3.1, the result follows obviously.

References

1. Buckley, F., Harary, F.: Distance in Graphs. Addison-Wesley, Redwood (1990)
2. Entringer, R.C., Jackson, D.E., Snyder, D.A.: Distance in graphs. Czechoslovak Math. J. **26**, 283–296 (1976)
3. Gutman, I., Polansky, O.E.: Mathematical Concepts in Organic Chemistry. Springer, Berlin (1986)
4. Gutman, I., Potgieter, J.H.: Wiener index and intermolecular forces. J. Serb. Chem. Soc. **62**, 185–192 (1997)
5. Gutman, I., Yeh, Y.N., Lee, S.L., Luo, Y.L.: Some recent results in the theory of the Wiener number. Indian J. Chem. **32A**, 651–661 (1993)
6. Hosoya, H.: Topological index. A newly proposed quantity characterizing the topological nature of structural isomers of saturated hydrocarbons. Bull. Chem. Soc. Jpn. **4**, 2332–2339 (1971)
7. Plesnik, J.: On the sum of all distances in a graph or digraph. J. Graph Theory **8**, 1–21 (1984)
8. Rouvray, D.H.: Should we have designs on topological indices. In: King, R.B. (ed.) Chemical Application of Topology and Graph Theory, pp. 159–177. Elsevier, Amsterdam (1983)
9. Rouvray, D.H.: Predicting chemistry from topology. Sci. Am. **255**(9), 40–47 (1986)
10. Rouvray, D.H.: The modelling of chemical phenomena using topological indices. J. Comput. Chem. **8**, 470–480 (1987)
11. Wiener, H.: Structural determination of paraffin boiling points. J. Am. Chem. Soc. **69**, 17–20 (1947)

Finite Point Method for the Time Fractional Convection-Diffusion Equation

Junchan Li and Xinqiang Qin(✉)

Department of Applied Mathematics, Xi'an University of Technology, Xi'an, China
278544643@qq.com, xqqin@xaut.edu.cn

Abstract. The paper presents a meshless finite point method for solving one-dimensional and two-dimensional time fractional convection-diffusion equations. The method is based on the combination of the moving least square shape function with the collocation method. Moreover, the process of forcing the stable term can effectively solve the convection dominant problem. The Caputo time fractional derivative is discretized by an approximation formula based on L1 interpolation. Furthermore, the stability related to the discretization of this approach is theoretically proven. At the same time, numerical examples are given to compare the errors of the proposed method and the L1 approximation finite difference method. The numerical results show that the proposed method has high accuracy and can effectively eliminate the oscillations.

Keywords: Meshless finite point method
Fractional convection-diffusion equation
Moving least square method · Collocation method

1 Introduction

The convection-diffusion model plays an important role in many practical applications, especially in fluid flow and heat transfer models such as ocean circulation, weather forecasting, and salt leaching in soil. In these applications, the convection terms are more important than the diffusion terms. The convection term is difficult to compute when the Peclet number is high, which is the source of oscillations. It is difficult to solve the steep boundary layer accurately by most numerical methods. The fractional convection-diffusion equation is an extension of the classical convection-diffusion equation. As a mathematical model, the fractional convection-diffusion equation can accurately describe the anomalous diffusion phenomena in complex heterogeneous aquifers and fractal geometry media, which makes up for the shortcomings of the classical model describing similar phenomena. However, the fractional derivative has historical memory and dependence, which is a quasi-differential operator with weak singular integrals. The difficulty of expressing the analytical solution of the differential equation with fractional derivative, and the complexity of numerical calculations result in the challenges for practical application and research. Therefore, numerical

N. Xiong et al. (Eds.): ISCSC 2018, AISC 877, pp. 28–36, 2019.
https://doi.org/10.1007/978-3-030-02116-0_4

algorithms are basic tools for studying fractional differential equations. Some numerical methods have been introduced to research on fractional convection-diffusion equation such as finite difference method (FDM) [1], finite element method [2], spectral method [3], wavelet method [4] and meshless method. In 1996, the scholars Onat and Idelsohn proposed the finite point method (FPM) [5]. The method is a real meshless method without any background grid. The meshless method has been applied in the fields of fluid mechanics and aerodynamics. In [6], the authors proposed an implicit meshless collocation scheme for solving the time fractional diffusion equation, and analyzed its stability and convergence. In [7], the authors also gave an implicit meshless scheme based on radial basis functions to simulate the time fractional diffusion equation. In [8], the authors proposed an implicit meshless method based on the MLS approximation and used a spline weight function to simulate the fractional convection-diffusion equation. In the paper, we mainly consider the following two-dimensional time fractional convection-diffusion equation.

$$\begin{cases} {}_0^C D_t^\alpha u(x,y,t) + b_1 \dfrac{\partial u(x,y,t)}{\partial x} + b_2 \dfrac{\partial u(x,y,t)}{\partial y} - a \cdot \Delta u(x,y,t) = f(x,y,t) \\ u(0,y,t) = g_1(y,t) \\ u(L,y,t) = g_2(y,t) \\ u(x,0,t) = h_1(x,t) \\ u(x,L,t) = h_2(x,t) \\ u(x,y,0) = u_0(x,y) \end{cases} \tag{1}$$

where $\Delta u(x,y,t) = \frac{\partial^2 u}{\partial x^2} + \frac{\partial^2 u}{\partial y^2}$, $(x,y,t) \in [0,L] \times [0,L] \times [0,T]$, b_1, b_2 and a are constants. Here the source function $f(x,y,t)$ is assumed to be sufficiently smooth; the time fractional derivative is defined as the Caputo fractional derivative and $0 < \alpha < 1$.

2 Finite Point Method Discrete Scheme

We can force the stable term [5,9] to (1)

$$ {}_0^C D_t^\alpha u(x,y,t) - r + \frac{h}{2|b|} \cdot b \cdot \nabla r = 0 \tag{2}$$

where h the feature length of the domain,

$$r = -b_1 \frac{\partial u}{\partial x} - b_2 \frac{\partial u}{\partial y} + a(\frac{\partial^2 u}{\partial x^2} + \frac{\partial^2 u}{\partial y^2}) + f$$

$$b = [b_1, b_2] \qquad \nabla r = (\frac{\partial r}{\partial x} \quad \frac{\partial r}{\partial y})^T.$$

Here the time fractional derivative is defined as the Caputo fractional derivative. We take the L1 approximation based on piecewise linear interpolation [10].

Letting $\tau = \frac{T}{n}$, $t_k = k\tau$, $0 \le k \le n$ and $a_l^{(\alpha)} = (l+1)^{1-\alpha} - l^{1-\alpha}$, $l \ge 0$, where τ is defined as the time step and n can assume any positive integer.
We can obtain [10]

$$ {}_0^C D_t^{\alpha} u_{pq}^{n+1} = \frac{\tau^{(-\alpha)}}{\Gamma(2-\alpha)} [a_0^{(\alpha)} u_{pq}^{n+1} - \sum_{k=1}^{n} (a_{n-k}^{(\alpha)} - a_{n-k+1}^{(\alpha)}) u_{pq}^{k} - a_n^{(\alpha)} u_{pq}^{0}] \quad (3) $$

where $u_{pq}^{n+1} = u(x_p, y_q, t_{n+1})$, $u_{pq}^k = u(x_p, y_q, t_k)$, $u_{pq}^0 = u(x_p, y_q, t_0)$.

The discretization of spatial direction is carried out by using the collocation method. Here we use the moving least square to construct approximate function:

$$ u(x_p, y_q) \cong u^h(x_p, y_q) = \sum_{i,j} \varphi_{pq,ij} \cdot (x_p, y_q) u(x_i, y_j) $$

where $\varphi_{pq,ij}$ represents the shape function formed by $(x_i, y_j)(i, j = 1, 2, \cdots, N)$ and $(x_p, y_q)(p, q = 1, 2, \cdots, N)$ which is the center of the influence domain. The weight function determines the continuity of the shape function, thus the selection of the weight function affects directly the result of the numerical simulation. Since Gaussian function has arbitrary order of continuous derivatives, we choose Gaussian weight function in this paper. We can be obtained the approximate expressions of the first partial derivatives of the unknown function at the collocation point (x_p, y_q) by semi-discretizing spatial variables.

$$ \frac{\partial u(x,y)}{\partial x}|_{x=x_p, y=y_q} = \sum_{i,j} \frac{\partial \varphi_{pq,ij}(x,y)}{\partial x} \cdot u(x_i, y_j)|_{x=x_p, y=y_q} $$

$$ \frac{\partial u(x,y)}{\partial y}|_{x=x_p, y=y_q} = \sum_{i,j} \frac{\partial \varphi_{pq,ij}(x,y)}{\partial y} \cdot u(x_i, y_j)|_{x=x_p, y=y_q} $$

Similarly, the approximation expressions of the second and third partial derivatives can be obtained. Substituting these expressions into (2)

$$ \begin{aligned} a_0^{(\alpha)} u_{pq}^{n+1} &+ \sum_{i,j} \beta_{pq,ij} u_{ij}^{n+1} = \sum_{k=1}^{n} (a_{n-k}^{(\alpha)} - a_{n-k+1}^{(\alpha)}) u_{pq}^{k} + a_n^{(\alpha)} u_{pq}^{0} \\ &+ \Gamma(2-\alpha)\tau^{\alpha} (f_{pq}^{n+1} - \frac{hb_1}{2|b|} \frac{\partial f_{pq}^{n+1}}{\partial x} - \frac{hb_2}{2|b|} \frac{\partial f_{pq}^{n+1}}{\partial y}) \end{aligned} \quad (4) $$

where

$$ \begin{aligned} \beta_{pq,ij} &= \Gamma(2-\alpha)\tau^{\alpha} [b_1 \frac{\partial \varphi_{pq,ij}(x,y)}{\partial x} + b_2 \frac{\partial \varphi_{pq,ij}(x,y)}{\partial y} \\ &- (a + \frac{hb_1^2}{2|b|}) \cdot \frac{\partial^2 \varphi_{pq,ij}(x,y)}{\partial x^2} - (a + \frac{hb_2^2}{2|b|}) \cdot \frac{\partial^2 \varphi_{pq,ij}(x,y)}{\partial y^2} \\ &- \frac{hb_1 b_2}{|b|} \cdot \frac{\partial^2 \varphi_{pq,ij}(x,y)}{\partial x \partial y} + \frac{ab_1 h}{2|b|} \cdot (\frac{\partial^3 \varphi_{pq,ij}(x,y)}{\partial x^3} + \frac{\partial^3 \varphi_{pq,ij}(x,y)}{\partial x \partial y^2}) \\ &+ \frac{ab_2 h}{2|b|} \cdot (\frac{\partial^3 \varphi_{pq,ij}(x,y)}{\partial y^3} + \frac{\partial^3 \varphi_{pq,ij}(x,y)}{\partial x^2 \partial y})]_{x=x_p, y=y_q} \end{aligned} $$

Equation (4) is written as the following matrix form:

$$Au_{pq}^{n+1} = \sum_{k=1}^{n}(a_{n-k}^{(\alpha)} - a_{n-k+1}^{(\alpha)})u_{pq}^{k} + a_{n}^{(\alpha)}u_{pq}^{0} + F_{pq}^{n+1} \tag{5}$$

where

$$F_{pq}^{n+1} = \Gamma(2-\alpha)\tau^{\alpha} \cdot (f_{pq}^{n+1} - \frac{hb_1}{2|b|}\frac{\partial f_{pq}^{n+1}}{\partial x} - \frac{hb_2}{2|b|}\frac{\partial f_{pq}^{n+1}}{\partial y})$$

$$A = \begin{pmatrix} \beta_{11,11} + a_0^{(\alpha)} & \beta_{11,12} & \cdots & \beta_{11,NN} \\ \beta_{12,11} & \beta_{12,12} + a_0^{(\alpha)} & \cdots & \beta_{12,NN} \\ \vdots & \vdots & \ddots & \vdots \\ \beta_{NN,11} & \beta_{NN,12} & \cdots & \beta_{NN,NN} + a_0^{(\alpha)} \end{pmatrix}$$

3 Stability Analysis of the Format

Lemma 1. *Let $\alpha \in (0,1), a_l^{(\alpha)} = (l+1)^{1-\alpha} - l^{1-\alpha}, l \geq 0$, then $1 = a_0^{(\alpha)} > a_1^{(\alpha)} > a_2^{(\alpha)} > \cdots > a_l^{(\alpha)} > 0$; when $l \to \infty, a_l^{(\alpha)} \to 0$.*

In (5), $A = (\beta_{pq,11}, \beta_{pq,12}, \cdots, \beta_{pq,pq} + a_0^{(\alpha)}, \cdots, \beta_{pq,NN})$, where $p, q = 1, 2, \cdots, N$. It is rare to see the case of $\beta_{pq,ij} = 0$ and $\beta_{pq,pq} < 0 (i, j = 1, 2, \cdots, N; ij \neq pq)$ occuring at the same time. Thus we can define approximately according to the infinite norm of the matrix

$$\begin{aligned} \|A\|_{\infty} &= \max_{1 \leq p,q \leq N} \sum_{i,j=1}^{N} |\beta_{pq,ij}| \\ &= \max_{1 \leq p,q \leq N} (|\beta_{pq,11}| + \cdots + |\beta_{pq,pq} + a_0^{(\alpha)}| + \cdots + |\beta_{pq,NN}|) \geq 1 \end{aligned}$$

At the same time, the result is verified by numerical experiments. Assuming that u_{pq}^{n+1} is a numerical approximation of $u(x_p, y_q, t_{n+1})$, $\tilde{u}_{pq}^{n+1}$ is an approximate numerical solution of (5), and the calculation of F_{pq}^{n+1} is exact, then the error $\varepsilon_{pq}^{n+1} = u_{pq}^{n+1} - \tilde{u}_{pq}^{n+1}$ satisfies:

$$\begin{aligned} &n = 0, A\varepsilon_{pq}^{1} = a_0^{(\alpha)} \cdot \varepsilon_{pq}^{0}; \\ &n = 1, A\varepsilon_{pq}^{2} = (a_0^{(\alpha)} - a_1^{(\alpha)}) \cdot \varepsilon_{pq}^{1} + a_1^{(\alpha)} \cdot \varepsilon_{pq}^{0}; \\ &n \geq 2, A\varepsilon_{pq}^{n+1} = \sum_{k=1}^{n}(a_{n-k}^{(\alpha)} - a_{n-k+1}^{(\alpha)}) \cdot \varepsilon_{pq}^{k} + a_n^{(\alpha)} \cdot \varepsilon_{pq}^{0}. \end{aligned}$$

Theorem 1. *The calculation format (5) is unconditionally stable.*

Proof. Use the mathematical induction to prove. Let $E^n = (\varepsilon_{11}^n, \varepsilon_{12}^n, \cdots, \varepsilon_{1N}^n, \varepsilon_{21}^n, \cdots, \varepsilon_{NN}^n)^T$, $\|A\|$ is the matrix norm on $C^{N^2 \times N^2}(R^{N^2 \times N^2})$, $\|\varepsilon_{pq}\|$ is the vector norm on $C^{N^2}(R^{N^2})$. Due to $\|A\varepsilon_{pq}\|$ is still a vector on $C^{N^2}(R^{N^2})$, $\|A\varepsilon_{pq}\| \leq \|A\| \cdot \|\varepsilon_{pq}\|$ $(p, q = 1, 2, \cdots, N)$. Let $|\varepsilon_l^1| = \max_{1 \leq p,q \leq N} |\varepsilon_{pq}^1|$ when $n = 1$, then

$$\begin{aligned} \|E^1\|_\infty = |\varepsilon_l^1| \leq \|A\|_\infty \cdot |\varepsilon_l^1| \leq a_0^{(\alpha)} \cdot |\varepsilon_l^0| \\ = a_0^{(\alpha)} \cdot \max_{1 \leq j \leq N} |\varepsilon_j^0| = a_0^{(\alpha)} \cdot \|E^0\|_\infty = \|E^0\|_\infty. \end{aligned}$$

Assume that $\|E^k\|_\infty \leq \|E^0\|_\infty$ hold when $k = 2, \cdots, n$.

Let $|\varepsilon_l^{n+1}| = \max_{1 \leq p,q \leq N} |\varepsilon_{pq}^{n+1}|$, then

$$\begin{aligned} \|E^{n+1}\|_\infty &= |\varepsilon_l^{n+1}| \\ &\leq \|A\|_\infty \cdot |\varepsilon_l^{n+1}| = \sum_{k=1}^{n} (a_{n-k}^{(\alpha)} - a_{n-k+1}^{(\alpha)}) \cdot |\varepsilon_l^k| + a_n^{(\alpha)} \cdot |\varepsilon_l^0| \\ &= \sum_{k=1}^{n} (a_{n-k}^{(\alpha)} - a_{n-k+1}^{(\alpha)}) \cdot \|E^k\|_\infty + a_n^{(\alpha)} \cdot \||E^0\|_\infty \\ &\leq [\sum_{k=1}^{n} (a_{n-k}^{(\alpha)} - a_{n-k+1}^{(\alpha)}) + a_n^{(\alpha)}] \cdot \|E^0\|_\infty \\ &= a_0^{(\alpha)} \|E^0\|_\infty = \|E^0\|_\infty \end{aligned}$$

The proof of Theorem 1 is completed.

4 Numerical Examples

In this section, we give two numerical examples. In order to demonstrate the validity and accuracy of the proposed algorithm, we calculate and compare the relative errors (Err) of the finite point method and L1 approximation finite difference method.

Example 1. Consider the following one-dimensional time fractional convection-diffusion equation

$${}_0^C D_t^\alpha u(x,t) + \frac{\partial u}{\partial x} - a \cdot \frac{\partial^2 u}{\partial x^2} = f(x,t)$$

where $0 \leq x < 1, 0 \leq t < T, \alpha = 0.5, a = 10^{-8}$. It is dominated by convection. With initial condition

$$u(x,0) = 0, 0 \leq x \leq 1$$

and boundary conditions

$$u(0,t) = u(1,t) = 0, 0 \leq t \leq T$$

The exact solution is

$$u_{exact}(x,t) = t^{2+\alpha}(x - x^2)$$

According to the exact solution of the equation, we can get the expression of the source term. We choose the linear basis function, and the support domain influence factor is 3.0.

From Tables 1 and 2 can be seen that the finite point algorithm holds comparable or even higher order accuracy by comparing with the finite difference method at different times, different number of nodes (Fig. 1).

From Fig. 2 can be seen that the finite point numerical solution converges to the exact solution and the image fits well.

Table 1. The comparison of calculation accuracy between FPM and FDM under different number of nodes

Time step	Time	The number of nodes	FPM(Err)	FDM(Err)
0.01	T = 1	11	4.7018e−04	5.2954e−04
		21	4.5594e−04	5.2205e−04
		51	4.2080e−04	5.1993e−04
		81	3.7903e−04	5.1967e−04
		101	3.5925e−04	5.1959e−04
		501	3.3459e−04	5.1833e−04

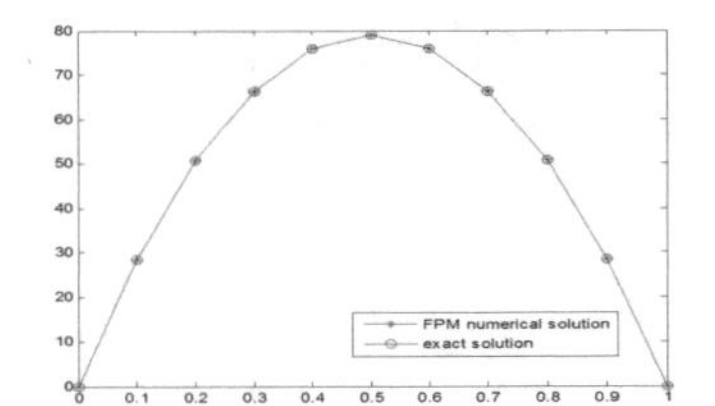

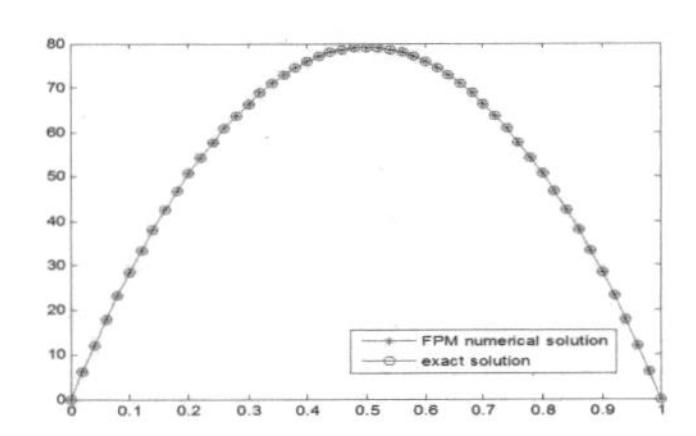

Fig. 1. $\tau = 0.01$. Comparison of numerical solution and exact solution for in T = 10 when the number of nodes is 11 (left) and 51 (right).

Example 2. We consider Eq. (1), where $\alpha = 0.5, b_1 = 1, b_2 = 1, a = 10^{-6}$. Here the exact solution is $u_{exact}(x,y,t) = t^{2+\alpha}(x-x^2)(y-y^2)$. Then all the boundary and initial values are 0.

From Table 3, it can be seen that compared the central processing unit (CPU) time of FPM and FDM when the number of different nodes. The proposed algorithm accuracy is better. It is obvious that the proposed algorithm CPU time is faster obviously than the finite difference method.

Table 2. Error comparison for different algorithms with different time steps when the number of nodes is 51.

The number of nodes	Time step	Time	FPM(Err)	FDM(Err)
51	0.01	T = 1	4.2080e−04	5.1993e−04
		T = 3	6.2451e−05	9.7143e−05
		T = 5	2.7043e−05	4.4930e−05
		T = 10	1.4754e−05	1.5846e−05
	0.1	T = 1	1.2454e−02	1.4881e−02
		T = 3	2.1094e−03	2.9073e−03
		T = 5	9.0235e−04	1.3625e−03
		T = 10	2.7461e−04	4.8675e−04

Form Table 4 can be seen that the accuracy of the proposed method is better than FDM. FPM still can reach the order of magnitude of e−02 when the time is calculated to T = 3. From Figs. 2 and 3 can be seen that the numerical solution of the proposed method is in good agreement with the exact solution, while the finite difference numerical solution has severe numerical oscillations. Therefore, FPM can eliminate oscillations to a certain extent, and in good agreement with the exact solution, which fully demonstrates the effectiveness of the algorithm.

Table 3. Comparison of error and computing time of FPM and FDM under different number of nodes.

Time step	Time	The number of nodes	FPM	CPU	FDM	CPU
0.001	T = 1	11 × 11	7.1575e−02	32.5	1.2380e−02	1051.4
		21 × 21	1.8950e−02	71.7	1.2201e−02	2433.4
		31 × 31	8.8807e−03	323.8	1.2174e−02	4682.1

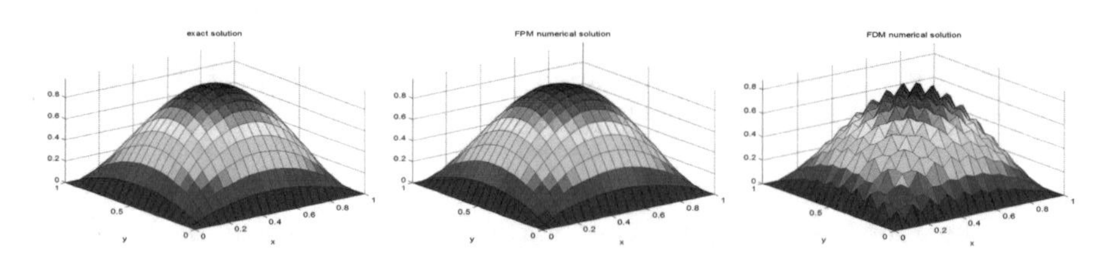

Fig. 2. Number of nodes is 21×21, $\tau = 0.001$. Comparison of FPM numerical solution, FDM numerical solution and exact solution in T = 3 time.

Table 4. The comparison of calculation accuracy between FPM and FDM at different times.

The number of nodes	Time step	Time	FPM(Err)	FDM(Err)
11×11	0.001	T = 0.1	3.9902e−02	4.5193e−01
		T = 1	7.1575e−02	1.2380e−02
		T = 3	8.3439e−02	1.4104e−01
31×31	0.001	T = 0.1	5.5290e−03	4.7271e−01
		T = 1	8.8807e−03	1.2174e−02
		T = 3	1.0417e−02	1.4402e−01

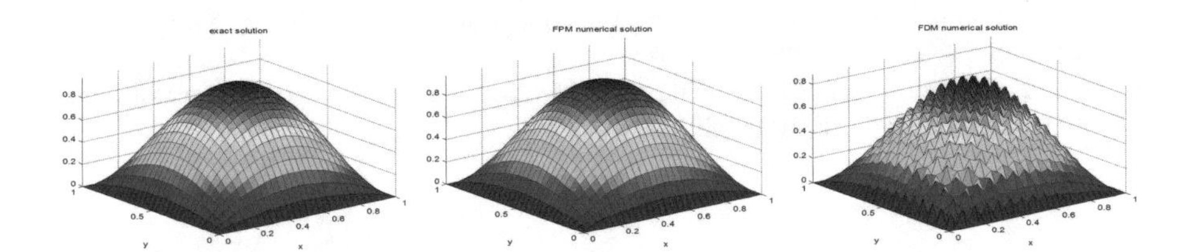

Fig. 3. Number of nodes is 31×31, $\tau = 0.001$. Comparison of FPM numerical solution, FDM numerical solution and exact solution in T = 3 time.

5 Conclusions

In this paper, we study the derivation of the finite point algorithm for the two-dimensional time fractional convection-diffusion equation, and analyze the stability of the method. Moreover, one-dimensional and two-dimensional numerical examples are given. The numerical results show our presented algorithm is efficient and simple. Compared with the traditional L1 approximation finite difference method, the finite point algorithm holds comparable or even higher order accuracy, and can eliminate oscillations better, especially for the two-dimensional convection dominant problem.

References

1. Cui, M.: A high-order compact exponential scheme for the fractional convection-diffusion equation. J. Comput. Appl. Math. **255**(285), 404–416 (2014)
2. Wang, J., Liu, T., Li, H., et al.: Second-order approximation scheme combined with H1-Galerkin MFE method for nonlinear time fractional convection-diffusion equation. Comput. Math. Appl. (2016)
3. Yu, Z., Wu, B., Sun, J.: A space-time spectral method for one-dimensional time fractional convection diffusion equations. Math. Methods Appl. Sci. **40**(7), 2634–2648 (2016)
4. Zhou, F., Xu, X.: The third kind Chebyshev wavelets collocation method for solving the time-fractional convection diffusion equations with variable coefficients. Appl. Math. Comput. **280**(C), 11–29 (2016)

5. Onate, E., Idelsohn, S., Zienkiewicz, O.C.: A stabilized finite point method for analysis of fluid mechanics problems. Comput. Methods Appl. Mech. Eng. **139**(1–4), 315–346 (1996)
6. Meerschaert, M.M., Tadjeran, C.: Finite difference approximations for fractional advection-dispersion flow equations. Elsevier Science Publishers B. V. (2004)
7. Dehghan, M., Shokri, A.: A meshless method for numerical solution of a linear hyperbolic equation with variable coefficients in two space dimensions. Numer. Methods Part. Differ. Equ. **25**(2), 494–506 (2010)
8. Zhuang, P., Gu, Y.T., Liu, F.: Time-dependent fractional advection-diffusion equations by an implicit MLS meshless method. Int. J. Numer. Methods Eng. **88**(13), 1346–1362 (2011)
9. Onate, E., Idelsohn, S., Zienkiewicz, O.C.: A finite point method in computational mechanics applications to convective transport and fluid flow. Int. J. Numer. Methods Eng. **39**(22), 3839–3866 (2015)
10. Sun, Z.Z., Gao, G.H.: Finite Difference Method for Fractional Differential Equations. Science Press, Beijing (2015)

The 13-Meter Radio Telescope Monitor and Control Software

Daxin Zhao(✉), Rongbing Zhao, Yongbin Jiang, Jinling Li, Cong Liu, and Hui Zhang

Shanghai Astronomical Observatory, 80 Nandan Road, Shanghai 200030, China
zhaodaxin@shao.ac.cn

Abstract. The 13-meter radio telescope monitor and control software, also called Client Graphics User Interface Library Tools (CGLT), is a set of graphical user interfaces provided to engineers and operators. It is a modular solution for monitoring and controlling the 13-meter radio telescope, which is a telescope under construction in Sheshan, Shanghai. Design philosophy, functionality and implementation of the software will be introduced in this article.

Keywords: 13-meter radio telescope
Client Graphics User Interface Library Tools · Monitor
Control · Software

1 Introduction

The 13-meter radio telescope–being used to observe and study radio waves from interesting sources–is a telescope under construction in Sheshan, Shanghai. It is a complex component of electronic and mechanical electronics units and consists of several subsystems like antenna system, receiver system, etc. All these subsystems work together to enable a successful astronomical observation. Telescope monitor and control software (M&C), which is also an essential subsystem, integrates all other subsystems and improves the observation efficiency. In the field of radio astronomy, requirements of the M&C are as follows. First of all, it would be highly desirable for parameters of all subsystems to be able to be periodically monitored. Secondly, numerous parameters and isolated physical location of subsystems make it a distributed system. Meanwhile, to allow a successful observation, all these subsystems must "set" as a user's requirements. To name a list, the antenna must track or scan the selected source, the receiver must be set to select the desired observation frequency band, the data acquisition equipment and data recording equipment must be set to the appropriate channel and data rate, etc. Finally, the health of subsystems should be monitored at all times so that should any subsystem fail, the fault information can be captured and pushed out, and also remedial actions could be taken to fix the wrong unit. The 13-meter radio telescope monitor and control software, also called Client Graphics User Interface Library Tools (CGLT), is a modular solution for

N. Xiong et al. (Eds.): ISCSC 2018, AISC 877, pp. 37–44, 2019.
https://doi.org/10.1007/978-3-030-02116-0_5

monitoring and controlling the 13-meter radio telescope. Believe or not, it contains all of the above-mentioned features. The CGLT allows one to (including but not limited to)

- Rotate the 13-meter radio telescope in azimuth and elevation, and to track or scan radio sources, satellites, and planets.
- Bring the required feed to the focus via the Feed Position System.
- Set the Up-Down Converter (UDC) including local oscillator, attenuation, and bandwidth.
- Set the receiver's Phase Calibration (PCal), band switching, noise injection.
- Select data acquisition terminal and data recording terminal parameters like channel bandwidth, data rate, etc.
- Coordinate antenna and power meter to record data in a Flexible Image Transport System (FITS)-format file at the same time.
- Monitor parameters of all subsystems and plot in real time when necessary.
- Register listener in health states of subsystems, catch and push exception whenever one's found.

2 Design of the Software

The design diagram of CGLT is shown in Fig. 1. All hardware devices are connected to the same network segment through routers. In the same network segment, the device manager connects to the physical device through the network port or serial port and retrieves all data. The CGLT and the device manager are connected by middleware–a computer software that connects software components and applications, through which device managers send out data and receive commands while the software receives data and sends commands. The main function packages or callable libraries that are used in the CGLT development process are as follows.

2.1 Device Manager

Each device corresponds to a manager in the system. The flow diagram of a device manager is shown in Fig. 2. Device managers are the interfaces between CGLT and devices. Those device managers collect parameters of devices and put them into data buffers, and CGLT calls common interface to obtain data from the buffers and finally display them. Meanwhile, device managers provide CGLT with methods for controlling devices, these methods are packaged to form callable device control commands which are issued by CGLT and delivered to devices by the device managers themselves. As we can see, device managers make devices transparent and provide unified interfaces to allow monitoring and controlling. Due to this design, device managers can be customized with additional devices.

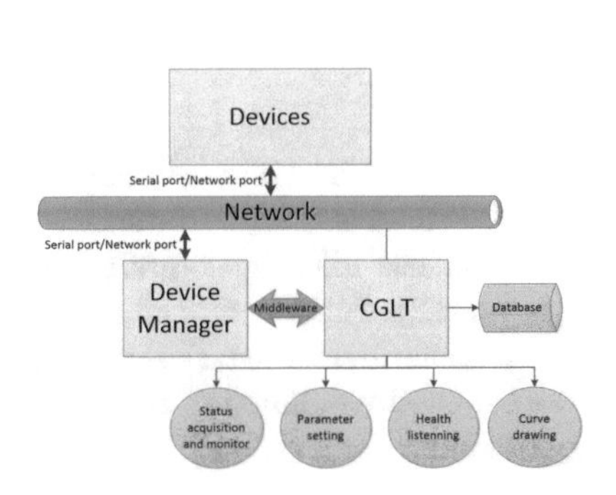

Fig. 1. Design diagram of CGLT

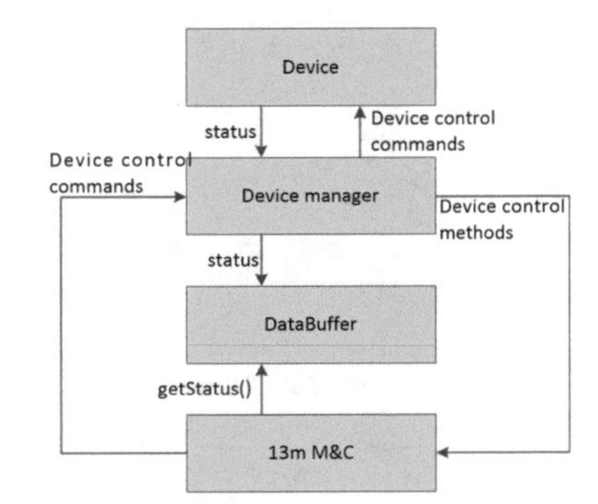

Fig. 2. Flow diagram of a device manager

2.2 Internet Communications Engine

As mentioned earlier, the 13-meter radio telescope is a complex component of electronic and mechanical electronics units. Isolation of physical locations of devices makes the control software distributed. It is difficult to design and implement such a huge system, so we introduced the distributed middleware. The main role of distributed middleware is to shield developers and users from the multiple heterogeneity between hardware and software, and to simplify the development work of developers. Throughout the field of radio astronomy control, distributed middleware is nothing more than Common Object Request Broker Architecture (CORBA) and Internet Communications Engine (ICE). Compared to the large and complex CORBA, ICE is easier to learn. ICE allows you to focus your efforts on your application logic, and it takes care of all interactions with low-level network programming interfaces. With ICE, there is no need to worry about details such as opening network connections, serializing and deserializing data for network transmission, or retrying failed connection attempts [1]. ICE supports multiple language mappings such as Java, C++, Python, etc. It is very suitable for multiple heterogeneous environments. What is said above is the reason why we chose ICE. ICE is powerful, yet we only used two of its services in CGLT, namely the Remote Protocol Call (RPC) service and the publish-subscribe service.

RPC Service. Server and client of application may appear on different machines, which makes it necessary to call interfaces through ICE RPC service. Before formalizing the steps of using RPC service, we have to introduce a concept–Special Lange for ICE (Slice). Slice is the fundamental abstraction mechanism for separating object interfaces from their implementations. Slice establishes a contract between client and server that describes the interfaces, operations and parameter types used by an application. This description is independent of the implementation language, so it does not matter whether the client is written in the same language as the server. Writing an ICE RPC application involves the following steps.

- Write a Slice definition and compile it.
- Write a server and compile it.
- Write a client and compile it.

For a more specific description, we take the antenna parameters acquisition as an example. First, we create a file called "acu.ice" and define the interfaces in the file as shown in Fig. 3. In directory of the file, use Slice to compile the file into a java interface. The related command is "slice2java acu.ice". After the compilation is completed, a folder named "Acu" is generated. If you're careful enough, you will find that its name is the same as the module name in "acu.ice". Copy the compiled folder to the project source directory.

```
module Acu {
    dictionary<string, string> dicparam;
        interface rpcACU
        {
                void getStatus (out dicparam para);
        };
};
```

Fig. 3. Interface defined in 'acu.ice'

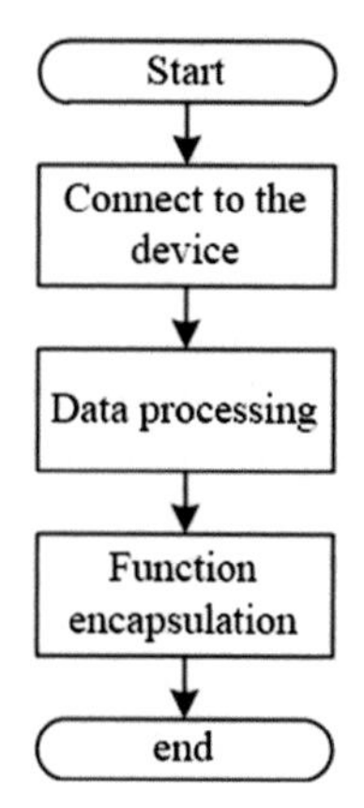

Fig. 4. Flow diagram of the server

```
public class Client {
  public static void main (String [] args) {
    try (com.zeroc.Ice.Communicator communicator =
        com.zeroc.Ice.Util.initialize(args)) {
        Map<String, String> map = new
            HashMap<String, String> ();
        com.zeroc.Ice.ObjectPrx base =
            communicator.stringToProxy(
              "Acu:default -p 14000");
        Acu.rpuACUPrx acu =
            Acu.rpuACUPrx.checkedCast(base);
        if (acu == null) {
            throw new Error ("Invalid proxy");
        }
        acu.getStatus(map);
    }
  }
}
```

Fig. 5. Full code of the client

Next, implement server-side code. Actually, this part is done in the device manager. Flow Diagram of the server is shown in Fig. 4. Data processing means retrieving data, processing data into a readable form and putting data into a buffer. Function encapsulation encapsulates one or more written functions through functions and classes, and provides only a simple function interface, enabling the client to get parameters and shielding the client from the way to achieve.

Finally, define the client, getting the parameters from the server and place them in a hashmap. The code is in full as shown in Fig. 5. The example gets parameters of the antenna once, but in reality, the "while" function has to be called to obtain the parameters and provide the CGLT periodically.

Publish-Subscribe Service. If ICE's RPC service can be regarded as the CGLT 'pulls' data from the server, its publish-subscribe service can be viewed as the server 'pushes' data to the CGLT, when the device has new parameters, the server sends them to the software. ICE's publish-subscribe service is implemented through its IceStorm service. IceStorm simplifies the collector implementation significantly by decoupling it from the monitors. As a publish-subscribe service, IceStorm acts as a mediator between the collector (the publisher) and the monitors (the subscribers). Similarly, before officially introducing the steps of using ICE's publish-subscribe service, we once again introduced a concept–Topic. An application indicates its interest in receiving messages by subscribing to a Topic. An IceStorm server supports any number of Topics, which are created dynamically and distinguished by unique names. Each Topic can have multiple publishers and subscribers. The publish-subscribe service is somewhat similar to the RPC service. Topics is basically equivalent to Slice, except that IceStorm transparently forwards each message to multiple recipients. Using IceStorm can be summarized as the following steps.

First, define the Topic interface and compile it. This is similar to the first step of the RPC service except for the module name, interface name, and method, we will skip the explanation here. For the convenience of the following description, there is a need declaring that the method defined in the interface is "void report (String measurements)".

Then, implement the publisher. To obtain three proxies may be the first step. The first one refers to the "TopicManager" which is the primary IceStorm object, used by both publishers and subscribers. The second one relates to the Topic defined above either by creating it if it does not exist, or retrieve the proxy for the existing Topic. The last one, which is provided for the purpose of publishing, involves the Topic's "publisher object". When three proxies are obtained, the publisher enters its main loop, collecting measurements and publishing them via the IceStorm publisher object.

Finally, implement the subscriber. To obtain a proxy for the "TopicManager" is still the work to be done here. Create an object adapter to host our Monitor servant as long as instantiate and activate it with the object adapter by implementing the Topic interface and override the method inside. After all these are done, it's clear to subscribe to the Topic and process"report" messages until shutdown.

2.3 JFreeChart

JFreeChart is an open-source Java chart library that makes it easy for developers to display professional quality charts in their applications. The library is used in the CGLT to:

- Plot power value of the power meter in real time all the time.
- Plot power value of each channel of the data acquisition equipment in real time when necessary.
- Plot curve of antenna pointing and Gaussian fitted curve of pointing.
- Plot device parameter curve in real time when necessary (see Sect. 2.4).

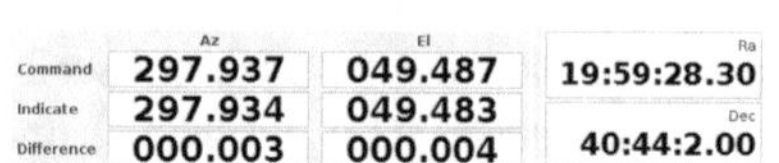

Fig. 6. Antenna parameters monitor with MonitorComponent

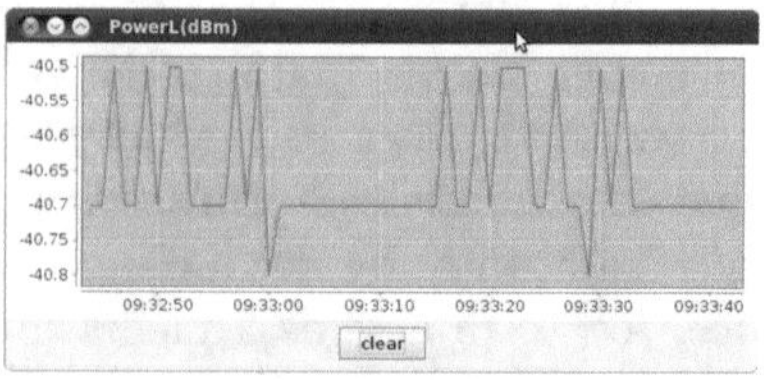

Fig. 7. Power value plot in MonitorComponent

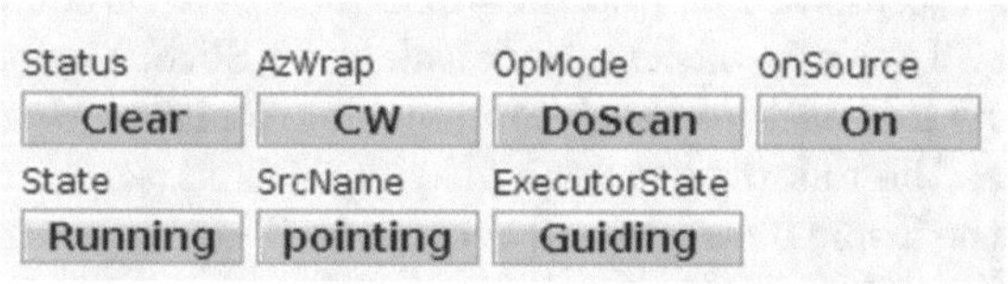

Fig. 8. Antenna status monitor with StatusComponent

2.4 MonitorComponent

MonitorComponent is a component that we developed and used to monitor numerical parameters. It is a hyperlink with or without a label as a heading. Figure 6 shows the situation where we use it to monitor antenna parameters. The main reason of MonitorComponent being a hyperlink rather than a label is that when we left click the component, a plot panel using JFreeChart to plot is showing. This panel is used to plot the acquired parameters of the device in real time. Figure 7 is the real-time plot panel that appears after clicking the MonitorComponent that monitors the power value of left circular polarization (LCP) amplifier of the receiver. A MonitorComponent can also monitor the normality of parameters, it is displayed in red when the parameters are abnormal.

2.5 StatusComponent

StatusComponent is a component that we have developed to monitor the device status or station status (appear as string). It is more like a LED with a string being its value. We can change the color of this component according to the value and judge the normality by the value as long as the color of the component. Figure 8 is a situation where we use StatusComponent to monitor the antenna status.

3 Main Window

We use the above packages or libraries (including but not limited to) to develop the CGLT. The main window of the software is shown in Fig. 9. As can be seen from the figure, the amount of data that the software needs to process is relatively large. The software has the ability to access almost any hardware, have hierarchical structure in monitoring and controlling, and it can be run manually or automatically, providing users with a graphical interface that is platform-independent, scalable, distributed, fast and reliable.

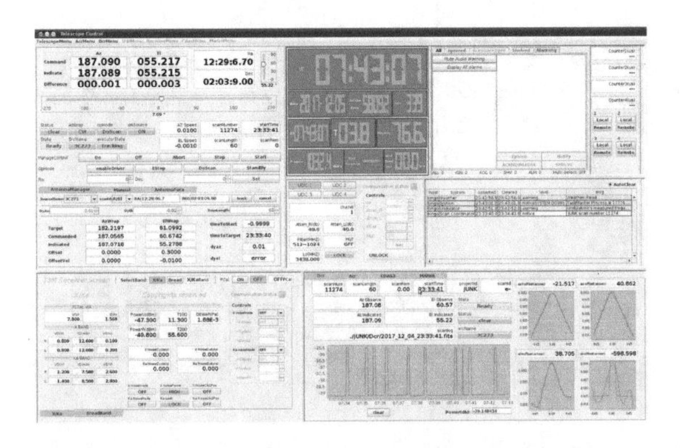

Fig. 9. Main window of the software

4 Implementation and Feasibility of the Software

The device managers are developed with C++ language while other parts of the CGLT are developed with Java language using packages and libraries mentioned above. The whole software can be operated in Ubuntu operation system as well as the Windows operation system.

We have conducted third-party testing of the software. After some comments are submitted, we made appropriate changes to the software and conducted the final third-party testing. The software is currently working well. Figure 10 shows the interference fringe obtained by 13-meter radio telescopes experimented with another antenna using the software. It can not only show that all subsystems of the 13-meter radio telescope are working properly, but also show the feasibility of the software.

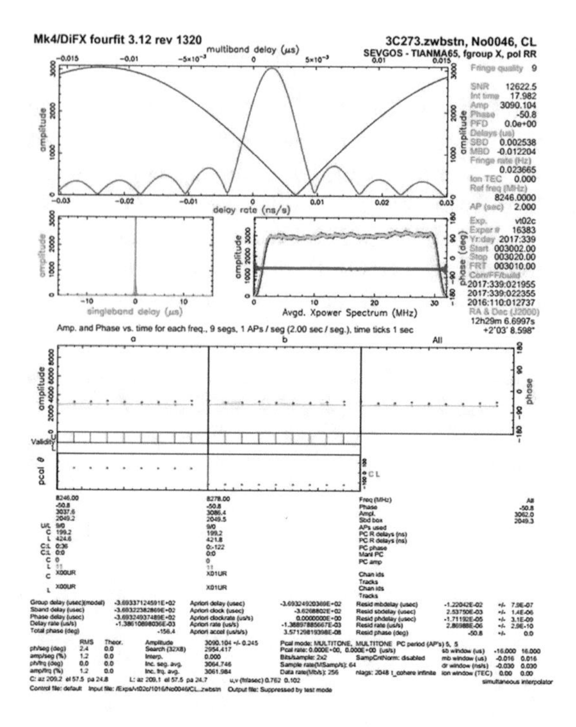

Fig. 10. Interference fringe obtained by 13-meter radio telescopes experimented with another antenna using CGLT

5 Future Maintenance and Upgrade

The development of web version and WeChat version of CGLT is in progress. In the near future, users can view the status of station system through Web pages or WeChat subscriptions. While remote control has become the goal of numerous telescope stations in the era of high automation requirements, yet several stations abroad have already done this, we are exploring on the way to remote control. The CGLT will also be applied to several other antennas of the same type, and it is expected that this software can be used to control multiple antennas in the future.

6 Summary

We have developed the 13-meter radio telescope monitor and control software, and it has been used for astronomical observations and various tests. We will continue to explore and add new features to the software to strive for further and further advancement in software automation.

Reference

1. Ice Manual. https://doc.zeroc.com/

The Method of Formulating Personalized Training Programs for New Employees Based on Knowledge Management

Chengqiang Wang, Hongfei Zhan(✉), Junhe Yu, and Rui Wang

Ningbo University, Ningbo 315000, China
zhanhongfei@nbu.edu.cn

Abstract. The training for new employees has always been a hot issue in the practice of human resources management. Based on the corporate research, this paper studies the method of formulating personalized training programs for new employees from the perspective of knowledge management. First, this paper puts forward the model of formulating personalized training programs for new employees. Then, a method of designing personalized training content based on business case is proposed. Finally, this paper constructs the framework of knowledge service system based on training program libraries. The purpose of this study is to provide a new pattern for the formulation of training programs, which can improve the targeted and scientific nature of the training for new employees and promote the reuse and inheritance of enterprise knowledge and experience.

Keywords: Knowledge management · Training programs · Business cases Training content · Knowledge services

1 Introduction

In the era of knowledge economy, knowledge resources have become the most important strategic resources of contemporary enterprises, and new employees, as an important part of this resource, must be effectively controlled and guided in order to make them become the core driving force for enterprise development [1]. To achieve this goal, companies should pay attention to the training of new employees. In the analysis of the survey data of nearly 300 different enterprises, the author found that 87% of the enterprises have business case libraries, but 61% of the new employees think that the biggest confusion is the lack of the reference to business cases. This data shows that most companies have established their own business case libraries, but they do not make full use these important knowledge resources in the training of new employees. From the perspective of knowledge management [2], new employees are the demand side of knowledge, the enterprise is the supply side of knowledge, and the business case library contains a lot of knowledge for solving business problems. So how to use the business case library to formulate training program is the starting point and innovation point of this paper.

N. Xiong et al. (Eds.): ISCSC 2018, AISC 877, pp. 45–55, 2019.
https://doi.org/10.1007/978-3-030-02116-0_6

As early as the beginning of the last century, the significance of employee training has been the focus of some researchers in developed countries. McGehee and Thayer [3] believes that training is a widely used human resource management function and is one of the important ways to organize human capital investment. The purpose is to improve employees' competency and production efficiency by influencing or changing employee attitudes, behaviors and skills. Meyer et al. [4] believe that from the training content, special skills training will enhance emotional commitment and sustained commitment; general skills training will enhance emotional commitment and reduce ongoing commitment. Mathieu et al. [5] found that the willingness to train is related to the employee's response to the training evaluation. Bartlett and Kang [6] believe that increasing training opportunities for employees is more effective than prior training methods that determine quantity and content.

In China, the concern for employee training began at the end of the last century. Most Chinese scholars use the method of empirical research to study this subject. Wang and Shi [7] put forward their own point of view when they learned and studied abroad on the evaluation of organizational training needs. They believed that follow-up training must be based on the results of the evaluation of training needs from three aspects: organization, task and personnel, which are complementary and indispensable. Qiao and Wan [8] tried to carry out in-depth research on the simulation methods commonly used in training, and carried out case experiments with specific research topics. Liu and Yang [9] cooperated to explore and discuss the western training model, focusing on nine kinds of patterns of training for foreign employees. Wu [10] and Su [11] carried out in-depth study in the training system. Zhang [12] discussed the strategies about the effective implementation of new employees training from four aspects.

To sum up, we can see that the research on the training for employees generally concerned about the basic theory and method. However, for the analysis of individual differences, domestic and foreign researches are very rare. Therefore, this paper attempts to study the method of formulating individual employee training programs based on the related theories of knowledge management, aiming to provide reference for the enterprise to train new employees.

2 The Model of Formulating Personalized Training Programs

The training for new employees is a matter that many companies need to do. On the one hand, effective new employee training program can assist new employees to quickly adapt to the new working environment and understand the knowledge and skills needed to master the work. On the other hand, it can help the production and operation of enterprises proceed smoothly. As a result, the formulation of new employee training programs is critical. So this paper builds the model of formulating personalized training programs for new employees, as shown in Fig. 1.

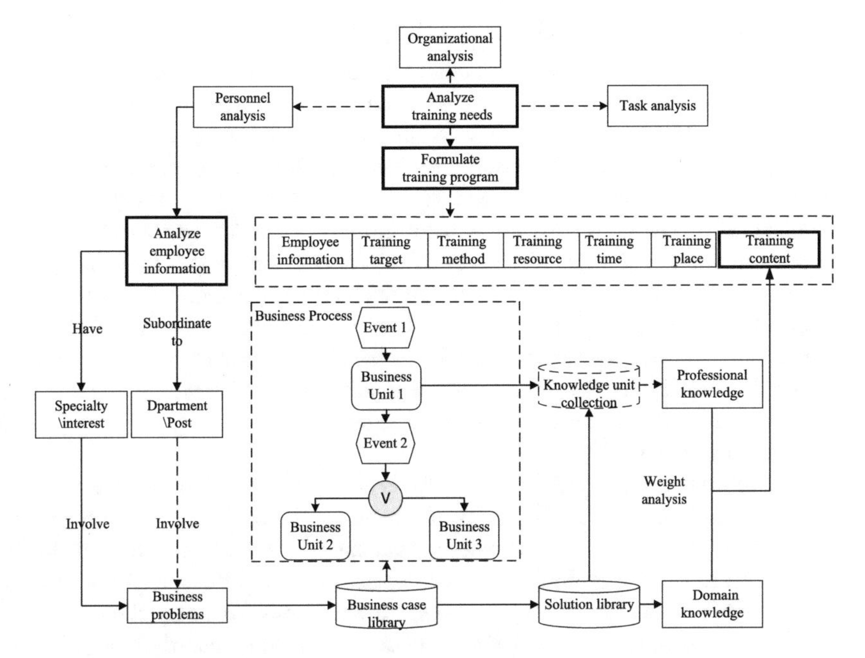

Fig. 1. The model of formulating personalized training programs.

2.1 The Analysis of Training Needs

The end of the last century, Goldstein et al. eventually modularized the analysis of training needs through unremitting efforts, and formed the famous Goldstein model [13]. Goldstein pointed out that training needs analysis should think about three aspects: organizational analysis, task analysis and personnel analysis. Organizational analysis is to determine which employees and departments in the organization need training. Task analysis can determine the various training tasks for the position, fine definition of the importance, frequency and difficulty of each task, and reveals successful completion of the task required training content such as knowledge, skills and attitudes. Personnel analysis is to analyze the gap between the existing situation and the ideal task requirements from the perspective of the actual situation of the employees, that is, the "target difference" in order to form the basis for training objectives and content. Through the comparison and synthesis of these three aspects of the analysis results, the most necessary knowledge, skills and attitudes of staff training can be revealed. Therefore, in order to better carry out training for new employees, the training needs of new employees should be analyzed from three aspects of organization, task, and personnel:

(1) **Organizational analysis:** The purpose of organizational analysis is to clarify the qualities, knowledge, and competencies that are available to enable new employees to do the job. Once new employees enter an enterprise, the company hopes they can adjust their minds as soon as possible from the following aspects. First, recognize themselves. Second, understand corporate history, development strategy, organizational structure, culture and values. Finally, comply with the rules and regulations of the enterprise. With new employees understanding the business initially and entering a department to work, the company hopes they can quickly become familiar with the functions of the department, personnel structure, the job responsibilities and work processes, with a certain ability to perform and communication coordination to blend into the work team as soon as possible. After new employees are familiar with the department and the post situation, company wants them to master the various knowledge and skills about their jobs, which enable them to work efficiently.

(2) **Task analysis:** The purpose of task analysis is to analyze the knowledge and skills that new employees should be equipped to carry out their jobs effectively. First of all, it's important for new employees to understand the basic situation of the job, including the job responsibilities of the department, the job responsibilities of the post, personnel structure, work processes, work relations. And they should have the abilities of teamwork, communication and execution. After understanding the basic information of the job, they will be able to carry out their own work skillfully and master knowledge and skills of the post by learning and working.

(3) **Personnel analysis:** The purpose of personnel analysis is to analyze the knowledge, qualities and skills that new employees should know and master. After entering the enterprise, due to the unfamiliar environment, new employees hope to be able to initially understand the organizational structure, corporate culture, departmental settings, products, corporate history, pay and benefits, rules, regulations and so on. With the understanding of the enterprise, they eager to understand the work, the career development, and whether they can be accepted by others. They want to understand the work of the team, job responsibilities, work objectives, interpersonal relationships and other content, and to master the industry or working language, which is conducive to cultivate good communication skills. Finally, new employees want to be able to continuously improve their professional knowledge and job skills, thus to get the organizational and others' recognition and affirmation.

2.2 The Analysis of Employee Information

After completing the training needs analysis, companies need to recruit suitable new employees and begin a new round of employee training. However, many companies often adopt the traditional collective unified training, ignoring the personalized information of employees, which undoubtedly greatly reduces the effect of training. Only a reasonable analysis of the relevant information of employees can formulate suitable training programs for employees. For example, judge the business problems that employees will involve in their work in the future by the department and post information of the employee, and then take the knowledge and skills needed to solve these problems as the content of employee training. In addition, based on the employee's specialty and interest information, the potential of employees can be tapped to determine the business areas in which they can be developed and the right career development path for employees can be assigned. Therefore, prior to the development of a training program, it is necessary to analyze the relevant information of employees so as to improve the pertinence and effectiveness of the training.

3 The Method to Design Personalized Training Content Based on Business Case Libraries

Projects in an enterprise are triggered by business problems that need to be solved in production and operation activities. They end with the successful application of the solution, and finally form a case that becomes the company's intellectual assets. Therefore, the solution reuse in historical cases for solving new business problems is a very important knowledge management activity in business operations. Case-based reasoning (CBR) technology is an important means of knowledge reuse. CBR-based knowledge systems can save users a lot of time and effort to solve problems [14]. For employee training, training content based on case-based reasoning can better meet the company's actual needs. Therefore, this paper will refine the knowledge in the business case, and propose a method for the development of personalized training content from the perspective of a more granular case knowledge reuse.

The business case in the enterprise is considered as a collection of business problems and solutions, and it is the precipitation of enterprise knowledge. Knowledge is used in business processes and there is a close relationship between business processes and knowledge. Knowledge is continuously input and output along with the execution of tasks in business processes. Business processes are the main axes of knowledge and knowledge services business. Therefore, there is a lot of knowledge required in the business case solving process in the business case, and this knowledge is the domain knowledge involved in employee work. From the viewpoint of system science, the knowledge in the business case is essentially a knowledge system [15], and it reflects its own knowledge and the structural relationship. The knowledge hierarchy is used to study the knowledge system in the business case, as shown in Fig. 2. The specific implementation is as follows:

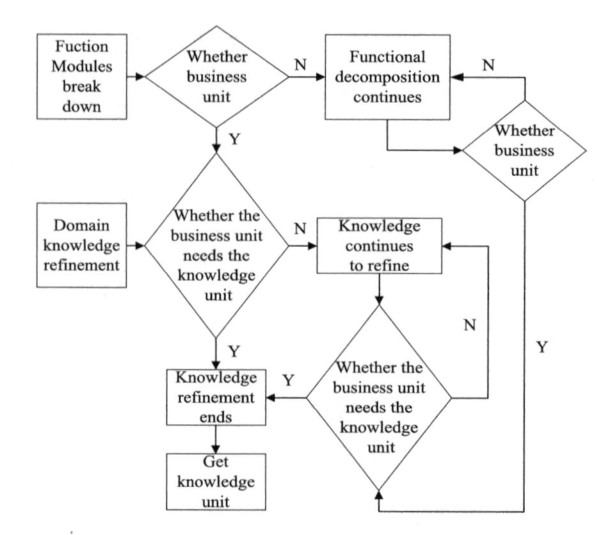

Fig. 2. Knowledge refinement based on business cases.

First, the knowledge in the enterprise is refined according to each function module of the business case (or according to the application needs). It can be divided into different knowledge areas, such as R&D knowledge, production knowledge, sales knowledge, management knowledge, product recycling knowledge and other knowledge areas.

Second, further decompose the function modules into sub-function modules. Since each functional module has large, abstract and complex features, we need to further decompose it in order to better complete each functional module. At the same time, each knowledge area is further refined into sub-fields accordingly. Each sub-field is called a knowledge sub-domain. For example, the above R&D knowledge area can be refined into business theory, policy and law, environment, process design, product design and other knowledge subdomains.

Finally, continue to decompose the sub-function modules, until they reach the operable, specific, and most basic level. Meanwhile, the corresponding knowledge subdomains are refined into smaller sub-areas of knowledge. For example, process design can be refined into smaller knowledge subfields such as design standards, resources, process options, and process models. There are two aspects to understand the operational and specific aspects: (1) the functional modules at this level are completed by the collaboration of a single employee and various individual resources (software, tools, etc.). (2) the function at each level has obvious differences.

The paper refers the operable, specific, and most basic level as the business unit level, the function at this level is called the business unit. At the same time the knowledge refinement is over, and the corresponding knowledge is called a knowledge unit. Through the above process, it can be seen that the business unit in the business case is completed by a series of knowledge units through mutual cooperation, and the knowledge unit collection is the professional knowledge that the staff needs to learn. According to new employee information and Pareto principle (that means 20% of domain knowledge and 80% of professional knowledge), company can determine the personalized training content of new employees.

4 The Construction of Knowledge Service System Framework Based on Training Program Libraries

4.1 The Model of Training Program

The training program is the guiding document that the company uses to implement training in the process of training employees. Modeling the training program helps to realize the reuse and inheritance of standardized training programs. This paper builds the training program model as shown in Fig. 3, among which the training program development process is the core.

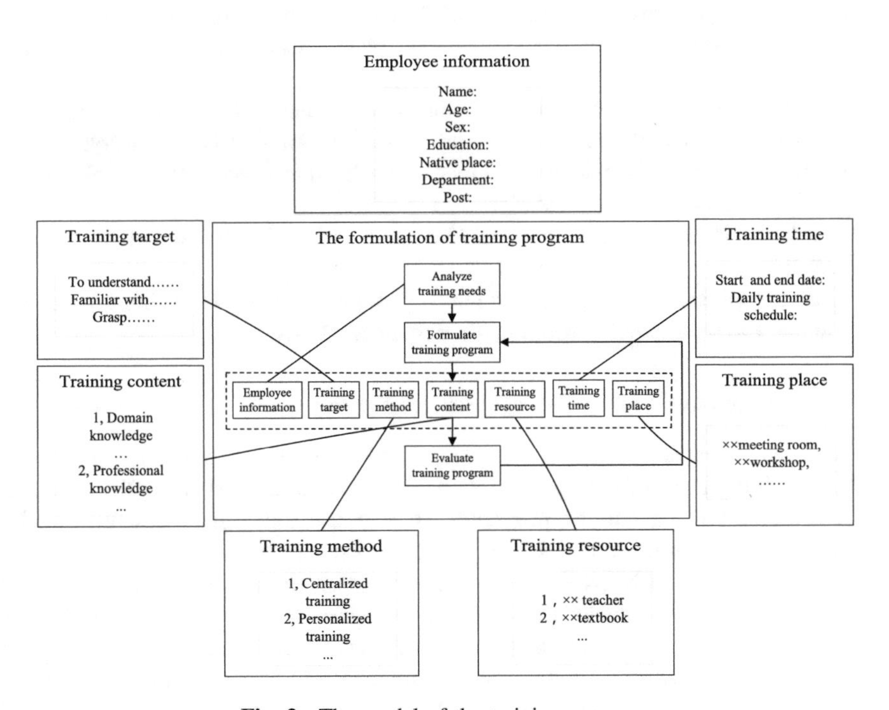

Fig. 3. The model of the training program.

The new employee training program is an organic combination of employee information, training target, training content, training resource, training method, training time and training place. The specific analysis is as follows:

(1) **Employee information:** Employee information is a description of the basic attributes of new employees, including name, age, sex, education, native place, department, position and other information.
(2) **Training target:** The training objective is a description of the company's requirements to be met by the trainee. The training objectives should set the overall goals and targets in a hierarchical manner, such as understanding of the company's history, a clear job responsibilities and content, etc.
(3) **Training content:** The training content is a description of the knowledge that new employees should be understood and grasped, including domain knowledge and professional knowledge. The professional knowledge is the training focus and also the individualized embodiment of the training content.
(4) **Training resource:** Training resource refers to the various support resources involved in the training period, generally including classrooms, conference rooms, work site, teaching materials, notebooks, pens, models, projectors, television, video, etc.
(5) **Training method:** The training methods are mainly divided into centralized training and personalized training. The centralized training is responsible for the training of domain knowledge, and the personalized training is responsible for the training of professional knowledge.
(6) **Training time:** Training time includes start and end date, and daily training schedule.
(7) **Training place:** Training place refers to the place where the training occurred, such as a certain meeting room and a certain workshop.

4.2 The Framework of Knowledge Service System

The company's valuable training knowledge and experience are often deposited in previous training programs. At present, companies do not have standardized methods for storing and reusing training programs. Therefore, a series of successfully implemented training programs are stored in the form of documents. They form a training program library, which helps the reuse and inheritance of corporate knowledge and experience. When new employees enter the company, the same or similar training programs are retrieved from the training plan database according to employee information, and corresponding training programs are invoked to obtain training content for new employees. The implementation of the training program needs to be evaluated and incorporated into the library management to provide reference for future staff training. Based on this, this paper constructs a knowledge service system framework based on the training program library. As shown in Fig. 4, it contains three levels: training model modeling layer, training program management layer, and training program application layer.

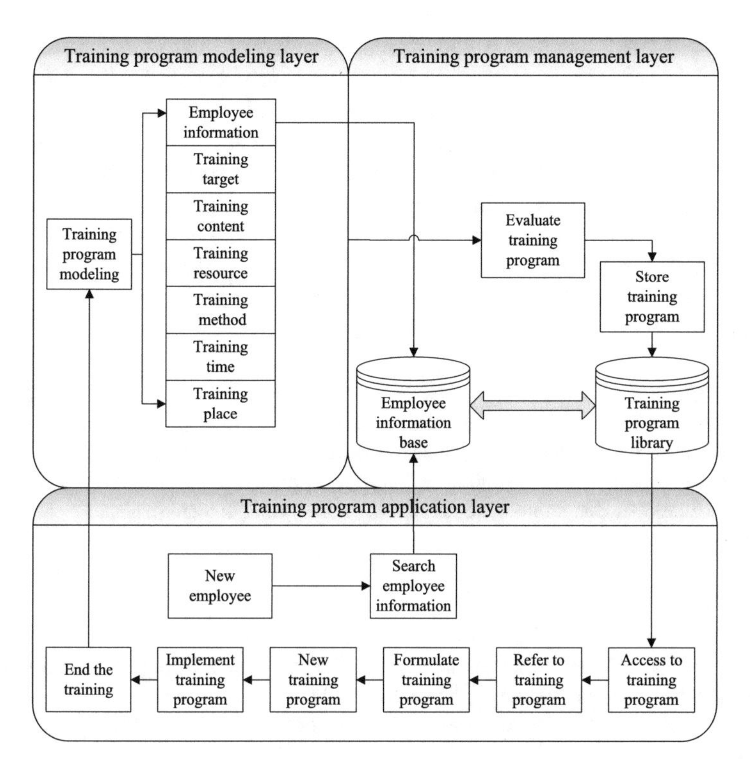

Fig. 4. The framework of knowledge service system.

(1) **Training program modeling layer:** The training program model constructed in this paper is composed of employee information, training target, training content, training resource, training method, training time and training place.
(2) **Training program management layer:** Training program management refers to evaluate and store training programs. Among them, the training program and employee information is a one-to-one relationship, that is to say, an employee corresponds to a training program. Therefore, access to training programs is available through employee information retrieval.
(3) **Training program application layer:** In the process of new employee training, enterprises access to the appropriate training program in accordance with the new employee information through employee information retrieval. And then refer to the obtained training programs and formulate new training programs which can improve the efficiency of formulating training programs.

5 Conclusions

In conclusion, the new employee training plays an increasingly important role in the development of enterprises. New employee training can enable them to achieve the change from the outsiders to the insiders as soon as possible, so that they can quickly adapt the new environment and identify the direction of their efforts. Besides, it can improve new employees constantly to create economic benefits for the enterprise.

Since new employees generally have different knowledge needs, the enterprises need to formulate personalized training programs for different new employees. This paper discusses the new employee training from the perspective of knowledge management. First, the model of formulating training program is constructed. And then the method of designing personalized training content is elaborated. Finally, this paper builds the training program model and the framework of knowledge service system based on training program libraries, which is helpful to the reuse and inheritance of corporate training knowledge and experience.

Acknowledgements. The research work of this paper is based on the project 71671097 which was supported by National Natural Science Foundation of China. This material is also based upon work funded by the Zhejiang Province Natural Science Foundation No. LY16G010004 and the Zhejiang Provincial Public Welfare Technology Application Research Project 2016C31047 and LGG18E050002. The authors would like to express their sincere thanks.

References

1. Lendzion, J.P.: Human resources management in the system of organizational knowledge management. Procedia Manuf. **3**, 674–680 (2015)
2. Sheng, X.: A review of domestic knowledge management research. J. Libr. Sci. China **3**, 60–64 (2002)
3. McGehee, W., Thayer, P.W.: Training in Business and Industry. Wiley, New York (1961)
4. Meyer, J.P., Bobocel, D.R., Allen, N.J.: Development of organizational commitment during the first year of employment: a longitudinal study of pre- and post-entry influences. J. Manag. **17**(4), 717–733 (1991)
5. Mathieu, J.E., Tannenbaum, S.I., Salas, E.: Influences of individual and situational characteristics on measures of training effectiveness. Acad. Manag. J. **35**(4), 828–847 (1992)
6. Bartlett, K.R., Kang, D.: Training and organizational commitment among nurses following industry and organizational change in New Zealand and the United States. Hum. Resour. Dev. Int. **7**(4), 423–440 (2004)
7. Wang, P., Shi, Z.: Research overview of training needs assessment. J. Dev. Psychol. **6**(4), 36–38 (1998)
8. Qiao, Z., Wan, D.: Analysis and application of management training simulation methods. Hum. Resour. Dev. China **11**(15), 49–52 (2004)
9. Liu, B., Yang, Q.: Summary of the enterprise training mode. Sci. Technol. Prog. Policy **4**, 141–143 (2004)
10. Wu, Y.: Construction of structured training system. Hum. Resour. Dev. China **3**, 50–51 (2004)
11. Su, W.: Discussion on effective training system. Hum. Resour. Dev. China **8**, 50–51 (2004)
12. Zhang, W.: New employee training effective implementation strategy. Hum. Resour. Dev. China **8**, 49–51 (2006)
13. Goldstein, H.W., Yusko, K.P., Braverman, E.P., Smith, D.B., Chung, B.: The role of cognitive ability in the subgroup differences and incremental validity of assessment center exercises. Pers. Psychol. **5**(1), 357–374 (1998)

14. Pal, S.K., Shiu, S.:: Foundations of Soft CASE-Based Reasoning, pp. 19–31. A Wiley-Interscience Publication, New York (2004)
15. Xi, Y., Dang, Y., Wu, J.: The complex network modeling method and its application to the knowledge system. In: International Federation for Systems Research, Kobe, Japan, pp. 421–431 (2005)

Refined Molecular Docking with Multi-objective Optimization Method

Ling Kang(✉)

Department of Software Engineering, Dalian Neusoft University of Information, Dalian 116023, China
kangling@neusoft.edu.cn

Abstract. Molecular docking is a typical method of structure-based drug design. It aims to identify a ligand that binds to a specific receptor binding site and determine its most favorable binding pose. A new multi-objective optimization method is introduced to improve the docking accuracy. Two diverse scoring functions, energy-based and contact-based, are applied to demonstrate the multi-objective strategy. The energy-based scoring function is a force field one which consists of the Coulomb and Van der Waals terms. The contact-based one is a simple summation of the number of heavy atom contacts between ligand and receptor. The flexibility of receptor is considered by modeling the residue groups movements that occur during the interaction. A new multi-objective docking program that accounts for protein flexibility has been developed. An adaptive multi-generation evolutionary algorithm is adopted to solve the optimization problem. Results show that the method can give more robust and reasonable conformation of ligand in molecular docking.

Keywords: Molecular docking · Optimization model · Multi-objective

1 Introduction

Structure-based drug design has become an accepted means in drug discovery [1]. Being a typical method of structure-based drug design, molecular docking is one of the most widely used in practice [2–4].

Molecular docking is an optimization problem which predicts the interaction energy while docking a ligand into the binding site of a receptor. During the optimization process, scoring functions are used to evaluate the binding affinities between the ligand and the receptor [5–7]. Since the first docking program DOCK by Kuntz et al. [8], numerous docking programs have been reported [9–15]. However, due to the huge difficulties to calculate the real binding energy, most conformational optimization methods in docking programs have adopted a single objective only, such as energy score, shape complementarity, or chemical complementarity. It is obvious that this evaluation is somewhat partial and very inaccurate [16–18]. Consensus scoring has been developed by combining existing score functions in order to reduce the deviations brought by single ones [19–21]. However, simple combination of scoring functions will lead to the discontinuities of energy curves. And it becomes more difficult to solve

N. Xiong et al. (Eds.): ISCSC 2018, AISC 877, pp. 56–63, 2019.
https://doi.org/10.1007/978-3-030-02116-0_7

the optimization problem. Thus it is reasonable to consider multiple score functions to improve the accuracy and efficiency.

Applying multiple score functions makes molecular docking to become a multi-objective optimization problem. In this paper, we respectively introduce two types of score functions which are energy score and contact score. Then a new multi-objective optimization method has been presented to solve the optimization problem. These two score functions follow different principles and focus on different aspects of ligand binding. The docking result indicates that the multi-objective strategy can enhance the pose prediction power of docking.

2 Multi-objective Optimization for Molecular Docking

2.1 Multi-objective Optimization Model

The multi-objective optimization problem can be formulated mathematically as

$$\begin{array}{ll} min & y = f(\mathbf{x}) = (f_1(\mathbf{x}), f_2(\mathbf{x}), \ldots, f_m(\mathbf{x})) \\ s.t. & g(\mathbf{x}) \le 0 \\ where & \mathbf{x} = \{\mathbf{x}_{lig}, \mathbf{x}_{rec}\}^T \\ & \mathbf{x}_{lig} = \{T_x, T_y, T_z, R_x, R_y, R_z, \alpha_1, \cdots, \alpha_n\}^T \\ & \mathbf{x}_{rec} = \{C_{1x}, C_{1y}, C_{1z}, \cdots, C_{px}, C_{py}, C_{pz}\}^T \end{array} \tag{1}$$

where the design vector x consists of two components $\mathbf{x}_{lig}$ and $\mathbf{x}_{rec}$. $\mathbf{x}_{lig}$ is made up of three parts: (T_x, T_y, T_z) and (R_x, R_y, R_z) are the state variables of translation and rotation of the entire ligand for the orientation search, respectively; and $\alpha_1, \cdots, \alpha_n$ are the torsion angles of the n rotatable bonds of the ligand for the conformation search. $\mathbf{x}_{rec}$ describes the conformation of the receptor. To consider the flexibility of the receptor, we classify the residues near the active site into several groups and then use the centers of these groups to describe the movements of the receptor. Here, $(C_{ix}, C_{iy}, C_{iz})(i = 1, 2, \cdots, p)$ is the center position coordinates of each residue group; p is the number of residue groups.

The constrain functions $g(\mathbf{x})$ can be represented as

$$\begin{cases} \underline{T_x} \le T_x \le \overline{T_x} \\ \underline{T_y} \le T_y \le \overline{T_y} \\ \underline{T_z} \le T_z \le \overline{T_z} \\ -\pi \le angle_L \le \pi, & angle_L = R_x, R_y, R_z \\ \underline{X}_{box} \le C_{ix} \le \overline{X}_{box} & i = 1, \ldots, p \\ \underline{Y}_{box} \le C_{iy} \le \overline{Y}_{box} & i = 1, \ldots, p \\ \underline{Z}_{box} \le C_{iz} \le \overline{Z}_{box} & i = 1, \ldots, p \\ -\pi \le angle_R \le \pi, & angle_R = \alpha_1, \cdots, \alpha_n \end{cases} \tag{2}$$

Here, the objective vector y consists of two parts, which are energy score $f_1(\mathbf{x})$ and contact score $f_2(\mathbf{x})$, respectively.

The objective function $f_1(\mathbf{x})$ is the interaction energy between the ligand and protein. It consists of the Coulomb and Van der Waals terms of force field functions:

$$f_1(\mathbf{x}) = \sum_{i=1}^{n_{lig}} \sum_{j=1}^{n_{rec}} \left(\frac{A_{ij}}{r_{ij}^a} - \frac{B_{ij}}{r_{ij}^b} + 332.0 \frac{q_i q_j}{D r_{ij}} \right) \quad (3)$$

where each term is a double sum over the ligand atom i and the receptor atom j, r_{ij} is the distance between atom i in the ligand and atom j in the receptor, A_{ij}, B_{ij} are van der Waals repulsion and attraction parameters, a, b are Van der Waals repulsion and attraction exponents, q_i, q_j are the point charges on atoms i and j, D is the dielectric function, and 332.0 is the constant that converts the electrostatic energy into kilocalories per mole.

The objective function $f_2(\mathbf{x})$ is contact score, which is a simple summation of the number of heavy atom contacts between the ligand and receptor. At the time of construction of the contact scoring grid, the distance threshold defining a contact is set. Atom VDW overlaps is penalized by checking the bump filter grid. It provides a simple assessment of shape complementarity. It can be useful for evaluating primarily non-polar interactions.

2.2 Multi-objective Evolutionary Strategy

A multi-objective evolutionary strategy was developed to solve the above multiple optimization problem. The optimization problem (1) can be transferred into following evaluation function model with a single objective

$$\begin{array}{ll} min & h(F) = \frac{1}{\rho} \ln \sum_{i=1}^{m} \exp(\rho \lambda_i f_i(\mathbf{x})) \\ s.t. & \mathbf{g}(\mathbf{x}) \leq 0 \\ where & \mathbf{x} = \{\mathbf{x}_{lig}, \mathbf{x}_{rec}\}^T \end{array} \quad (4)$$

where $\lambda_i \geq 0$ is the weight of each objective, and $\sum_{i=1}^{m} \lambda_i = 1 (i = 1, 2, \ldots, m)$; By varying the weights, the set of the pareto solutions was generated. ρ is a large positive real number, and it is set to be 103 in this work.

An improved genetic algorithm with multi-population strategy and an adaptive evolutionary strategy is adopted to solve the above optimization problem [22]. A multi-population genetic strategy is used to maintain diversity among different populations and to avoid premature problems to some extent. Namely, different populations with certain number of individuals are generated randomly with the same searching space. Selection and mutation operations for each population are performed independently, while a crossover operation is carried out between different populations.

3 Results and Discussion

3.1 Preparation of Test Dataset

A subset of 37 protein-ligand complexes is chosen from the public GOLD test dataset [9] for evaluation of reproduction. The rotatable bonds of the ligands range widely from 1 to 16. According to the biological interaction pairs, each complex was separated into a probe molecule and a ligand. Each protein molecule was obtained by excluding all the structural water molecules and the ligand from the receptor PDB file. Next, a mol2 file of the protein was generated by adding hydrogen atoms and Kallman charge. Then, the active site was defined as the area which consists of those residues around the ligand within a radius of 6.5 Å. The ligand was prepared by adding hydrogen atoms and Gasteiger-Marsili atomic charges.

Usually, the weight value of each objective must be provided as initial data. However, it is very difficult to set these two values. Different values will lead to different results. Here one of the weights is assigned to be the added design variables to overcome the difficulty in confirming the values. The lower and upper limits of it can be defined in a reasonable region, here $0 < \lambda_i < 1$, $i = 1, 2$ $\lambda_1 + \lambda_2 = 1$.

3.2 Docking Accuracy

In general, the critical criterion for evaluating docking methods is docking accuracy [12]. Root-mean-square deviation (RMSD) value, which is defined as the locations of all heavy atoms in the model from those of the crystal structure, is applied to evaluate the docking accuracy. Generally, the docking accuracy is acceptable if the RMSD value between the docked pose and native structure is less than 2.0 Å.

According the docking results by using the proposed method, a statistics has been performed. The multi-objective docking method yielded 46% docking solutions with RMSD values below 1.0 Å. Nearly 81% cases got the results with RMSD values under 2.0 Å. The average RMSD of this study is only 1.27 Å.

Two docking methods have chosen to make comparisons, which are GAsDock [14] and Frdock [15]. Both of the two methods adopt single-objective score function. GAsDock is performed with flexible ligands and rigid receptor supposition, while Frdock is performed with flexible ligands and flexible receptor.

Thymidylate Synthase (PDB entry: 1TSD) in complex with 1843U89 was chosen for evaluation of reproduction (Fig. 1)

The RMSD value of the multi-objective docking of 1TSD is 0.89 and the docked pose is given in Fig. 2. The molecule shown in blue is the native pose derived from the crystal structure and the molecule shown in yellow is docked pose of multi-objective docking method.

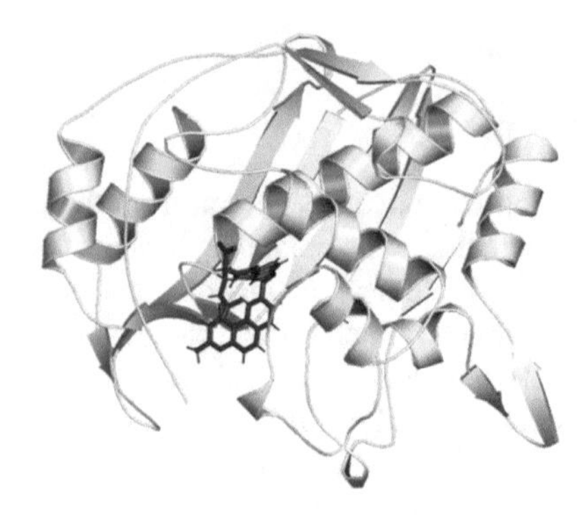

Fig. 1. Thymidylate Synthase (PDB entry: 1TSD)

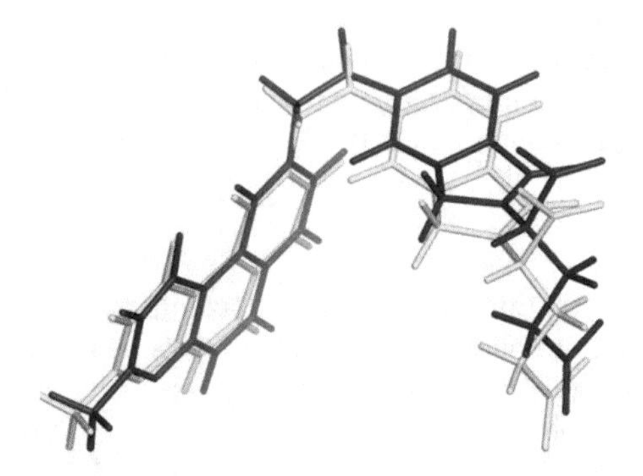

Fig. 2. The native pose and docked ligand pose with the multi-objective docking strategy (blue: native pose, yellow: docked pose)

Table 1 shows the docking results of these three methods, which consist of two parts: energy score and RMSD value. Figures 3 and 4 show the docked conformations by using GAsDock and Frdock respectively.

Table 1. Comparison of docking accuracy with other methods

Docking programs	Energy score (kcal/mol)	RMSD (Å)
GAsDock	−47.88	3.88
Frdock	−31.60	1.06
Proposed	−28.99	0.89

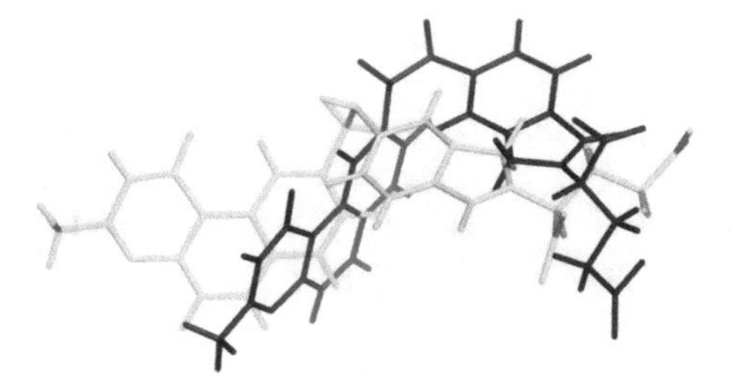

Fig. 3. The native pose and docked ligand pose with GAsDock (blue: native pose, yellow: docked pose)

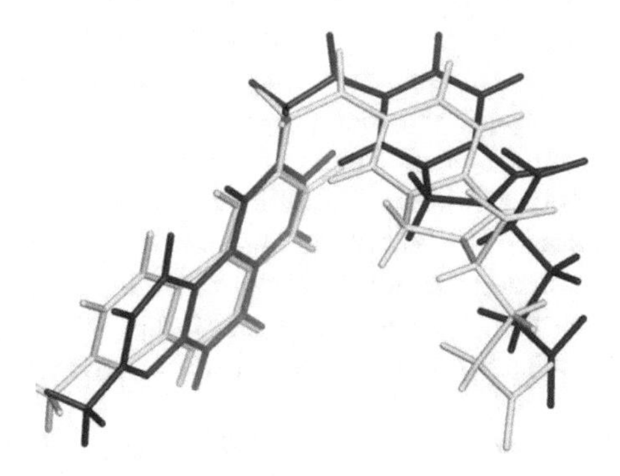

Fig. 4. The native pose and docked ligand pose with Frdock; (blue: native pose, yellow: docked pose)

3.3 Multi-objective Versus Single-Objective

The performance of the multi-objective strategy against the single objectives is discussed first.

The RMSD value of the two single-objective docking methods, GAsDock and Frdock, are 3.88 and 1.06 respectively. For the proposed one, the RMSD value compared with the conformation of X-ray complex is 0.98.

It is obvious that the multi-objective optimization method can decrease the deviations derived from individual score function and find a pose more close to the native one.

3.4 Semi-flexible Versus Flexible

GAsDock is a rigid-flexible docking program, while the proposed one is flexible-flexible docking method.

As shown in Table 1, the RMSD value of the proposed one is quite better than the semi-flexible one. The flexible-receptor docking performs better than rigid-receptor docking. From this point of view, flexible docking can decrease the ratio of wrong prediction by considering the conformational change of the receptor during the receptor-ligand binding process.

However, the binding score of the proposed method is −28.99 while the score of the semi-flexible one is −47.88. From the view of this case, the multi-objective strategy will find a pose with balance between energy distribution and contact distribution and then a more reasonable solution will be found.

4 Conclusions

This paper has presented a multi-objective molecular docking optimization model. To reduce the correlations between each other, two different kinds of scoring functions—energy-based and contact-based, are treated as the objectives during the docking optimization. Instead simple combination with predefined weight factors, an aggregate function is adopted to combine the multiple objectives and to approximate the real solution of the original multi-objective and multi-constraint problem. Besides, receptor flexibility has been considered during the optimization process. Finally, an improved genetic algorithm is introduced to solve the optimization problem.

The multi-objective strategy is not restricted to these scoring functions. Different scoring functions can be combined with this multi-objective strategy to research the best combinations, and this will be the direction of our next work.

Acknowledgement. The author gratefully acknowledges financial support for this work from the Foundation of Liaoning Education Ministry (No. L2015037).

References

1. Cheng, T., Li, Q., Zhou, Z., Wang, Y., Bryant, S.H.: Structure-based virtual screening for drug discovery: a problem-centric review. ASPS J. **14**(1), 133–141 (2012)
2. Kitchen, D.B., Decornez, H., Furr, J.R., Bajorath, J.: Docking and scoring in virtual screening for drug discovery: methods and applications. Nat. Rev. Drug Discovery **3**(11), 935–949 (2004)
3. Ripphausen, P., Nisius, B., Peltason, L., Bajorath, J.: Quo vadis, virtual screening? A comprehensive survey of prospective applications. J. Med. Chem. **53**(24), 8461–8467 (2010)
4. Meng, X.Y., Zhang, H.X., Mezei, M., Cui, M.: Molecular docking: a powerful approach for structure-based drug discovery. Curr. Comput. Aided Drug Des. **7**(2), 146–157 (2011)
5. Chung, H.W., Cho, S.J.: Recent development of scoring functions on small molecular docking. J. Chosun Nat. Sci. **3**(1), 49–53 (2010)
6. Truchon, J.F., Bayly, C.I.: Evaluating virtual screening methods: good and bad metrics for the “early recognition” problem. J. Chem. Inf. Model. **47**(2), 488–508 (2007)
7. Wang, R., Yipin Lu, A., Wang, S.: Comparative evaluation of 11 scoring functions for molecular docking. J. Med. Chem. **46**(12), 2287–2303 (2003)
8. Kuntz, I.D., Blaney, J.M., Oatley, S.J., Langridge, R., Ferrin, T.E.: A geometric approach to macromolecule-ligand interactions. J. Mol. Biol. **161**(2), 269–288 (1982)
9. Jones, G., Willett, P., Glen, R.C., Leach, A.R., Taylor, R.: Development and validation of a genetic algorithm for flexible docking. J. Mol. Biol. **267**(3), 727–748 (1997)

10. Friesner, R.A., Banks, J.L., Murphy, R.B., Halgren, T.A., Klicic, J.J., Mainz, D.T., et al.: Glide: a new approach for rapid, accurate docking and scoring. 1. method and assessment of docking accuracy. J. Med. Chem. **47**(7), 1739–1749 (2004)
11. Kramer, B., Rarey, M., Lengauer, T.: Evaluation of the FlexX incremental construction algorithm for protein–ligand docking. Proteins: Struct. Funct. Bioinf. **37**(2), 228–241 (1999)
12. Jain, A.N.: Surflex: fully automatic flexible molecular docking using a molecular similarity-based search engine. J. Med. Chem. **46**(4), 499–511 (2003)
13. Cosconati, S., Forli, S., Perryman, A.L., Harris, R., Goodsell, D.S., Olson, A.J.: Virtual screening with AutoDock: theory and practice. Expert Opin. Drug Discov. **5**(6), 697–707 (2010)
14. Li, H., Li, C., Gui, C., Luo, X., Chen, K., Shen, J., et al.: GAsDock: a new approach for rapid flexible docking based on an improved multi-population genetic algorithm. Bioorg. Med. Chem. Lett. **14**(18), 4671–4676 (2004)
15. Kang, L., Li, H., Zhao, X., Jiang, H., Wang, X.: A novel conformation optimization model and algorithm for structure-based drug design. J. Math. Chem. **46**(1), 182–198 (2009)
16. Perola, E., Walters, W.P., Charifson, P.S.: A detailed comparison of current docking and scoring methods on systems of pharmaceutical relevance. Proteins: Struct. Funct. Bioinf. **56**(2), 235–249 (2004)
17. Garcíasosa, A.T., Hetényi, C., Maran, U.: Drug efficiency indices for improvement of molecular docking scoring functions. J. Comput. Chem. **31**(1), 174–184 (2010)
18. Esmaielbeiki, R., Nebel, J.C.: Scoring docking conformations using predicted protein interfaces. BMC Bioinform. **15**(1), 1–16 (2014)
19. Charifson, P.S., Corkery, J.J., Murcko, M.A., Walters, W.P.: Consensus scoring: a method for obtaining improved hit rates from docking databases of three-dimensional structures into proteins. J. Med. Chem. **42**(25), 5100–5109 (1999)
20. Feher, M.: Consensus scoring for protein–ligand interactions. Drug Discovery Today **11**(9), 421–428 (2006)
21. Yang, J.M., Chen, Y.F., Shen, T.W., Kristal, B.S., Hsu, D.F.: Consensus scoring criteria for improving enrichment in virtual screening. J. Chem. Inf. Model. **45**(4), 1134–1146 (2005)
22. Kang, L., Li, H., Jiang, H., Wang, X.: An improved adaptive genetic algorithm for protein-ligand docking. J. Comput. Aided Mol. Des. **23**(1), 1–12 (2009)

Classifying Motor Imagery EEG Signals Using the Deep Residual Network

Zilong Pang[1], Jie Li[1(✉)], Yaoru Sun[1], Hongfei Ji[1], Lisheng Wang[1], and Rongrong Lu[2]

[1] Tongji University, CaoAn NO. 4800, Shanghai, China
nijanice@163.com
[2] Fudan University, HanDan NO. 220, Shanghai, China

Abstract. Classification of motor imagery electroencephalogram (EEG) signals is widely applied in the non-invasive brain–computer interface (BCI) field. Studying the feature and correlations is important for the classification of motor imagery. In this paper, we propose a method based on Deep Residual Network (DRN) for classifying the EEG signals of motor imagery. Firstly, the raw EEG data are represented in the time-frequency field using continuous wavelet transformation, and then a DRN model is built to classify MI tasks. Results show improved accuracy of the proposed method compared with other classification methods (SCP + SVM).

Keywords: EEG · Motor imagery · BCI · Deep residual network

1 Introduction

Brain Computer Interface (BCI) is a system that provides means of communication between the human brain and computers [1]. The most popular signal used for BCI is the scalp recorded electroencephalogram, because it is a noninvasive measurement that recording procedure is safe and easy to apply, and it is potentially applicable to almost all people including those who are seriously amputated and paralyzed.

The widely used mental control model in BCI is the motor imagery task [2]. Specifically, it has been known that imagining motor motions yields event-related synchronization (ERS) and event-related desynchronization (ERD) of the brain rhythmic activity within specific frequency band over centro-parietal lobes. The main control strategy of MI is to control machines using only the brain signal without actual movement of the subject's body.

In recent years, the deep learning method has shown good performance in various research fields such as language recognition and computer vision. Convolution neural networks (CNN) is one of the deep learning methods, and it is widely used in image recognition with high accuracies. Compared with the traditional CNN, the deep residual network (DRN) is easier to optimize, and it can increase the accuracy by increasing the equivalent depth.

Here, we propose a new method combining DRN and time-frequency representation to perform the classification of MI.

N. Xiong et al. (Eds.): ISCSC 2018, AISC 877, pp. 64–68, 2019.
https://doi.org/10.1007/978-3-030-02116-0_8

2 Method

2.1 Experiment Setup and Data Acquisition

The datasets collected during our BCI experiments were used to verify the effectiveness and robustness of the proposed tensor based scheme. Four healthy male subjects, aged from 21 to 30, participated in the experiment. They all gave informed consent as approved by the Ethics Committee. Sixty-two channels of EEG were recorded with a 64 Channel EEG System (SynAmps2, Neuroscan). In the data collection stage, each subject was asked to seat in an armchair, keeping their arms on the chair arms with their hands relaxed, and the eyes were requested to look at 1 m in front of the subject at eyes level.

The dataset was collected from four subjects in the motor imagery task, which has been widely used in BCI systems. Each subject was instructed to imagine a movement of the right or left hand for about 2 s to control a cursor movement on the computer screen. EEG signals were recorded, sampled at 500 Hz and band pass filtered between 0.1 Hz and 100 Hz. For each subject, 100 left and 100 right trials were acquired.

2.2 Residual Neural Networks

The convolution neural network adopts biological heuristic design and uses spatial local correlation to compensate completely connected neural networks to form a network with local connectivity between adjacent layer neurons. In recent years, CNNs have shown good results when applied to data types containing spatio-temporal information, such as human speech [3], music [4] and videos [5]. CNN can extract low/mid/high-level features. The more layers in the network, the richer the features that can be extracted in different levels. In addition, the deeper the features extracted by the network are, the more abstract the features are and the more semantic information is involved. However, if the CNN increases only in the depth, it will cause a gradient dispersion or gradient explosion, and deeper neural networks are more difficult to train. Therefore, to make the optimization of such a deep model tractable, we use residual connections and batch normalization [6]. Based on CNN, we insert shortcut connections which turn the network into its counterpart residual version called Resnet.

We adopt residual learning to every few layers. A building block is shown in Fig. 1. Formally, we consider a building block defined as:

$$y = F(x, \{W\_i\}) + x \tag{1}$$

where x and y are the input and output vectors. The function F(x, {W_i}) is the residual mapping to be learned. For the example in Fig. 1 that has two layers, F = W_2 σ(W_1 x) in which σ denotes. This shortcut connection does not introduce extra parameter or computation complexity. If The dimensions of x and F are not equal, we can perform a linear projection W_s x by the shortcut connections to match the dimensions:

$$y = F(x, \{W\_i\}) + W\_sx \tag{2}$$

W_s x is only used when matching dimensions.

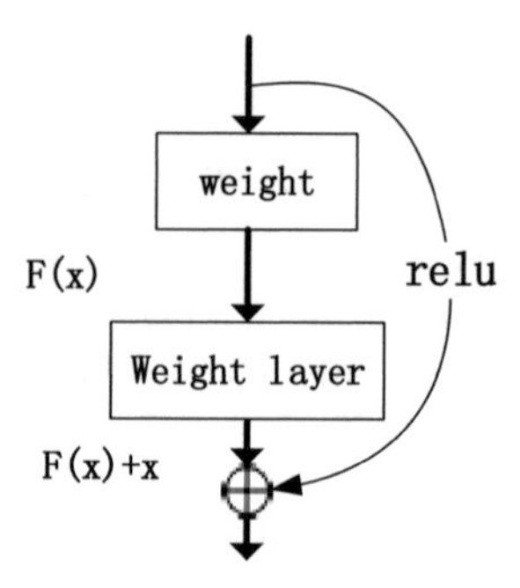

Fig. 1. Residual learning: a building block

2.3 Data Processing

In order to retrieve temporal and spatial characteristics from the dataset, the original data was time-frequency converted and spatially stitched. In detail, by wavelet transform, 2-way EEG spectrums maps (frequency * time) were obtained from 1-way data (time) acquired in every channel. Then, arrange the data according to the lead position (Fig. 2) stepped by 30 ms. Each trial acquired 21 samples and each sample consisted of a 2D matrix with the dimension of 240×207.

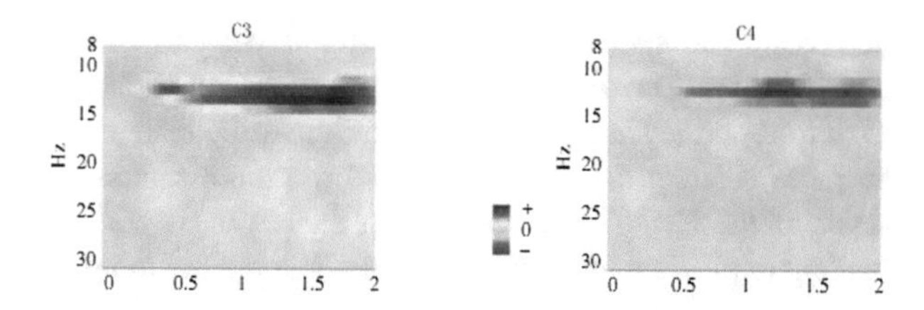

Fig. 2. C3 and C4 channels 2-way maps for Sub. 3

We use convolution neural networks for the learning task. The architecture of the network is shown in Fig. 3. The network inputs the raw EEG signal and outputs predictions, which present left hand or right hand movement imagination. In order to make the optimization of such a network tractable, we employ shortcut connections in a similar manner to those found in the Residual Network architecture. The shortcut connections between neural network layers optimize training by allowing information to propagate well in deep neural networks.

The network consists of 1 basic block and 16 residual blocks. The basic block has two convolutional layers. The convolutional layers uses 64 kernels of size 5×5 with a stride of one pixel. The residual blocks have one convolutional layer per block. The convolutional layer uses 64 kernels of size 7×7 with a stride of 2 pixel, followed by a max pooling layer that uses a 3×3 kernel and stride of 2. When a residual block subsamples the input, the corresponding shortcut connections also subsample their input using a Max Pooling operation with the same subsample factor.

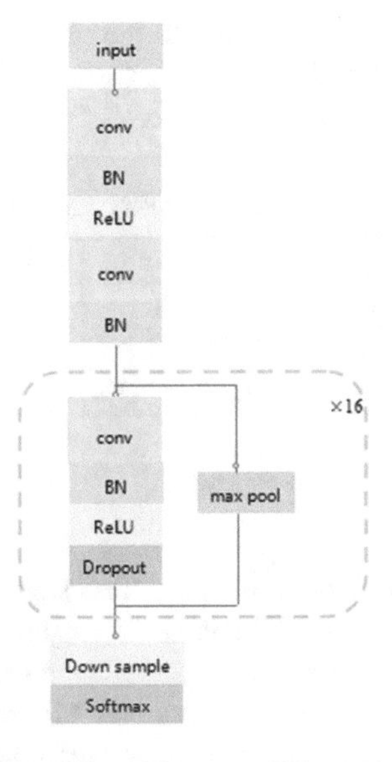

Fig. 3. The architecture of the network

Before each convolution layer we apply Batch Normalization [7] and a rectified linear activation, adopting the pre-activation block design [8]. The first and last layers of the network are special cased due to this pre-activation block structure. We also apply Dropout [8] between the convolution layers and after the non-linearity. The final fully connected layer and softmax activation produce a distribution over the 2 output classes for each time-step. We train the networks from scratch, initializing the weights of the convolution layers as in [9]. We use the Adam [10] optimizer with the default parameters and reduce the learning rate by a factor of 10 when the validation loss stops improving. We save the best model as evaluated on the validation set during the optimization process.

3 Result

In this study, the experiments were conducted in matlab and caffe which is a deep learning framework developed by Berkeley AI Research (BAIR) and by community contributors on an Intel 4.00 GHz Core i7 PC with 16 GB of RAM. Matlab was used for Input data processing and caffe was used for designing and testing the proposed networks.

The training dataset was used in the study is comprised of 4200 samples for each subject. 80% of this data was separated into training set and remaining 10% was separated into test set using 10×10 folds cross-validation. DRN was performed with a batch size of 50 and an epoch of 300.

We also applied CSP + SVM classifier to the same data set in order to investigate the role of deep networks in classification performance. Table 1 shows The classification accuracy of the CSP and DRN scheme.

Table 1. The classification accuracies of CSP, DRN and optimal number of spatial patterns.

Subject	Sub 1	Sub 2	Sub 3	Sub 4
Patterns number	2	8	4	6
CSP accuracy	52.9%	52.9%	45%	51.4%
DRN accuracy	96.6%	63.5%	69.0%	64.3%

4 Conclusions

We have presented a method of motor imagery EEG signal classification using the deep residual network. Firstly, the raw EEG data are represented in the time-frequency field using continuous wavelet transform and then a DRN model is built to classify MI tasks. Comparison with SCP method which is common MI classification algorithms, DRN classification method has showed the promising performance in MI classification. As future works, we are working on optimizing the network parameters and developing more effective neural networks schemes on MI classification researches.

References

1. Allison, B.Z., Wolpaw, E.W., Wolpaw, J.R.: Brain–computer interface systems: progress and prospects. Expert Rev. Med. Devices **4**, 463–474 (2007)
2. Grosse-Wentrup, M., Gramann, K., Buss, M.: Adaptive spatial filters with predefined region of interest for EEG based brain-computer-interfaces. In: Advances in Neural Information Processing Systems (2007)
3. Graimann, B., Allison, B., Pfurtscheller, G.: Brain–computer interfaces: a gentle introduction. In: Brain-Computer Interfaces, pp. 1–27. Springer, Heidelberg (2009)
4. Graves, A., Mohamed, A., Hinton, G.: Speech recognition with deep recurrent neural networks. In: 2013 IEEE International Conference on Acoustics, Speech and Signal Processing (ICASSP). IEEE (2013)
5. Karpathy, A., et al.: Large-scale video classification with convolutional neural networks. In: Proceedings of the IEEE Conference on Computer Vision and Pattern Recognition (2014)
6. Dieleman, S., Brakel, P., Schrauwen, B.: Audio-based music classification with a pretrained convolutional network. In: 12th International Society for Music Information Retrieval Conference (ISMIR-2011)
7. He, K., et al.: Deep residual learning for image recognition. In: Proceedings of the IEEE Conference on Computer Vision and Pattern Recognition (2016)
8. He, K., et al.: Identity mappings in deep residual networks. In: European Conference on Computer Vision. Springer, Cham (2016)
9. Kinga, D., Ba Adam, J.: A method for stochastic optimization. In: International Conference on Learning Representations (2015)
10. Srivastava, N., et al.: Dropout: a simple way to prevent neural networks from overfitting. J. Mach. Learn. Res. **15**(1), 1929–1958 (2014)

Factors Influencing Financial Flexibility of Enterprises

Take the Manufacturing Enterprises in Jiangsu and Zhejiang Provinces as an Example

Xiushan Zhou[1(✉)] and Jingming Li[2]

[1] Washburn University, Topeka, KS 66621, USA
xiushan.zhou@washburn.edu
[2] Wuhan University of Sci and Tech, Wuhan, Hubei 430000, China
ljm500@wust.edu.cn

Abstract. In recent years, the financial development is tending to be more flexible instead of rigid and the scope covers the financing, management, investment and profit distribution of financial activities in four aspects. Based on such appeal, this paper firstly analyzes the concept and characteristics of financial flexibility theoretically. Then, from Jiangsu and Zhejiang provinces manufacturing enterprise financial flexibility and EVA rate, their correlation in this paper is not only the theory of knowledge analysis but more deeply into the relationship between the variables. Multivariate regression analysis is applied in this process, and the method of robustness analysis is also adopted. Finally, based on the results of the study, more reasonable and operable suggestions are given to the financial flexibility in the practice of manufacturing enterprises in Jiangsu and Zhejiang provinces.

Keywords: Financial flexibility · EVA rate · Multiple regression analysis

1 Introduction

1.1 The Research Background of Manufacturing Enterprises in Jiangsu and Zhejiang Provinces

The financial sectors, financial personnel's professional degrees and financial systems of Jiangsu and Zhejiang provinces manufacturing enterprise will be affected by the scale of Jiangsu and Zhejiang provinces manufacturing enterprise, the manufacturing enterprise itself, the development stage and the value of finance.

The manufacturing enterprises of Jiangsu and Zhejiang provinces in our country is in an economic knowledge, network, and information integration age. So, manufacturing enterprises need to change in the rapidly changeable environment. Besides, it is also important to change the financial character because the capital is the lifeblood of development and survival for manufacturing enterprises [1].

N. Xiong et al. (Eds.): ISCSC 2018, AISC 877, pp. 69–77, 2019.
https://doi.org/10.1007/978-3-030-02116-0_9

1.2 The Research Purpose of Manufacturing Enterprises in Jiangsu and Zhejiang Provinces

The manufacturing enterprises of Jiangsu and Zhejiang provinces are changed from financial rigidity into financial flexibility. Because the traditional finance has no longer adapted to the development of the modern manufacturing enterprise. mechanized finance become stiffer and rigid which cannot be timely meets the needs of manufacturing enterprises in the financial and cannot create more value for the manufacturing industries in Jiangsu and Zhejiang provinces. So, at this point, it is necessary to change the traditional finance. The opposite concept of rigidity is flexible [2]. Financial flexibility to some extent will be able to solve the traditional rigid financial abuses in manufacturing enterprises in Jiangsu and Zhejiang provinces, can be more adapted to the development level of manufacturing enterprises in Jiangsu and Zhejiang provinces, and cope with its internal and external changing environment.

2 The Meaning and Characteristics of Financial Flexibility

2.1 The Meaning of Financial Flexibility

Financial flexibility refers to many unstable financial factors which are faced in the enterprise, and the financial management also needs to change. Financial flexibility is more and more in line with the development needs of manufacturing enterprises in Jiangsu and Zhejiang provinces. So, it is important to further study the financial flexibility measurement, evaluation, model construction and management measures.

2.2 The Characteristics of Financial Flexibility

The financial flexibility of manufacturing enterprises in Jiangsu and Zhejiang provinces has the characteristics of timeliness, flexibility, and systematisms.

- It is about timeliness. Jiangsu and Zhejiang provinces manufacturing enterprises should develop a suitable financial plan and make financial decisions in a short time as the environment changes instead of the rigid finance that traditionally makes relevant financial activities according to the internal plan of the manufacturing enterprises.
- It is about flexibility. In contrast to rigid and procedural, the new financial is flexible, and not programmed. They are more flexible in terms of the financial system, financial way, financial ideas or financial personnel [3].
- It is about systematisms. Analysis of financial flexibility from financial financing, investment and distribution perspectives. Capital structure, debt management, and flexible combination, their reasonable collocation can reduce the risk of financing [4]. Enterprises can adjust their income distribution flexibly on the premise of balancing the interests of all parties and environmental differences.

3 The Influence Mechanism of Financial Flexibility

3.1 Explained Variable of Financial Flexibility

Financial flexibility is financial preparation based on the financial situation of the past and the present company's internal financial situation and the external environment of the enterprise [5]. At the same time, the evaluation index of EVA value is trying to use historical data to measure the results and the managers' compensation which will make the enterprise management behavior of managers affects the future financial flexibility, enterprise value and so on.

From a series of calculation formulas of EVA rate, we can see that EVA rate completely considers the opportunity cost of capital which is also the cost of capital in the formula. They are unlike earnings per share and net income which do not consider the opportunity cost and the authenticity [6]. They are also unlike the accounting profit measured by a traditional accounting indicator. The introduction of EVA rate instead of EVA is aimed at the difference of enterprise size in Jiangsu and Zhejiang provinces, because of the different enterprise size which would lead to the deviation of the absolute value of EVA. So, we scientifically study the economic added value of each unit capital cost of different size enterprises (EVA ratio = and EVA/the capital cost).

3.2 Explanatory Variable of Financial Flexibility

The selection and measurement of explainable variables are from the four financial activities which are financing, operation, investment, and distribution. We introduce two concepts to make a better logic arrangement before we analyze variables. First, the leverage index which is based on the financing and allocation of two financial activities which is a measure of the solvency of creditors or shareholders [7]. The two angles are measured in four specific variables: short-term solvency, comprehensive solvency, cash dividend payment capability and comprehensive payment capability. Second, the capital index is from operation and investment financial activities which mainly involve the inflows and outflows of cash flow and is also a reasonable capacity measure of capital input and output. The two angles are measured in terms of cash holdings, total asset cash recovery, investment activity financing ratio and cash reinvestment ratio (Table 1).

Table 1. Variable definition summary

Explained variable	Explanatory variable				
One measure	Two index	Four financial activities	Eight elements	Variable abbr.	Function formula
Financial flexibility	Lever index	Financing activity	Short term solvency	STS	= Net cash flows/Current Liabilities
			Comprehensive solvency	CS	= Net cash flows/Total Liabilities
		Operating activity	Cash dividend paid	CDP	= Net cash flow/Cash dividend
			Comprehensive paid	CP	= Net cash flows/Equity
	Financial index	Investment activity	Cash holdings	CH	= (Cash + Cash equivalents)/Total assets
			Rate return on all assets	RRAA	= Net operating cash flow/Total assets
		Distribution activity	Investment activity financing ratio	IAFR	= Net cash flows from investing activities/(Net cash flows from operating activities + Net cash flows from financing activities)
			Cash reinvestment ratio	CRR	= Net cash flow-cash dividend-interest/Reinvestment quota

4 Research Hypothesis and Empirical Analysis of Financial Flexibility

4.1 Sample Selection and Data Collection

In this paper, the related date of Jiangsu and Zhejiang manufacturing companies during recent 3 years were selected from CSMAR database as the research object, and the data are appropriately screened. (1) 107 Jiangsu and Zhejiang manufacturing companies that went bankrupt or closed from 2015 to 2018 were eliminated. (2) Exclude 56 ST-type manufacturing companies in Jiangsu and Zhejiang Provinces. (3) 89 manufacturing companies from Jiangsu and Zhejiang provinces that are missing data were excluded. According to the above selection, 427 samples were obtained.

4.2 Variable Description Statistical Analysis

For the explained variables and explanatory variables listed above, the variable description statistical analysis method can be used to measure whether the two datasets are in a line. In another word, it would measure the linear relationship between the fixed distance variables.

In the table below, the letters are the abbreviations for variables in Table 2. Under the condition of the significant level of 5%, the EVA rate is negatively correlated with the short-term solvency and the comprehensive solvency. When the short-term solvency is higher, the EVA rate will be lower, and vice versa. The EVA rate was positively correlated with the other six factors at a significant level of 5%.

Table 2. Variable description statistical analysis table

	ER	STS	CS	CDP	CP	CH	RRAA	IAFR	CRR
ER	1	−0.5756	−0.6307	0.5652	0.5802	0.4371	0.3332	0.5652	0.5802
STS		1	0.5763	−0.1758	−0.2639	−0.164	−0.4401	−0.3433	−0.5763
CS			1	−0.6308	−0.7705	−0.2307	−0.1317	0.8491	−0.8641
CDP				1	0.1496	0.1634	0.1415	0.1304	0.1454
CP					1	0.1493	0.0503	0.1284	0.1434
CH						1	0.035	0.068	0.083
RRAA							1	0.242	0.257
IAFR								1	0.4521
CRR									1

4.3 The Multiple Regression Analysis

The Multiple regression analysis is to analyze the correlation between independent variables and dependent variables. In uncertain relationships, we would like to find out as much as possible about the mathematical expression between them and predict and control the relationship between them or the analysis of the factors. Using the variables explained in the previous text and the relationship of financial flexibility, the analysis of multiple regression analysis is used to further analyze (Table 3).

Therefore, this multiple linear regression equation is:

$$\begin{aligned} Y = & -0.0247X1 - 0.0292X2 + 0.0130X3 + 0.0144X4 + 0.0211X5 \\ & + 0.0309X6 + 0.0292X7 + 0.0307X8 \end{aligned} \tag{1}$$

The third line model for the above table is 0.6760. Under 5% significance level, the variation of the EVA rate can be explained by multiple regression equation of the ratio of 67.60% to the eight factors which are short-term solvency, comprehensive solvency, cash dividend payment capability, comprehensive payment capability, cash holdings, total asset cash recovery, investment activity financing ratio, cash reinvestment ratio.

Table 3. Multiple regression analysis

Multiple regression analysis	Multiple regression analysis			
	Coefficients	Standard deviation	T-state	P-value
Intercept	0.4487	0.1164	3.8184	0.0001
Short-term solvency	−0.0247	0.0241	−1.9859	0.0406
Comprehensive solvency	−0.0292	0.0249	0.6277	0.0282
Comprehensive solvency	−0.0292	0.0249	0.6277	0.0282
Cash dividend payment capability	0.0130	0.0056	1.9586	0.0503
Comprehensive payment capability	0.0144	0.0071	1.9601	0.0518
Cash holdings	0.0211	0.0195	3.177	0.0012
Total asset cash recovery	0.0309	0.0164	1.9997	0.0329
Investment activity financing ratio	0.0292	0.0038	−5.7692	0
Cash reinvestment ratio	0.0307	0.0053	−5.7707	0
Adj R2	0.6760			
F Value	54.4729			
Sign F	0			

4.4 The Robustness Analysis

Robustness testing is an adjustment and change by using a different standard of classification, other variables, or measuring method to replace a parameter. After a series of repeated experiments, the empirical results still maintain a relatively consistent explanation. If the empirical results are not stable. The cause needs to be found and retest will be needed, too. The relationship between EVA rate and eight factors is established by multiple linear regression equation. Then, we need to test the regression coefficients of the regression equation to be further determined. For any Bi (I = 1, 2, 3, 4, 5, 6, 7, 8)

H0: Hi = 0, H1: Bi does not equal 0.

In a significant level of 5%, If $P < 0.05$ in p-value, the factors can influence the EVA rate.

In terms of financing, the coefficient of short-term solvency is −0.0247. In a significant level of 5%, if other conditions remain unchanged, for every 1 increase in short-term solvency, the EVA rate would be reduced by 0.0224. In other words, the short-term solvency ratio is inversely proportional to the EVA rate. So, when the enterprise reduces the coefficient of short-term solvency, the rate of EVA will be appropriately increased, and it also improves and optimize the enterprise. The comprehensive solvency coefficient is −0.0292, which is reflects the inverse relationship between the comprehensive solvency and financial flexibility in 5% significant level.

In terms of investment, the ratio of cash dividend payment capability is 0.0130, indicating that the ability to pay cash dividends is positively correlated with financial flexibility. And the results are consistent with previous statistics. The ratio of comprehensive payment capability is 0.0144. Under the same significance level, we can know that it has more influence on EVA rate compared with the ratio of cash dividend payment capability because of its' higher ratio than cash dividend payment capability

[8]. At the same time, if other conditions of financial flexibility remain unchanged, the average rate of EVA increased by 0.0144 for every 1 increase in the enterprise's comprehensive payment capability.

In terms of operation, the coefficient of cash holdings is 0.0211. Although it means that cash holdings of the enterprise are positively correlated with the EVA rate, the linear relationship between them is not obvious. Thus, Other factors should be considered to analyze the EVA rate. The coefficient of total asset cash recovery is 0.0309, which means that in 5% level significant when the total asset cash recovery is higher, the EVA rate of enterprises is higher. In this situation, the liquidity of the enterprise assets is relatively fast, and obviously, there will be more funds to improve its operation.

In terms of profit distribution, the coefficient of investment activity financing ratio is 0.0292 and the ratio of cash reinvestment ratio is 0.0307 in the same 5% significance level. Based on the above data, we can draw the following conclusion. First, investment activity financing ratio and financial flexibility is a positive correlation. So, if the ratio of investment activity financing ratio is higher, the EVA rate will be higher which result in financial flexibility degree is higher. Thus, manufacturing enterprises need to gradually raise investment activity financing ratio to improve the degree of financial flexibility. Second, combined with the previous analysis and coefficient data, we can know that the stability of the cash reinvestment ratio is significant to the degree of financial flexibility.

5 The Implementation of Financial Flexible Management Measures in Jiangsu and Zhejiang Provinces

- In terms of financial functions, we need to redefine the concept of financial work at the beginning. What's more, the connotation of traditional financial functions cannot be confused with the meaning of accounting function which only emphasizes budget, accounting functions, and the financial work scope is limited to the level of simple data processing such as planning and control. So, we should turn to the level of collections, integration, and analysis of effective information to support decision making [9]. Then, we need to improve the adaptability and the strategy of traditional financial to the external environment. We need to get a reasonable market investigation, the details of corporate strategy, corporate liquidity and financing channel and so on. So, we can adjust the mechanized financial functions to be more flexible and forward-looking. Finally, it is necessary to increase the collaboration between the finance department and other departments, especially the sales and marketing departments. Only in this way can performance increase efficiency.
- In terms of organizational structure level, the previous organizational hierarchy often brought the tension between the upper and lower levels. The collection and communication of information in the internal and between internal and external will become more slowly which will result in high cost of financial management and inefficient financial management [10]. As the reaction rate of market demand is

increasing, the vertical financial organization structure appears more and more unsuited, the transverse is necessary to replace the old organization structure gradually.

- In terms of management, enterprise for financial staff do not take the often makes these financial clerks repeated accounting work or to be satisfied with the job. If financial leaders cannot be integrated into the enterprise decision-making plan, they often do not have a global perspective, and cannot lead the team to work better. With the globalization of the environment and the deepening of the network, the application of foreign languages and computers will be the inevitable course of integration with international finance. So, this requires financial staff to have the necessary foreign language and computer skills and to be able to use it in financial operations. At the same time, there are some educational activities and training that can improve the knowledge reserve of financial personnel. When we create a better learning environment and pay attention to build a learning organization of the financial department and, this financial personnel will not be disconnected from the needs of the enterprise development. Meanwhile, it makes the financial personnel feel enough care which will create more value for the enterprises. In addition, enterprises can properly import overseas talents to inject international forces into the enterprise itself and bring an international perspective.
- In terms of the management mode, the previous management model achieves the goal of financial tasks through the authority of the superior to the subordinate. Sometimes, we can control the finances well on some levels. But for a long time, it will make financial personnel be lack of autonomy, creativity, and vitality instead of making manufacturing enterprises be more flexible in an increasingly complex and changeable environment. So, the management mode should be more humanized, and more incentives. What's more, we need to pay attention to the department optimization, concern employee behavior and so on. Rewarding the employees with material incentives and emotional rewards.
- In terms of capital, financial personnel need to be flexible in the management of funds. For financing, combined with the international framework, we use the right means of financing channels and consider the cost of funding risk factors to get relatively low financial risks and costs and gain higher financial capital to adapt to the needs of enterprises development. For investment, we should know about the investment market in advance and make the corresponding investment decisions. At the same time, it is necessary to adjust the proportion of investment projects and the duration of investment with the change of the investment market. For operating capital, according to the different factors of operating capital, flow cycle, flow rate, flow cost in a specific situation, we could analyze the operating ratio of current assets and liabilities and reduce the number of idle assets as much as possible after forecasting the ability of the asset to borrow. For profit distribution, we not only need to predicate and plan on corporate profits accurately. But more importantly, we need to involve the financial staff in the process. In this case, the financial personnel's enthusiasm can be improved, and the benefits of the Jiangsu and Zhejiang provinces manufacturing enterprises will get more.

6 The Conclusions and Prospects

This paper establishes the basic and thick theoretical framework, which involves the concept, the characteristics, and a deeper understanding of financial flexibility. Multivariate regression analysis and robustness analysis are used for more scientific research and analysis. Finally, the management of the enterprise financial flexibility is mainly about the following aspects. First, the change in management thinking. Second, the appropriate changes from three aspects-people, wealth, and goods. The first aspect of people relates to the training, incentive method, talent introduction and other measures of the organization personnel. The second aspect of wealth is refined separately into four links of financing, investment, operating capital and profit distribution. The last aspect changes three angles from the financial function, organization structure, management mode.

This paper still exists some shortcomings in this paper, because of the limitation of individual research ability, the restriction of documents channel, the lack of time in writing papers, and the hysteresis of social practice research and application. First, although the financial flexibility is measured from multiple perspectives, the relationship between the variables and the degree of influence of each other is unclear. Second, the theory applied to the practice. When we communicate with managers in Jiangsu and Zhejiang provinces manufacturing enterprises, we have not been enough detailed and specific management plans about the later improvement of enterprise financial flexibility.

References

1. Weilai Zhang, F.: Monetary tightening, financial flexibility, and enterprise risk bearing. J. Contemp. Financ. **384**(11), 45–56 (2016)
2. Weiqiu Yuan, F., Haijiao Wang, S., Xu Huang, T.: Financial flexibility, enterprise investment and business performance – empirical research based on the background of monetary tightening. J. Invest. Res. **35**(4), 57–73 (2016)
3. Xiaoxu Liu, F., Hu Xiang, S.: Environmental uncertainty, enterprise characteristics, and financial flexibility. Macroecon. Res. **355**(78), 127–134 (2014)
4. Manchu Wang, F., Xiujuan Sha, S., Minhao Tian, T.: Financial flexibility, corporate governance, and excessive financing. Econ. Econ. J. **33**(6), 102–106 (2016)
5. Hartzell, D.F., Howton Shawn, D.S., Howton, S.T.: Financial flexibility and at-the-market (atm)equity offerings: evidence from real estate investment trusts. Real Estate Econ. **540**(8), 20–42 (2016)
6. Takami, S.F.: Preserving and exercising financial flexibility in the global financial crisis period: the Japanese example. Corp. Account. Financ. **245**(3), 218–225 (2016)
7. Jong, A.F., Verbeek, M.S., Verwijmeren, T.: Financial flexibility reduce investment ability and firm value: evidence from firms with spare debt capacity. J. Financ. Res. **48**(2), 243–259 (2012)
8. Ferrado, A.F., Marchica Mura, R.S.: Financial flexibility and investment ability across the euro area and the UK. Eur. Financ. Manag. **23**(4), 214–219 (2017)
9. Daniel, N.D.F., Denis, D.J.S., Naveen, L.T.: Sources of Financial Flexibility: Evidence from Cash Flow Shortfalls, 2nd edn. Drexel University, Philadelphia (2010)
10. Byoun, S.F.: Financial Flexibility and Capital Structure Decision, 2nd edn. Baylor University, Waco (2011)

Prime LI-ideal Topological Spaces of Lattice Implication Algebras

Chunhui Liu(✉)

School of Mathematics and Statistics, Chifeng University, Chifeng 024000, Inner Mongolia, China
chunhuiliu1982@163.com

Abstract. In this paper, topological spaces based on prime LI-ideals and maximal LI-ideals of a lattice implication algebra are constructed. We conclude that the topological space based on prime LI-ideals is a compact T_0 space and the topological space based on maximal LI-ideals is a compact T_2 space.

Keywords: Fuzzy logic · Lattice implication algebra · LI-ideal Prime LI-ideal · Topological space

1 Introduction

With the developments of mathematics and computer science, non-classical logic has been actively studied [1]. So far, many-valued logic has become an important part of non-classical logic. In order to study the many-valued logical system whose propositional value is given in a lattice, Xu [2] proposed the concept of lattice implication algebras and discussed some of their properties. Since then, this logical algebra has been extensively investigated by many authors [3–13]. Among them, Jun [5] introduced the notion of LI-ideals in lattice implication algebras and investigated their properties. Liu [6] introduced the notions of ILI-ideals and prime LI-ideals. Chen [8] constructed a topological space based on LI-ideals of lattice implication algebras and discussed the topological properties of them. It's a common sense that topology, logic and order theory are closely linked with each other. On one hand, the algebra corresponding to the logic of finite observations is a kind of generalized topologies (i.e., frames). On the other hand, topologies can be viewed as geometric logic. Topology is also a useful tool in studying logic and logic algebras. For instance, the Stone's representation theorem for Boolean algebras and the Priestley's representation theorem for distributive lattices are expressed by the prime ideal spaces for the two algebras (cf., [14]). In order to seek more combined points of topology and non-classical mathematical logic, in this paper, topological spaces based on prime LI-ideals and maximal LI-ideals of a lattice implication algebra are constructed. It is proved that the topological space constructed on the prime LI-ideals is a compact T_0 space and the topological space constructed on the maximal LI-ideals is a compact T_2 space. Note that the topology on the prime LI-ideals to be constructed

N. Xiong et al. (Eds.): ISCSC 2018, AISC 877, pp. 78–85, 2019.
https://doi.org/10.1007/978-3-030-02116-0_10

in this paper is similar to but different from the classical topology on the prime ideals for distributive lattices. The key difference of the two topologies is that the topology to be constructed is a one-side topology and the classical topology is a two-side topology w.r.t. the set inclusion order of the prime ideals.

2 Preliminaries

In this section we recall some basic notions and results which will be needed in the sequel. For notions about topology, please refer to [15].

Definition 1. [2]. *Let $(L, \vee, \wedge, ', \rightarrow, O, I)$ be a bounded lattice with an order-reversing involution $'$, where I and O are the greatest and the smallest element of L respectively, $\rightarrow: L \times L \rightarrow L$ is a mapping. Then $(L, \vee, \wedge, ', \rightarrow, O, I)$ is called a lattice implication algebra if the following conditions hold for any $x, y, z \in L$:*

(I_1) $x \rightarrow (y \rightarrow z) = y \rightarrow (x \rightarrow z)$;
(I_2) $x \rightarrow x = I$;
(I_3) $x \rightarrow y = y' \rightarrow x'$;
(I_4) $x \rightarrow y = y \rightarrow x = I$ *implies* $x = y$;
(I_5) $(x \rightarrow y) \rightarrow y = (y \rightarrow x) \rightarrow x$;
(L_1) $(x \vee y) \rightarrow z = (x \rightarrow z) \wedge (y \rightarrow z)$;
(L_2) $(x \wedge y) \rightarrow z = (x \rightarrow z) \vee (y \rightarrow z)$.

A lattice implication algebra $(L, \vee, \wedge, ', \rightarrow, O, I)$ is denoted by L in short.

Lemma 1. [2,9]. *Let L be a lattice implication algebra, then for any $x, y, z \in L$ the following assertions hold:*

(1) $O \rightarrow x = I$, $x \rightarrow I = I$ *and* $I \rightarrow x = x$;
(2) $x \leqslant y$ *if and only if* $x \rightarrow y = I$;
(3) $x' = x \rightarrow O$;
(4) $x \rightarrow y \leqslant (y \rightarrow z) \rightarrow (x \rightarrow z)$;
(5) $x \vee y = (x \rightarrow y) \rightarrow y$;
(6) $x \leqslant y$ *implies* $y \rightarrow z \leqslant x \rightarrow z$ *and* $z \rightarrow x \leqslant y \rightarrow x$;
(7) $x \rightarrow (y \vee z) = (x \rightarrow y) \vee (x \rightarrow z)$;
(8) $x \rightarrow (y \wedge z) = (x \rightarrow y) \wedge (x \rightarrow z)$.

Definition 2. [5]. *Let L be a lattice implication algebra. An LI-ideal A is a non-empty subset of L such that for any $x, y \in L$,*

(LI1) $O \in A$;
(LI2) $(x \rightarrow y)' \in A$ *and* $y \in A$ *imply* $x \in A$.

By Definition 2, $\{O\}$ and L are trivial LI-ideals of a lattice implication algebra L. If $\mathcal{A}$ is a non-empty family of LI-ideals of a lattice implication algebra L, then it is easy to show that $\cap\mathcal{A}$ is also an LI-ideal of L.

Let L be a lattice implication algebra. Define a binary operation $\oplus : L \times L \to L$ such that $x \oplus y = x' \to y$. Then $\oplus$ is commutative and associative. In addition, one can prove that $x \vee y \leqslant x \oplus y$ holds for any $x, y \in L$. With these properties of operation $\oplus$, it is not difficult to show the following characterization theorem for LI-ideals.

Lemma 2. [9]. *Let L be a lattice implication algebra. A non-empty subset A of L is an LI-ideal if and only if the following conditions hold for any $x, y \in L$:*

(1) *$x \leqslant y$ and $y \in A$ implies $x \in A$;*
(2) *$x, y \in A$ implies $x \oplus y \in A$, i.e., A is closed with $\oplus$.*

Let A be a subset of a lattice implication algebra L. The least LI-ideal containing A is called the LI-ideal generated by A, denoted by $\langle A \rangle$. If $A = \{a\}$, we write $\langle A \rangle$ as $\langle a \rangle$. For the sake of simplicity, we shall write $[a_1, a_2, \cdots, a_n, x]$ for $a_1 \to (a_2 \to (\cdots \to (a_n \to x) \cdots))$, and write $[a^n, x]$ if $a_1 = a_2 = \cdots = a_n = a$.

Lemma 3. [5]. *Let L be a lattice implication algebra and $\emptyset \neq A \subseteq L$. Then $\langle A \rangle = \{x \in L |\ [a_1', a_2', \cdots, a_n', x'] = I$ for some $a_1, a_2, \cdots, a_n \in A\}$.*

Corollary 1. *Let L be a lattice implication algebra and $\emptyset \neq A \subseteq L$, then $\langle A \rangle = \{x \in L |\ x \leqslant a_1 \oplus a_2 \oplus \cdots \oplus a_n$ for some $a_1, a_2, \cdots, a_n \in A\}$.*

Lemma 4. [5,6]. *Let L be a lattice implication algebra. Let $a, b \in L$ and $\emptyset \neq A, B \subseteq L$. Then*

(1) *$A \subseteq B$ implies $\langle A \rangle \subseteq \langle B \rangle$;*
(2) *$a \leqslant b$ implies $\langle a \rangle \subseteq \langle b \rangle$;*
(3) *If A is an LI-ideal of L, then $\langle a \rangle \subseteq A$ if and only if $a \in A$;*
(4) *$\langle a \rangle \cap \langle b \rangle = \langle a \wedge b \rangle$ and $\langle \langle a \rangle \cup \langle b \rangle \rangle = \langle a \vee b \rangle$.*

Definition 3. [6]. *Let L be a lattice implication algebra. A proper LI-ideal of L is said to be a prime LI-ideal if $x \wedge y \in A$ implies $x \in A$ or $y \in A$ for any $x, y \in L$. We shall denoted the set of all prime LI-ideal by $\mathcal{P}_{LI}(L)$.*

Lemma 5. [6]. *Let L be a lattice implication algebra and A a proper LI-ideal of L. Then the following are equivalent:*

(1) *A is a prime LI-ideal;*
(2) *$(x \to y)' \in A$ or $(y \to x)' \in A$ for all $x, y \in L$;*
(3) *$x \wedge y = 0$ implies $x \in A$ or $y \in A$ for all $x, y \in L$;*
(4) *$A_1 \cap A_2 \subseteq A$ implies $A_1 \subseteq A$ or $A_2 \subseteq A$ for all LI-ideals A_1 and A_2 of L.*
(5) *$\langle a \rangle \cap \langle b \rangle \subseteq A$ implies $a \in A$ or $b \in A$ for all $a, b \in L$.*

Lemma 6. (Prime *LI*-ideal theorem) [6]. *Let L be a lattice implication algebra and A an LI-ideal of L. Let S be a filtered subset of L and $A \cap S = \varnothing$. Then there exists a prime LI-ideal B such that $A \subseteq B$ and $B \cap S = \varnothing$.*

Corollary 2. *The following conclusions for a lattice implication algebra L hold:*

(1) *If $a \notin A \in \mathcal{P}_{LI}(L)$, then there is $B \in \mathcal{P}_{LI}(L)$ such that $A \subseteq B$ and $a \notin B$;*
(2) *If $x \in L$ and $x \neq 0$, then there exists $B \in \mathcal{P}_{LI}(L)$ such that $x \notin B$;*
(3) *If A is a proper LI-ideal of L, then $A = \cap\{B \in \mathcal{P}_{LI}(L) | A \subseteq B\}$.*

3 Topological Spaces of Prime LI-ideals

Definition 4. *Let L be a lattice implication algebra and X a subset of L. Define $\|X\|^*$ on $\mathcal{P}_{LI}(L)$ as following:*

$$\|X\|^* = \{A \in \mathcal{P}_{LI}(L) | X \nsubseteq A\}. \tag{3.1}$$

Proposition 1. *Let L be a lattice implication algebra. Let X, Y be subsets of L and $\{X_\alpha\}_{\alpha \in \Gamma}$ a family of subsets of L. Then*

(1) *$X \subseteq Y$ implies $\|X\|^* \subseteq \|Y\|^*$;*
(2) *$\cup_{\alpha\in\Gamma}\|X_\alpha\|^* = \| \cup_{\alpha\in\Gamma} X_\alpha\|^*$;*
(3) *$\|X\|^* \cap \|Y\|^* = \|\langle X\rangle \cap \langle Y\rangle\|^*$;*
(4) *$\|X\|^* = \mathcal{P}_{LI}(L)$ if and only if $\langle X\rangle = L$;*
(5) *$\|X\|^* = \emptyset$ if and only if $X = \emptyset$ or $X = \{O\}$;*
(6) *$\|X\|^* = \|\langle X\rangle\|^*$;*
(7) *$\|X\|^* = \|Y\|^*$ if and only if $\langle X\rangle = \langle Y\rangle$.*

Proof.(1) Suppose $A \in \|X\|^*$. Then $A \in \mathcal{P}_{LI}(L)$ and $X \nsubseteq A$. So we have $A \in \mathcal{P}_{LI}(L)$ and $Y \nsubseteq A$ by $X \subseteq Y$. Thus $A \in \|X\|^*$, i.e., $\|X\|^* \subseteq \|Y\|^*$.

(2) Obviously, $\|X_\alpha\|^* \subseteq \| \cup_{\alpha\in\Gamma} X_\alpha\|^*$ for any $\alpha \in \Gamma$ by (1), and so we have that $\cup_{\alpha\in\Gamma}\|X_\alpha\|^* \subseteq \| \cup_{\alpha\in\Gamma} X_\alpha\|^*$. On the other hand, suppose $A \in \| \cup_{\alpha\in\Gamma} X_\alpha\|^*$. Then $A \in \mathcal{P}_{LI}(L)$ and $\cup_{\alpha\in\Gamma}X_\alpha \nsubseteq A$. Thus there exists $\alpha_0 \in \Gamma$ such that $X_{\alpha_0} \nsubseteq A$ and $A \in \|X_{\alpha_0}\|^*$. So, we have that $\| \cup_{\alpha\in\Gamma} X_\alpha\|^* \subseteq \cup_{\alpha\in\Gamma}\|X_\alpha\|^*$ and $\cup_{\alpha\in\Gamma}\|X_\alpha\|^* = \| \cup_{\alpha\in\Gamma} X_\alpha\|^*$.

(3) Suppose $A \in \|X\|^* \cap \|Y\|^*$. Then $A \in \mathcal{P}_{LI}(L)$, $X \nsubseteq A$ and $Y \nsubseteq A$ by (3.1). Thus $\langle X\rangle \nsubseteq A$ and $\langle Y\rangle \nsubseteq A$ by Lemma 4(1). If $\langle X\rangle \cap \langle Y\rangle \subseteq A$, then by Lemma 5(3) we have $\langle X\rangle \subseteq A$ or $\langle Y\rangle \subseteq A$, a contradiction. So $\langle X\rangle \cap \langle Y\rangle \nsubseteq A$. This means that $A \in \|\langle X\rangle \cap \langle Y\rangle\|^*$ and so $\|X\|^* \cap \|Y\|^* \subseteq \|\langle X\rangle \cap \langle Y\rangle\|^*$. Conversely, let $A \in \|\langle X\rangle \cap \langle Y\rangle\|^*$. Then $A \in \mathcal{P}_{LI}(L)$ and $\langle X\rangle \cap \langle Y\rangle \nsubseteq A$. So $\langle X\rangle \nsubseteq A$ and $\langle Y\rangle \nsubseteq A$. From this and the definition of generated LI-ideals we get $X \nsubseteq A$ and $Y \nsubseteq A$. Thus $A \in \|X\|^* \cap \|Y\|^*$ and $\|\langle X\rangle \cap \langle Y\rangle\|^* \subseteq \|X\|^* \cap \|Y\|^*$. Therefore $\|X\|^* \cap \|Y\|^* = \|\langle X\rangle \cap \langle Y\rangle\|^*$.

(4) Let $\|X\|^* = \mathcal{P}_{LI}(L)$. If $\langle X\rangle \neq L$, then $\langle X\rangle$ is a proper LI-ideal of L. Thus there exists $B \in \mathcal{P}_{LI}(L)$ such that $\langle X\rangle \subseteq B$ by Corollary 2. By $X \subseteq \langle X\rangle$ we have $X \subseteq B$, i.e., $B \notin \|X\|^*$, contradicting to $\|X\|^* = \mathcal{P}_{LI}(L)$. So $\langle X\rangle = L$. Conversely, if $\langle X\rangle = L$, then it's easy to show that $\|X\|^* = \mathcal{P}_{LI}(L)$ by Definition 4 and the definition of generated LI-ideals.

(5) Suppose $\|X\|^* = \emptyset$. If $X \neq \emptyset$ and $X \neq \{O\}$, then there exists $a \neq O$ such that $a \in X$. It follows from Corollary 2 that there exists $B \in \mathcal{P}_{LI}(L)$ such that $a \notin B$. So $X \nsubseteq B$ and $B \in \|X\|^*$. This means that $\|X\|^* \neq \emptyset$, a contradiction. Thus $X = \emptyset$ or $X = \{O\}$. Conversely, let $X = \emptyset$ or $X = \{O\}$. Since for any $A \in \mathcal{P}_{LI}(L)$ we have $O \in A$ and $\emptyset \subseteq A$, $\|\{O\}\|^* = \emptyset$ and $\|\emptyset\|^* = \emptyset$, i.e., $\|X\|^* = \emptyset$.

(6) Suppose $A \in \mathcal{P}_{LI}(L)$. Then $A \in \|X\|^*$ if and only if $X \nsubseteq A$ if and only if $\langle X\rangle \nsubseteq A$ if and only if $A \in \|\langle X\rangle\|^*$. So $\|X\|^* = \|\langle X\rangle\|^*$.

(7) Suppose $\langle X\rangle = \langle Y\rangle$. Then $\|\langle X\rangle\|^* = \|\langle Y\rangle\|^*$. So by (6) we have $\|X\|^* = \|Y\|^*$. Conversely, let $\|X\|^* = \|Y\|^*$. If $\langle X\rangle \neq \langle Y\rangle$, then there exists $B \in \mathcal{P}_{LI}(L)$ such that $\langle X\rangle \subseteq B$ but $\langle Y\rangle \nsubseteq B$, or $\langle Y\rangle \subseteq B$ but $\langle X\rangle \nsubseteq B$. This implies $B \in \|Y\|^*$ but $B \notin \|X\|^*$, or $B \in \|X\|^*$ but $B \notin \|Y\|^*$. Each of them contradicts to condition $\|X\|^* = \|Y\|^*$. So $\langle X\rangle = \langle Y\rangle$.

If $X = \{x\}$, we write $\|X\|^*$ as $\|x\|^*$ and $\|x\|^* = \{A \in \mathcal{P}_{LI}(L) | x \notin A\}$. By Lemma 4 and Proposition 1 one can easily prove the following proposition:

Proposition 2. *Let L be a lattice implication algebra and $x, y \in L$. Then*

(1) $x \leqslant y$ *implies* $\|x\|^* \subseteq \|y\|^*$;
(2) $\|x\|^* \cap \|y\|^* = \|x \wedge y\|^*$;
(3) $\|x\|^* \cup \|y\|^* = \|x \vee y\|^* = \|x \oplus y\|^*$.

Definition 5. *Let L be a lattice implication algebra. Let $\mathcal{T} \subseteq \mathcal{P}_{LI}(\mathcal{L})$ such that*

$$\mathcal{T} = \{\|X\|^* | X \subseteq L\}. \tag{3.2}$$

Then by Proposition 1 we have

(1) $\emptyset, \mathcal{P}_{LI}(L) \in \mathcal{T}$;
(2) $\|X\|^*, \|Y\|^* \in \mathcal{T}$ *implies* $\|X\|^* \cap \|Y\|^* \in \mathcal{T}$;
(3) $\{\|X\|^*_\alpha\}_{\alpha\in\Gamma} \subseteq \mathcal{T}$ *implies* $\cup_{\alpha\in\Gamma}\|X\|^*_\alpha \in \mathcal{T}$.

So, $\mathcal{T}$ is a topology on $\mathcal{P}_{LI}(L)$, called the prime LI-ideal topology of L. The topological space $(\mathcal{P}_{LI}(L), \mathcal{T})$ is called the prime LI-ideal topological space of L.

Theorem 1. *Let L be a lattice implication algebra. Then the set $\{\|x\|^* | x \in L\}$ is a base for prime LI-ideal topology $\mathcal{T}$ of L.*

Proof. Suppose $X \subseteq L$, then $\|X\|^*$ is an open set of prime LI-ideal topological space $(\mathcal{P}_{LI}(L), \mathcal{T})$ and by Proposition 1 we have

$$\|X\|^* = \| \cup_{x\in X} \{x\}\|^* = \cup_{x\in X}\|x\|^*.$$

This means that open sets of LI-ideal topological space $(\mathcal{P}_{LI}(L), \mathcal{T})$ can be expressed as unions of some elements of $\{\|x\|^* | x \in L\}$, i.e., $\{\|x\|^* | x \in L\}$ is a base for topology $\mathcal{T}$.

Theorem 2. *Let L be a lattice implication algebra. Then for any $x \in L$, $\|x\|^*$ is a compact subset of prime LI-ideal topological space $(\mathcal{P}_{LI}(L), \mathcal{T})$ of L.*

Proof. For any $x \in L$, by Theorem 1 we can assume $\{\|x_\alpha\|^*\}_{\alpha\in\Gamma}$ is an arbitrary open covering of $\|x\|^*$, i.e., $\|x\|^* = \cup_{\alpha\in\Gamma}\|x_\alpha\|^*$. Then by Proposition 1,

$$\|x\|^* = \cup_{\alpha\in\Gamma}\|x_\alpha\|^* = \| \cup_{\alpha\in\Gamma} \{x_\alpha\}\|^*.$$

Furthermore, $\langle x\rangle = \langle \cup_{\alpha\in\Gamma}\{x_\alpha\}\rangle$ and $x \in \langle \cup_{\alpha\in\Gamma}\{x_\alpha\}\rangle$. It follows from Corollary 1 that exists $x_{\alpha_1}, x_{\alpha_2}, \cdots, x_{\alpha_n} \in \cup_{\alpha\in\Gamma}\{x_\alpha\}$ such that

$$x \leqslant x_{\alpha_1} \oplus x_{\alpha_2} \oplus \cdots \oplus x_{\alpha_n}.$$

By Proposition 2 we have that

$$\begin{aligned}\|x\|^* \subseteq& \|x_{\alpha_1} \oplus x_{\alpha_2} \oplus \cdots \oplus x_{\alpha_n}\|^* \\ =& \|x_{\alpha_1}\|^* \cup \|x_{\alpha_2}\|^* \cup \cdots \cup \|x_{\alpha_n}\|^* \subseteq \|x\|^*,\end{aligned}$$

i.e., $\|x\|^* = \cup_{k=1}^{n} \|x_{\alpha_k}\|^*$. This shows that $\|x\|^*$ is compact in $(\mathcal{P}_{LI}(L), \mathcal{T})$.

Theorem 3. *Let L be a lattice implication algebra. Then prime LI-ideal topological space* $(\mathcal{P}_{LI}(L), \mathcal{T})$ *of L is a compact* T_0*-space.*

Proof. Since $\mathcal{P}_{LI}(L) = \|I\|^*$ and $I \in L$, by Theorem 2, $(\mathcal{P}_{LI}(L), \mathcal{T})$ is a compact space. In order to prove that $(\mathcal{P}_{LI}(L), \mathcal{T})$ is a T_0-space, we need to prove for any pair of $A, B \in \mathcal{P}_{LI}(L)$ with $A \neq B$, there is an open set $U \subseteq \mathcal{P}_{LI}(L)$ such that $A \in U$ but $B \notin U$, or $B \in U$ but $A \notin U$. In fact, if $A \neq B$, then $A \nsubseteq B$ or $B \nsubseteq A$. Assume $A \nsubseteq B$ without loss of generality. Then there is $x \in A$ such that $x \notin B$. Take $U = \|x\|^*$. Then $A \notin U$ but $B \in U$. This shows that $(\mathcal{P}_{LI}(L), \mathcal{T})$ is a T_0-space.

Theorem 4. *Let L be a lattice implication algebra. Then prime LI-ideal topological space* $(\mathcal{P}_{LI}(L), \mathcal{T})$ *of L is a* T_1*-space if and only if* $A \neq B$ *implies* $A \nsubseteq B$ *and* $B \nsubseteq A$ *for any* $A, B \in \mathcal{P}_{LI}(L)$.

Proof. Suppose $(\mathcal{P}_{LI}(L), \mathcal{T})$ is a T_1-space and $A, B \in \mathcal{P}_{LI}(L)$ with $A \neq B$. Then exists two open sets $U, V \subseteq \mathcal{P}_{LI}(L)$ such that $A \in U$, $B \notin U$ and $B \in V, A \notin V$. Thus $A \nsubseteq B$ and $B \nsubseteq A$. Conversely, if $A \neq B$ implies $A \nsubseteq B$ and $B \nsubseteq A$ for any $A, B \in \mathcal{P}_{LI}(L)$, then exists $x \in A$ such that $x \notin B$ and exists $y \in B$ such that $y \notin A$. Let $U = \|x\|^*$ and $V = \|y\|^*$ we obtain that $B \in U, A \notin U$ and $A \in V, B \notin V$, i.e., $(\mathcal{P}_{LI}(L), \mathcal{T})$ is a T_1-space.

Definition 6. [6]. *Let L be a lattice implication algebra. A maximal LI-ideal of L is defined to be a proper LI-ideal of L which is not a proper subset of any proper LI-ideal of L. We shall denote the set of all maximal LI-ideals of L by* $\mathcal{M}_{LI}(L)$.

Lemma 7. [6]. *In a lattice implication algebra, maximal LI-ideals are all prime.*

By Lemma 7 we know that in a lattice implication algebra L, any maximal LI-ideal must be a prime LI-ideal. So, if we restrict the prime LI-ideal topology $\mathcal{T}$ of L to the set $\mathcal{M}_{LI}(L)$ then we can obtain a topological subspace $(\mathcal{M}_{LI}(L), \mathcal{T}|_{\mathcal{M}_{LI}(L)})$ and call it maximal LI-ideal topological space of L. For any $X \subseteq L$ we write $\|X\|^* \cap \mathcal{M}_{LI}(L)$ as $\|X\|_M^*$. Then $\|X\|_M^* \in \mathcal{T}|_{\mathcal{M}_{LI}(L)}$ and $\{\|x\|_M^* | x \in L\}$ is a base of topology $\mathcal{T}|_{\mathcal{M}_{LI}(L)}$. In order to explore more properties of maximal LI-ideal topological space $(\mathcal{M}_{LI}(L), \mathcal{T}|_{\mathcal{M}_{LI}(L)})$ of L, we prove the following lemma first.

Lemma 8. *In a lattice implication algebra L, then* $\forall x, y \in L$, $(x \to y)' \wedge (y \to x)' = O$.

Proof. By Lemma 1 (2) and (8) we have

$$x \to (x \wedge y) = (x \to x) \wedge (x \to y) = x \to y.$$

Thus according to Definition 1 and Lemma 1 we obtain that

$$\begin{aligned}
&(x \to y)' \wedge (y \to x)' \\
=&((x \to y) \vee (y \to x))' \\
=&(((x \to y) \to (y \to x)) \to (y \to x))' \\
=&(((x \to (x \wedge y)) \to (y \to (x \wedge y))) \to (y \to x))' \\
=&((y \to ((x \to (x \wedge y)) \to (x \wedge y))) \to (y \to x))' \\
=&((y \to (x \vee (x \wedge y))) \to (y \to x))' \\
=&((y \to (x \wedge y)) \to (y \to x))' \\
=&((y \to x) \to (y \to x))' \\
=&1' = 0.
\end{aligned}$$

This completes the proof.

Theorem 5. *Let L be a lattice implication algebra. Then the maximal LI-ideal topological space $(\mathcal{M}_{LI}(L), \mathcal{T}|\mathcal{M}_{LI(L)})$ of L is a compact T_2-space, hence both regular and T_4-space.*

Proof. At first, we prove that $(\mathcal{M}_{LI}(L), \mathcal{T}|\mathcal{M}_{LI(L)})$ is a compact space. Suppose that $\{\|x_\alpha\|_M^*\}_{\alpha\in\Gamma}$ is an arbitrary open cover of $\mathcal{M}_{LI}(L)$. By Proposition 1 we have

$$\begin{aligned}
\|I\|_M^* = \mathcal{M}_{LI}(L) = \cup_{\alpha\in\Gamma} \|x_\alpha\|_M^* \\
=\| \cup_{\alpha\in\Gamma} \{x_\alpha\}\|_M^* = \|\langle\cup_{\alpha\in\Gamma}\{x_\alpha\}\rangle\|_M^*,
\end{aligned}$$

hence, $I \in \langle\cup_{\alpha\in\Gamma}\{x_\alpha\}\rangle$. By Corollary 1, exists $x_{\alpha_1}, x_{\alpha_2}, \cdots, x_{\alpha_n} \in \cup_{\alpha\in\Gamma}\{x_\alpha\}$ such that $I \leqslant x_{\alpha_1} \oplus x_{\alpha_2} \oplus \cdots \oplus x_{\alpha_n}$. By Proposition 2 we get

$$\begin{aligned}
\mathcal{M}_{LI}(L) = \|I\|_M^* =&\|x_{\alpha_1} \oplus x_{\alpha_2} \oplus \cdots \oplus x_{\alpha_n}\|_M^* \\
=&\|x_{\alpha_1}\|_M^* \cup \|x_{\alpha_2}\|_M^* \cup \cdots \cup \|x_{\alpha_n}\|_M^*.
\end{aligned}$$

This shows that $(\mathcal{M}_{LI}(L), \mathcal{T}|\mathcal{M}_{LI(L)})$ is a compact space.

Next, we prove that $(\mathcal{M}_{LI}(L), \mathcal{T}|\mathcal{M}_{LI(L)})$ is a T_2-space. Let $A, B \in \mathcal{M}_{LI}(L)$ and $A \neq B$, then by the maximality of A and B there exist $a \in A \setminus B$ and $b \in B \setminus A$. Let $x = (b \to a)'$ and $y = (a \to b)'$. Then by the definition of LI-ideal we have $x \notin A$ and $y \notin B$, for otherwise there will be a contradiction that $b \in A$ or $a \in B$. This shows that $A \in \|x\|_M^*$ and $B \in \|y\|_M^*$. By Proposition 2 and Lemma 8 we have

$$\begin{aligned}
\|x\|_M^* \cap \|y\|_M^* =&\|x \wedge y\|_M^* \\
=&\|(b \to a)' \wedge (a \to b)'\|_M^* = \|O\|_M^* = \emptyset.
\end{aligned}$$

Thus $(\mathcal{M}_{LI}(L), \mathcal{T}|\mathcal{M}_{LI(L)})$ is a T_2-space.

4 Conclusion

In this paper, two topologies based on prime LI-ideals and maximal LI-ideals of lattice implication algebras are constructed. Compactness and some separation properties of the two topologies are obtained. This work presents interactions between Non-Classical Mathematical Logic and General Topology. It should be noticed that other topologies and topological properties can also be considered in this framework. So, it is hopeful that more research topics of Non-Classical Mathematical Logic will arise with this work.

References

1. Wang, G.J.: Non-Classical Mathematical Logic and Approximate Reasoning. Science in China Press, Beijing (2003)
2. Xu, Y.: Lattice implication algebras. J. Southwest Jiaotong Univ. **28**(1), 20–27 (1993)
3. Xu, Y., Qin, K.Y.: Lattice H implication algebras and lattice implication classes. J. Hebei Min. Civ. Eng. Inst. **3**, 139–143 (1992)
4. Xu, Y., Qin, K.Y.: On filters of lattice implication algebras. J. Fuzzy Math. **1**(2), 251–260 (1993)
5. Jun, Y.B., Roh, E.H., Xu, Y.: LI-ideals in lattice implication algebras. Bull. Korean Math. Soc. **35**(1), 13–24 (1998)
6. Liu, Y.L., Liu, S.Y., Xu, Y.: ILI-ideals and prime LI-ideals in lattice implication algebras. Inf. Sci. **155**, 157–175 (2003)
7. Song, Z.M.: Implication filter spaces. J. Fuzzy Math. **8**(1), 263–266 (2000)
8. Chen, S.W., Jiang, B.Q., Yang, X.W.: LI-ideal spaces of lattice implication algebras. Chin. Quart. J. Math. **22**(4), 504–511 (2007)
9. Xu, Y., Ruan, D., Qin, K.Y., Liu, J.: Lattice-Valued Logics. Springer, Berlin (2003)
10. Liu, C.H.: Prime fuzzy LI-ideals and its spectrum space of lattice implication algebras. Appl. Math. J. Chin. Univ. (Ser. A) **29**(1), 115–126 (2014)
11. Liu, C.H.: LI-ideals lattice and its prime elements characterizations in a lattice implication algebra. Appl. Math. J. Chin. Univ. (Ser. A) **29**(4), 475–482 (2014)
12. Liu, C.H.: Extended LI-ideals in lattice implication algebras. Appl. Math. J. Chin. Univ. (Ser. A) **30**(3), 306–320 (2015)
13. Liu, C.H.: On(, q(,))-fuzzy LI-ideals in lattice implication algebras. J. Shandong Univ. (Natural SScience) **53**(2), 65–72 (2018)
14. Davey, B.A., Priestly, H.A.: Priestley Introduction to lattices and Order. Cambridge University Press, Cambridge (2002)
15. Kelley, J.L.: General Topology. Springer, New York (1991)

Application of Big Data on Self-adaptive Learning System for Foreign Language Writing

Jiang-Hui Liu(✉), Ling-Xi Ruan, and Ying-Yu Zhou

Guangdong University of Foreign Studies,
Guangzhou, Guangdong 510006, People's Republic of China
247031690@qq.com

Abstract. Personalized learning is a hot topic in education study in recent years. How to evaluate the students and find out the problem, is a problem that has to be solved in the study of individualized learning. Although the traditional education platform enables students to make full use of network resources, it does not really realize students' personalized learning. This study analyzes the general reference model of the adaptive learning system to understand the characteristics of each part. And we established an adaptive learning system model for foreign language writing to realize students' individual independent learning. In order to provide reference for the future research and application of adaptive learning technology, this study also makes an in-depth analysis of the adaptive learning technology and its application. It shows that adaptive learning technology is involved in a wide range of fields, but its application fields tend to be concentrated. The improvement of learning style model and the application of education big data and adaptive engine provide more accurate and intelligent personalized learning services for learners.

Keywords: Big data · Self-adaptive learning · Micro learning
Foreign language writing

1 Preface

The term Big Data originated in 1997. It means that it is a massive, high-growth and diversified information assets having a more powerful decision-making force, insight and process optimization capability through the new processing mode. The key to big data is not "massive", but "effective". It uses cloud computing, data mining, artificial intelligence and other methods to extract useful information and make decisions based on data. [1, 2] Big data application needs to go through three necessary stages, including data collection, data analysis and data visualization. Compared with traditional education data, big data has two aspects of values, predictive value and scientific research value [3]. Predictive value is the core value of big data. Mining algorithm is perfectly applied to mass data to predict the possibility of things happening [4].

Adaptive learning is a software and platform that can automatically adjust to learners' personalized learning needs while they learning [5]. It belongs to the category of ITAI, specifically, the individualized learning type of ITAI. Adaptive learning

N. Xiong et al. (Eds.): ISCSC 2018, AISC 877, pp. 86–93, 2019.
https://doi.org/10.1007/978-3-030-02116-0_11

technology, fully considers the learner's own habits and demand. It adjusts the teaching material by individual user data. It also predicts learners' needs of what kind of contents at a specific point to make timely follow up [6]. It constantly guides the learners' learning path and provides learners with a smoother study way in the process of learning. It makes learners concentrate on the study content more easily and improves the efficiency of individual learning as well. In general, adaptive learning technology enables the platform to adapt to learners rather than learners to adapt to the platform.

2 The Research of Adaptive Learning System

In the field of personalized adaptive learning for foreign language writing, foreign countries started early. Peter Brusilovsky, a professor at the University of Pittsburgh in the United States, constructs user modeling based on students' learning background, interest preference and knowledge level. In order to meet the personalized learning needs of learners in the process of interaction between learners and the system, adaptive learning systems such as InterBook, elm-art, KnowledgeSea, AnnotatEd and TaskSieve have been developed successively. Many of the results of subsequent studies were improvements and supplements on its basis. Professor DeBra, Eindhoven University of Technology, Professor Wolf, Royal Melbourne Institute of Technology University and Professor Papanikolaou, University of Athens also developed AHA!, iWeaver, INSPIRE and other personalized education hypermedia systems [7].

Domestic research on personalized adaptive learning is still in the stage of a theoretical discussion and small-scale attempt. Among them in the field of education technology, Beijing normal university professor Sheng-Quan Yu studied adaptive learning system early. He published an academic paper about adaptive learning, "adaptive learning – the development trend of education distance learning". Adaptive learning model was proposed from three key links, learning diagnosis, learning strategy and dynamic organization of learning content [8]. Dr. Pin-De Chen of south China normal university has completed his doctoral dissertation on adaptive learning, research on web-based adaptive learning support system. In this paper, the prototype system of the a-tutor (Adaptive Tutor) is designed, and questions such as "why should I adapt" and "what can be adapted" were put forward [9]. Professor Jian-Ping Zhang of Zhejiang University has also made a deep research on the adaptive learning support system. He has published a series of academic achievements. One of the works about adaptive learning, research on network learning and adaptive learning support system, the book elaborated the conceptual research about user model, learning ability, adaptive testing and knowledge visualization [10].

Adaptive learning technology provides a way to realize personalized learning. Whether by using semantic Web ontology technology to build the domain model, using the Web to use and to identify the learning interest of text mining, or present recommendation algorithm is adopted to improve the adaptive learning content and the formation of adaptive learning navigation, adaptive learning technology is always for the personalized learning services, in order to achieve personalized study provides an effective way.

3 The Design of the Adaptive Learning System

3.1 Functional Requirements and System Characteristics

The current online learning characteristics of Chinese students (for example, low level of learning strategies, poor autonomy, dependence on teachers, emphasis on memory and neglect of learning process, exam-driven learning and lack of learning self-motivation) show that online learning has not really happened. Even if it happens, it is only superficial learning, not deep learning. Therefore, an improved adaptive learning system is needed to solve the above problems. Adaptive learning system resources construction makes diverse learning resources for a same knowledge. (such as text, video, pictures, etc.) It can meet the needs of different users. The resource construction can meet the needs of different users. Establish an adaptive learning system that can provide adaptive learning services for learners.

3.2 Design of the Personalized Adaptive Learning System

Based on the above analysis, design the basic architecture of personalized adaptive learning system. The system is divided into five levels: perception layer, data layer, information layer, control layer and application layer, as shown in Fig. 1.

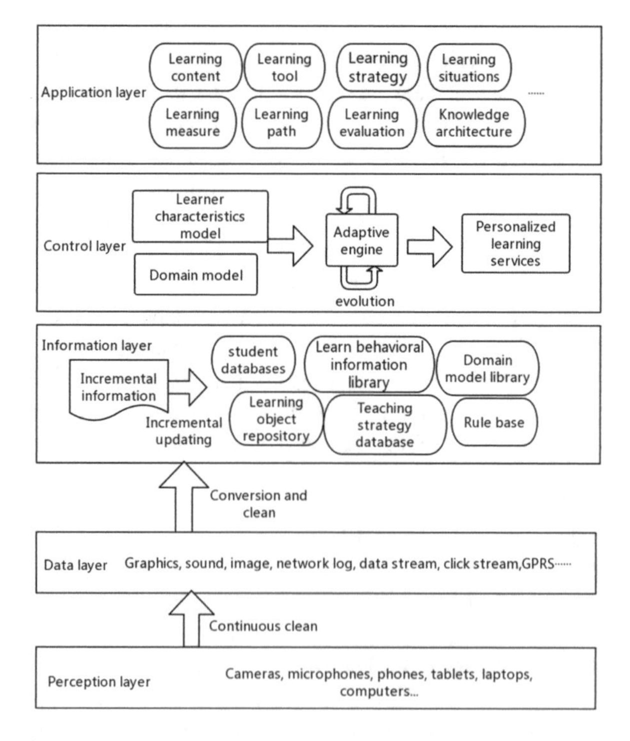

Fig. 1. The personalized adaptive learning system framework.

The perception layer. The perception layer includes various terminals, such as mobile phones and IPad. It can perceive learners' learning status and obtain various learning data. Data sources also include student information systems, performance management systems, etc.

The Data layer. Data layer collects data from the perception layer. It forms images, sounds, images, foreign language writing adaptive system network log, data stream, etc. And the original data pool is formed as well.

The Information layer. Through ETL and other technologies, the information layer integrates, transforms and cleans the data of the data layer. It converts the data into information and stores it in various databases. According to the continuous data collection and transformation, incremental information is constantly formed to update the contents in various databases.

The Control layer. Control layer is the core of personalized adaptive learning system. The adaptive engine provides personalized learning services according to the learner's characteristic model and domain model, and constantly updates the engine rules according to the feedback in the service process.

The Application layer. The application layer is a collection of business services for users, including learning content presentation, learning tools, learning situations, learning strategies and so on.

In conclusion, the perception layer continuously collects all kinds of learning data in the process of using the system. They are temporarily stored in the data layer and are stored in each database after being cleaned by transformation. The system applies the learner feature model and domain model in the control layer and continuously evolves itself through the adaptive engine. Therefore, it provides personalized learning services, which are finally reflected in the application layer.

4 Case Analysis

The knowledge map is shown in Fig. 2 used for Rule one analysis.

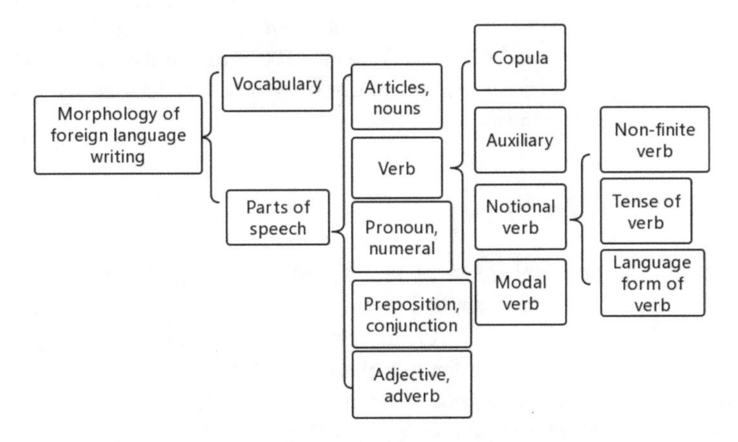

Fig. 2. The part of the knowledge map of foreign language writing.

If learners master “Non-finite verb” and have a cognitive thinking level of M0 for “Non-finite verb”, then the learning support for “Notional verb” is S = F(M0) × S0. Similarly, the learning support of “Tense of verb” and “Language form of verb” for “Notional verb” is S = F(M1) × S1 and S = F(M2) × S2, respectively. Rule two. For knowledge “Notional verb”, there are the following two cases:

If suitable learning c $\mathrm{Min}(1, \sum_{i=0}^{n} F(Mi) \times Si) = 1$, then the knowledge points can be learned.

If suitable learning c $\mathrm{Min}(1, \sum_{i=0}^{n} F(Mi) \times Si) \geq 0.683$, and similar peers in similar circumstances as well. Recommend the learner to learn this points directly.

5 System Implementation

The relevant technical tools used in this paper are as follows:

VBScript technology. VBScript is an important WEB development technology.

It is the basis of IIS, ASP and CGI program design and an important part of Microsoft Active X and COM technology. It is not only a part of HTML documents, but more importantly, it is based on HTML. VBScript USES an Active X (R) script to talk to the host application. Browsers and other host applications no longer require special integration code for each Script part because they use Active X Script.

SQL and ADO database access technologies. Another essential aspect of database access is Microsoft’s latest database access technology, ADO, which can be used in any ODBC-enabled relational database. SQL is an instruction that allows the database to perform data operations quickly after the data source and data objects are established.

ASP technology. ASP is a server-side scripting environment for developing dynamic web pages. It can support multiple scripting languages, such as JavaScript and Script, and can embed dynamic elements into pages. ASP can be fully integrated with Html files and can be extended with controls. The resulting page is not only highly interactive, but also capable of complex data manipulation. In addition, since this technology is a script executed on the server side. The source program will not be sent to the browser side, so the source program can be avoided to be used by others. This will improve the system security.

6 Experimental Result

In order to test the practical effect of the self-adaptive learning system for foreign language writing, the system is applied to a class in our school. Another class adopts the traditional teaching method. Try to make the external conditions, hardware equipment and teaching content of the two classes basically the same. Both classes start from scratch. After the same period, students in each class were finally selected for unified test to make relevant comparison (excluding students with relevant knowledge foundation before the experiment). The comparison is as shown in Table 1:

Table 1. Comparison of experimental conditions.

External conditions	Class A	Class B	Remarks
Teacher situation	Teach	Guide	
Classroom	48-seat multimedia classroom with projection equipment	48-seat multimedia classroom with equipment with adaptive learning system software	
Students	First semester of first grade	First semester of first grade	Random classes of 44 people per class
Number of study weeks	6	6	
Test topic	Some people believe that job satisfaction is more important than job security. Others believe that people cannot always enjoy their jobs and that a permanent job is more important. Discuss both views and give your own opinion	Some people believe that job satisfaction is more important than job security. Others believe that people cannot always enjoy their jobs and that a permanent job is more important. Discuss both views and give your own opinion	

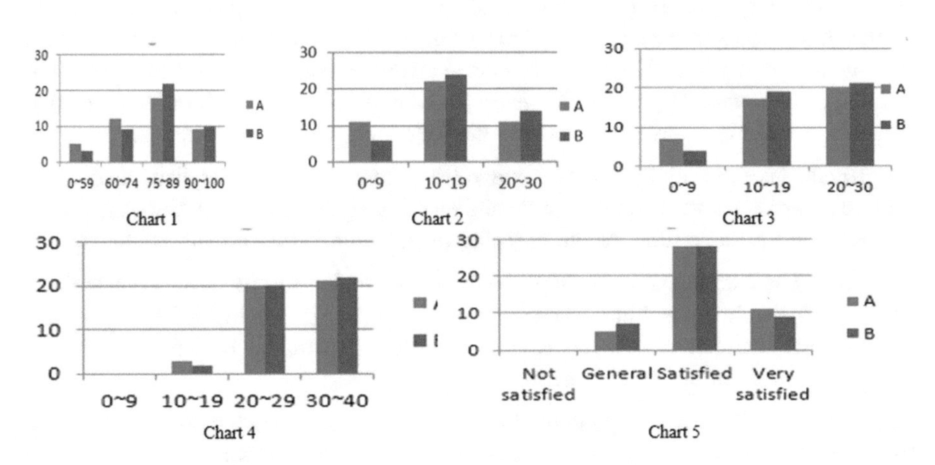

Fig. 3. Comparison of test results.

As is shown in Fig. 3, total score distribution (100) is in Chart 1. Achievements in the Application of Vocabulary and sentence patterns (30) is in Chart 2. Grammar performance (30) is in Chart 3. Key points cover achievement (40) is in Chart 4. Degree of students' satisfaction with the adaptive learning system for foreign language writing is in Chart 5.

Through the analysis of charts in Fig. 3, it can be seen that after six weeks' study, students who have used the self-adaptive learning system for foreign language writing (Class B) have achieved better results in total scores and the use of vocabulary, sentence patterns and grammar than those who have not. This proves that the adaptive learning system for foreign language writing is effective in practical application, especially in the vocabulary and sentence patterns part and the grammar part, which contributes to the total score difference between Class B and Class A.

However, it can also be found that while using adaptive learning system to learn, the satisfied sense of experience is relatively less than that of teachers. We need to consider more about the users' experience of the human machine interface module. This will be the direction that needs to be researched and improved in the future.

7 Main Research Work in the Future

People's emotional state and their changes will have an important influence on the performance of learning activities. In network learning environment, the human-computer interaction is different from the social reality, learners in the learning process prone to mood disorders such as depression, anxiety, loneliness, loss, which will affect the learning efficiency. How to solve the network learning environment, learners' emotional loss, and improve the learning efficiency, is the inevitable requirement of adaptive learning depth development. The development of the adaptive learning system cannot be separated from the improvement of the learner model, and the learner model, as a key component to realize the personalized adaptive function, cannot be short of the learning style model. At present, the theoretical and practical research of learning style model is constantly improving, showing the following development trends.

Integration. The learning style model integrates various elements of stylistic theory and is more comprehensive. Cognitive style and learning style are taken as the elements of the model at the same time, or the mapping relationship between them is studied.

Dynamics. The learning style model not only allows students to fill in the style scale to determine the style types, but also can judge the learners' preference information from the real-time interaction between students and the department system. And it can update the model information dynamically.

Materialization. It mainly refers to the localization of learning style model research, the combination with specific grades and disciplines, the consideration of the differences between subjects and learners of different ages, and the enrichment and modification of our learning style model according to the teaching objectives and learning strategies of subjects.

Systematic. The learning style model is only a sub-model of the learner model, and future studies will give more consideration to the relationship with other sub-models of the learner model (such as the knowledge state model) and the interaction and connection with other models of the adaptive learning system (such as the domain model).

8 Conclusion

The original purpose of education is to make students' learning process more independent and efficient. However, the existing education platform cannot fully realize students' effective independent learning due to teachers, students, technology or other reasons. Starting from these problems, this paper analyzes the characteristics of the reference model of the adaptive learning system, and discusses the supporting elements needed for the personalized learning environment, and proposes a personalized learning system based on the adaptive learning system. This model complements the deficiency of the existing education platform on the Internet, enabling the system to update course resources dynamically. Students can choose their own learning content according to their own conditions. This gives full play to students' subjective initiative and also enables students to obtain personalized learning services in the learning process.

References

1. Qiang, J., Wei, Z., Pengjiao, W.: Personalized adaptive online learning analysis model and implementation based on big data. China Educ. Technol. **1**, 85–92 (2015)
2. Zhiting, Z., Demei, S.: New paradigm of education technology research based on big data. E-educ. Res. **10**, 5–13 (2010)
3. Xiaoji, Y.: Research on the construction of personalized teaching information service platform based on big data application. Inf. Sci. **11**, 53–56 (2015)
4. Zhiting, Z., Bin, H., Demei, S.: Reverse innovation in information-based education. Educ. Res. (3), 5–12+50 (2014)
5. Brusilovsky, P., Millán, E.: User models for adaptive hypermedia and adaptive educational systems. In: Brusilovsky, P., Kobsa, A., Nejdl, W. (eds.) The Adaptive Web. Lecture Notes in Computer Science, vol. 4321, pp. 3–53. Springer, Heidelberg (2007)
6. Qiang, J., Wei, Z., Pengjiao, W.: Study on the architecture of adaptive learning system based on GALSRM model. Mod. Distance Educ. **1**, 71–77 (2013)
7. Ma, X., Zhong, S., Xu, D.: Research on support model and implementation mechanism of personalized adaptive learning system from the perspective of big data. China Educ. Technol. **4**, 97–102 (2017)
8. Bian, L., Xie, Y.: Research on the adaptive strategy of adaptive learning system. In: Zhang X., Zhong S., Pan Z., Wong K., Yun R. (eds.) Entertainment for Education. Digital Techniques and Systems. Edutainment. Lecture Notes in Computer Science, vol. 6249, pp. 203–214. Springer, Heidelberg (2010)
9. Jia-Jiunn, L., Ya-Chen, Ch., Shiou-Wen, Y.: Designing an adaptive web-based learning system based on students' cognitive styles identified online. Comput. Educ. **58**(1), 209–222 (2012)
10. Mulwa, C., Lawless, S., Sharp, M., Wade V.: A web-based framework for user-centred evaluation of end-user experience in adaptive and personalized e-learning systems. In: WI-IAT 2011 Proceedings of the IEEE/WIC/ACM International Conferences on Web Intelligence and Intelligent Agent Technology, pp. 351–356. IEEE (2011)

Perturbation Analysis of the Fully Fuzzy Linear Systems

Kun Liu[✉], Hong-xia Li, and Ying Guo

College of Mathematics and Statistics, Longdong University,
Qingyang 745000, Gansu, China
liukunws@163.com
http://sxx.ldxy.edu.cn/

Abstract. The main aim of this paper is to investigate perturbation analysis of the fully fuzzy linear systems (shown as FFLS) $\tilde{A} \otimes \tilde{x} = \tilde{b}$, where $\tilde{A}$ and $\tilde{b}$ are respectively a fuzzy matrix and a fuzzy vector. For showing how the perturbation of the right hand vector impact the fuzzy approximate solution vector to FFLS. We first transform the original fully fuzzy linear systems into three crisp linear systems by the Dehghan's method. Next, the situation that the right hand side is slightly perturbed while the coefficient matrix remains unchanged is discussed; Finally, the relative error bounds of perturbed fully fuzzy linear systems are obtained in terms of the distance of LR-type triangular fuzzy vector.

Keywords: LR-type fuzzy numbers · The fully fuzzy linear systems
Perturbation analysis · FFLS

1 Introduction

Systems of simultaneous linear equations are an essential mathematical tool in science and technology. Whenever a relation exists between two sets of variables, the simplest model to try is one in which the dependent variables are sums of constant times the independent variables. When we want to determine the values of constants or to reserve the role of dependent and independent variables, we must solve a system of linear equations. However, in practice, uncertainty is often involved in any model design process. It is known that fuzzy sets provide a widely appreciated tool to deal with these uncertainty. In many applications, some or all of the parameters of the system may be represented by fuzzy quantities rather than crisp ones, and hence it is important to develop mathematical theory and numerical schemes to treat systems of fuzzy and fuzzy linear equations occurred in many fields, such as control problems, information, physics, statistics, engineering, economics, finance and even social sciences (see [1–6]).

It's worth noting that when we use numerical methods to solve fuzzy or fully fuzzy linear systems, round-off error is inevitable. If we desire a perfect numerical result then we must measure the error between the computational result and the exact one. In addition, perturbation analysis can assert whether

N. Xiong et al. (Eds.): ISCSC 2018, AISC 877, pp. 94–101, 2019.
https://doi.org/10.1007/978-3-030-02116-0_12

the problem is well-conditioned or ill-conditioned and then help us find efficient numerical methods. Therefore, perturbation analysis is very important issues in numerical methods for solving fuzzy or fully fuzzy linear systems. Tang [7] consider the perturbation problems of a fuzzy matrix equation by perturbation methods [8]. Wang et al. [9] analyzed the perturbation of an $n \times n$ fuzzy linear system $Ax = \tilde{b}$. They focused their discussion only on fuzzy linear systems with the right hand side being triangular fuzzy numbers and their results were only valid for numerical approaches based on the embedding method. Tian et al. [10] define a metric in fuzzy vector space and using the embedding method transform the original fuzzy linear system $Ax = \tilde{b}$ into a functional system. By solving this functional linear system, they gave the relative error bounds for the perturbed system in terms of the metric of the fuzzy vectors and the spectral norm of the real matrix. However, they focused their discussion only on the fuzzy linear systems of the form $Ax = \tilde{b}$ by the embedding method.

In this paper, we transform the original fully fuzzy linear systems into three crisp linear systems by the Dehghan's method [11]. By solving three crisp linear systems we give the fuzzy approximate solutions to the fully fuzzy linear systems and the perturbed fully fuzzy linear systems. Moreover, we deduce the relative error bounds for perturbed fully fuzzy linear system based on the distance of LR-type triangular fuzzy vector.

2 Notation and Basic Definitions

In this section, we give some definitions and introduce the notation which will be used throughout the paper.

Let us denote by $F(R)$ the class of fuzzy subsets of the real axis (i.e. $u : R \to [0,1]$). An arbitrary fuzzy number is defined as an ordered pair of functions $(\underline{u}(r), \overline{u}(r))$, $0 \le r \le 1$, which satisfies the following requirements (see [12]):

(1) $\underline{u}(r)$ is a bounded monotonic increasing left continuous function over [0,1],
(2) $\overline{u}(r)$ is a bounded monotonic decreasing left continuous function over [0,1],
(3) $\underline{u}(r) \le \overline{u}(r)$, $0 \le r \le 1$.

A special type of representation for fuzzy numbers which increases computational efficiency is LR-type fuzzy numbers.

Definition 1 see [13–15]. *A fuzzy number M is said to be an LR-type fuzzy number if*

$$\mu_M(x) = \begin{cases} L(\frac{m-x}{\alpha}), \; x \le m, \; \alpha > 0, \\ R(\frac{x-m}{\beta}), \; x \ge m, \; \beta > 0, \end{cases} \tag{1}$$

where m is the mean value of M, and α and β are left and right spreads, respectively, and a function $L(\cdot)$ the left shape function satisfying:

(i) $L(x) = L(-x)$,
(ii) $L(0) = 1$ *and* $L(1) = 0$,
(iii) $L(x)$ *is non-increasing on* $[0, \infty)$.

Naturally, a right shape function $R(\cdot)$ is similarly defined as $L(\cdot)$. Using its mean value and left and right spreads, and shape functions, such an LR-type fuzzy number M is symbolically written $M = (m, \alpha, \beta)_{LR}$. □

The fuzzy arithmetic operations of fuzzy numbers have grown in importance during recent years as an advanced tool in numerical calculation of all kinds of fuzzy problems. Based on the extension principle, Dubois and Prade designed exact formulas for $\oplus$ and $\ominus$ together with the approximate formulas for $\otimes$ and scalar multiplication to LR-fuzzy numbers, whose advantage comes from the fact that arithmetic operations with them can be performed in a relatively easy manner (see [13,14]).

Let $F_{LR}(R)$ denote the set of all LR-type fuzzy numbers. Hung [16] defined the distance for any M and N in $F_{LR}(R)$ with $M = (m, \alpha, \beta)_{\mathrm{LR}}$ and $N = (n, \gamma, \delta)_{\mathrm{LR}}$ as follows:

$$d^2_{LR}(M,N) = \frac{1}{3}\{(m-n)^2 + ((m-l\alpha)-(n-l\gamma))^2 + ((m+r\beta)-(n+r\delta))^2\}, \quad (2)$$

where $l = \int_0^1 L^{-1}(\omega)d\omega$ and $r = \int_0^1 R^{-1}(\omega)d\omega$.

For an LR-type fuzzy number $M = (m, \alpha, \beta)_{\mathrm{LR}}$, if $L(\cdot)$ and $R(\cdot)$ are of the form

$$T(x) = \begin{cases} 1 - x, 0 \leq x \leq 1, \\ 0, \quad \text{otherwise}, \end{cases} \quad (3)$$

then M is called a triangular fuzzy number and is represented by triplet $M = (m, \alpha, \beta)_{\mathrm{T}}$, with the membership function defined by the expression

$$\mu_M(x) = \begin{cases} 1 - (\frac{m-x}{\alpha}), \; x \leq m, \; \alpha > 0, \\ 1 - (\frac{x-m}{\beta}), \; x \geq m, \; \beta > 0. \end{cases} \quad (4)$$

Consider two triangular fuzzy numbers $M = (m, \alpha, \beta)_{\mathrm{T}}$ and $N = (n, \gamma, \delta)_{\mathrm{T}}$ of the space $F_T(R)$ which represents the collection of all triangular fuzzy numbers, the distance on $F_T(R)$ can be defined as

$$\begin{aligned} d^2_T(M,N) &= \frac{1}{3}\{(m-n)^2 + ((m-\frac{1}{2}\alpha)-(n-\frac{1}{2}\gamma))^2 + ((m+\frac{1}{2}\beta)-(n+\frac{1}{2}\delta))^2\} \\ &= \frac{1}{3}\{(m-n)^2 + ((m-n)-\frac{1}{2}(\alpha-\gamma))^2 + ((m-n)+\frac{1}{2}(\beta-\delta))^2\}. \quad (5) \end{aligned}$$

Denotes $F^n_T(R)$ n-dimensional fuzzy vectors space, which consists of all n-dimensional triangular fuzzy number vectors. Let $U, V \in F^n_T(R)$ and $U = (u_1, u_2, \cdots, u_n)^T$, $V = (v_1, v_2, \cdots, v_n)^T$. We define the distance of LR-type fuzzy vector U and V as follows:

$$d^2_T(U,V) = \frac{1}{3}\sum_{i=1}^{n}\{(u_i-v_i)^2 + ((u_i-v_i)-\frac{1}{2}(\alpha_i-\gamma_i))^2 + ((u_i-v_i)+\frac{1}{2}(\beta_i-\delta_i))^2\} \quad (6)$$

where $u_i = (u_i, \alpha_i, \beta_i)_{\mathrm{T}}, v_i = (v_i, \gamma_i, \delta_i)_{\mathrm{T}}, i = 1, 2, \cdots, n$.

Naturally, the norm of fuzzy vector U is defined as $\|U\| = (d_T^2(U, \tilde{0}))^{\frac{1}{2}}$, i.e. the metric of U and zero fuzzy vector $\tilde{0}$. By zero fuzzy vector we mean that each of its entries is fuzzy number zero whose membership is $\chi_{\{0\}}$.

In addition, the Euclidian norm of a real vector $x = (x_1, x_2, \cdots, x_n)^T \in C^n$ is defined as $\|x\| = (\sum_{i=1}^{n} |x_i|^2)^{\frac{1}{2}}$.

The spectral norm of a real matrix $A = (a_{ij}) \in C^{n\times n} (1 \leq i, j \leq n)$ is defined as $\|A\|_2 = (\lambda_{\max(A^H A)})^{\frac{1}{2}}$.

A matrix $\tilde{A} = (\widetilde{a_{ij}})$ is called a fuzzy matrix, if each element of $\tilde{A}$ is a fuzzy number (see [13–15]).

$\tilde{A}$ will be positive (negative) and denoted by $\tilde{A} > 0$ ($\tilde{A} < 0$) if each element of $\tilde{A}$ be positive (negative). Similarly non-negative and non-positive fuzzy matrices will be defined.

Up to rest of this paper, we use LR-type triangular fuzzy numbers and Multiplication formula, in place of exact $\otimes$. We may represent $n \times n$ fuzzy matrix $\tilde{A} = (\widetilde{a_{ij}})_{n\times n}$, that $\widetilde{a_{ij}} = (a_{ij}, \alpha_{ij}, \beta_{ij})_{\mathrm{T}}$, with new notation $\tilde{A} = (A, M, N)$, where $A = (a_{ij})$, $M = (\alpha_{ij})$ and $N = (\beta_{ij})$ are three crisp $n \times n$ matrices, with the same size of $\tilde{A}$.

Definition 2 (see [11]). *Let $\tilde{A} = (\widetilde{a_{ij}})$ and $\tilde{B} = (\widetilde{b_{ij}})$ be two $m \times n$ and $n \times p$ fuzzy matrices. Then their approximate multiplication was defined as $\tilde{A} \otimes \tilde{B} = \tilde{C} = (\widetilde{c_{ij}})$ which is $m \times p$ matrix where*

$$\widetilde{c_{ij}} = \sum_{k=1,\cdots,n}^{\oplus} \widetilde{a_{ik}} \otimes \widetilde{b_{kj}}. \tag{7}$$

□

Definition 3 (see [11]). *Consider the $n \times n$ linear system of equations:*

$$\begin{cases} \tilde{a}_{11} \otimes \tilde{x}_1 \oplus \tilde{a}_{12} \otimes \tilde{x}_2 \oplus \cdots \oplus \tilde{a}_{1n} \otimes \tilde{x}_n = \tilde{b}_1, \\ \tilde{a}_{21} \otimes \tilde{x}_1 \oplus \tilde{a}_{22} \otimes \tilde{x}_2 \oplus \cdots \oplus \tilde{a}_{2n} \otimes \tilde{x}_n = \tilde{b}_2, \\ \cdots\cdots\cdots\cdots\cdots\cdots\cdots\cdots\cdots \\ \tilde{a}_{n1} \otimes \tilde{x}_1 \oplus \tilde{a}_{n2} \otimes \tilde{x}_2 \oplus \cdots \oplus \tilde{a}_{nn} \otimes \tilde{x}_n = \tilde{b}_n. \end{cases} \tag{8}$$

The matrix form of the above equations is

$$\tilde{A} \otimes \tilde{x} = \tilde{b} \tag{9}$$

or simply $\tilde{A}\tilde{x} = \tilde{b}$, where the coefficient matrix $\tilde{A} = (\tilde{a}_{ij}), 1 \leq i, j \leq n$ is an $n \times n$ fuzzy matrix and $\tilde{x}_i, \tilde{b}_i \in F_T(R), 1 \leq i \leq n$. This system is called a fully fuzzy linear systems (in short, FFLS). Also, if each element of $\tilde{A}$ and $\tilde{b}$ is an non-negative LR-type fuzzy number, then the Eq. (2) is called an non-negative FFLS. □

Definition 4 (see [11]). *The fuzzy vector* $\tilde{x} = (x, y, z) = ((x_1, y_1, z_1), (x_2, y_2, z_2), \cdots, (x_n, y_n, z_n))^{\mathrm{T}}$ *denotes a solution vector to non-negative fully fuzzy linear systems* (2)*, if and only if*

$$Ax = b, Ay + Mx = g, Az + N = h,$$

where the membership function of each element of $\{x \mid \mu_{\tilde{x}}(x) > 0\}$ *can be defined with the same functions* $L(\cdot)$ *and* $R(\cdot)$ *which used in* $\tilde{A}$ *and* $\tilde{b}$*. Moreover, if* $y \geq 0, z \geq 0$ *and* $x - y \geq 0$, $\tilde{x} = (x, y, z)$ *be called a consistent solution of non-negative fully fuzzy linear systems* (2)*. Otherwise, it will be called a dummy solution.* □

3 Perturbation Analysis of the Fully Fuzzy Linear Systems

In this section, we will show how the perturbations of the right hand vector impact the fuzzy approximate solution vector to FFLS (2). We assume that the fuzzy vector $\tilde{b}$ in the fully fuzzy linear systems $\tilde{A} \otimes \tilde{x} = \tilde{b}$ is slightly perturbed to another fuzzy vector $\tilde{b'}$ (the perturbed fully fuzzy linear systems is $\tilde{A} \otimes \tilde{x^*} = \tilde{b'}$) while the fuzzy coefficient matrix $\tilde{A}$ remain unchanged. Assuming that A is an non-singular crisp matrix, using approximate multiplication, $\tilde{A} \otimes \tilde{x} = \tilde{b}$ and $\tilde{A} \otimes \tilde{x^*} = \tilde{b'}$ can be written as $(Ax, Ay + Mx, Az + Nx) = (b, g, h)$ and $(Ax^*, Ay^* + Mx^*, Az^* + Nx^*) = (b', g', h')$, respectively. Thus we have

$$\begin{cases} Ax = b, \\ Ay + Mx = g, \\ Az + Nx = h, \end{cases} \tag{10}$$

and

$$\begin{cases} Ax^* = b', \\ Ay^* + Mx^* = g', \\ Az^* + Nx^* = h'. \end{cases} \tag{11}$$

So we easily get

$$\begin{cases} x = A^{-1}b, \\ y = A^{-1}g - A^{-1}MA^{-1}b, \\ z = A^{-1}h - A^{-1}NA^{-1}b, \end{cases} \tag{12}$$

and

$$\begin{cases} x^* = A^{-1}b', \\ y^* = A^{-1}g' - A^{-1}MA^{-1}b', \\ z^* = A^{-1}h' - A^{-1}NA^{-1}b'. \end{cases} \tag{13}$$

By Eqs. (12) and (13), the fuzzy approximately solution vector to two fully fuzzy linear systems can be written as

$$\tilde{x} = (x, y, z) = (A^{-1}b, A^{-1}g - A^{-1}MA^{-1}b, A^{-1}h - A^{-1}NA^{-1}b), \tag{14}$$

$$\tilde{x^*} = (x^*, y^*, z^*) = (A^{-1}b^{'}, A^{-1}g' - A^{-1}M^{-1}b', A^{-1}h' - A^{-1}N^{-1}b'). \tag{15}$$

We investigate a relative error bound for the fuzzy solution of a fully fuzzy linear system in the following.

Theorem 1. *Let A be an non-singular crisp matrix, $\tilde{x} = (\tilde{x}_1, \tilde{x}_2, \cdots, \tilde{x}_n)^T$ and $\tilde{x^*} = (\tilde{x_1^*}, \tilde{x_2^*}, \cdots, \tilde{x_n^*})^T$ be the fuzzy approximate solution vectors to the fully fuzzy linear systems $\tilde{A} \otimes \tilde{x} = \tilde{b}$ and $\tilde{A} \otimes \tilde{x^*} = \tilde{b'}$, respectively. Also, let $\delta b = b - b', \delta g = g - g', \delta h = h - h'$. Then*

(a)

$$\frac{d(x^*, x)}{\|x\|} \leq (\|A\|_2 \|A^{-1}\|_2) \frac{d(b, b')}{\|b\|};$$

(b)

$$\frac{d(y^*, y)}{\|y\|} \leq (\|A\|_2 \|A^{-1}\|_2) \frac{\|\delta g - MA^{-1}\delta b\|}{\|g - MA^{-1}b\|};$$

(c)

$$\frac{d(z^*, z)}{\|z\|} \leq (\|A\|_2 \|A^{-1}\|_2) \frac{\|\delta h - NA^{-1}\delta b\|}{\|h - NA^{-1}b\|}.$$

Moreover

$$\frac{d(\tilde{x^*}, \tilde{x})}{\|\tilde{x}\|} \leq (\|A\|_2 \|A^{-1}\|_2) \max\{\frac{d(b, b')}{\|b\|}, \frac{\|\delta g - MA^{-1}\delta b\|}{\|g - MA^{-1}b\|}, \frac{\|\delta h - NA^{-1}\delta b\|}{\|h - NA^{-1}b\|}\}.$$

□

Proof. By using the Dehghan's method, we can transform $\tilde{A} \otimes \tilde{x} = \tilde{b}$ and $\tilde{A} \otimes \tilde{x^*} = \tilde{b'}$ into three crisp linear systems (12) and (13), respectively. Let $x = (x_1, x_2, \cdots, x_n)$ and $x^* = (x_1^*, x_2^*, \cdots, x_n^*)$ are solution vectors to $Ax = b$ in (12) and $Ax^* = b'$ in (13), respectively. It follows that

$$x^* - x = A^{-1}(b' - b) = A^{-1}\delta b.$$

Taking supremum norm, we have

$$\|x^* - x\| \leq \|A^{-1}\|_2 \|\delta b\|.$$

Since $\|b\| \leq \|A\|_2 \|x\|$, so

$$\frac{\|x^* - x\|}{\|x\|} \leq \frac{\|A^{-1}\|_2 \|A\|_2 \|\delta b\|}{\|b\|}.$$

According to the definition of vector norm in Sect. 2, we obtain the desired inequality

$$\frac{d(x^*, x)}{\|x\|} \leq (\|A\|_2 \|A^{-1}\|_2) \frac{d(b, b')}{\|b\|}.$$

Hence the inequality (a) holds.

Next we prove (b). According to Eqs. (12) and (13), It follows that

$$\begin{aligned} y^* - y &= A^{-1}(g' - g) - A^{-1}MA^{-1}(b' - b) \\ &= A^{-1}(\delta g - MA^{-1}\delta b). \end{aligned}$$

Taking supremum norm, we have

$$\|y^* - y\| \leq \|A^{-1}\|_2 \|\delta g - MA^{-1}\delta b\|.$$

Since $\|g - Mx\| \leq \|A\|_2 \|y\|$, so

$$\frac{\|y^* - y\|}{\|y\|} \leq (\|A\|_2 \|A^{-1}\|_2) \frac{\|\delta g - MA^{-1}\delta b\|}{\|g - MA^{-1}b\|}.$$

According to the definition of vector norm in Sect. 2, we obtain the desired inequality

$$\frac{d(y^*, y)}{\|y\|} \leq (\|A\|_2 \|A^{-1}\|_2) \frac{\|\delta g - MA^{-1}\delta b\|}{\|g - MA^{-1}b\|}.$$

This completes the proof of (b).

In the same way, we have the inequality (c).

It is obvious that

$$\frac{d(\tilde{x^*}, \tilde{x})}{\|\tilde{x}\|} \leq (\|A\|_2 \|A^{-1}\|_2) \max\{\frac{d(b, b')}{\|b\|}, \frac{\|\delta g - MA^{-1}\delta b\|}{\|g - MA^{-1}b\|}, \frac{\|\delta h - NA^{-1}\delta b\|}{\|h - NA^{-1}b\|}\}.$$

□

4 Conclusion

In this paper, the perturbation analysis of the fully fuzzy linear systems shown as FFLS) $\tilde{A} \otimes \tilde{x} = \tilde{b}$ is investigate. Based on LR-type triangular fuzzy numbers and its approximate arithmetic operators, the original fully fuzzy linear systems is converted to three crisp linear equations. And using the distance of LR-type triangular fuzzy vector and the Dehghan's method, the relative error bounds for perturbed fully fuzzy linear systems are obtained.

Acknowledgements. Thanks to the support by Provincial Science and Technology Program Foundation of Gansu (18JR3RM238) and PhD Scientific Research Start-up Funded Projects of Longdong University (XYBY05).

References

1. Friedman, M., Ma, M., Kandel, A.: Fuzzy linear systems. Fuzzy Sets Syst. **96**, 201–209 (1998). https://doi.org/10.1016/S0165-0114(96)00270-9
2. Ezzati, R.: Solving linear systems. Soft Comput. **3**, 193–197 (2011). https://doi.org/10.1007/s00500-009-0537-7
3. Allahviranloo, T., Mikaeilvand, N., Kiani, N.A., Shabestari, R.M.: Signed decomposition of fully fuzzy linear systems. Int. J. Appl. Appl. Math. **3**, 77–88 (2008). https://www.researchgate.net/publication/229006519
4. Mikaeilvand, N., Allahviranloo, T.: Solutions of the fully fuzzy linear systems. AIP Cpmf. Proc. **1124**, 234–243 (2009). https://doi.org/10.1063/1.3142938

5. Dehghan, M., Hashemi, B.: Solution of the full fuzzy linear systems using the decomposition procedure. Appl. Math. Comput. **182**, 1568–1580 (2006). https://doi.org/10.1016/j.amc.2006.05.043
6. Dehghan, M., Hashemi, B., Ghatee, M.: Solution of the full fuzzy linear systems using iterative techniques. Chaos Solitons Fractals **34**, 316–336 (2007). https://doi.org/10.1016/j.chaos.2006.03.085
7. Tang, F.C.: Perturbation teechniques for fuzzy matrix equations. Fuzzy Sets Syst. **109**, 363–369 (2000). https://doi.org/10.1016/S0165-0114(98)00021-9
8. Nayfeh, A.H.: Perturbation methods. In: IEEE Transactions on Systems, Man, and Cybernetics, pp. 417–418. IEEE Press, New York (1978). https://doi.org/10.1109/TSMC.1978.4309986
9. Wang, K., Chen, G., Wei, Y.: Perturbation analysis for a class of fuzzy linear systems. J. Comput. Appl. Math. **224**, 54–65 (2009). https://doi.org/10.1016/j.cam.2008.04.019
10. Tian, Z.F., Hu, L.J., Greenhalgh, D.: Perturbation analysis of fuzzy linear systems. Inf. Sci. **180**, 4706–4713 (2010). https://doi.org/10.1016/j.ins.2010.07.018
11. Dehghan, M., Hashemi, B., Ghatee, M.: Computational methods for solving fully fuzzy linear systems. Appl. Math. Comput. **179**, 328–343 (2006). https://doi.org/10.1016/j.amc.2005.11.124
12. Goetschel, R., Voxman, W.: Elementary calculus. Fuzzy Sets Syst. **18**, 31–43 (1986). https://doi.org/10.1016/0165-0114(86)90026-6
13. Dubois, D., Prade, H.: Operations on fuzzy numbers. Int. J. Syst. Sci. **9**, 613–626 (1978). https://doi.org/10.1080/00207727808941724
14. Dubois, D., Prade., H.: Fuzzy Sets and Systems: Theory and Applications, New York (1980). https://doi.org/10.2307/2581310
15. Gong, Z.T., Zhao, W.C., Liu, K.: A straightforward approach for solving fully fuzzy linear programming problem with LR-type fuzzy numbers. J. Oper. Res. Soc. Jpn. **61**, 172–185 (2018). http://www.orsj.or.jp
16. Hung, W.L., Yang, M.S., Ko, C.H.: Fuzzy clustering on $LR-$ type fuzzy numbers with an application in Taiwanese tea evaluation. Fuzzy Sets Syst. **150**, 561–577 (2005). https://doi.org/10.1016/j.fss.2004.04.007

Anisotropy of Earth Tide from Deepwell Strain

Anxu Wu(✉)

Beijing Earthquake Agency, Beijing 100080, China
wu-anxu@163.com

Abstract. The drilling strain observation techniques have been played a very important role in the research field of geodynamics and earthquake prediction, and have been gotten wide applications. We have achieved a number of observational solid tide data from the station without the influence and interference from ground. The apparatus could record clear and stable solid tide and strain wave with coseism. Through the solid tide model of Nakai, we calculated and gained solid tide proportion factors in the directions of deepwell, from deepwell strain observation, to verify that the tide response to the theoretical tidal factor model in Baishan station is a very large dispersion and a regular azimuthal anisotropy. The high precision deepwell solid tide observation and tidal response anisotropy phenomenon is expected to be on the dynamics, earthquake prediction, to provide a reference to improve the tide displacements of observation data in space geodesy.

Keywords: Deepwell strain mete · Solid tide · Nakai model · Anisotropy

1 Introduction

The drilling strain observation techniques started early and have been playing an important role in the researching of seismic prediction. However, the deep-well observation technologies have been restrained by specific technologies, and as a result the strain observation and researching have not been developed in china. In October of 2008, The Institute of Crustal Dynamics of CEA, Beijing Earthquake Agency and ChangPing Distric Earthquake Agency carried out a cooperation to install the self-developed deep-well deformation observing system at BaiShan of ChangPing Distric. Moreover, there are a number of faults around the station, which is significant to the study of activity of fault, earthquake prediction and geodynamics.

The observation of RZB-3 deep well strain filled this gap and made it possible to carry out further study in some problems of geodynamics and seismology. Based on the latest observations, we took the advantages of multi-directions to carry out the differentiation characteristic analysis about the response direction of the earth tide and study the characteristics of the directional distribution of solid tidal response for this type of instruments in this well.

N. Xiong et al. (Eds.): ISCSC 2018, AISC 877, pp. 102–108, 2019.
https://doi.org/10.1007/978-3-030-02116-0_13

2 Multicomponent Observation and Data

2.1 Deepwell Strain Meter

The RZB-3 system consists of 8 components of horizontal strain, 1 component of vertical strain, 2 components of bias and ground temperature, 3 components of observation of seismic waves. It uses the measurement technology of capacitance displacement with high-precision, integration technology of deep well integrated observing system, deep well sealing and testing technology underground multi-channel measurement and data collection and transmission technology, deep well cement consolidation and installation technology.

2.2 Observation Records

The RZB-3 type system started the trial operation on 5th Oct., 2008. Since then, several data missing cases occurred in the two years. All of the reasons for the data missing are power supplying interruption or a failure to download data, which are all ground matters. For the normal record, the data are fine in quality and are able to record clear earth tide and coseismic variation, which proved that the instrument accuracy meets the related demand.

3 Anisotropy Analysis of Earth Tide

The anisotropy features of the dielectric exist commonly in the crust. Researching of the anisotropy features in seismology is common, but the analysis on the different response ability of earth tide in each orientation is rarely. It is mainly because there were no such observation and research conditions before [1].

In the past two years, more than one borehole strain instruments with shallow wells were installed in china. During the installation of these instruments, the limitation and difficulty of the install orientation are normally different. The analysis about the new installed 4 components borehole strain apparatus, Chi et al. [1] show that: different orientations have different tidal factors. Therefore, he believed that the anisotropy exists in the response ability of earth tide.

3.1 Algorithm for Anisotropy of Strain Earth Tide

To get the install location of the RZB-3 type deep-well strain meter and the orientation distribution of the anisotropy of ground response to the tide, we need at least 4 components of borehole strain to calculate the ground strain in each orientation and get the tidal response factor. Then, we can plot the orientation rose maps of the tidal response on the location where the instruments installed.

This paper is mainly based on the average response of earth tide. We applied NAKAI earth tide simulation formula [2] to calculate the average amplitude response factor and analyze the response ability of the earth tide in different orientation. For the relation between the earth tide observation and the impact factor, we can simulate it from the NAKAI formula. Through this formula, we can get tidal, non-tidal, rate and some other changing information. For the data y(t) observed at time t, can be expressed as [2]

$$y(t) = \alpha R(t) - \beta \frac{dR(t)}{dt} + K_0 + K_1 t + K_2 t^2(t) \quad (1)$$

where $R(t)$ is theoretical earth tide; α is the tidal scaling factor; β is a unknown number related to phase lag of the tide observation; K_0, K_1, K_2 are respectively constants, velocity related to non-tidal, unknown number related to acceleration.

There are many methods to calculate the theoretical value of the earth tide and its differential coefficient. In this paper, we applied a more accurate and fast method to calculate them about the tidal strain [3] (Figs. 1 and 2).

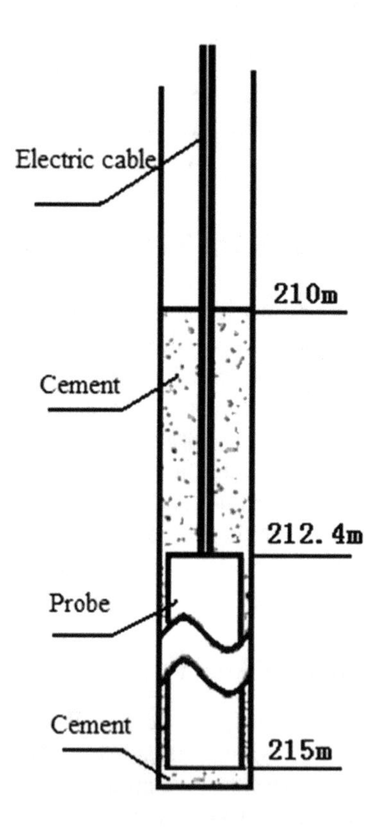

Fig. 1. Sketch map for the location of installed probe of the RZB-3 type deep-well strain meter and the borehole in BaiShan station of ChangPing district.

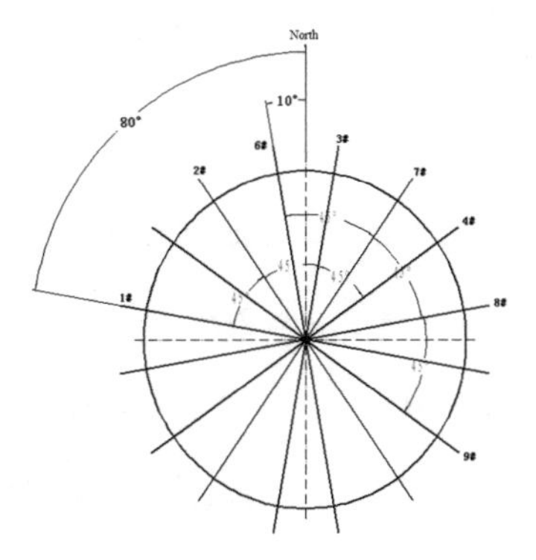

Fig. 2. Sketch map for the installing angles of the measuring components in RZB-3 type deep-well strain meter in BaiShan station of ChangPing district.

3.2 Strain Calculating in Any Direction

Assume μ_A is the displacement of the borehole surface in A direction of the ground surface to the horizontal direction, σ_1, σ_2 are main ground stress.

According to elastic analysis [1], the hole in the vertical direction, in plane stress state, μ_A, σ_1 and σ_2 have a relationship as:

$$\mu_A = \frac{a}{E}((\sigma_1 + \sigma_2) + 2(\sigma_1 - \sigma_2)\cos 2(A - \psi)) \tag{2}$$

where E is young modulus; ψ is the angle of the main stress σ_1.

In case of $\sigma_{zz} = 0$, the main stress and main strain of the ground have a relationship as:

$$\sigma_1 = E(e_1 + ve_2)/(1 - v^2) \tag{3}$$

$$\sigma_2 = E(e_2 + ve_1)/(1 - v^2) \tag{4}$$

where e_1, e_2 are main strains on ground; v is poison's ration and took 0.25 as the v value in this calculation.

Put the Eqs. (3) and (4) to (2), we can get

$$\varepsilon_A = \mu_A/a = (e_1 + e_2)/(1 - v) + 2(e_1 - e_2)\cos 2(A - \psi)/(1 + v) \tag{5}$$

If ε_A is the data ε_1 observed in direction #1, then the value in #1, #2, #3, #4 direction of YRY-4 type multi-component borehole strain meter have a relationship to e_1 and e_2 as:

$$\varepsilon_1 = (e_1 + e_2)/(1 - v) + 2(e_1 - e_2)\cos 2(A - \psi)/[1 + v] \tag{6}$$

$$\varepsilon_2 = (e_1 + e_2)/(1 - v) - 2(e_1 - e_2)\sin 2(A - \psi)/[1 + v] \tag{7}$$

$$\varepsilon_3 = (e_1 + e_2)/(1 - v) - 2(e_1 - e_2)\cos 2(A - \psi)/[1 + v] \quad (8)$$

$$\varepsilon_4 = (e_1 + e_2)/(1 - v) + 2(e_1 - e_2)\sin 2(A - \psi)/[1 + v] \quad (9)$$

From Eq. (6–9), we can calculate the main strain e_1, e_2 and main angle ψ.

$$\varepsilon_1 + \varepsilon_3 = \varepsilon_2 + \varepsilon_4 = 2(e_1 + e_2)/(1 - v) \quad (10)$$

$$\varepsilon_1 - \varepsilon_3 = 4(e_1 - e_2)\cos 2(A - \psi)/(1 + v) \quad (11)$$

$$\varepsilon_2 - \varepsilon_4 = -4(e_1 - e_2)\sin 2(A - \psi)/(1 + v) \quad (12)$$

$$\psi = A - \left[tg^{-1}(\varepsilon_4 - \varepsilon_2)/(\varepsilon_1 - \varepsilon_3)\right]/2 \quad (13)$$

The strain e_Φ of the ground in direction Φ can be determined by e_1, e_2 and ψ

$$e_\Phi = (1/2)(e_1 + e_2) + (1/2)(e_1 - e_2)\cos 2(\Phi - \psi) \quad (14)$$

3.3 Earth Tide Factor and Response Ability

Based on the formula (14), with a sampling interval of 3° in whole 360, we carried out the comparison between the simulation of theoretical value and actual value. From the Nakai formula (1) [1] and mirror fitting method, for the data observed in BaiShan station in Changping, we calculated the factor of the tidal response in different directions (Figs. 3, 4, 5 and 6).

Referring to the faults distribution map, we calculated the inspecting observe data using the horizontal observations from 1#(280°), 2#(325°), 3#(10°), 4#(55°). At the same time, we get the theoretical earth tide value and its differential coefficient value. Then we fitted out the earth tide from Nakai formula (1) to get out the amplitude factor (Fig. 3). For the comparison, we also calculated out all the theoretical earth tide value and its differential coefficient value in four directions.

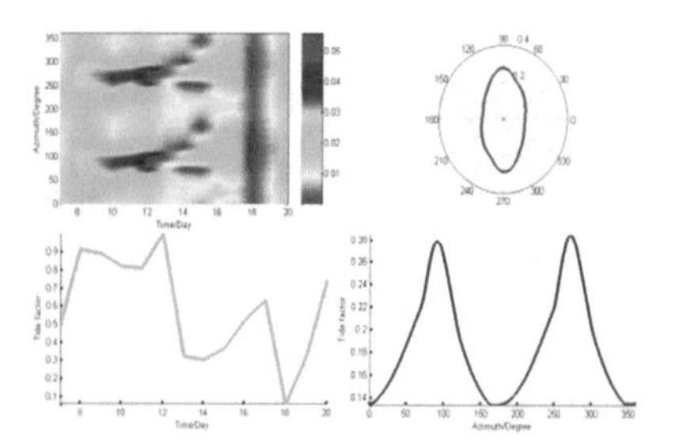

Fig. 3. Anisotropy response of the solid tidal scaling factor from observations for line #1, #2, #3 and #4.

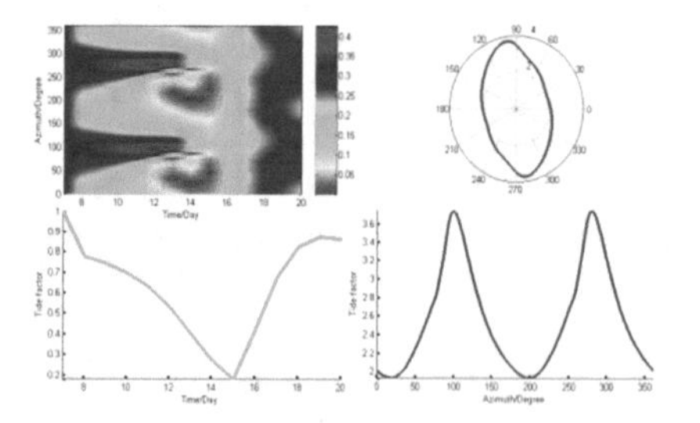

Fig. 4. Anisotropy response of the solid tidal scaling factor from theoretical value for line #1, #2, #3 and #4.

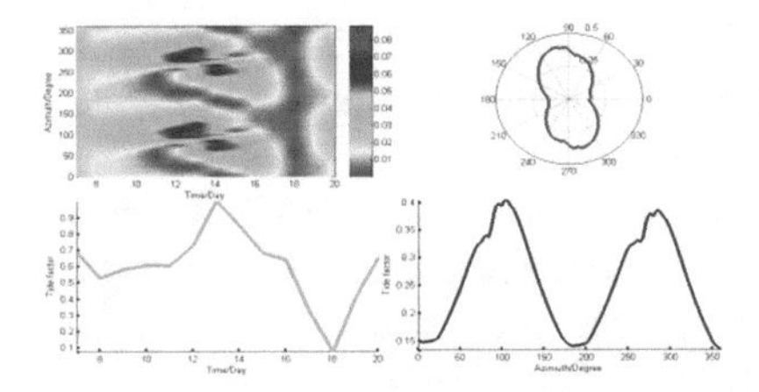

Fig. 5. Anisotropy response of the solid tidal scaling factor from observations for line #6, #7, #8 and #9.

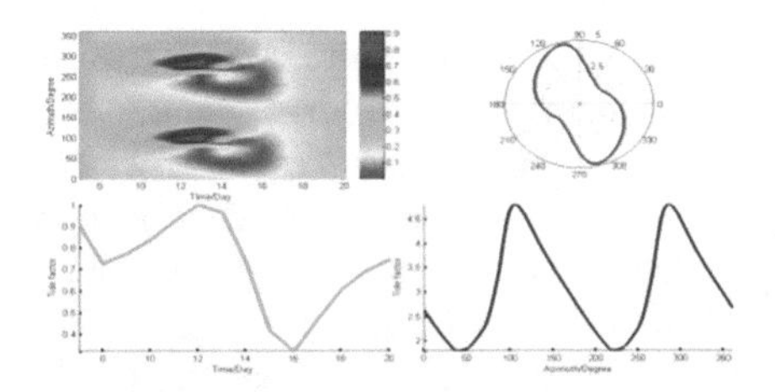

Fig. 6. Anisotropy response of the solid tidal scaling factor from theoretical value for #6, #7, #8 and #9.

Take the value calculated from Nakai formula as observation, we fitted out the related amplitude factors (Fig. 4). For comparison, take the same method as above, we also processed the other four strain records form direction 6#(350°), 7#(35°), 8#(80°), 9#(125°) to get the related scaling factor of earth tide amplitude.

Figures 3 and 4 show that, the solid tidal response reached the maximum value at the angle of 90°, which stands for east (its 0° when the direction is north). Considering the distribution of the faults, in the location of this station, the earth tide response was not blocked-up by the faults and could also get 'outside information' easily.

Then we can say that the direction of 90° is the best direction of earth tide response. This direction is basically along the main fault striking. The result of the theoretical value will not be impacted by the striking of the fault or the blocking-up of the faults.

Figures 5 and 6 also shows that the response of earth tide reaches the maximum value in the direction of 100°, which is the direction of a little deviate to the south from east. It is more tend to be parallel to the fault striking. It is obviously shown from Fig. 1 that, this direction is basically the same as the direction where were not blocked up by the faults.

However, the theoretical value is not also impacted by the fault striking or its activity, which shows that, the simulation of the earth tide theoretical value is based on the uniform model of the earth. Even though with the different observation group, the result is the same.

4 Analysis and Conclusion

4.1 Result Analysis

Comparing the Fig. 3 with Fig. 6, from the results calculated from different observation groups, we can find that: The earth tide simulated from the theoretical value is evidently not impacted by the distribution of the faults; Although the shapes of the observation group vary, the best direction is the same. However, these two results obtained from the calculation of actual observations shows that the best directions are not the same, but they are all mainly tend to the direction parallel to the faults.

This result shows that: The solid tidal response is impacted by the faults striking, the activity of the faults and the blocking-up of faults. In ChangPing district, the impacts of these are more complex; In addition, the nonuniform of the medium also may be an influence factor. Then, the results obtained from different observation group will vary naturally. But the difference is not distinct. This can be proved by the Fig.; Furthermore, the difference between each orientation seems not obvious between the deep well observation and shallow well observation. Perhaps, for the RZB-3 Type deep well observation system, the impaction from the ground decreased and then the anisotropy features resulted from the ground false information decreased. But it is not sufficient to prove this conclusion.

4.2 Preliminary Understanding

The tidal strain obtained from the borehole strain instruments that installed on each side of the faults or faults blocking-up area come from two sources. One source is the tidal strain from the bottom of the earth and the other is the decreased tidal strain from the adjacent plates of the faults. If the fault, where the instrument be installed, has been active recently and the fault has not been filled and not be 'weld', then the medium is not 'continuous'. So the tidal strain that vertical to the fault striking may not be received and the tidal scaling factor will be very small.

Based on the results above, we can say that the anisotropy of this instrument in BaiShan station is not decreased in the deep well and have an obviously direction characteristic. The result may be impacted by the striking and blocking-up of the faults. Therefore, it is should be noted in the study of geodynamics and seismic prediction. The fact revealed in this paper shows a guideline for improving the methods to correct the displacement in the space geodetic data.

Acknowledgment. This work is supported by projects of Spark Program of Earthquake Science and Technology (XH16003), Three Combination of China Earthquake Administration(CEA-JC/3JH-170203) and National Natural Science Foundation (NNSF) of China under Grant 41474087.

References

1. Chi, S.L., Wu, H.L., Luo, M.J.: Discussion on strain tidal factor separation and anisotropy—analysis of first data of borehole component strain-meter of China's digital seismological observational networks. Prog. Geophys. **22**(6), 1746–1753 (2007). (in Chinese)
2. Nakai, S.: Pre-processing of tidal data. BIM **76**, 4334–4340 (1977)
3. Jiang, Z., Zhang, Y.B.: A differential analytical-representation of the theoretical value of earth tide and the fit-testing of earth tide data. Acta Geophysica Siniva **37**(6), 776–786 (1994). (in Chinese)
4. Zhang, Y.B.: Applicatlon of numerical differential in Nakai process of tidal data. Crustal Deform. Earthq. **9**(4), 79–83 (1989). (in Chinese)
5. Tang, J.A.: The zenith distance formulas of the first order differential quotient for the theoretic value of the earth tides applied to the fit-testing of the gravity, tilter and line strain earth tides. Crustal Deform. Earthq. **10**(2), 1–8 (1990). (in Chinese)
6. Feng, K., Zhang, J.Z., Zhang, Q.X.: Numerical Computing Methods. Defense Industry Press, Beijing (1978). (in Chinese)

Azimuth Determination of Borehole Strain Meter

Anxu Wu(✉)

Beijing Earthquake Agency, Beijing 100080, China
wu-anxu@163.com

Abstract. Determination the azimuth of borehole strain is very significant, and it is hard or impossible to determine the azimuth of borehole strain by azimuth measuring apparatus under the condition that the borehole strain meter is firmly fixed in the deep well. However, based on the characteristic that solid tide's theoretic value could be calculated, generally considered observation data and its major influence factors, and used extended Nakai solid tide fitting model, adopted solid tide strain model's theoretic value, we proceeded effective simulation, with the Borehole strain observation data of Gubeikou station and the nonlinear particle swarm optimization method, we obtained the apparatus's azimuth angle of the station, then, the inversion model real test with the other time observation data and the obtained optimizing solution was done, the result shows well and some problems in it were discussed.

Keywords: Borehole strain · Azimuth · Solid tide · Nakai · Inversion

1 Introduction

Beijing Earthquake Agency has installed five borehole strain meters in Beijing. All the five instruments have recorded clear solid tidal information and the observations are all of great quality (Fig. 1). They are playing an important and active role in the prediction of earthquake. But restrained by techniques at that time, the azimuth angle was not measured. As a result the observations can't be widely applied. In fact, the observation obtained from high-precision borehole strain meters contains huge earth's interior information, such as the earth deformation and the structure of the earth. The information can be used to confirm the change of the earth tidal parameters in the strain changing function. But for the five borehole strain meters in Beijing, we can't get the direction angle below the well for various reasons so far. Therefore, the solid tidal data can't be well used. This brought trouble to analyze the solid tidal observations. Then, it is regrettable to failure finding out more earth tidal information.

For this reason, we tried to carry out the mixture inversion of the azimuth angle. The inversion procedure follows some certain forward inversion model. During the processing, we tried to use the feature that the solid tidal theoretical value can be calculated in advance and considered all the possible impact factors. Using the possibility and rationality of the researching model of the solid tidal theoretical value and taking the inversion result from other time interval, we confirmed the responsibility of the result and the rationality of the inversion model. For the stations whose azimuth

N. Xiong et al. (Eds.): ISCSC 2018, AISC 877, pp. 109–114, 2019.
https://doi.org/10.1007/978-3-030-02116-0_14

angles have not been fixed, it is an experience accumulation to carry out a further detailed inversion of the borehole strain meter angle. It also founded a good theoretical foundation to inverse all angles in each station reasonably. From this research, we can provide a real pregnant analyzing material with the earth tide for the study of earthquake prediction and geodynamics.

2 Method and Model

2.1 Inversion Model

There are many factors that can impact the borehole strain observation, such as theoretical strain value, its differential coefficient and linear and non-linear quadratic drift. It also could be affected by the temperature, pressure, water level, rainfall and other factors. Meanwhile, there may be some coupling between these impact factors. Some factors may have some time lag. Therefore, the impact factors in the strain observation are of great complex.

To simplify the problem, we can take no account of rainfall and temperature. To avoid the affect of the rainfall, we can use the data in no rain days. And to avoid the affect of the temperature, considering the temperature in certain depth of the well are almost constant, we can assume the impact from the temperature as a constant. Therefore, for the strain observation, we can just consider the theoretical strain value and its differential coefficient, three drift factor, assistant water level and pressure.

For the relationship between the observation and the impact factors, it can be simulated by the Nakai formula [1]. But this formula does not contain the water level and pressure in traditional processing. If assuming the observation has relationship with these two factors, then Nakai formula can be extended by adding the factor of water level and pressure. The advantage of choosing this formula is that the information obtained from this formula can be directly used for harmonic analyzing.

Nakai formula is a more classic tidal data processing method, which is based on the tidal theory. It is founded on the basis of tidal theory to simulate and test abnormal data from the observations. Through this formula, we can get tidal, non-tidal, rate and much other changing information (Jiangjun et al. 1994). Adding the water level and air pressure, the data $y(t)$ observed at time t can be expressed as:

$$y(t) = \alpha R(t) - \beta \frac{dR(t)}{dt} + K_0 + K_1 t + K_2 t^2 + wW(t) + pP(t) \tag{1}$$

where, $R(t)$ is theoretical value of the solid tide; α is the tidal scale factor, which is the focus of this study. It is used to analyze the response ability of the solid tide in different orientation; β is an unknown number related to phase lag of the tide observation; K_0, K_1, K_2 are respectively constants, velocity related to non-tidal, and unknown number related to acceleration. W is the impact factor of water lever, p is the impact of pressure. There are many methods to calculate the theoretical value of the solid tide and its differential coefficient.

In this paper, we applied a more accurate and fast method to calculate the theoretical value of drilling tidal strain and its derivative [2].

2.2 Observation Records

The inversion strategy of the azimuth angle needs no linear process for the non-linear method. The theoretical value of earth tide and its derivative can be calculated in real-time. And then, the equation can be solved based on the extended Nakai formula and the mirror method [5]. Under the principle of minimum residual and maximum correlation coefficient, we can obtain the azimuth angle of the borehole strain meter. There are a lot of non-linear inversion methods.

In this paper, we applied the particle swarm optimization (PSO) [6, 7] to invert the azimuth angle. PSO is a new inversion method, similar to the genetic algorithm. It does not need to calculate the differential coefficient of the inversion parameter, but with more advantages compared to the genetic algorithm. The PSO algorithm has advantages of less parameter, no need of differential coefficient, high efficiency and can be applied convenience and quickly. The principle of PSO algorithm [6, 7] is as follows:

The particle swarm optimization (PSO) was put forward by Kennedy and Eberhart [6] inspired by the foraging behavior of birds in 1995. In the particle swarm optimization (PSO), all particles have a fitness that decided by the fitting function and a displacement whose times of iterations determined by the speed $V_i = (v_{i1}, v_{i2}, \ldots, v_{id})$. The PSO is initialized to a group of random particles (random solutions), in which the position of the i particle in d dimensions of solution space is expressed as: $X_i = (x_{i1}, x_{i2}, \ldots, x_{id})$.

According to the following formula, the particle updates the speed and the position.

$$v_{id} = w \times v_{id} + c_1 \times rand() \times (p_{id} - x_{id}) + c_2 \times rand() \times (g - x_{id}) \quad (2)$$

$$x_{id} = x_{id} + v_{id} \quad (3)$$

in the above formula (2) and (3), v_{id} is the speed of the particle i in d dimension, $v_d \in [-v_{d\,\max}, +v_{d\,\max}]$. x_{id} is the current position of the particle i in d dimension. w is the inertia weight and initialized with 0.92 and the times of iterations is decreased to 0.41.

This way, the algorithm can initially focus on the global search to convergence the search space in to a certain region and then obtain the high-precision solution by local refined search. Rand() are random numbers uniformly distributed in (−1, 1); c_1 and c_2 are learning factors. For the learning factors, if it is too small, the particle may be far away from the target region and if it is too large the particle may fly to the target region in a sudden. Usually, we make $c_1 = c_2 = 2$. The flying speed of the particle has a maximum limitation $v_{d\,\max}$. Assuming that the d dimension of the search space is defined in intervals $[-x_{d\,\max}, +x_{d\,\max}]$, then usually $v_{d\,\max} = kx_{d\,\max}$ and $0.1 < k < 1.0$.

3 Result Analysis

3.1 Azimuth Determination

In order to illustrate the feasibility of idea and model of the inversion and the reliability of the results, we only select the borehole strain observations from Gubeikou station ($N40.68°, E117.15°$), from which station the theoretical value of earth tide and observations are thought to be better to carry out the experiment. In the experiment, we carried out the inversion and analyzing for the fixed azimuth angle and studied the feasibility and reliability of the inversion method.

The inversion strategy to solve the azimuth are: setup the azimuth angle by using the PSO algorithm [6, 7], obtain the theoretical strain value and its differential coefficient [2], form equations according to formula (1), solve the over determined system of equations with the use of more effective mirror method [5], calculate the minimum variance of the theoretical value and observation, return to the PSO optimization system to decide whether to select. Through repeated optimization, the best azimuth angle can be derived. Since only the azimuth angle was took as the unknown inversion value and the parameters in Nakai were solved by using mirror method, then the constraints of the inversion will be stronger and the optimal solution is easier to be obtained. This solving process is called mixed inversion process.

According to this demand, we chose several days to carry out the inversion. All the inversion angles were analyzed to find out the most reasonable azimuth angle. At the end, we got the best orientation angle of 46.80449°, for which angle the minimum variance of the fitting is 4.60134×10^{-9}. The correlation coefficient between the observation and fitting data is 0.9985. The inversion results from the other days were similar. The maximum error reached ±0.87°. At that time the best parameters of PSO were set to $w = 0.5$, $c_1 = c_2 = 2$, $k = 0.7$. Figure 2 is the inversion result from the observation obtained on 4th jun., 2017.

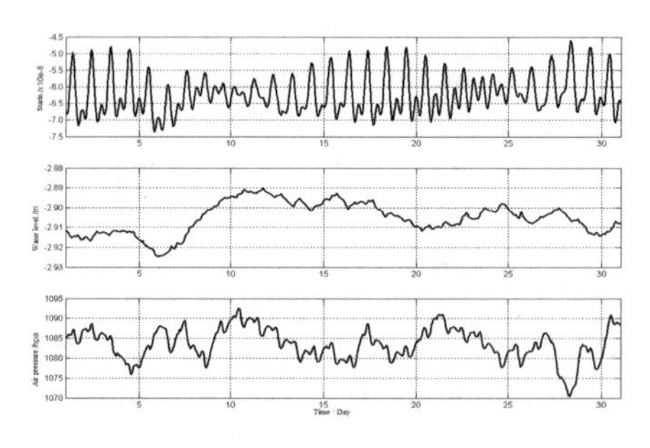

Fig. 1. The observation recording of the borehole strain meter, water and atmospheric pressure at Gubeikou station.

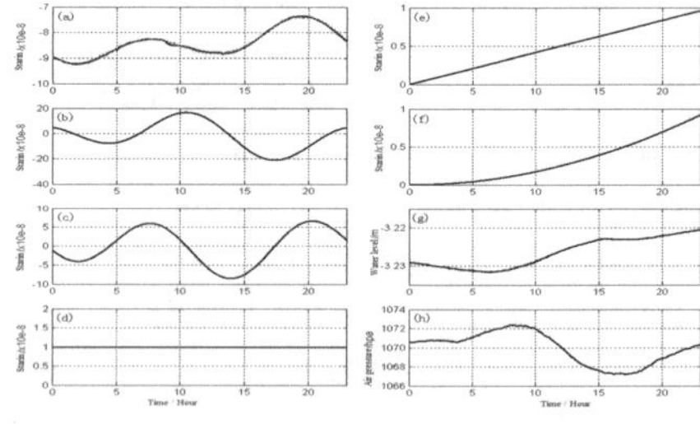

Fig. 2. Curves of observation, fitting, theory, derivative, excursion, water and atmospheric pressure at Gubeikou station on 4th Oct., 2017.

3.2 Data Testing in Normal Time

Among the more stable data of the borehole strain observation from Gubeikou station, we randomly selected the data on 3rd Oct., 2017 to test the inversion model (Fig. 3). Figure 3 shows that the accuracy of the experiment is high and the fitting variance reached 4.90×10^{-9}. The correlation coefficient between the observation and fitting data is 0.99. The fitting result has a good effect. For the deformities, however, it can't be completely simulated. It is a normal phenomenon, because this model was supposed to reflex the trend shape of the earth tide but not to get the detailed changing. Otherwise, though the fitting result has good effect, the result may be deviate from the actuality. It is reasonable to apply normal observations to test the inversion model and the inversion azimuth angles are acceptable.

3.3 Data Testing in the Coseismic Effect Period

The borehole strain meter located in Gubeikou station clearly recorded an occurrence of strong earthquake in Japan. The observations in this time can be chosen to the test (Fig. 4). By analyzing the results we can get the conclusions that, although there are great wave changes in Fig. 4, the accuracy of the test result is high. The fitting variance reached 5.66×10^{-9} and the correlation coefficient reached 0.98. In the period with no seismic, the prediction values are consistent with the observations. In the period with seismic, the trend is consistent. Applying this model, the affect on records from seismic wave shape can be eliminated easily. In return the seismic wave data can be derived to be used in other researches.

The testing results both from normal observing period and period containing coseismic effect show that, the fitting results from the testing samples are of high accuracy. The shape of the fitting curve reflected the trend of the observations. It indicates that the inversion azimuth angle should be reasonable or at least approach the true azimuth.

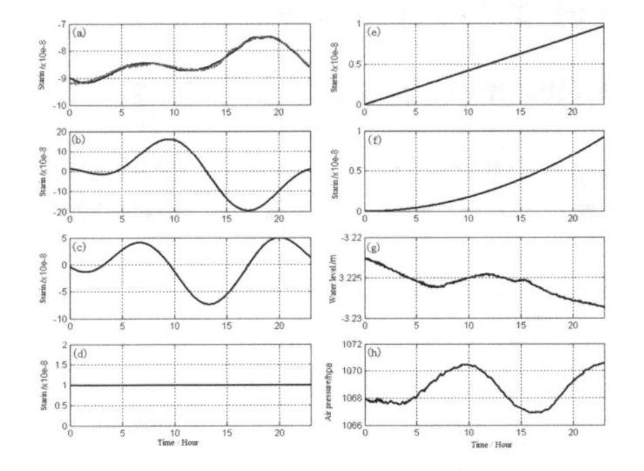

Fig. 3. Curves of observation, fitting, theory, derivative, excursion, water and atmospheric pressure at Gubeikou station on 3rd Oct., 2017.

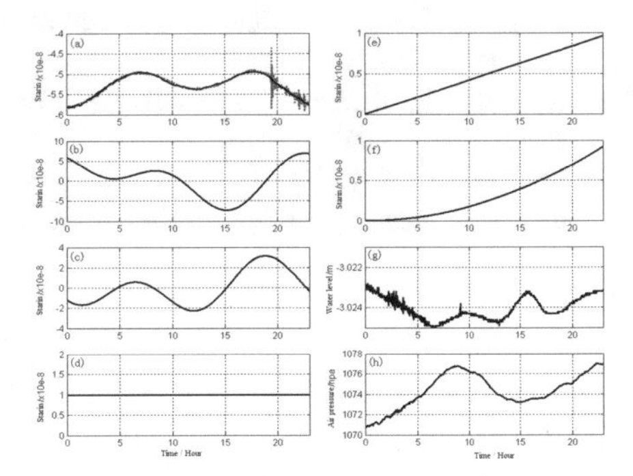

Fig. 4. Curves of observation, fitting, theory, derivative, excursion, water and atmospheric pressure at Gubeikou station.

4 Concluding Remarks

The inversion of the azimuth angle needs the attention on the materials selection. In order to simplify the inversion model and improve the inversion accuracy, it is better to select the observations with clear solid tidal information with non-interference, no rain and no seismic as far as possible before the inversion.

The inversion idea to integrate the PSO algorithm and extended Naikai model ought to be reasonable. It can derive the furthest optimum solutions that close to the true solution. The accuracy of the inversion results, proved by different type of information, is acceptable. The inversion result can reflect the true condition of the observations. It is of great practical significance for excavating the strain solid tidal information in this borehole. Of course, this method can provide the necessary services for the geodynamics and earthquake prediction.

Acknowledgment. This work is supported by projects of Spark Program of Earthquake Science and Technology (XH16003), Three Combination of China Earthquake Administration (CEA-JC/3JH-170203) and National Natural Science Foundation (NNSF) of China under Grant 41474087.

References

1. Nakai, S.: Pre-processing of tidal data. BIM **76**, 4334–4340 (1977)
2. Jiang, Z., Zhang, Y.B.: A differential analytical-representation of the theoretical value of earth tide and the fit-testing of earth tide data. Acta Geophys. Siniva **37**(6), 776–786 (1994). (in Chinese)
3. Zhang, Y.B.: Application of numerical differential in Nakai process of tidal data. Crustal Deform. Earthq. **9**(4), 79–83 (1989). (in Chinese)
4. Tang, J.A.: The zenith distance formulas of the first order differential quotient for the theoretic value of the earth tides applied to the fit-testing of the gravity, tilter and line strain earth tides. Crustal Deform. Earthq. **10**(2), 1–8 (1990). (in Chinese)
5. Feng, K., Zhang, J.Z., Zhang, Q.X.: Numerical Computing Methods. Defense Industry Press, Beijing (1978). (in Chinese)
6. Kennedy, J., Eberhart, R.: Particle swarm optimization. IEEE Int. Conf. Neural Netw. **4**(27), 1942–1948 (1995)
7. Eberhart, R., Kennedy, J.: A new optimizer using particle swarm theory. In: Proceedings of the Sixth International Symposium on Micro Machine and Human Science, pp. 39–43 (1995)

Analysis and Calculation of the Electronic Evidence Admissibility Model

Deqiang Chen(✉)

East China University of Political Science and Law, Shanghai 201620, China
dqchen@ecupl.edu.cn

Abstract. In this paper, the problem of how to measure and predict the credibility of collected electronic evidence is discussed. By introducing Markov chain as a measure of admissibility model, this paper fully considers the dependence in the electronic evidence system. Empirical result shows that the transfer matrix can well evaluate and predict the admissibility of the electronic evidence system.

Keywords: Markov chain model · Electronic evidence · Admissibility model

1 Introduction

Compared with traditional evidence [1], electronic evidence has a distinct system, and systematization is a major feature of electronic evidence. Any electronic evidence is not isolated, in particular, some electronic evidence is based on information network, then the same behavior will leave the relevant electronic evidence in different network nodes. Some of the electronic evidence is based on stand-alone space, then the same behavior can produce the content evidence (such as WORD document, EXCEL document, PPT document, PDF document, picture file, e-mail, etc.), those evidences will form some subsidiary information synchronously (such as document creation time, modification time, access time, save person, type, Format, and so on), and leave a series of associated traces (such as a WINDOWS system log file, hibernation file, paging file, delete pointer, or data store rule). In any case, it is theoretically possible to find mutual proof of electronic evidence and form a system.

Credibility is a quantitative indicator of the admissibility model. The true and false of evidence always exist in three situations: evidence is true, false, and uncertain. If there is an evidence that the electronic evidence is false, it cannot be used to build a crediting system. If it proves that there is a certain degree of false possibility of electronic evidence, the credit system can be constructed by reducing its credibility. Therefore, the credibility of electronic evidence is a very important measure [2].

The rest of the paper is organized as follows: Sect. 2 presents the Markov chain model and how to use the Markov chain model in our experiments. Section 3 uses the transfer matrix to describe the dependence in electronic evidence system. Section 4 presents the results of our experiments and analyses them in details. Section 5 makes our conclusions.

N. Xiong et al. (Eds.): ISCSC 2018, AISC 877, pp. 115–122, 2019.
https://doi.org/10.1007/978-3-030-02116-0_15

2 Materials and Methodology

In the traditional evidence appraisal, some scholars have also proposed the "weighted" quantitative feature scoring standard [3], the addition of scientific criteria such as the minimum number of conformity feature standards [4], area and feature quality standards [5]. However, the dependence in the electronic evidence cannot be objectively reflected by the above methods.

When judging the verification system of electronic evidence, three basic steps should be followed: the first step, Evaluate the credibility of each electronic evidence (including a small amount of conventional evidence), and simply translate it into initial probability values; the second step is to calculate the transfer matrix; the third step, according to the Markov chain rule, calculate the credibility of all electronic evidence (including a small amount of traditional evidence). The evidence model is analyzed and calculated.

As the amount of evidence and the credibility of a single piece of evidence increases, the probative power of the relevant evidence increases correspondingly, and the probability that the evidence is adopted is also increased. These are obvious.

We know that to describe the random phenomenon of a certain period of time, such as whether certain electronic evidence is admissibility, we can use a random variable X_n, but to describe the overall prediction of the entire electronic evidence system, we need a series of random variables $X_1, X_2, \cdots, X_n, \cdots$, We call $\{X_t,\ t \in T,\ T$ is a parameter set$\}$ a random process, and the set of $\{X_t\}$ values is called the state space. If the parameters of the random process $\{X_n\}$ are non-negative integers, X_n is a discrete random variable, and $\{X_n\}$ has no post-validity (or Markov character), this random process is called a Markov chain (abbreviated as Markov chain). The so-called no-response, intuitively speaking, is that if you consider the parameter n of $\{X_n\}$ as time, what value it takes in the future depends only on its current value, regardless of what value was taken in the past.

For a Markov chain with N states, describe its probabilistic nature, and the most important is the one-step transition probability that it transitions to state j at the next moment in state i at time n:

$$P(X_{n+1} = j | X_n = i) = p_{ij}(n) \qquad i,j = 1,\ 2,\ \cdots,\ N \tag{1}$$

If it is assumed that the above formula is not related to n, that is, it can be recorded as (in this case, the process is called smooth), and

$$P = \begin{pmatrix} p_{11} & p_{12} & \cdots & p_{1N} \\ p_{21} & p_{22} & \cdots & p_{2N} \\ \cdots & \cdots & \cdots & \cdots \\ p_{N1} & p_{N2} & \cdots & p_{NN} \end{pmatrix} \tag{2}$$

This is called the transition probability matrix. The transition probability matrix has the following properties:

$$(1)\, p_{ij} \geq 0, \quad i,j = 1,\ 2,\ \cdots,\ N. \quad (2) \sum_{j=1}^{N} p_{ij} = 1, \quad i = 1,\ 2,\ \cdots,\ N.$$

If we consider the case of multiple transitions of the state, then there is a k-step transition probability that the process transitions to state j at time $n + k$, but at time n in state i:

$$P(X_{n+k} = j | X_n = i) = p_{ij}^{(k)}(n) \qquad i,j = 1,2,\cdots,N. \tag{3}$$

Also by the stability, the probability of the above formula is independent of n, can be written as,

$$P^{(k)} = \begin{pmatrix} p_{11}^{(k)} & p_{12}^{(k)} & \cdots & p_{1N}^{(k)} \\ p_{21}^{(k)} & p_{22}^{(k)} & \cdots & p_{2N}^{(k)} \\ \cdots & \cdots & \cdots & \cdots \\ p_{N1}^{(k)} & p_{N2}^{(k)} & \cdots & p_{NN}^{(k)} \end{pmatrix}. \tag{4}$$

This is called the k-step transition probability matrix.

$$p_{ij}^{(k)} \geq 0, \quad i,j = 1,\ 2,\ \cdots,\ N; \sum_{j=1}^{N} p_{ij}^{(k)} = 1, \quad i = 1,\ 2,\ \cdots,\ N. \tag{5}$$

In general, if there is a one-step transfer matrix, the k-step transfer matrix will be written as,

$$P^{(k)} = \begin{pmatrix} p_{11}^{(k)} & p_{12}^{(k)} & \cdots & p_{1N}^{(k)} \\ p_{21}^{(k)} & p_{22}^{(k)} & \cdots & p_{2N}^{(k)} \\ \cdots & \cdots & \cdots & \cdots \\ p_{N1}^{(k)} & p_{N2}^{(k)} & \cdots & p_{NN}^{(k)} \end{pmatrix}. \tag{6}$$

In the Markov prediction method, the estimation of the transition probability of the system state is very important. There are usually two methods for estimating: First, the subjective probability method, which is based on people's long-term accumulated experience and knowledge of predicted events, and a subjective estimate of the likelihood of event occurrence. This method is generally in the absence of the data's history. The second is the statistical estimation method, which is introduced through examples.

This example records the identification of 24 electronic evidences in the same system and is given in Table 1. We want to find the transition probability matrix of its trusted state. Among them, state 1 indicates that the electronic evidence is correctly identified, and state 2 indicates that the electronic evidence is erroneous and is not accepted.

Table 1. 2-state Markov chain model

No.	Status	No.	Status	No.	Status	No.	Status
1	1	7	1	13	1	19	2
2	1	8	1	14	1	20	1
3	2	9	1	15	2	21	2
4	1	10	2	16	2	22	1
5	2	11	1	17	1	23	1
6	2	12	2	18	1	24	1

The data in the Table 1 included 15 true, 9 false, 5 continuous true, and the number of from true to false and from false to true is 7. The number of consecutive false was 2. As a result, the following transition matrix can be obtained. Now calculate the transition probability. To replace the probability with frequency, you can get the probability of continuous true:

$$p_{11} = \frac{\text{The number of consecutive "true" occurrences}}{\text{The number of "true" occurrences}} = 0.5$$

The number in the denominator is 15 minus 1 because the 24th evidence is identified as “true” and there is no follow-up record, so it needs to be reduced by 1.

Similarly, the probability of changing from “true” to “false”

$$p_{12} = \frac{\text{The number of occurrences from "true" to "false"}}{\text{The number of "true"}} = 0.5$$

$$p_{21} = \frac{\text{The number of occurrences from "false" to "true"}}{\text{The number of "false"}} = 0.78$$

$$p_{22} = \frac{\text{The number of consecutive "false" occurrences}}{\text{The number of "false" occurrences}} = 0.22$$

In summary, the state transition probability matrix for the identification result is

$$P = \begin{pmatrix} p_{11} & p_{12} \\ p_{21} & p_{22} \end{pmatrix} = \begin{pmatrix} 0.5 & 0.5 \\ 0.78 & 0.22 \end{pmatrix}$$

3 Analysis and Calculation

3.1 Calculate the State Transition Probability Matrix

Assume that there are n possible states for the development of an event, namely $S_1, S_2, \cdots S_n$. Denoted P_{ij} as the state transition probability from state S_i to state S_j, then the transfer matrix is as follow,

$$P = \begin{pmatrix} P_{11} & P_{12} & \cdots & P_{1n} \\ P_{21} & P_{22} & \cdots & P_{2n} \\ \vdots & \vdots & \vdots & \vdots \\ P_{n1} & P_{n2} & \cdots & P_{nn} \end{pmatrix}.$$

In the actual situation, we do not need to consider so many states, generally take three kinds of states. Now we consider the three states of the identification result of electronic evidence, namely "true", "uncertain" and "false." Remember that S_1 is in the "true" state, S_2 is the "uncertain" state, and S_3 is the "false" state. Table 2 gives the status of the results of the identification of 40 relevant electronic evidences for this system. We need to calculate the state transition probability matrix that identifies the overall credibility.

Table 2. 3-state Markov chain model

No.	1	2	3	4	5	6	7	8	9	10
Status	S_1	S_1	S_2	S_3	S_2	S_1	S_3	S_2	S_1	S_2
No.	11	12	13	14	15	16	17	18	19	20
Status	S_3	S_1	S_2	S_3	S_1	S_2	S_1	S_3	S_3	S_1
No.	21	22	23	24	25	26	27	28	29	30
Status	S_3	S_3	S_2	S_1	S_1	S_3	S_2	S_2	S_1	S_2
No.	31	32	33	34	35	36	37	38	39	40
Status	S_1	S_3	S_2	S_1	S_1	S_2	S_2	S_3	S_1	S_2

As can be seen from Table 2, there are 15 states that proceeded from S_1 (transferred out), 3 states are transferred from S_1 to S_1 ($1 \rightarrow 2$, $24 \rightarrow 25$, $34 \rightarrow 35$), 7 states are transferred from S_1 to S_2 ($2 \rightarrow 3$, $9 \rightarrow 10$, $12 \rightarrow 13$, $15 \rightarrow 16$, $29 \rightarrow 30$, $35 \rightarrow 36$, $39 \rightarrow 40$), and 5 states are transferred from S_1 to S_3 ($6 \rightarrow 7$, $17 \rightarrow 18$, $20 \rightarrow 21$, $25 \rightarrow 26$, $31 \rightarrow 32$).

$$P_{11} = P(S_1 \rightarrow S_1) = P(S_1|S_1) = 0.200, P_{12} = P(S_1 \rightarrow S_2) = P(S_2|S_1) = 0.467$$

$$P_{13} = P(S_1 \rightarrow S_3) = P(S_3|S_1) = 0.333, P_{21} = P(S_2 \rightarrow S_1) = P(S_1|S_2) = 0.539$$

$$P_{22} = P(S_2 \rightarrow S_2) = P(S_2|S_2) = 0.154, P_{23} = P(S_2 \rightarrow S_3) = P(S_3|S_2) = 0.308$$

$$P_{31} = P(S_3 \rightarrow S_1) = P(S_1|S_3) = 0.364, P_{32} = P(S_3 \rightarrow S_2) = P(S_2|S_3) = 0.455$$

Calculate according to the same method as above,

$$P_{33} = P(S_3 \rightarrow S_3) = P(S_3|S_3) = 0.182.$$

Therefore, the state transition probability matrix is

$$P = \begin{pmatrix} 0.200 & 0.467 & 0.333 \\ 0.539 & 0.154 & 0.308 \\ 0.364 & 0.455 & 0.182 \end{pmatrix}.$$

3.2 Predictive Calculation

The state probability $\pi_j(k)$ represents the probability that the event is in the state S_j at the k-th time (period) after k times of state transitions under the condition that the initial ($k = 0$) state is known. According to the nature of the probabilities, there are obviously:

$$\sum_{j=1}^{n} \pi_j(k) = 1 \tag{7}$$

From the initial state, the state reaches S_j after k times of state transition. This state transition process can be seen as the first state after $k - 1$ state transitions, and then from the first state S_i transition to the state S_j. According to Markov process's non-post-effect and Bayes' conditional probability formula, there are

$$\pi_j(k) = \sum_{i=1}^{n} \pi_j(k-1)P_{ij} \quad (j = 1, 2, \cdots, n) \tag{8}$$

If a vector is $\pi(k) = [\pi_1(k), \pi_2(k), \cdots, \pi_n(k)]$, the recursive formula for the successive calculation of the state probability can be obtained.

$$\begin{cases} \pi(1) = \pi(0)P \\ \pi(2) = \pi(1)P = \pi(0)P^1 \\ \quad \vdots \\ \pi(k) = \pi(k-1)P = \cdots = \pi(0)P^k \end{cases} \tag{9}$$

$\pi(0) = [\pi_1(0), \pi_2(0), \cdots, \pi_n(0)]$ is the initial state probability vector.

4 Experiments and Results

The experimental data comes from Table 2. Through the calculation and analysis, we use the 15 and 40 states as the initial vector and predict backwards 10 steps to obtain the following forecasting map of the identification system.

As can be seen from the two figures, the state fluctuates at the beginning, but gradually stabilizes with time. The "true" state is higher than the "uncertain" state, and the "false" state (Figs. 1 and 2).

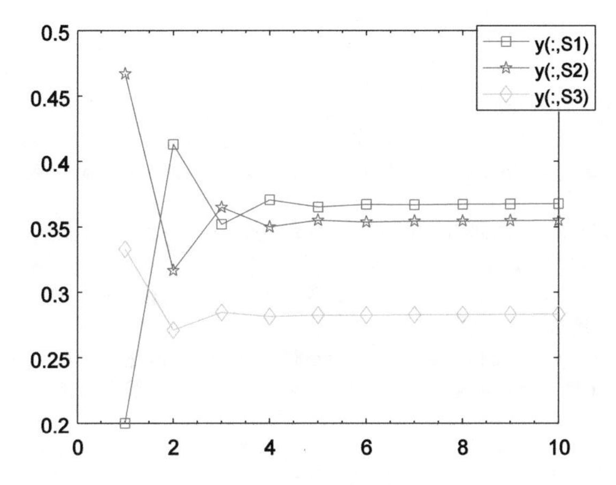

Fig. 1. 15 states as the initial vector and predict backwards 10 steps.

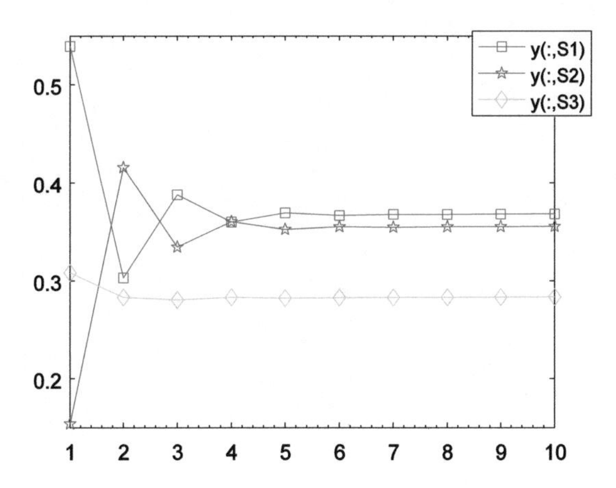

Fig. 2. 40 states as the initial vector and predict backwards 10 steps.

5 Conclusion

For the adoption of multiple electronic evidences of the same system, if a homogeneous and stable transfer matrix can be obtained by using the Markov chain model, it can be shown that the evaluation of the electronic evidence in this system is more reliable, it can more accurately reveal the correlation among the various electronic evidences. This provides a new idea and new method for quantifying the evaluation of electronic evidence in judicial practice and adopting it reasonably.

Acknowledgements. This research is supported by the East China University of Political Science and Law's research projects (A-0333-18-139015) and Beijing Qihoo science and Technology Co., Ltd. cooperation project (C-6901-18-023).

References

1. Kwan, M., et al.: Reasoning about evidence using Bayesian networks. In: IFIP International Conference on Digital Forensics, pp. 275–289. Springer, Boston (2008)
2. Tribe, L.H.: Trial by mathematics: precision and ritual in the legal process. Harvard Law Rev. **84**(6), 1329 (1971)
3. Bonnici, T., et al.: Evaluation of the effects of implementing an electronic early warning score system: protocol for a stepped wedge study. BMC Med. Inform. Decis. Mak. **16**(1), 19 (2015)
4. Mahajan, R.: The naked truth: appearance discrimination, employment, and the law. J. Am. Acad. Audiol. **14**(1), 56 (2007)
5. Garrie, D.B., Morrissy, J.D.: Digital forensic evidence in the courtroom: understanding content and quality. Northwest. J. Technol. Intellect. Prop. **12**(2) (2014)

The Research of Multi-objective Optimization Based on ACO

Limei Jia[1(✉)] and Mengtian Cui[2]

[1] Normal College, Chifeng University, Chifeng 024000, Inner Mongolia, China
happyzg2@163.com

[2] Computer Science and Technology College, Southwest Minzu University, Chengdu 610041, China
hangkongzy@163.com

Abstract. To solve the problem of multi-objective optimization, the novel methods based on ant colony optimization algorithm (ACO) is proposed in the paper. The essential advantages of ACO were discussed. Considering the shortcomings of the traditional ACO, colony fitness is improved. Cross operation and mutation of genetic algorithms were introduced into the ACO in order to improve its searching ability and to realize the dynamically adjusting the influence of each ant to the trail information updating and the selected probabilities of the paths. In the end, the experimental results show that the methods proposed compared with traditional methods can improve the speed of converge and can get the optimal solution quickly.

Keywords: Multi-objective optimization · Ant Colony Optimization (ACO) Optimal solution

1 Introduction

The shortcoming Multi-objective optimization is its difficulty in solving NP problem [1]. The optimal solution is usually not a single solution, but multiple solutions, and it is very hard to compare with the various solutions. In recent years, there have been a lot of good multi-objective optimization algorithms, such as genetic algorithms, fish algorithm, PSO and its improved algorithms [2]. But there is still slow convergence of these algorithms, and it is easy to fall into local optimal solution and other issues. So it is need be further improved. In order to optimize data functions with multi-variable and multi-modal data functions, Maniezzo et al. [3] proposed an ant heuristic for the frequency assignment problem in 2000. Wang et al. [4] proposed the algorithm for multimedia multicast routing based on ant colony optimization in 2002. Liu et al. [5] proposed the QoS routing algorithm based on the combination of genetic algorithm and ant colony algorithm in 2007. In the same year, Cui et al. [6] proposed the optimization algorithm of multi-level discrete fuzzy neural networks for solving global stability. The algorithm is effective to solve global stability via applying the *if-then* rule of fuzzy inference system and hard *C* Means clustering and evolutionary algorithm to obtain the fuzzy neural networks's eventual relationship among the fuzzy sets and to optimize the model. Karaboga et al. [7] proposed a comparative study of

N. Xiong et al. (Eds.): ISCSC 2018, AISC 877, pp. 123–129, 2019.
https://doi.org/10.1007/978-3-030-02116-0_16

artificial bee colony algorithm in 2009. Farzi [8] also proposed the efficient job scheduling in grid computing with modified artificial fish swarm algorithm in 2009. Fang et al. [9] proposed the genetic hybrid particle swarm algorithm in the power plant unit load in the application of combinatorial optimization and Xu et al. [10] proposed the weight of a nonlinear adaptive particle swarm optimization in 2010.The algorithms are simulated group of bees foraging intelligent algorithm, according to their division of labor for different activities to achieve the exchange of information a group of bees sharing the body in order to find the optimal solution. Function optimization results show it has better optimization performance compared with than the genetic algorithm, particle swarm optimization and different evolution algorithm. In the paper, the novel methods based on ant colony optimization algorithm (ACO) is proposed in order to solve the problem of multi-objective optimization.

2 Multi-objective Optimization Problem

Considering the following multi-objective optimization problem

$$\min f(x) = (f_1(x), f_2(x), \cdots, f_m(x)) \text{ , s.t. } x \in [a, b]. \tag{1}$$

where, the decision vector $x \in R^{SN}$, $x = (x_1, x_2, \cdots, x_{SN})$, the target vector $f(x) \in R^m$.

There are many constraints among the various goals in multi-objective optimization. The optimized goals are often at the expense of other goals [1, 2]. In the paper, in order to solve the problem of multi-objective optimization, the novel methods based on ant colony optimization algorithm (ACO) is proposed in the paper. The essential advantages of ACO including cooperativity, obustness and positive feedback and distributed nature were discussed. Considering the shortcomings of the traditional ACO, colony fitness is improved. Cross operation and mutation of genetic algorithms were introduced into the ACO in order to improve its searching ability and to realize the dynamically adjusting the influence of each ant to the trail information updating and the selected probabilities of the paths. In the end, experimental results show that the methods proposed compared with traditional methods can improve the speed of converge and can get the optimal solution quickly.

3 Multi-objective Optimization Based on ACO

3.1 Individual Fitness

In this paper, the sort method and adaptive method is used. In order to obtain the fitness value of the individual based on dominating relationship, first of all, to sort individual using the algorithm of ACO. Then, to calculate the value of adaptive double density according to the congestion around the density values. In the end, to identify adaptation values comprehensively. The method in detail is as follows.

Step1: Computing Individual Sort $R^{'}$ (i) in groups Q

$$R'(i) = |\{j \mid j \in Q, j \succ i\}| \quad \forall i \in Q \tag{2}$$

where, $R'(i)$ means dominating relationship, which express the number of dominating individual i on the current group Q.

Step2: Individual sorting

$$R(i) = R'(i) + \sum_{j \in Q, i < j} R'(j) \qquad \forall i \in Q \tag{3}$$

It is shown that ordering number R (i) of individual i is equal to the number of sorting. And it can control the number of all of the individual.

Step3: The target space is divided into n grids.

According to the population size, dimensional space n_e for the number of grids of each target is set. Set ting the integer part of $\sqrt[k]{SN}$g is a, and the fractional part is r, then

$$n_e = \begin{cases} a & r = 0 \\ a + 1 & r \neq 0 \end{cases} \tag{4}$$

Step4: Individual fitness value is as follows.

$$fit_i = \frac{1}{\exp(R(i) * \rho(i))} \tag{5}$$

where, $R(i)$ is the rank of each individual I and $\rho(i)$ is the individual density values.

3.2 The Novel Ant Colony Algorithm and Genetic Algorithm Description

Ant colony algorithm is that ant is always looking for food sources with the shortest path between nests. The algorithm has a positive feedback, coordinated, parallel and robustness, which is easy to integrate with other methods. However, the algorithm generally require a longer time for searching and prone to the phenomenon of precocious and stagnation which result the lack of global optimal solution. However, Ant colony algorithm take some advantage of improvement the global searching ability and searching speed, where the use of population-based genetic algorithm optimization techniques to prevent the population of convergence and stagnation at any local optimal point.

Genetic algorithms have a choice, exchange and variation [1, 2] the number of three basic operations. Select an operation from the current generation of generating a new set of stocks, which determines which individuals are involved in generating the next generation of individuals, that is, the next generation of the mating pool process. Chains according to their individual genetic adaptation function values are copied in the mating pool. So the algorithms mainly including the choice of strategy, updating the amount of the local information the local update, finding a local optimal solution of the local search algorithm and update the four parts of the overall information. The specific implementation principle is as follows.

(I) In order to reduce the computation time, the larger transition probability is taken in the first iteration.
(II) In order to prevent falling into local optimal solution and missing the global optimal solution, crossover and mutation of the genetic algorithm can be carried out in order to adjust and narrow path of information after searching a certain time on the best and the worst the gap between the amount due to increase the probability of randomly selected, so that a more complete search space to expand your search.
(III) In the pheromone update, all of better characteristics will be accounted through changing the update strategy.

In this paper, the individual's fitness value will be selected by the probability of ant information choice of probability, which is calculated as follows.

$$p_i = \frac{fit_i}{\sum_{n=1}^{SN} fit_n} \tag{6}$$

where, fit_n is the fitness value of individual i and SN is the number of searching path or the number of information. In order to produce a new position from the memory location, ACO algorithm is as follows.

$$v_{ij} = x_{ij} + \phi_{ij}(x_{ij} - x_{kj}) \tag{7}$$

where, $k \in \{1, 2, \cdots SN)$, $j \in \{1, 2, \cdots, D\}$ are selected randomly, and $k \neq i$; ϕ is random number between [−1,1]; It controls the new field x_{ij} of ants and bees production within the scope of the two visual comparison of two ants location, it is obvious to find from Eq. (7) that the smaller the disturbance on the location while the narrower the gap between x_{ij} and x_{kj}. So the solution space with the optimal solution was approaching, steps will be reduced accordingly.

In the artificial colony algorithm, if the position x_i after a limited number of ants cycle still cannot be improved, then the ants collected path at a scouts, the location of the ants will be randomly generated within the solution space replaced by a location x_i the scouts found a new ant source and replace them. The operation is as follows.

$$x_i^j = x_{\min}^j + rand[0, 1](x_{\max}^j - x_{\min}^j) \tag{8}$$

where, $j \in \{1, 2, \cdots, D\}$.

4 Experimental Verification

In order to verify the validity of the proposed method, reference [7] testing function is used before performance.

(I) Sphere function is as follows.

$$f(x) = \sum_{i=1}^{n} x_i^2 \quad -100 \leq x_i \leq 100 \tag{9}$$

It is a continuous, unimodal convex function, the minimum of global function is 0. It is to say that

$$\vec{x}_{opt} = (x_1, x_2, \cdots x_n) = (0, 0, \cdots, 0) \tag{10}$$

(II) Origin function is as follows.

$$f(x) = \sum_{i=1}^{n} (x_i^2 - 10\cos(2\pi x_i) + 10) \quad -5.12 \leq x_i \leq 5.12 \tag{11}$$

The value of global optimization is equal to 0. The optimize the results is as follows.

$$\vec{x}_{opt} = (x_1, x_2, \cdots, x_n) = (0, 0, \cdots, 0) \tag{12}$$

(III) Rosen brock function is as follows.

$$f(x) = \sum_{i=1}^{n-1} 100\left(x_{i+1} - x_i^2\right)^2 + (x_i - 1)^2 \quad -10 \leq x_i \leq 10 \tag{13}$$

The value of global optimum is 0, and the results after optimizing is as follows.

$$\vec{x}_{opt} = (x_1, x_2, \cdots, x_n) = (1, 1, \cdots, 1) \tag{14}$$

The maximum of cycles is 2000 before and after the improvement of the algorithm. In order to statistic the average error and average of convergence algorithms, experiments were done 30 times for each testing function. Computing time per cycle observed ants and information were collected 50% of the population. Tables 1, 2 and 3 are for different populations of ants before and after results improvement algorithms respectively.

It can be seen from the Tables 1, 2 and 3, the convergence speed has been greatly improved by suggested method comparing with the traditional methods. In addition, this approach is not available in the case of small populations, but it is available in the case of the population increasing in the number, and the algorithm in the case of the larger population is better than artificial colony algorithm.

Table 1. Testing results of Sphere function

Sphere function	Population size 10		Population size 30		Population size 50	
	Traditional methods	Suggested methods	Traditional methods	Suggested methods	Traditional methods	Suggested methods
Mean	0.000750316	1.85669e−06	5.57868e−10	1.04213e−013	2.2408e−15	4.45573e−17
Mean	0.000750316	1.85669e−06	5.57868e−10	1.04213e−013	2.2408e−15	4.45573e−17
Std	0.000855695	1.39426e−05	7.79432e−10	2.83441e−15	3.54077e−15	2.89256e−17
Std	0.000855695	1.39426e−05	7.79432e−10	2.83441e−15	3.54077e−15	2.89256e−17

Table 2. Testing results of Rastrigin function

Rastrigin function	Population size 10		Population size 30		Population size 50	
	Traditional methods	Suggested methods	Traditional methods	Suggested methods	Traditional methods	Suggested methods
Mean	2.48241	3.31653	0.524612	0.241352	0.008027	1.07643e−06
Mean	2.48241	3.31653	0.524612	0.241352	0.008027	1.07643e−06
Std	2.373	2.97476	1.86301	0.97305	0.04359	1.47197e−005
Std	2.373	2.97476	1.86301	0.97305	0.04359	1.47197e−005

Table 3. Testing results of Rosen-brock function

Rosen-brock function	Population size 10		Population size 30		Population size 50	
	Traditional methods	Suggested methods	Traditional methods	Suggested methods	Traditional methods	Suggested methods
Mean	0.921806	0.180521	0.098569261	0.016460	0.033730235	0.0039679
Mean	0.921806	0.180521	0.098569261	0.016460	0.033730235	0.0039679
Std	0.574671844	0.139630	0.096541546	0.020842	0.023992851	0.010462
Std	0.574671844	0.139630	0.096541546	0.020842	0.023992851	0.010462

5 Conclusions

The novel methods based on ant colony optimization algorithm (ACO) is proposed in the paper in order to solve the problem of multi-objective optimization. Considering the shortcomings of the traditional ACO, colony fitness algorithm was improved. Crossing operation and mutation of genetic algorithms were introduced into the ACO in order to improve its searching ability and to realize the dynamically adjusting the influence of each ant to the trail information updating and the selected probabilities of the paths. Finally, experimental results on three basic testing functions proved that the methods proposed can improve the speed of converge and obtain the optimal solution quickly compared with traditional methods.

Acknowledgment. The authors deeply appreciate support for this paper by the National Natural Science Foundation of China (Grant No. 61379019), the Scientific Research Foundation for Returned Scholars Fellowship from Sichuan Province.

References

1. Michalewicz, Z., Schoenauer, M.: Evolutionary algorithms for constrained parameter optimization problems. J. Evol. Comput. **4**(1), 1–32 (1996)
2. Van, D.A., Gary, V., Lamont, B.: Multi-objective evolutionary algorithm research: a history and analysis evolutionary computation. J. Evol. Comput. **8**(2), 125–147 (1998)
3. Maniezzo, A., Carbon, V., Aro, A.: An ant heuristic for the frequency assignment problem. J. Future Gener. Comput. Syst. **16**(8), 927–935 (2000)
4. Wang, Y., Xie, J.Y.: Algorithm for multimedia multicast routing based on ant colony optimization. J. Shanghai Jiaotong Univ. **36**(4), 526–531 (2002)
5. Liu, P., Gao, F., Yang, Y.: QoS routing algorithm based on the combination of genetic algorithm and ant colony algorithm. J. Appl. Res. Comput. **35**(9), 224–227 (2007)
6. Cui, M.T., Fu, L.X., Zhao, H.J., et al.: Optimization algorithm of multi-level discrete fuzzy neural networks for solving global stability. J. Univ. Electr. Sci. Technol. China **36**(3), 628–631 (2007)
7. Karaboga, D., Akay, B.: A comparative study of artificial bee colony algorithm. J. Appl. Math. Comput. **214**(1), 108–132 (2009)
8. Farzi, S.: Efficient job scheduling in grid computing with modified artificial fish swarm algorithm. Int. J. Comput. Theory Eng. **1**(1), 13–18 (2009)
9. Fang, Y.T., Peng, C.H.: Genetic hybrid particle swarm algorithm in the power plant unit load in the application of combinatorial optimization. J. Electr. Power Autom. Equip. **10**, 22–26 (2010)
10. Xu, G., Yang, Y.K., Wong, N.D.: The weight of a nonlinear adaptive particle swarm optimization. J. Comput. Eng. Appl. **35**, 49–51 (2010)

Linear Programming Model of Communication Capability Configuration

Xinshe Qi[(✉)], Weiwei Zhao, Qingzheng Xu, Na Wang, Yuyi Li, and Xin Wang

College of Information and Communication, National University of Defense Technology, 710106 Xi'an, China
tyqxs@163.com

Abstract. Based on a regional network, a linear programming model for minimum capability redundancy is established firstly. According to the principle of subordinate relationship and minimum capability, the capability data integration model is proposed and optimized, and then the communication scheme with the minimum total capability can be obtained. Finally, according to the minimum redundancy, the security capabilities of communication line is analyzed systematically and deeply.

Keywords: Linear programming model · Capability matching
Communication guarantee

1 Introduction

The existing communication sites and lines of a regional network are shown in Fig. 1. The circles in the figure indicate communication site; the number in each circle and the letter next to the circle indicate the configuration capability (GB) and the name of the site respectively. Each site can be divided into three levels according to the communication task attribute and then constitute a subordinate relationship in the communication service. There are two first-level central sites (O and AF), five second-level sites (I, L, U, Y and AC; the first two sites affiliated site O, and the last three affiliated site AF), and 27 third-level sites (subordinate sites I: B, C, D, E, F, G and H; subordinate sites L: J, K, M, N and P; subordinate sites U: Q, R, S, T and V; subordinate sites Y: W, X, Z, AA and AG; subordinate sites AC: AB, AD, AE, AH and AI). The communication traffic per unit time is 1 GB between the first-level central sites, 2 GB between the first-level central sites and the subordinated second-level sites, and 0.6 GB between the second-level sites and the subordinate third-level sites. The traffic between two sites can be split into the smallest unit 0.1 GB. Now the capability of the communication network equipment is need to be effectively matched.

N. Xiong et al. (Eds.): ISCSC 2018, AISC 877, pp. 130–136, 2019.
https://doi.org/10.1007/978-3-030-02116-0_17

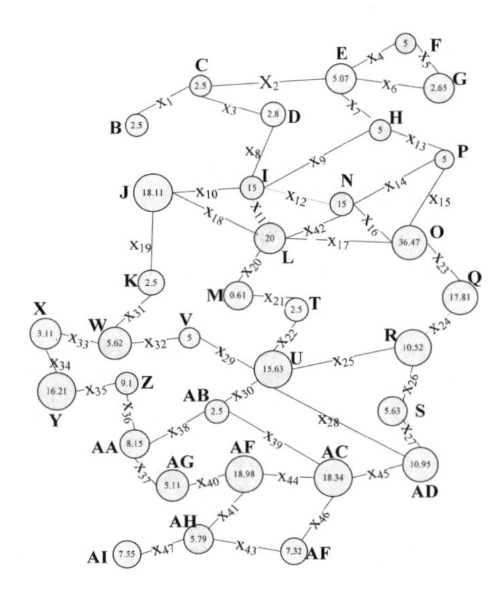

Fig. 1. A regional communication site and distribution diagram.

2 Introduction

The communication network equipment capability matching means that the communication between two related sites need to occupy the same capability. The amount of the redundancy at each site is the main basis for judging the rationality of communication network equipment [1, 2]. The guarantee of communication equipment is mainly to ensure that the station on each line have enough matching capability to communicate. For ease of study, it is assumed that each line has a minimum amount of matching capability.

2.1 Model Building

Let us suppose x_i the matching capability of the first line, q_i the fixed capability of the site, S_i the redundancy of the site, and R_n the fixed capability of the site n. As seen from Fig. 1, there is only one line between any two associated sites. Because the capability configuration is fixed, it is easy to calculate the total capability of the communication network. Then the minimum of the total redundancy can be determined by setting the objective function.

Meanwhile, the capability configuration must meet the requirement that the sum of the capability of the site and other sites cannot exceed the total capability of the site, being $\min S = \min \sum_{i=1}^{47} S_i = \sum_{i=1}^{47} q_i - 2\sum_{i=1}^{47} x_i$, in which $0 \leq \sum_{n} x_i \leq R_n$, and n denotes any site in B, C, D, …, AI.

2.2 Model Solving and Optimization

With Lingo software, theoretical values for matching capability on each line can be obtained when the redundancy barrier is minimized [3], as shown in Table 1. But through data analysis, it has been found that a large number of capability values are zero for each line matching capability, which does not meet the actual needs of communication. Therefore, the stated capability value only theoretically meets the matching to achieve the minimum total redundancy.

Table 1. Theoretical value and adjustment value of matching capability of each line.

Line	Capability	Adjustment capability	Line	Capability	Adjustment capability
B-C	2.5	2.3	H-P	0	0.1
C-E	0	0.1	N-P	0	0.1
C-D	0	0.1	P-O	5	4.8
E-F	3.71	3.61	O-N	15	14.7
F-G	1.29	1.39	O-L	0.78	1.48
G-E	1.36	1.26	J-L	14.13	13.58
E-H	0	0.1	J-K	1.88	1.88
D-I	2.8	2.7	L-M	0	0.1
H-I	5	4.8	M-T	0.61	0.51
I-J	2.11	2.66	T-U	1.89	1.99
I-L	5.1	4.75	O-Q	15.69	15.49
I-N	0	0.1	Q-R	2.12	2.32
R-U	8.4	8.1	AA-AG	0	0.1
R-S	0	0.1	AA-AB	2.5	2.3
S-AD	5.63	5.53	AB-AC	0	0.1
U-AD	5.32	5.32	AG-AF	5.11	5.01
U-V	0	0.1	AF-AH	0	0.1
U-AB	0	0.1	N-L	0	0.1
K-W	0.62	0.62	AH-AE	0	0.1
W-V	5	4.9	AC-AF	11.02	10.92
W-X	0	0.1	AC-AD	0	0.1
X-Y	3.11	3.01	AE-AC	7.32	7.22
Y-Z	3.45	3.35	AH-AI	5.79	5.59
Z-AA	5.65	5.75	MinS(GB)	14.16	12.96

To ensure the smooth of the communication line, the value of the line matching capability needs to be improved, being $x_i > 0(i = 1, 2, 3, \ldots, 47)$. Therefore, as long as there is such a line between the two sites, it is necessary to ensure the best of the mother line $x_i \geq 0.1$. Recalculation can draw the adjustment value of matching capability of each line, as shown in Table 1.

3 The Establishment of Capability Integration Model

In accordance with the reduction of waste of communication resources and the improvement of the working efficiency of communication networks, the capability matching of each line and data integration of each station needs to be done. It can be finished through these three steps.

First, the data for each site's capability processing, according to the requirement that the minimum amount of traffic can be split between the two sites is 0.1 GB. And the last decimal place of the fixed capability of each site can be removed. Secondly, combined with the affiliation diagram, calculating the capability of matching circuits based on the traffic requirements per unit of time. Finally, establishing a linear programming mathematical model.

3.1 Fixed Matching Capability Processing

In the integration process, the principle of the tail-off method is mainly followed, that is to say, remove the decimal value on the percentile, where the M-site data integration follows principle of proximity. The reason is as following.

First, the site L and the site M belong to the communication between the second-level site and the third-level site, and the least capability of 0.6 GB should be guaranteed. Second, the site M and site T belong to the communication between the third-level site. To ensure the communication smoothness of the same level site, a given communication capability is required. Being the communication capability is at least 0.1 GB. Therefore, the capability of the site M is integrated to 0.7 GB.

3.2 Adjustment of Minimum Matching Capability

According to the affiliation and communication requirements, and taking into account that there is no affiliation between sites at the same level, a given capability of 0.1 GB is needed to ensure the smooth flow of the line. The channel capability for a site includes not only the capability of the directly connected site, but also the capability of the reserved channel of the communication with the connected station. Therefore, the minimum communication capability of each line can be obtained.

3.3 Capability Integration Model Establishment

In order to give a communication implementation scheme that minimizes the total equipment capability, the shortest path can be selected firstly. There must be two types of shortest paths from one vertex to other vertices and the shortest path between each pair of vertices in the graph [4]. Following the principle of proximity and large, that is to say, the more nodes the communication passes through, the more equipment it occupies, it is reasonable to choose a higher level site to reduce the capability consumption.

By using the broken circle method, the circuit in the figure is opened so that there is only one communication channel. The communication circuit diagram can be implemented. Then a linear programming model that occupied the total equipment capability can be established.

The objective function is defined as:

$$\min S = \min \sum_{i=1}^{47} S_i = 2 \sum_{i=1}^{47} x_i \tag{1}$$

$$0 \leq \sum_{n} x_i \leq RI_n \tag{2}$$

In the formula, n represents any site, and RI_n means fixed capability R_n retains a decimal value of n sites.

3.4 Solution and Analysis of the Model

The Lingo software is used to solve the linear programming model to obtain the optimal configuration of each line capability and the minimum value of the equipment capability in the occupancy. The line matching capability is shown in Table 2. The total equipment capability value min S = 117.6 GB.

Table 2. Line capability best matching table.

Line	Capability	Line	Capability	Line	Capability	Line	Capability
B-C	0.7	H-P	0.1	R-U	2.8	AA-AG	2.7
C-E	0.1	N-P	0.7	R-S	0.7	AA-AB	0.1
C-D	1.3	P-O	0.1	S-AD	0.1	AB-AC	0.6
E-F	0.7	O-N	0.1	U-AD	3.1	AG-AF	2.1
F-G	0.1	O-L	4	U-V	0.6	AF-AH	0.1
G-E	0.7	J-L	1.2	U-AB	0.1	N-L	0.1
E-H	1.9	J-K	0.7	K-W	0.1	AH-AE	1.3
D-I	1.8	L-M	0.6	W-V	0.1	AC-AF	5
H-I	2.4	M-T	0.1	W-X	0.7	AC-AD	3.6
I-J	0.1	T-U	0.6	X-Y	1.2	AE-AC	1.8
I-L	2.1	O-Q	1.1	Y-Z	3.8	AH-AI	0.7
I-N	1.2	Q-R	1.7	Z-AA	3.3	minS	117.6

This result is based on the distribution path of the communication site in the absence of the affiliation of the communication site with a capability of 0.1 GB. From the capability matching table of the line and sites distribution, the model can come true that the communication path between any two sites is the shortest path, and the occupied total backup capability is as small as possible.

4 In-Depth Analysis of Security Capabilities of Communication Line

The network communication security capability between the central site O and its subordinate sites is taken as an example for analysis and modeling, and then a communication line configuration capability adjustment implementation plan is proposed.

First, to determine the objective function and its constraint conditions to obtain a matching value range of the channel capability between each adjacent site. Second, according to the upper limit of the matching capability, a value that guarantees to realize the connection without disruption of each channel can be obtained. Third, based on the distribution of traffic between sites at all levels, the probability that the inter-site channel achieves communication can be obtained, and then the equipment capability of each site is optimally configured.

4.1 Communication Line Matching Capability Range

In the central site O area, the objective function that can match the maximum capability and the minimum capability on the communication line can be expressed as $\max x_i, \min x_i$, both to meet the sum of the capability values of the lines allocated to the site does not to exceed the capability of the site, but also to meet the traffic value per unit of time required by the sites in the model of the capability data integration.

According to the actual situation, if the minimum value is selected, the communication network cannot be fully realized. Therefore, the maximum value of x_i should be selected as possible and then to determine the boundary value between two adjacent sites, so as to improve the network communication security.

4.2 Communication Line Reliability Calculation

The traffic between first-level site O and the subordinate second-level site is subject to the normal distribution $N(2, 0.5^2)$, the second-level site and the subordinate third-level site is subject to the normal distribution $N(0.6, 0.2^2)$. After calculation, it can be concluded that the probability of communication and maintenance of the purlin line: $\mu = 2, \sigma = 0.5$ between the first-level site O and the subordinate site L; $\mu = 0.6, \sigma = 0.2$ between the second-level site D and the subordinate third-level site I.

$$P(x_{17} \leq 12.9) = \Phi(\frac{12.9 - 2}{0.5}) = \Phi(21.8), \tag{3}$$

$$P(x_8 \leq 0.6) = \Phi(\frac{0.6 - 0.6}{0.2}) = \Phi(0) \tag{4}$$

The same can be obtained for other lines guaranteed rate. According to the analysis, the conclusion can be obtained that the larger the Φ value, the higher the security rate, and the stronger the communication capability of the line.

5 Conclusion

The line capability matching model has examined the least redundancy and the largest occupancy under the constraint that it does not exceed the fixed capability of the site and ensure the smooth communication network with an increase the utilization of communication networks. The results show that after this configuration, the reliability of the information transmission of the line has been greatly improved, and the capability of the channel has been utilized to the maximum extent. So it has a practical application prospect.

References

1. Tan, Y.J., Lv, X., Wu, J., Deng, H.Z.: On the invulnerability research of complex networks. Syst. Eng. Theory Pract. **28**(suppl), 116–120 (2008). (in Chinese)
2. Shi, C.H.: Research on Survivability of Complex Network Topology. University of Electronic Science and Technology of China, Chengdu, China (2012). (in Chinese)
3. Ding, K.S., Zhang, X.Y., Liang, X.J.: The reliability of communication network meaning and its comprehensive measure indicators. J. Commun. **20**(10), 75–78 (1999). (in Chinese)
4. Xu, J., Xi, Q.M., Wang, Y.L.: Systemic nuclear and nuclear theory (II) - optimized design and reliable communication network. J. Syst. Eng. **9**(1), 1–11 (1994). (in Chinese)

Quality Survey and Analysis of China's Key Service Industries in 2017

Yawei Jiang and Huali Cai(✉)

Quality Management Branch,
China National Institute of Standardization, Beijing, China
352362986@qq.com, 583451349@qq.com

Abstract. Service industries represent an important part of modern economic system while quality a key pillar for the development of service industries. With the further enhanced leading role of China's service industries in economic growth, service quality occupies an important position in satisfying the increasing demands of the public for a better life and facilitating high-quality economic growth. In order to understand the current development of service quality, this paper makes an analysis of China's key service industries in terms of customer satisfaction, consumer complaint and brand value in 2017, which is of great significance for improving service quality.

Keywords: Service quality · Key service industries · Customer satisfaction
Consumer complaint · Brand value

1 Introduction

In 2017, the added value of service industries occupied a percentage of 51.6% among the total GDP of China, making the industries a leader in the tertiary industry for 5 consecutive years. The growth of service industries contributed 58.8% to the growth of national economy and 4.0% to the growth of the national GDP [1]. The leading role of service industries in economic growth has been further enhanced and service quality has become a key support for the development of service industries. In order to better understand the development of service industries, and based on customer satisfaction model, this paper makes an analysis of China's key service industries, including automobile insurance, online tourism, supermarket, etc., in terms of customer satisfaction, consumer complaint and brand value, etc.

2 Customer Satisfaction

The Customer satisfaction evaluation model structure is shown in Fig. 1.

Where, circles indicate latent variables; the arrows between circles indicate the causal relationship among latent variables; rectangles indicate observable variables; and the arrows between rectangles and circles indicate the observable variable-latent variable response relationship.

N. Xiong et al. (Eds.): ISCSC 2018, AISC 877, pp. 137–143, 2019.
https://doi.org/10.1007/978-3-030-02116-0_18

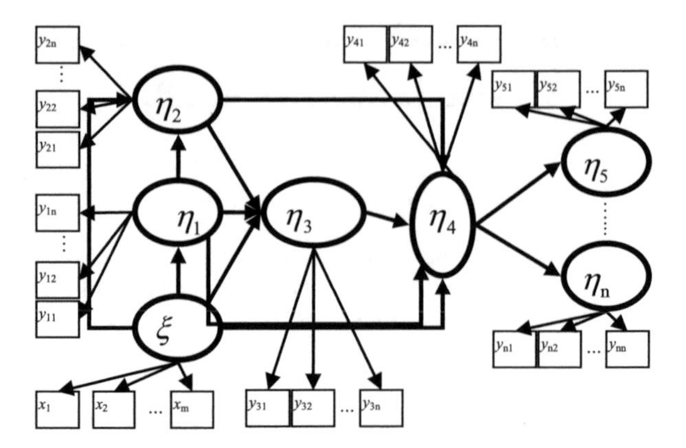

Fig. 1. Example of customer satisfaction evaluation model.

The example of mathematical form for structural equation is shown as follows:

$$\eta = B\eta + \Gamma\xi + \zeta \tag{1}$$

Where, η means endogenous latent variable; ξ means exogenous latent variable; B means relationship between endogenous latent variables; Γ means impact of exogenous latent variable on endogenous latent variable; and ζ means residual of structural equation, which reflects the parts that are unable to be interpreted in equation and subject to independent normal distribution with average value being zero.

The example of mathematical form measuring the equation is shown as follows:

$$X = \Lambda x\xi + \delta \tag{2}$$

$$Y = \Lambda y\eta + \varepsilon \tag{3}$$

Where, X means the vector formed by exogenous indexes; Y means the vector formed by endogenous indexes; Λx means the relationship between exogenous and endogenous indexes, or factor loading matrix of exogenous index on exogenous latent variable; Λy means the relationship between endogenous index and endogenous latent variable or factor loading matrix of endogenous index on endogenous latent variable.

By using this model, we conducted a survey of 250 major cities in China, and each industry have 2000 effective sample under 80% confidence. Through our calculation, the customer satisfaction with China's service industries was 74.75 in 2017 (Fig. 2), 2.86% up compared to the year 2016, and remained within the "Relatively satisfied" range. As is shown in Table 1, the scores of customer satisfaction in automobile insurance (75.24), supermarket (76.7), securities (72.72), online shopping (74.36), household broadband (73.95), online payment (73.66), life insurance (73.01), civil aviation service (75.22), online tourism (73.3), express hotel (72.12), mobile communications (74.47), web portal (76.16) and express delivery (75.7) [2] increased to varying extents compared to the previous year. However, the score of customer satisfaction in automobile after-sales service (76.0) [3] decreased by 1 in contrast to the year 2016.

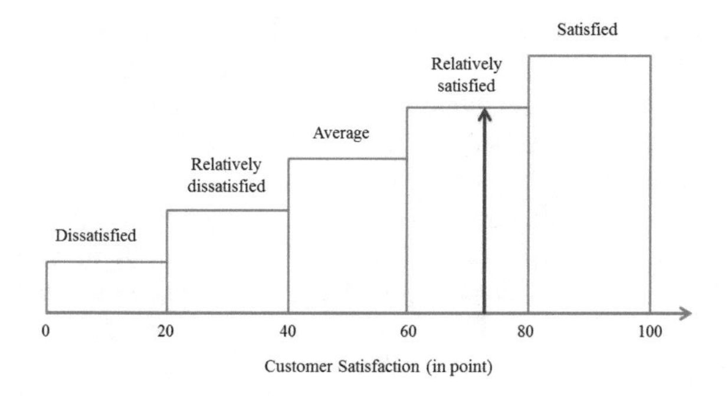

Fig. 2. Customer satisfaction ratings

Table 1. Customer satisfaction in service industries from 2016 to 2017.

Industry	2017	2016	Change
Automobile insurance	75.24	69.79	↑5.45
Supermarket	76.70	71.31	↑5.39
Securities	72.72	68.85	↑3.87
Online shopping	74.36	70.73	↑3.63
Household broadband	73.95	70.48	↑3.47
Online payment	73.66	70.70	↑2.96
Life insurance	73.01	70.37	↑2.64
Civil aviation service	75.22	72.63	↑2.59
Online tourism	73.30	70.99	↑2.32
Express hotel	72.12	70.02	↑2.10
Mobile communications	74.47	72.62	↑1.85
Web portal	76.16	74.33	↑1.83
Express delivery	75.70	74.70	↑1.00
Automobile after-sales service	76.00	77.00	↓1.00

Among all factors influencing customer satisfaction, brand image and perceived quality have a greater influence, with the average influence coefficient of 0.58 and 0.57, beyond perceived value (0.33) and expected quality (0.32). In securities, online shopping, household broadband, life insurance, civil aviation service, express hotel, mobile communications, web portal and other fields, brand image produces the largest influence over customer satisfaction, which reflects consumers' increased focus on brand cognition when they purchase or use goods. In the fields of supermarket, online payment and online tourism, perceived quality has the most influence over customer satisfaction, indicating that consumers focus more on the quality of products and services. In the automobile insurance field, perceived value produces the largest influence over customer satisfaction, indicating consumers' greater focus on price performance ratio.

Among customer satisfaction monitoring indicators, service characteristics, rich service items, attitude of service staff, etc. have low scores, showing the great homogeneity of service industries in China. It is required that service innovation should be further facilitated and the professional level and comprehensive accomplishments of some service staff are to be improved. As is shown in Table 2, in the fields such as household broadband, mobile communications, web portal and civil aviation service, China enjoys greater customer satisfaction than the USA; while in the fields featuring high integration with the Internet such as online shopping and online tourism as well as living services such as express delivery, life insurance, express hotel and supermarket, China still lags behind developed countries.

Table 2. Comparison between China and the USA in terms of customer satisfaction in some service fields.

Fields	China	USA	Gap
Household broadband	73.95	64	9.95
Mobile communications	74.47	72	2.47
Web portal	76.16	75	1.16
Civil aviation service	75.22	75	0.22
Supermarket	76.7	79	−2.3
Express hotel	72.12	76	−3.88
Life insurance	73.01	78	−4.99
Express delivery	75.7	81	−5.3
Online tourism	73.3	79	−5.7
Online shopping	74.36	82	−7.64

3 Consumer Complaint

Consumer complaint can be deemed as one side that reflects the quality of industry development. By analyzing the changes in the number of complaints per 10,000 consumers in the manufacturing and service fields, we can determine the change in number of quality issues, find out key factors that constrain consumption and define the focus of quality improvement. The number of complaints per 10,000 consumers is calculated as follows:

Number of complaints per 10,000 consumers = total number of complaints/total population of China

Where, the total number of complaints and total population of China are sourced from China Consumers Association [4] and the Statistical Communiqué of the People's Republic of China on the 2017 National Economic and Social Development released by the National Bureau of Statistics of the People's Republic of China [5].

Figure 3 showed the trend of the number of complaints per 10,000 consumers. Through our calculation, in 2017, the number of complaints per 10,000 consumers in service fields (3.03) was on a rise, higher than that in the manufacturing fields (2.20) again. The service fields have become a hotspot talked about by consumers and a focus

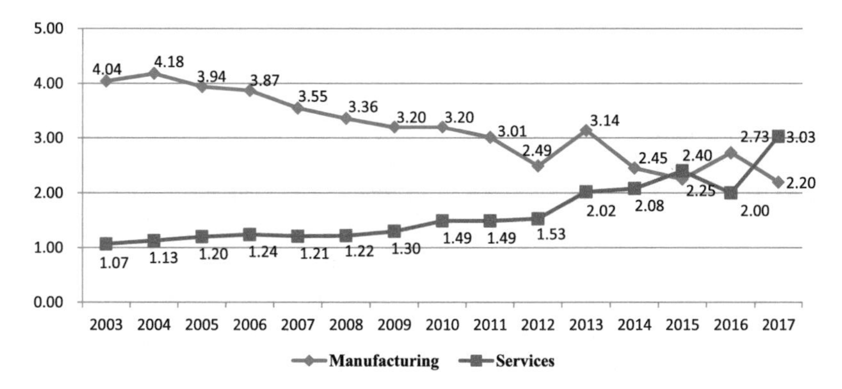

Fig. 3. Number of Complaints per 10,000 consumers

for quality improvement. According to the statistics of China Consumers Association, the ratios concerning contract, after-sales service, false publicity and human dignity were on a slight rise. It reflects the weak points of a small number of emerging high-tech enterprises in contract, after-sales service, publicity and other fields due to the constant emergence of new business models and new marketing means. Among service complaints, Internet service, sales service, living and social service, telecommunication service and culture, amusement and sports service ranked top 5. Compared with the year 2016, the number of Internet service complaints witnessed a significant increase, by 330.86%. The number of remote shopping complaints, with online shopping as focus, remained far ahead in service complaints. Some emerging bicycle sharing enterprises had difficulty in refunding deposits to their users. For e-commerce platform and individual online businesses represented by We-Chat business and TV shopping, there are serious issues concerning inferior quality of commodity and service. Insurance service, health care service, financial service had a large drop in the number of consumer complaints, down by 64.28%, 44.70% and 24.19% respectively on a year-on-year basis. It indicates that we should pay attention to the regulated development of emerging industries such as Internet.

4 Brand Value

Brand value represents a combination of financial indicators including enterprise revenues and profitability and consumers' brand cognition and an important aspect measuring quality level. By analyzing the growth rate in brand value, we may measure the quality efficiency and sustainability of service industry development. The growth rate in brand value is expressed in the following formula:

Growth rate in brand value = (brand value of the current year-brand value of the previous year)/brand value of the previous year * 100%

Where, the brand value data is sourced from the 2017 Top 100 Most Valuable Global Brands and 2017 Top 100 Most Valuable Chinese Brands [6], released by BrandZ, a world brand laboratory.

According to our statistics, among 2017 Top 100 Most Valuable Chinese Brands, 47 service brands appeared in the ranking, 6 brands more than 2016. The total value of service brands reached USD 458.197 billion, 77.94% of all brands listed in the ranking, and 13.23% up on a year-on-year basis. Most service industry brands maintained a growth in value. Tourism, education, retail and other fields increased the most in brand value, by 31.77%, 31.65% and 17.75% respectively; and insurance, banking and other traditional economic brands dropped in value, by 6.16% and 5.92% respectively. It reflects the fact that with the improvement of people's living standards, service brands are shifting towards meeting and improving people's needs for a better life (Fig. 4).

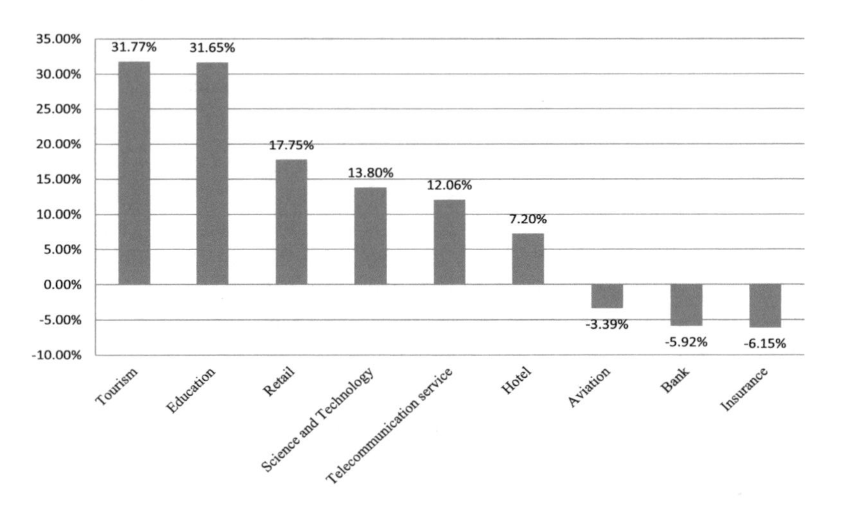

Fig. 4. Growth rate in the value of Chinese service brands

5 Conclusions

This paper analyzes the quality of China's key service industries in 2018 from three perspectives: customer satisfaction, consumer complaint and brand value growth. The survey indicates that the customer satisfaction in China's service industries was 74.75 in 2017, 2.86 up compared to the previous year, and remained within the "relatively satisfied" range. The customer satisfaction in 13 service industries including automobile insurance, supermarket, securities, online shopping, household broadband, online payment, life insurance, civil aviation service, online tourism, express hotel, mobile communications, web portal and express delivery witnessed an increase in contrast to the previous year. In 2017, the brand value in tourism, education, retail and other fields of China had a large growth and the value in traditional economic brands such as insurance and banking witnessed a drop, which reflects the fact that with the improvement of people's living standards, service brands are shifting towards meeting and improving people's needs for a better life. However, in terms of consumer complaint, the number of complaints per 10,000 consumers in service fields tended to increase in 2017, higher than that of the manufacturing fields (2.20) again. The service fields have become a hotspot talked about by consumers and a focus for quality improvement. Online service, sales service, living and social service, telecommunication service and culture, entertainment, and sports service ranked top 5 in terms of consumer complaints. Overall, in 2017, China's key service industries witnessed a growth in quality while maintaining stability and gradual increase in brand value, despite of rising service complaints. In the future, we should attach great importance to the improvement of consumption environment so as to effectively reduce consumer complaints.

References

1. National Bureau of Statistics of the People's Republic of China. http://www.stats.gov.cn/tjsj/sjjd/201801/t20180119_1575485.html. Accessed 10 May 2018
2. State Post Bureau of the People's Republic of China. http://www.askci.com/news/chanye/20180129/171314117201.shtml. Accessed 10 May 2018
3. China Association for Quality. http://www.caq.org.cn/html/zltj/yhmygcdtxx/6467.html. Accessed 10 May 2018
4. China Consumers Association. http://www.cca.org.cn/tsdh/detail/27876.html. Accessed 10 May 2018
5. National Bureau of Statistics of the People's Republic of China. http://www.stats.gov.cn/tjsj/zxfb/201802/t20180228_1585631.html. Accessed 10 May 2018
6. China-10.com. http://www.china-10.com/news/486523.html. Accessed 21 June 2018

Subdivision Algorithm of Quartic Q-Ball Surfaces with Shape Parameters

Gang Hu(✉), Dan Lv, and Xinqiang Qin

Department of Applied Mathematics, Xi'an University of Technology, Xi'an, China
{hg_xaut,xqqin}@xaut.edu.cn, 18229035865@163.com

Abstract. Based on a set of quartic Ball basis functions, the quartic Q-Ball surfaces with shape parameters are proposed, which not only inherits the main properties of classical cubic Ball surfaces, but also have excellent shape adjustability. Then it is introduced some basic properties of the Ball basis function and quartic Q-Ball surfaces. With the aim to solve the problem that local adjustment of complex surface shape in modeling design, the coefficient subdivision algorithm of quartic Q-Ball surfaces are investigated. It is not only deduced the coefficient subdivision algorithm for the quartic Q-Ball surfaces, gave the specific subdivision steps, but also geometric modeling examples. The modeling examples show that the proposed algorithm is effective and easy to implement, which greatly enhances the ability to adjusting the local shape of the Q-Ball surfaces. In addition, surfaces subdivision and local shape adjustment have wide range of applications in shape design and shape adjustment of complex surface in engineering.

Keywords: Quartic Q-ball surface · Shape parameters
Subdivision algorithm · Coefficient subdivision

1 Introduction

The Ball curves are free-form parametric curves constructed by the Ball basis function. As the distinctive and excellent properties of Ball basis functions, they become one of the most practical methods to express the parametric curves in CAGD. In recent years, many scholars have proposed different generalized Ball curves and studied them in depth [1–5]. Their studies show that the generalized Ball curves have many nice analytic properties like Bézier curves, such as partition of unity, convex hull property, terminal property, transformation invariance and calculation stability. In addition, generalized Ball curves are far superior to the Bézier curves in computing speed of the degree elevation and reduction [4–6]. Since the curves hold these excellent properties and its widely application value in shape design, the generalized Ball curves have attracted scholars more and more attention.

In [3], Wu et al. considered that the shape adjustment of quartic Ball curve with shape parameter is superior to the quartic Bézier curve with shape parameter. At present, there are few researches on the algorithms of generalized Ball

N. Xiong et al. (Eds.): ISCSC 2018, AISC 877, pp. 144–151, 2019.
https://doi.org/10.1007/978-3-030-02116-0_19

curves with parameters, though, many scholars have proposed different generalized Ball curves in recent years [4–6]. Among them, the quartic Q-Ball curves proposed by Liu et al. [7] not only have the characteristics of cubic Ball curves, but also have excellent shape adjustability and better approximation. Therefore, it is significant to study the correlation algorithm of the this parametric curves. Fan et al. [8] first proposed subdivision algorithm of C-Bézier curves, which effectively enhanced the ability of C-Bézier methods to control and expressed curves shape. And then they popularized the subdivision of curves to surfaces, an arbitrary subdivision algorithm for C-Bézier surfaces are proposed in [9]. Tan et al. [10] proposed subdivision algorithm of cubic H-Bézier curves, which have greatly increased the flexibility of H-Bézier curves. Subsequently in 2011, an arbitrary subdivision algorithm for H-Bézier surfaces is given in [11]. Moreover, many researchers have studied the subdivision algorithms of different Bézier curves and surfaces in recent years. But no researchers studied the generalized Ball curves and surfaces subdivision algorithms. In order to improve the flexibility and ability of local adjustment of the quartic Q-Ball surfaces in this paper, here we investigate the subdivision technology of the quartic Q-Ball surfaces.

2 Quartic Q-Ball Surfaces

2.1 Definition of Ball Basis Function

Definition 1. Let $\lambda, \mu \in [-2, 4]$, for any $t \in [0, 1]$, the following polynomial functions with respect to t

$$b_{i3}(t) = \begin{cases} b_{03}(t) = (1 - \lambda t + \lambda t^2)(1-t)^2 \\ b_{13}(t) = (2 + \lambda - \lambda t)(1-t)^2 t \\ b_{23}(t) = (2 + \mu t)(1-t)t^2 \\ b_{33}(t) = (1 - \mu t + \mu t^2)t^2 \end{cases} \tag{1}$$

$b_{i3}(t)$ are called quartic Ball basis functions associated with the shape parameters λ, μ.

It can be easily proved that the quartic Ball basis functions have the following properties:

(a) Non-negativity. Let $\lambda, \mu \in [-2, 4]$, for any $t \in [0, 1]$, $b_{i3}(t) \geq 0 (i = 0, 1, 2, 3)$.
(b) Degeneracy. In particular, when the shape parameters $\lambda = \mu = 0$, the basis function are just the cubic Ball basis functions.
(c) Symmetry. When $\lambda = \mu, b_{i3}(t)(i = 0, 1, 2, 3)$, are symmetric

$$b_{03}(1-t) = b_{33}(t), b_{13}(1-t) = b_{23}(t)$$

(d) Terminal properties. $b_{i3}(t)(i = 0, 1, 2, 3)$ meet the following properties at the endpoints:

$$b_{03}(0) = b_{33}(1) = 1, b_{13}(0) = b_{23}(1) = 0$$

(e) Linear independence. For any $\lambda, \mu \in [-2, 4]$, the quartic Ball basis functions $b_{i3}(t)(i = 0, 1, 2, 3)$ are linearly independent.

2.2 Definition of the Quartic Q-Ball Surfaces

Definition 2. Given surfaces representation of the form

$$\mathbf{S}(u,v)=\sum_{i=0}^{3}\sum_{j=0}^{3}b_{j3}(u;\lambda_j,\mu_j)b_{i3}(v;\lambda_i,\mu_i)\mathbf{Q}_{ij} \tag{2}$$

$\mathbf{S}(u,v)$ are called the quartic Q-Ball surfaces, where the $\mathbf{Q}_{ij}(i,j=0,1,2,3)$ are control mesh points, and the basis functions $b_{j3}(u), b_{i3}(v)$ are defined by (1), any $(u,v)\in[0,1]\times[0,1]$, the shape parameters $\lambda_j,\mu_j,\lambda_i,\mu_i\in[-2,4]$.

From the properties of quartic Ball basis functions, the quartic Q-Ball surfaces have the following properties. And given the modeling in Fig. 1.

(a) Non-negativity. For all $\lambda_j,\mu_j,\lambda_i,\mu_i\in[-2,4]$ and(u,v), we have

$$b_{j3}(u;\lambda_j,\mu_j)b_{i3}(v;\lambda_i,\mu_i)\geq 0$$

(b) Convex hull property. The entire quartic Q-Ball surfaces $\mathbf{S}(u,v)$ are contained in the convex hull of its control points (the $\mathbf{Q}_{ij}$).
(c) Endpoint interpolation properties. The surfaces $\mathbf{S}(u,v)$ interpolates the four corner control points.

$$\mathbf{S}(0,0)=\mathbf{Q}_{00},\mathbf{S}(0,1)=\mathbf{Q}_{03}$$

$$\mathbf{S}(1,0)=\mathbf{Q}_{30},\mathbf{S}(1,1)=\mathbf{Q}_{33}$$

(d) symmetry. The quartic Q-Ball surface defined by control mesh points $\mathbf{Q}_{ij}(i,j=0,1,2,3)$ are same as the surface defined by $\mathbf{Q}_{(3-i)(3-j)}(i,j=0,1,2,3)$.
(e) Approximation property. When triangulated, the control net forms a planar polyhedral approximation to the surface.
(f) Shape adjustable properties. Fixing the control net, the shape of the quartic Q-Ball surfaces can still be adjusted by changing the shape parameters $\lambda_i,\mu_i,\lambda_j,\mu_j$

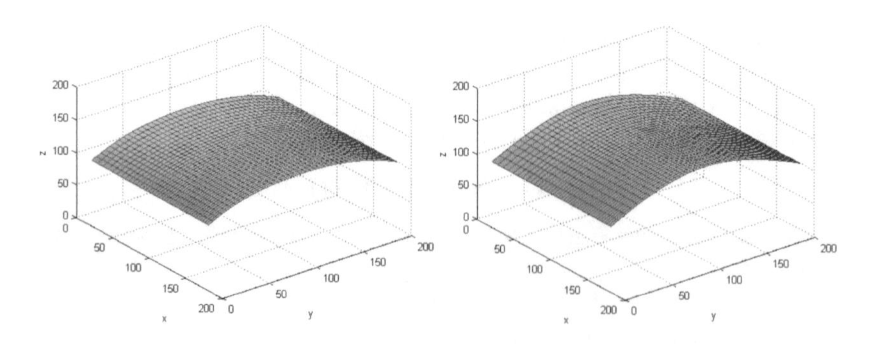

Fig. 1. The effects of the shape parameters on quartic Q-Ball surfaces

3 Coefficient Subdivision of the Quartic Q-Ball Surfaces

3.1 The Coefficient Subdivision Steps of Sub-surfaces in v Direction

Coefficient subdivision can utilize the principle of remaining the shape of the original surface unchanged and making the expression of the original surface same to the expression of the sub-surfaces, and then computing the control mesh points of the sub-surfaces by using the same coefficients of the corresponding term of the surface expression. Assume the original surfaces $\mathbf{S}(u, v)$ defined by (2), two sub-surfaces $\mathbf{S}_1(u, v)$ and $\mathbf{S}_2(u, v)$ can be obtained by dividing the original surface $\mathbf{S}(u, v)$ at any division lines $\mathbf{C}(u, v^*)$, which corresponding to the parameter v^*. And so $\mathbf{S}_1(u, v^*) = \mathbf{C}(u, v^*)$. Now, we reconstructed the quartic Q-Ball surfaces formula after subdivision.

$$\begin{aligned}\mathbf{S}_1(u, v) &= \sum_{i=0}^{3}\sum_{j=0}^{3} b_{j3}(u; \lambda_j, \mu_j) b_{i3}(v; \lambda_i^1, \mu_i^1)\mathbf{Q}_{ij}^1 \\ &= \sum_{i=0}^{3}\sum_{j=0}^{3} b_{j3}(u; \lambda_j, \mu_j) b_{i3}((v' - 0)/(v^* - 0); \lambda_i^1, \mu_i^1)\mathbf{Q}_{ij}^1 \\ &= \sum_{i=0}^{3}\sum_{j=0}^{3} b_{j3}(u; \lambda_j, \mu_j) b_{i3}(v/v^*; \lambda_i^1, \mu_i^1)\mathbf{Q}_{ij}^1\end{aligned} \quad (3)$$

Where, $0 \le v \le 1, 0 \le v' \le v^*$,and $v'' = v'/v^*$, so $0 \le v'' \le 1$. According to the principle of the coefficient subdivision and the endpoint interpolation properties of the quartic Q-Ball surfaces, we obtained the sub-surface $\mathbf{S}_1(u, v)$ meet the following condition:

$$\mathbf{S}(u, v') = \mathbf{S}_1(u, v'') \quad (4)$$

Let the coefficients of $\mathbf{S}(u, v')$ are equal to the corresponding coefficients of $\mathbf{S}_1(u, v)$, we can obtain the control points of sub-surface $\mathbf{S}_1(u, v)$ as follows:

$$\begin{cases} \mathbf{Q}_{i0}^1 = \mathbf{Q}_{i0}, \mathbf{Q}_{i3}^1 = \mathbf{Q}_i^* \\ \mathbf{Q}_{i1}^1 = \mathbf{Q}_{i0}^1 + \dfrac{(\mathbf{Q}_{i1} - \mathbf{Q}_{i0})(2 + \lambda_i)}{2 + \lambda_i^1} * v^* \\ \mathbf{Q}_{i2}^1 = \dfrac{3\lambda_i(\mathbf{Q}_{i0} - \mathbf{Q}_{i1}) + \mathbf{Q}_{i0} - 4\mathbf{Q}_{i1} + 2\mathbf{Q}_{i2} + \mathbf{Q}_{i3}}{2} * v^{*2} \\ \quad - \dfrac{3\lambda_i^1(\mathbf{Q}_{i0}^1 - \mathbf{Q}_{i1}^1) + \mathbf{Q}_{i0}^1 - 4\mathbf{Q}_{i1}^1 + \mathbf{Q}_{i3}^1}{2} \end{cases} \quad (5)$$

Where $\lambda_i, \mu_i, \mathbf{Q}_{ij}, \lambda_i^1, \mu_i^1, \mathbf{Q}_{ij}^1$ $(i, j = 0, 1, 2, 3)$are shape parameters and control mesh points of the original surface $\mathbf{S}(u, v)$ and first sub-surface $\mathbf{S}_1(u, v)$.

In order to calculate the control points of sub-surface $\mathbf{S}_2(u, v)$, we use the symmetry of the quartic Ball basis functions and the quartic Q-Ball surfaces to facilitate the process of computation. The symmetrical basis functions are given by (6) and correspondingly, the original surface transformed into the form as (7):

$$c_{i3}(t) = \begin{cases} c_{03}(t) = (1 - \mu t + \mu t^2)(1-t)^2 \\ c_{13}(t) = (2 + \mu - \mu t)(1-t)^2 t \\ c_{23}(t) = (2 + \lambda t)(1-t)t^2 \\ c_{33}(t) = (1 - \lambda t + \lambda t^2)t^2 \end{cases} \tag{6}$$

$$\mathbf{S}(u,v) = \sum_{i=0}^{3}\sum_{j=0}^{3} c_{j3}(u;\lambda_j,\mu_j)c_{i3}(v;\lambda_{3-i},\mu_{3-i})\mathbf{Q}_{(3-i)j} \tag{7}$$

Where $(u,v) \in [0,1] \times [0,1]$. Analogously, we need reconstructing the quartic Q-Ball sub-surfaces formula after subdivision.

$$\begin{aligned} \mathbf{S}_2(u,v) &= \sum_{i=0}^{3}\sum_{j=0}^{3} c_{j3}(u;\lambda_j,\mu_j)c_{i3}(v;\lambda_{3-i}^2,\mu_{3-i}^2)\mathbf{Q}_{(3-i)j}^2 \\ &= \sum_{i=0}^{3}\sum_{j=0}^{3} c_{j3}(u;\lambda_j,\mu_j)c_{i3}((v'-0)/(1-v^*-0);\lambda_{3-i}^2,\mu_{3-i}^2)\mathbf{Q}_{(3-i)j}^2 \\ &= \sum_{i=0}^{3}\sum_{j=0}^{3} c_{j3}(u;\lambda_j,\mu_j)c_{i3}(v'/(1-v^*);\lambda_{3-i}^2,\mu_{3-i}^2)\mathbf{Q}_{(3-i)j}^2 \end{aligned} \tag{8}$$

Where, $0 \le v \le 1, 0 \le v' \le 1 - v^*$, and $v'' = v'/(1-v^*)$, so $0 \le v'' \le 1$. Similarly, we have $\mathbf{S}_2(u,(1-v^*)) = \mathbf{C}(u,v^*)$ in division lines. And then using the principle of coefficient subdivision, we obtained the sub-surface $\mathbf{S}_2(u,v)$ satisfy the following condition:

$$\mathbf{S}(u,v') = \mathbf{S}_2(u,v'') \tag{9}$$

It follows that the control mesh points of $\mathbf{S}_2(u,v)$ are

$$\begin{cases} \mathbf{Q}_{i0}^2 = \mathbf{Q}_{(3-i)3}, \mathbf{Q}_{i3}^2 = \mathbf{Q}_{(3-i)}^* \\ \mathbf{Q}_{i1}^2 = \mathbf{Q}_{(3-i)0}^2 + \dfrac{(\mathbf{Q}_{(3-i)2} - \mathbf{Q}_{(3-i)3})(2+\mu_{(3-i)})}{2+\mu_i^2} * (1-v^*) \\ \mathbf{Q}_{i2}^2 = \dfrac{3\mu_{(3-i)}(\mathbf{Q}_{(3-i)3} - \mathbf{Q}_{(3-i)2}) + \mathbf{Q}_{(3-i)3} - 4\mathbf{Q}_{(3-i)2} + 2\mathbf{Q}_{(3-i)1} + \mathbf{Q}_{(3-i)0}}{2} \\ \quad * (1-v^*)^2 - \dfrac{3\lambda_i^2(\mathbf{Q}_{i0}^2 - \mathbf{Q}_{i1}^2) + \mathbf{Q}_{i0}^2 - 4\mathbf{Q}_{i1}^2 + \mathbf{Q}_{i3}^2}{2} \end{cases} \tag{10}$$

Where $\lambda_{3-i}, \mu_{3-i}, \mathbf{Q}_{(3-i)j}, \lambda_i^2, \mu_i^2, \mathbf{Q}_{ij}^2$ $(i,j = 0,1,2,3)$ are shape parameters and control mesh points of the original surface $\mathbf{S}(u,v)$ and second sub-surface $\mathbf{S}_2(u,v)$.

3.2 Coefficient Subdivision of Surfaces in u Direction

Analogously, we can obtain the expression of sub-surfaces $\mathbf{S}_3(u,v)$, which are divided in u direction. And the algorithm steps just as the above process, so here simplify proof process. Due to the expression of the sub-surfaces have the same expression before and after subdivision, it followed that

$$\mathbf{S}(u', v) = \mathbf{S}_3(u'', v) \tag{11}$$

Where, $0 \le u \le 1, 0 \le u' \le u^*$, and $u'' = u'/u^*$, so $0 \le u'' \le 1$. From the upper equation and $\mathbf{S}_3(u^*, v) = \mathbf{C}(u^*, v)$ in division lines, it can derive that

$$\begin{cases} \mathbf{Q}_{0j}^1 = \mathbf{Q}_{0j}, \mathbf{Q}_{3j}^1 = \mathbf{Q}_j^* \\ \mathbf{Q}_{1j}^1 = \mathbf{Q}_{0j}^1 + \dfrac{(\mathbf{Q}_{1j} - \mathbf{Q}_{0j})(2 + \lambda_j)}{2 + \lambda_j^1} * u^* \\ \mathbf{Q}_{2j}^1 = \dfrac{3\lambda_j(\mathbf{Q}_{0j} - \mathbf{Q}_{1j}) + \mathbf{Q}_{0j} - 4\mathbf{Q}_{1j} + 2\mathbf{Q}_{2j} + \mathbf{Q}_{3j}}{2} * u^{*2} \\ \quad - \dfrac{3\lambda_j^1(\mathbf{Q}_{0j}^1 - \mathbf{Q}_{1j}^1) + \mathbf{Q}_{0j}^1 - 4\mathbf{Q}_{1j}^1 + \mathbf{Q}_{3j}^1}{2} \end{cases} \tag{12}$$

Where $\lambda_j, \mu_j, \mathbf{Q}_{ij}, \lambda_j^1, \mu_j^1, \mathbf{Q}_{ij}^1$ $(i, j = 0, 1, 2, 3)$ are shape parameters and control mesh points of the original surface $\mathbf{S}(u, v)$ and first sub-surface $\mathbf{S}_3(u, v)$.

In this steps, the principle of the coefficient subdivision can be combined with symmetry and the endpoint interpolation properties of the quartic Q-Ball surfaces, then we can obtain the $\mathbf{S}_4((1 - u^*), v) = \mathbf{C}(u^*, v)$ in division lines, and $\mathbf{S}_4(u, v)$ satisfy the following condition:

$$\mathbf{S}(u', v) = \mathbf{S}_4(u'', v) \tag{13}$$

Where, $0 \le u \le 1, 0 \le u' \le 1 - u^*$,and $u'' = u'/(1 - u^*)$, so $0 \le u'' \le 1$. And we can derive the control mesh points of sub-surface $\mathbf{S}_4(u, v)$ from above equation, and as follows:

$$\begin{cases} \mathbf{Q}_{0j}^2 = \mathbf{Q}_{3(3-j)}, \mathbf{Q}_{3j}^2 = \mathbf{Q}_{(3-j)}^* \\ \mathbf{Q}_{1j}^2 = \mathbf{Q}_{3(3-j)}^2 + \dfrac{(\mathbf{Q}_{2(3-j)} - \mathbf{Q}_{3(3-j)})(2 + \mu_{(3-j)})}{2 + \mu_j^2} * (1 - u^*) \\ \mathbf{Q}_{2j}^2 = \dfrac{3\mu_{3-j}(\mathbf{Q}_{3(3-j)} - \mathbf{Q}_{2(3-j)}) + \mathbf{Q}_{3(3-j)} - 4\mathbf{Q}_{2(3-j)} + 2\mathbf{Q}_{1(3-j)} + \mathbf{Q}_{0(3-j)}}{2} \\ \quad * (1 - u^*)^2 - \dfrac{3\lambda_j^2(\mathbf{Q}_{0j}^2 - \mathbf{Q}_{1j}^2) + \mathbf{Q}_{0j}^2 - 4\mathbf{Q}_{1j}^2 + \mathbf{Q}_{3j}^2}{2} \end{cases} \tag{14}$$

Where $\lambda_{3-i}, \mu_{3-i}, \mathbf{Q}_{(3-i)j}, \lambda_i^2, \mu_i^2, \mathbf{Q}_{ij}^2$ $(i, j = 0, 1, 2, 3)$ are shape parameters and control mesh points of the original surface $\mathbf{S}(u, v)$ and second sub-surface $\mathbf{S}_4(u, v)$.

3.3 Modeling Examples

In this section, we carry out the coefficient subdivision in v direction and in u direction respectively in Fig. 2. There are four sub-surfaces $\mathbf{S}_1(u, v)$, $\mathbf{S}_2(u, v)$, $\mathbf{S}_3(u, v)$ and $\mathbf{S}_4(u, v)$.

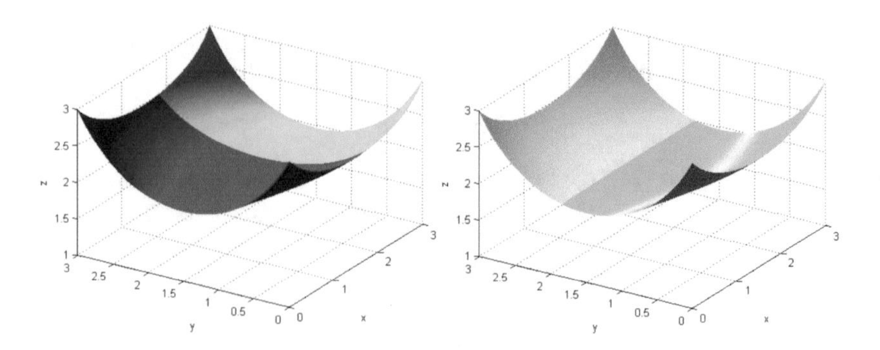

Fig. 2. The coefficient subdivision of quartic Q-Ball surfaces

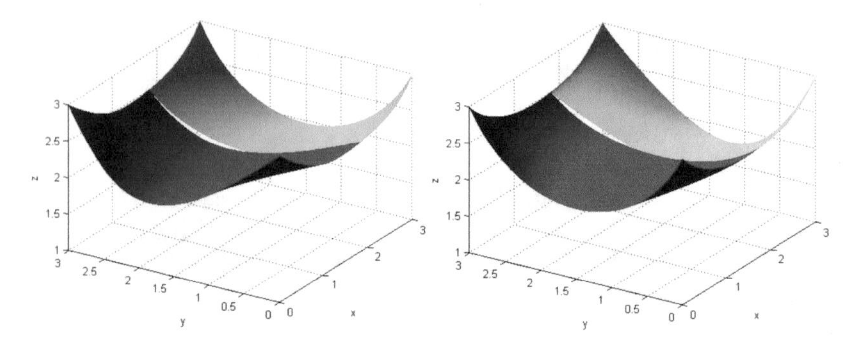

Fig. 3. Local shape adjustment of sub-surfaces $\mathbf{S}_1(u,v)$ and $\mathbf{S}_2(u,v)$

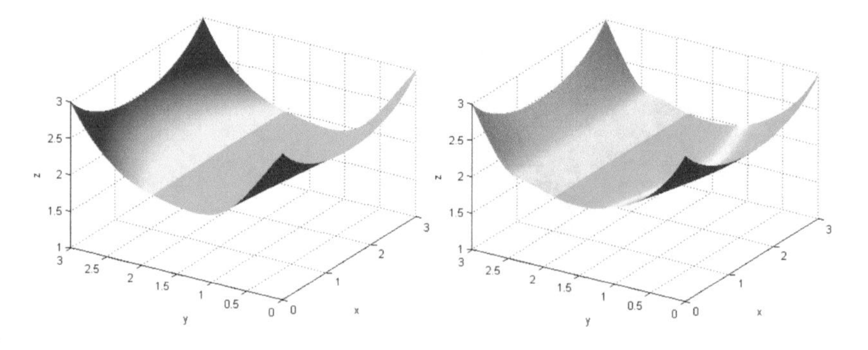

Fig. 4. Local shape adjustment of sub-surfaces $\mathbf{S}_3(u,v)$ and $\mathbf{S}_4(u,v)$

According to the examples of modeling in Fig. 2, we realize the coefficient subdivision of quartic Q-Ball surfaces. And then, the local shape adjustment of sub-surfaces $\mathbf{S}_1(u,v)$ and $\mathbf{S}_2(u,v)$ present in Fig. 3 by altering the shape parameters of sub-surfaces respectively. Besides, the shape of two sub-surfaces $\mathbf{S}_3(u,v)$ and $\mathbf{S}_4(u,v)$ in Fig. 4 can also be adjusted by changing their shape parameters. We know that when the shape parameters of sub-surfaces take different values, the shape of quartic Q-Ball surfaces can be adjusted. Therefore, we achieve the local shape adjustment of the sub-surfaces in v, u direction and do not affect the shape of other surfaces.

4 Conclusion

In this paper, we have mainly studied coefficient subdivision algorithm of quartic Q-Ball surfaces, and then the arbitrary subdivision algorithm in v direction and u direction both are presented. The modelling examples show that the presented subdivision algorithms are simple and effective in aspect of the algorithm implementation, and the shape parameters can be flexibly controlled sub-surfaces. In addition, the shape adjustment of sub-surfaces have no influence on other parts, which means the local shape adjustment of the quartic Q-Ball surfaces are realized completely.

References

1. Hu, S.M., Wang, G.J., Jin, T.G.: Properties of two types of generalized Ball curves. Comput.-Aided Des. 125–133 (1996)
2. Ding, Y.: Generalized ball curves properties and its application. Acta Math. Appl. Sin. **23**(4), 580–595 (2000)
3. Xiaoqin, W., Han, X.: Shape analysis of quartic Ball curve with shape parameter. Acta Math. Appl. Sin. **34**(04), 671–682 (2011)
4. Wang, G.J., Jiang, S.: The algorithms for evaluating two new types of generalized Ball curves/surfaces and their applications. Acta Math. Appl. Sin. **2004**(01), 52–63 (2004)
5. Wang, C.: Extension of cubic ball curve. J. Eng. Graph. **29**(1), 77–81 (2008)
6. Yan, L., Liang, J., Rao, Z.: Two new extension of cubic curve. J. Eng. Graph. **32**(05), 20–24 (2011)
7. Huayong Liu, L., Li, D.Z.: Quartic ball curve with multiple shape parameters. J. Shandong Univ. (Eng. Sci.) **41**(02), 23–28 (2011)
8. Fan, J., Yijie, W., Lin, X.: Subdivision algorithm and G1 condition for C-Bézier curves. J. Comput.-Aided Des./Comput. Graph. **05**, 421–424 (2002)
9. Fan, J., Luo, G., Wang, W.: Subdivision algorithm and G1 condition and application for C-Bézier surfaces. J. Eng. Graph. **03**, 133–138 (2002)
10. Tan, J., Wang, Y., Li, Z.: Subdivision algorithm, connection and application of cubic H-Bézier curves. J. Comput.-Aided Des./Comput. Graph. **21**(5), 584–588 (2009)
11. Zhang, J., Tan, J.: Subdivision algorithm and connection for H-Bézier surfaces. Comput. Eng. Appl. **47**(09), 152–155 (2011)

Calculation of Financing Gap of Agricultural Modernization and New-Type Urbanization Based on ARIMA Model

Shuyu Hu and Ming Huang(✉)

Hunan Radio and Television University, Changsha 410004, China
632340962@qq.com

Abstract. Along with the constant advance of agricultural modernization and new-type urbanization construction, the financing demand increases continuously, and a huge financing gap appears. This paper firstly introduces the development history of China's agricultural modernization and urbanization, forecasts the development tendency adopting the logic growth curve model, and then calculates the financing gap of agricultural modernization and new-type urbanization construction based on ARIMA model.

Keywords: Calculation of financing gap · Agricultural modernization New-Type urbanization · ARIMA model

1 Overview and Development Tendency of Agricultural Modernization and New-Type Urbanization

1.1 Overview of Agricultural Modernization

The 18th National Congress of the CPC puts forward the strategic deployment of "adhering to the new industrialization, informatization, urbanization and agricultural modernization road with Chinese characteristics … and promoting the synchronous development of industrialization, informatization, urbanization and agricultural modernization". The engine of "new four modernizations" is the new industrialization, urbanization, the power or locomotive is the informatization, and the foundation is the agricultural modernization. However, the level of China's agricultural modernization is far below China's overall modernization level. As evaluated by He Chuanqi and other experts, "in 2008, the agricultural modernization of China is about 8 years lower than its modernization level, about 10% lower".

In 2009, Wang Jingxin once referred domestic and foreign similar researches, and constructed a simple evaluation index system composed of 8 first-level indexes and 11 second-level indexes in the Rural Reform and Village-Level Economic Transition in Yangtze River Delta (China Social Sciences Publishing House, 2009), evaluated based on the 2nd agriculture census and relevant statistical data, and obtained the conclusion: in accordance with the target raised by Premier Zhou at the 4th National People's Congress of realizing the mechanization, irrigation, electrification, fertilization, variety improvement and other purposes, at the end of 2007, China's agricultural

N. Xiong et al. (Eds.): ISCSC 2018, AISC 877, pp. 152–166, 2019.
https://doi.org/10.1007/978-3-030-02116-0_20

modernization completed about 2/3, among which: the realization degree of agricultural production equipment (labor and farm machinery power) reached 61%, the mechanized production and socialized service replacing labor become the agricultural production method of many areas; as of 2005, the proportion of effective irrigation area in the total cultivated area was only 30.1%, far behind the standard value of 80% farmland irrigation rate, the realization degree only reached 37.6%; the applying quantity of chemical fertilizer reached 3786.8 ton/10,000 hectares, lower than it of Germany (3837 ton/10,000 hectares), and higher than it of the US (1173 ton/10,000 hectares) and Japan (2521 ton/10,000 hectares); the realization degree of contribution rate of agricultural technology reached 70%. The total grain output of China realized "11 consecutive growths", agricultural mechanization and contribution of agricultural technology played an important role. In addition, the realization degree of rural non-agriculturization (rural labor force non-agricultural employment and urban population proportion) was separately 50.6% and 75%; the realization degree of peasants' intellectualization (education level of rural labor, culture education expenditure ratio) was separately 58.3% and 71.9%. However, according to requirements of new-type modern agriculturalization focusing on knowledgeablization, informatization and ecologicalization, China's agricultural modernization just completed about 1/3; the realization degree of overall agricultural productivity (gross output of farming, forestry, husbandry and fishing per agricultural labor force) was 31.1%; the realization degree of peasant living quality (peasant per capita net income) just reached 39.7%; the side effect of agricultural modernization - environmental pollution and ecological damage was not changed materially, the food safety, quality of agricultural product and food risk still confused and tortured us from time to time.

1.2 Overview of New-Type Urbanization

1.2.1 Development History of China's Urbanization

Since the establishment of the People's Republic of China, China's urbanization development has changed for several times, from the slow development at early days of new China to rapid development after the reform and opening-up. The division of China's urbanization course for over 60 years has been a hot spot and difficulty all the time. According to different bases, China's urbanization course can be divided into different stages, such differences are neither superior nor inferior to each other, and every kind of division is valuable for knowing and researching China's urbanization.

Considering the division of China's urbanization course should neither be too coarse not too fine, in this paper, the urbanization course of China for over 60 years is divided into three stages: the slow starting stage from early days of new China to Chinese economic reform, the pluralistic development stage from Chinese economic reform to the end of the 20th century, and the rapid development stage since the 21st century.

(1) The slow starting stage from early days of new China to Chinese economic reform (1949–1978). Early of New China, the urbanization rate was just 10.64%. After three years of recovery, China entered its "first five-year plan", starting its large scale industrialization construction and urban construction. During this period,

China adopted the "key developing" urban development policy, the urbanization had been promoted steadily, the urbanization improved from 10.6% to 15.4% during 1949–1957, with an annual growth of 0.6%. However, it was followed by the turbulent "Great Leap Forward" and "Great Cultural Revolution", which made the urbanization construction remain stagnant and even become retrogressive. In 1962, China's urbanization rate was 17.33%, while in 1978, China's urbanization rate was just 17.92%, during 16 years, China's urbanization rate just improved 0.89%, the number of cities increased from 132 to 193, only 61 new cities emerged during this period. This fact is resulted from China's choice of focusing on industrialization, anxiety for success, and low urbanization level starting points (Table 1).

Table 1. 1949–1978 China's urbanization course

Year	National population	Urban population	Urbanization rate	Yearly urbanization rate improvement percentage [unit: 10,000 persons/%]
1949	54167	5765	10.64	–
1950	55196	6169	11.18	0.54
1955	61465	8285	13.48	0.46
1960	66207	13073	19.75	1.254
1965	72538	13046	17.99	−0.352
1970	82992	14424	17.38	−0.122
1978	96259	17245	17.92	0.108

Data source: China statistical yearbook

(2) The pluralistic development stage from Chinese economic reform to the end of the 20th century (1978–2000). From 1978 to 2000, China's urbanization level was improved from 17.92% to 36.22%, with 0.83% of annual growth, the number of cities increased from 193 to 636, and the number of organic towns increased from 2173 to 20312. After Chinese economic reform, China's urbanization course obviously accelerates, the first reason is that the work priority of "focusing on economic construction" is defined; the second reason is that the industrialization strategy is transformed to focus on light textile industry; the third reason is that China vigorously implements the development strategy of driving the region by the city, the national policy supports the development of the eastern costal region, big coastal cities become the key of urbanization development, many small and medium cities as well as small towns spring up in the eastern coastal region. The urbanization development of this period is characterized by restorative, diversity and acceleration.

(3) The rapid development stage since the 21st century (2000 up to now). After entering the 21st century, the overall strategy for accelerating the urbanization development is formally formulated. Since the Sixteenth National Congress of the Communist Party of China, China's urbanization has developed rapidly. From 2002 to 2011, China's urbanization rate had developed at the speed of annual 1.35%, and the urban population had increased 209610,000 persons every year. In

2013, the urban population proportion reached 53.73%, 14.64% higher than it in 2002; the urban population was 7311,110,000 persons, 2289,910,000 persons more than 2002. In 2014, China's urbanization rate reached 54.77%. It is estimated that in 2020, China's urbanization rate will reach 60%.

1.2.2 Connotation and Meaning of New-Type Urbanization

The concept of new-type urbanization is corresponding to the urbanization. It is different from the previous urbanization mode with low cost and rapid expansion, the new-type urbanization is the connotative and benefit-oriented growth paying more attention to the quality and efficiency, emphasizing "urban and rural overall development, urban-rural integration, industry-urban interaction, energy saving & intensification, pleasant living environment, and harmonious development". The core of new-type urbanization is the urbanization of "people". The urbanization is to transform rural population to towns, realize the change from peasants to citizens, not just to create plazas and build tall building; the key it to realize the equalization of public service and integration of urban and rural infrastructure, promote the rapid economic development, and realize the common prosperity not at the expense of sacrificing ecological environment and agriculture.

The connotation of new-type urbanization can be divided into four "new", namely "new urban region", "new citizen", "new agriculture" and "new industry". While comparing with the new-type urbanization, the traditional urbanization can be summarized as one "new", namely the construction of new cities and towns. The transformation from the traditional urbanization to the new-type urbanization also refers to the transformation from one "new" to four "new".

The "new urban region" mainly refers to that during the urbanization construction, the construction of newly increased urban area and infrastructure construction surrounding the old town and shantytowns transformation. To be specific, the construction includes the infrastructure construction of power supply, water supply, gas, road and sewage treatment, as well as land comprehensive improvement and industrial park. The "new citizen" refers to the population who move from villages to cities and transfer to be citizen during the new-type urbanization construction that enjoys infrastructures and fundamental public services same to original residents of cities and towns. The "new agriculture" refers to the "large-scale agriculture" formed because the agricultural production develops gradually to the direction of modernization and intensification along with the new-type urbanization construction. The "new industry" refers to the new-type urbanization construction promotes the course of rural industrialization and drives the development of some middle and small-sized enterprises and township enterprises.

Currently, domestic economic growth slows, overseas market demand decreases and domestic demand is sufficient. Facing difficulties of economic development, the government hopes to simulate the domestic demand by promoting the new-type urbanization construction, further to promote the economic development. On the one hand, the new-type urbanization construction drives the investment demand of the industrial engineering and service industry via the town infrastructure and property construction; on the other hand, the transformation from peasant to citizen forms a more wide consumption market, the total demand increases and promotes comprehensive economic development relying on the multiplier effect.

1.3 Tendency of New-Type Urbanization Development

China's urbanization rate reached 54.77% in 2014. It is pointed by National new-Type Urbanization Planning (2014–2020) that in 2020, the permanent population urbanization rate will reach about 60%, and the registered population urbanization rate will reach about 45%, and we will strive to realize about 0.1 million agricultural population transfer and other permanent population settle in cities and towns. American urban geographer (Ray M. Northam) finds out the staged development law of urbanization is not up straight and can be summarized as one elongated S-form curve, through observing the overall tendency of urbanization development of various countries in the world since 1800.

Combining with the universal law of world urbanization development, it is indicated that China is at the middle stage of urbanization development, and has entered the critical development period. In the future years, China's urbanization development level and speed can be estimated with the logical growth curve model.

The growing development of biological identities and the development of some technical and economic characteristics experience three stages. In the occurrence stage, the change speed is slow; in the development stage, the change speed accelerates; to the mature stage, the change speed tends to be low again. The object change development curve obtained as per the development law is called as the growth curve or logical growth curve, it is also names as S curve because its shape is similar to S,

(1) The generic curve of Pier growth curve is:

$$y = \frac{K}{1 + e^{f(x)}} \tag{1}$$

$$f(x) = a_0 + a_1x + a_2x^2 + \cdots + a_nx^n \tag{2}$$

Thereinto, K is the constant.

The common form of Pier growth curve model:

$$y = \frac{K}{1 + be^{-ax}}(a > 0, b > 0) \tag{3}$$

The MATLAB nonlinear regression order method can be used to solve the unknown constant a, b and K.

(2) The common form of Lindnord growth curve model (Table 2):

$$N(t) = \frac{L}{1 + \left(\frac{L}{N} - 1\right)e^{-rt}}(t \geq 0) \tag{4}$$

Table 2. China's urbanization rate during 1995–2016

Year	Urbanization rate	Year	Urbanization rate [Unit: year/%]
1995	29.04	2006	43.90
1996	29.37	2007	44.94
1997	29.92	2008	45.68
1998	30.4	2009	46.59
1999	30.89	2010	49.68
2000	36.22	2011	51.27
2001	37.66	2012	52.57
2002	39.09	2013	53.73
2003	40.53	2014	54.77
2004	41.76	2015	56.1
2005	42.99	2016	57.35

Data source: China statistical yearbook

Through the MATLAB programming (appendix), the Pier growth curve model is established as shown below:

$$y = \frac{75.11}{1 + 1.8462 \quad e^{-0.0798 \quad x}} \tag{5}$$

The forecast result is obtained as below (Table 3):

Table 3. 2015–2020 China's urbanization rate forecast

Year	Urbanization rate	Year	Urbanization rate [unit: year/%]
2015	56.1	2021	61.87
2016	57.35	2022	62.71
2017	58.02	2023	63.52
2018	59.05	2024	64.28
2019	60.03	2025	64.99
2020	60.97	–	–

It is estimated that China's urbanization rate will reach 60.97% in 2020, which is similar to the target that the permanent population urbanization rate will reach about 60% in 2020 as stipulated in the National New-Type Urbanization Planning (2014–2020), indicating that the model has a good simulating effect, and the estimated urbanization level is accurate (Fig. 1).

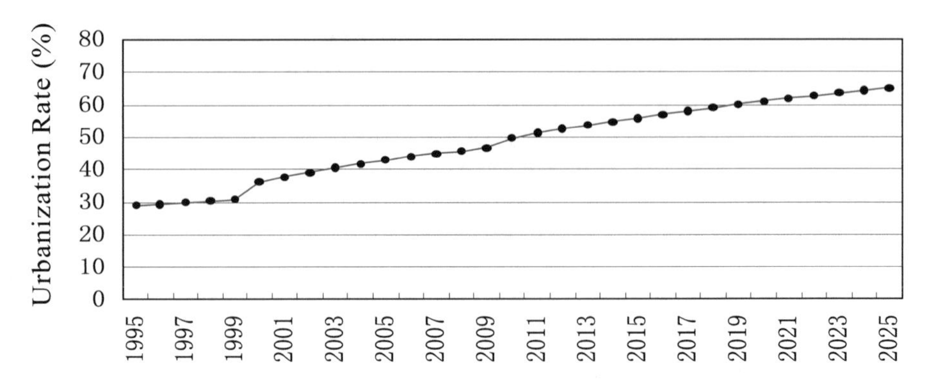

Fig. 1. China's urbanization development tendency

2 Calculation of Construction Financing Gap

2.1 Introduction of ARIMA Model

The ARIMA method is one common and effective method regarding the time series forecast. It explains the variable Yt using variable random error and its own time-delay terms, different from the method using k external variables, X1, X2.... Xk, to example the general regression model of Yt. ARIMA method is widely used in economic and financial field forecasts, under the condition that the data mode is unknown, a proper data survey model will be found.

2.1.1 ARMA Model

If the model owns the structure as shown below, this model will be called as the autoregressive movement average model, namely ARMA (p, q):

$$\begin{cases} x_t = \phi_1 x_{t-1} + \cdots + \phi_p x_{t-p} + \varepsilon_t - \theta_1 \varepsilon_{t-1} - \theta_q \varepsilon_{t-q} \\ \phi_p \neq 0, \ \theta_q \neq 0 \\ E(\varepsilon_t) = 0, Var(\varepsilon_t) = \sigma_\varepsilon^2, E(\varepsilon_t \varepsilon_s) = 0, t \neq s \\ E(x_s \varepsilon_t) = 0, \forall s < t \end{cases} \tag{6}$$

The delay operator is introduced, the ARMA (p, q) model is recorded briefly as $\Phi(B)x_t = \Theta(B)\varepsilon_t$.

Thereinto, $\Phi(B) = 1 - \phi_1 B - \cdots - \phi_p B^p$ is the autoregression coefficient multinomial of p order.

$\Theta(B) = 1 - \theta_1 B - \cdots \theta_q B^q$ is the movement average coefficient multinomial of q order.

It can been seen that AR (p) model and MA (q) model separately refer to special ARM A(p, q) model examples when p = 0 and q = 0.

The stationarity condition of model is: the root of $\Phi(B) = 0$ is beyond the unit round, namely it must be larger than 1; and the invertibility condition is that: the root of $\Theta(B) = 0$ is beyond the unit round, namely it must be larger than 1. The stationarity of ARMA (p, q) model is decided by AR section, and the invertibility is decided by MA section. When the root of $\Phi(B) = 0, \Theta(B) = 0$ is beyond the unit round, the ARMA (p, q) model is stationary and invertible.

In order to define the ARMA model order and recognize the model, firstly, the autocorrelation coefficient and partial autocorrelation coefficient should be calculated, and the fitting is carried out as per natures expressed by them. The basic principle is as shown below:

$\hat{\rho}_k$	$\hat{\phi}_{kk}$	Model order determination
Trailing	P order truncation	AR (p) model
Q order truncation	Trailing	MA (q) model
Trailing	Trailing	ARMA (p, q) model

2.1.2 ARIMA Model

It is difficult to forecast non-stationary series. The ARIMA model raised by Box-Jenkins can effectively solve the forecast of non-stationary series. After the difference calculation, may time series will become stationary series, fit stationary series receiving the difference calculation according to the ARMA model will obtain the ARIMA model, and the structure is as shown below:

$$\begin{cases} \Phi(B)\nabla^d x_t = \Theta(B)\varepsilon_t \\ E(\varepsilon_t) = 0, Var(\varepsilon_t) = \sigma_\varepsilon^2, E(\varepsilon_t\varepsilon_s) = 0, t \neq s \\ E(x_s\varepsilon_t) = 0, \forall s < t \end{cases} \tag{7}$$

Thereinto, $\nabla^d = (1 - B)^d$; ARIMA can be recorded briefly as: $\nabla^d x^t = \frac{\Phi(B)}{\Theta(B)}\varepsilon_t$

ε_t is the zero mean value, white noise series.

It is observed that ARIMA model is the combination of difference calculation and ARMA model.

2.2 Forecast of Construction Financing Demand

The construction of agricultural modernization and new-type urbanization is a system engineering needing huge capital investment. The agricultural modernization is combined with the new-type urbanization usually; therefore, the new-type urbanization is used as the example to analyze the construction financing demand. Along with the change of external conditions and internal motivation of agricultural modernization and urbanization, as well as increase of urban population and a large number of external population rushing into the city, more and higher fund demands are put forward on the construction of urban public utilities, infrastructure and social security system. A large number of investments are required for remolding laggard conditions of existing urban infrastructures, huge funds are needed for new planned cities and cities upgraded from towns, and capitals are necessary as the guarantee for solving “three rural” problems during the urbanization process.

The construction of agricultural modernization and new-type urbanization is a complex systematic engineering, involving all aspects of the society. There are a lot of factor affecting the financing demand for the construction of agricultural modernization and new-type urbanization, difficult to be estimated. In this paper, we use the predicted value of urban fixed assets investment to express the financing demand of construction of agricultural modernization and new-type urbanization.

China's urban fixed assets investment amount during 1995–2014 is as shown below. We can find that the urban fixed assets investment amount increase every year, and its growth tendency is stronger and stronger. There are a lot of factors affecting the urban fixed assets investment, such as the population increase, economic foundation, environment, resources and various factors which influence its composition, and the relationship among these factors is very complicated; therefore, it is difficult to analyze and forecast the urban fixed assets investment applying the structural causal model. We can try to makes the urban fixed assets investment experienced as a time series, to analyze the change law, further to establish the forecast model (Table 4).

Table 4. China's urban fixed assets investment amount during 1995–2016

Year	Urban fixed assets investment	Year	Urban fixed assets investment [unit: 0.1 billion Yuan]
1995	15, 643.70	2005	75, 095.10
1996	17, 567.20	2007	117, 464.47
1997	19, 194.20	2008	148, 738.30
1998	22, 491.40	2009	193, 920.39
1999	23, 732.00	2010	243, 797.79
2000	26, 221.80	2011	302, 396.06
2001	30, 001.24	2012	364, 854.15
2002	35, 488.76	2013	435, 747.43
2003	45, 811.70	2014	502, 004.90
2004	59, 028.19	2015	551, 590
2006	93, 368.68	2016	596, 501

Data source: State statistics bureau website

Through the stationary test of urban fixed assets investment, no time stationary series is found, and the integration order is 2. The ARIMA model identification is carried out according to the autocorrelation coefficient and partial autocorrelation coefficient figure of the urban fixed assets investment data, to determine ARIMA ((1, 2), 2, 2). The urban fixed assets investment amount is shown as X, the difference calculation is carried out against the data during 1995–2013 twice, the order 0 integration variable DDX is obtained; among which, DDX represents that the difference calculation is carried out against X twice, DX represents that the difference calculation is carried out against X once. The ARIMA ((1, 2), 2, 2) model can be fitted applying Eviews, as shown below.

$$\mathrm{DDXt} = 7308.7 - 0.2857 * \mathrm{DDXt} - 1 + 0.8532 * \mathrm{DDXt} - 2 - 0.9803 * \varepsilon \mathrm{t} - 2 \quad (8)$$

As shown by the above model, the urban fixed assets investment amount during 2014–2025 can be gradually estimated, and the forecast result is as shown in Table 5.

Table 5. Forecast of urban fixed assets investment amount during 2014–2025

Year	Urban fixed assets investment	Year	Urban fixed assets investment [unit: 0.1 billion Yuan]
2014	503598.19	2020	1104515.81
2015	587425.71	2021	1232395.00
2016	675205.14	2022	1362570.28
2017	773567.43	2023	1505371.01
2018	875438.82	2024	1649684.44
2019	988498.12	2025	1807498.99

Among which, the predicted value of urban fixed assets investment in 2014 is 50359.819 billion Yuan, which is close to the historical data of 2014 for 50200.490 billion Yuan as published by the State Statistics Bureau, indicating that the model has a good simulating effect, and the forecast of the State Budgetary Appropriation on the urban fixed assets investment is accurate.

In Table 6, the estimated urban fixed assets investment amount is the aggregate-value, which yearly increase amount is calculated as shown in Table 6.

Table 6. Forecast of 2015–2025 urban fixed assets investment amount absolute

Year	Urban fixed assets investment	Year	Urban fixed assets investment [unit: 0.1 billion Yuan]
2015	85420.71	2021	130175.28
2016	87779.43	2022	142800.73
2017	98362.30	2023	144313.44
2018	101871.39	2024	157814.55
2019	113059.29	2025	602510.81
2020	127879.19	Total	1305493.99

As shown in Table 6, the urban fixed assets investment demand during 2015–2020, namely the new-type urbanization financing demand forecast, is 60251.081 billion Yuan, and the financing demand forecast during 2015–2025 is 130549.399 billion Yuan.

2.3 Forecasting of Urban Fixed Assets Investment Supply

The urban fixed assets investment mainly has five capital sources: state budgetary appropriation, foreign investment, domestic loan, self-raised fund and other funds. Through the stationary test of urban fixed assets investment supply data from various channels, no time stationary series is found, and the integration order of every series is as shown below (Table 7).

Table 7. Urban fixed assets investment fund

Variable name	Fixed assets investment (excluding peasant household) state budgetary appropriation	Fixed assets investment (excluding peasant household) domestic loan	Fixed assets investment (excluding peasant household) foreign investment	Fixed assets investment (excluding peasant household) self-raised fund	Fixed assets investment (excluding peasant household) other funds
Variable code	X1	X2	X3	X4	X5
Integration order	1	1	0	2	1

Note: Since 2011, the urban fixed assets investment data release has been changed to be fixed assets investment (excluding peasant household), fixed assets investment (excluding peasant household) is equal to the original urban fixed assets investment plus project investment of rural enterprise organizations.

In this paper, the integration order of five fund supply data will be used to confirm the proper ARIMA model for the forecast. The difference calculation is carried out against variable X1, X2, X4 and X5, and we will separately obtain the 0 order integration variable DX1, DX2, DDX4 and DX5 will be obtained. Substitute DX1, DX2, X3, DDX4 and DX5 into the model equation, 5 ARIMA models will be fitted, which can be used to separately forecast the fund supply of various fund channels, finally, the fund supply forecast of five channels is summarized, and then the forecast result of the urban fixed assets investment fund supply is obtained (Table 8).

Table 8. Selection of urban fixed assets investment supply forecasting model

Variable name	Fixed assets investment (excluding peasant household) state budgetary appropriation	Fixed assets investment (excluding peasant household) domestic loan	Fixed assets investment (excluding peasant household) foreign investment	Fixed assets investment (excluding peasant household) self-raised fund	Fixed assets investment (excluding peasant household) other funds
Variable code	X1	X2	X3	X4	X5
Model selection	Arima (0, 1, 3)	Arima (3, 1, 0)	Arma (1, 1)	Arima (1, 2, (2, 3))	Arima (3, 1, 4)

According to the division of capital source channels, main channels and capital limit of urban fixed assets investment capital channels during 1995–2014 is as shown below (Table 9).

Table 9. Summary of capital source funds for urban fixed assets investment

Year	State budgetary appropriation	Domestic loan	Foreign investment	Self-raised fund	Other funds [unit: 0.1 billion Yuan]
1995	568.98	3511.86	2114.05	7940.79	2013.66
1996	576.4	3903.19	2475.6	7748.18	3308.9
1997	631.68	4136.68	2424.49	8722.33	3597.73
1998	1108.72	4918.03	2377.89	9885.48	4512.13
1999	1613.81	5249.8	1832.21	10042.86	4893.13
2000	1795.02	6245.82	1526.15	11227.52	5619.95
2001	2261.74	6672.49	1570.5	13708.53	6561.44
2002	2750.81	8167.51	1825.83	16567.7	7723.91
2003	2360.14	11223.89	2211.7	23617.35	9448.21
2004	2855.6	12842.9	2706.59	32196.07	12514.5
2005	3637.87	15363.86	3386.41	44154.51	14369.67
2006	4438.74	18814.82	3811.05	56547.51	18147.02
2007	5464.13	22136.08	4549.02	74520.88	24073.34
2008	7377.01	25466.01	4695.79	97846.45	23194.42
2009	11493.63	37634.14	3983.55	127557.67	38117.69
2010	13104.67	45104.7	4339.64	165751.97	44823.61
2011	14843.29	46 034.83	5061.99	220860.23	50094.76
2012	18958.66	51292.37	4468.78	268560.22	56555.02
2013	22305.26	59056.31	4319.44	324431.5	70953.34
2014	25, 410.89	64, 092.24	4, 042.09	370, 015.99	67, 271.94

Data source: State statistics bureau website

Take the state budgetary appropriation forecast as an example: the state budgetary appropriation is expressed by X1, the difference calculation is done against X1 data during 1995–2013 once, and then the single integer variable DX1 of stage 0 is obtained. The ARIMA (0, 1, 3) model can be fitted by applying the Eviews, as shown below.

$$DX1t = 1441.342 + 0.8746 * \varepsilon t - 3 \quad (9)$$

According to the above mentioned model, the state budgetary appropriation fund supply for urban fixed assets investment during 2014–2025 is estimated gradually. The forecast result is as shown in Table 10.

Table 10. Forecast of state budgetary appropriation fund supply for urban fixed assets investment

Year	State budgetary appropriation	Year	State budgetary appropriation [unit: 0.1 billion Yuan]
2014	25452.47	2020	34100.52
2015	26893.81	2021	35541.87
2016	28335.15	2022	36983.21
2017	29776.50	2023	38424.55
2018	31217.84	2024	39865.89
2019	32659.18	2025	41307.24

Among which, the state budgetary appropriation forecast of urban fixed assets investment in 2014 is estimated as 2545.247 billion Yuan, which is extremely similar to the historical data of 2541.089 billion Yuan of 2014 as published by the State Statistics Bureau, indicating that the model has a good simulating effect, and the forecast of the State Budgetary Appropriation on the urban fixed assets investment is accurate.

As for four aspects of the urban fixed assets investment fund source, separately foreign investment, domestic loan, self-raised and other funds, the forecast is carried out adopting the same method with the state budgetary appropriation, the prediction result of five aspects of the fund supply is summarized, the prediction result of the urban fixed assets investment supply is obtained (Table 11).

Table 11. Urban fixed assets investment supply forecasts of all channels

Year	Fixed assets investment (excluding peasant household) state budgetary appropriation	Fixed assets investment (excluding peasant household) domestic loan	Fixed assets investment (excluding peasant household) foreign investment	Fixed assets investment (excluding peasant household) self-raised fund	Fixed assets investment (excluding peasant household) other funds	Total [unit: 0.1 billion Yuan]
2015	26893.81	68847.48	4051.78	442971.41	70990.73	613755.21
2016	28335.15	73275.14	4064.78	496646.91	76182.05	678504.04
2017	29776.50	77732.64	4076.00	553454.56	81327.26	746366.95
2018	31217.84	82188.63	4085.68	613394.36	86481.04	817367.55
2019	32659.18	86649.87	4094.04	676466.31	91683.16	891552.56
2020	34100.52	91126.47	4101.26	742670.40	96863.75	968862.40
2021	35541.87	95602.29	4107.50	812006.64	102048.34	1049306.64
2022	36983.21	100080.82	4112.88	884475.03	107255.51	1132907.45
2023	38424.55	104567.26	4117.52	960075.56	112452.62	1219637.52
2024	39865.89	109053.30	4121.53	1038808.25	117651.60	1309500.58
2025	41307.24	113540.74	4125.00	1120673.08	122861.12	1402507.17

The urban fixed assets investment and domestic budget fund, domestic loan, foreign capital, self-raised fund and other capital, as fixed assets investment's source, are the accumulated data. Their respective yearly increase amount is calculated, as shown in Table 12.

Table 12. Urban fixed assets investment supply forecasting absolute

Year	2015	2016	2017	2018	2019	2020 [unit: 0.1 billion Yuan]
Urban fixed assets investment supply	82922.06	64748.82	67862.92	71000.59	74185.02	77309.84
Year	2021	2022	2023	2024	2025	Total
Urban fixed assets investment supply	80444.23	83600.81	86730.07	89863.06	93006.59	871674.02

Add the fixed assets investment fund supply of all channels, and then we will find that the total supply of urban fixed assets investment during 2015–2020 is estimated as 49649.56743 Billion Yuan, and the urban fixed assets investment supply during 2015–2020 is estimated as 110034.87 Billion Yuan.

2.4 Calculation of Gap of Supply and Demand of Construction

According to the estimated value of financing demand and supply of agricultural modernization and new-type urbanization construction during 2015–2025, the financing gap is as shown in Table 13.

Table 13. Forecast of financing gap of agricultural modernization and new-type urbanization construction during 2015–2025

Year	Urban fixed assets investment amount	Urban fixed assets investment supply	Financing gap [unit: 0.1 billion Yuan]
2015	85420.71	82922.06	2498.65
2016	87779.43	64748.82	23030.61
2017	98362.30	67862.92	30499.38
2018	101871.39	71000.59	30870.79
2019	113059.29	74185.02	38874.28
2020	116017.70	77309.84	38707.85
Total of 2015–2020	602510.81	438029.25	164481.56
2021	127879.19	80444.23	47434.96
2022	130175.28	83600.81	46574.47
2023	142800.73	86730.07	56070.65
2024	144313.44	89863.06	54450.38
2025	157814.55	93006.59	64807.95
Total of 2015–2025	1305493.99	871674.02	433819.97

As shown by the table above, it is estimated that the financing gap of agricultural modernization and new-type urbanization during 2015–2020 will be 16448.156 billion Yuan, and the financing gap during 2015–2020 will be 43381.997 billion Yuan.

Funds. The research is supported by Hunan Provincial Social Science Achievement Review Committee General Subjects (XSP17YBZC052) and Hunan Provincial Department of Education Scientific Research Projects (15C0932).

References

1. Stopher, P.R.: Financing urban rail projects: the case of Los Angeles. Transp. (20), 229–250 (1993)
2. Kim, K.-H.: Housing finance and urban infrastructure finance. Urban Stud. **34**, 1597–1630 (1997)
3. Cho, S.H., Wu, J., Boggess, W.G.: Measuring interactions among urbanization, land use regulations, and public finance. Am. Agric. Econ. Assoc. **85**(4), 988–999 (2003)
4. Chang, M.: Urban water investment and financing in China. Water **21**(10), 14–18 (2004)
5. Koch, C., Buser, M.: Emerging met a governance as an institutional framework for public private partnership networks in Denmark. Int. J. Project Manage. **24**(7), 548–556 (2006)

The Discussion and Prospect on Key Problems for Vision Accurate Inspection of Wear Debris

Guicai Wang(✉) and Donglin Li

HeNan University of Technology, Zhengzhou, China
wangguicai@tom.com

Abstract. Due that the wear debris is the critical information carrier and wear mechanism criterion in frictional and wear process. The status and attendant problems for wear debris vision inspection technology have been researched in analysis of the paper. On the basis combination of wear debris requirement and the practical situation, the paper has taken the research wear debris real-time inspection problem as foundation and has made it as target to around existed vision methods of wear debris are not effectively solved the three key problems: diversity and time variability of image features, uncertainty and complexity species classification, multi-restrictions and accuracy of wear debris density auto-counting. The objective of the paper is to establish a new kind theory and technique based on the vision accurate inspection by deeply research the vision inspection theory for wear debris.

Keywords: Wear debris · Image analysis · Accurate inspection

1 Introduction

Wear is the main failure form of mechanical equipment, abrasive wear in the process plays an indispensable role [1]. Tribological study shows that wear debris' morphology (including shape, surface texture, color, etc.), color and size etc. has close relationship with the types of wear and wear process. Meanwhile, there is a direct relationship between the wear debris density and wear degree and rate of friction surface. In addition, with the development of computer and image processing technology and color images contain more information than gray images are widely so that can be applied to wear debris Thus, it is necessary that accurate inspection of wear debris [2]. Roylance [3] put forward using machine vision and pattern recognition technology to wear debris inspection and develop computer aided vision engineer system, which has the advantages of high accuracy, small amount of labor and results visualization. It has opened up a new way for wear debris inspection and classification. It is of great significance to the analysis of wear debris. Due to wear debris existing outside layer of abrasive, vision inspection is more suitable for wear debris. For these reasons, vision inspection is a research focus and main technological means research field of wear debris inspection. Deepen research on wear debris feature extraction, classification and density estimation has important practical significance to shorten analysis time, improve recognition rate and lighten the workload in [4, 5]. The vision inspection

N. Xiong et al. (Eds.): ISCSC 2018, AISC 877, pp. 167–176, 2019.
https://doi.org/10.1007/978-3-030-02116-0_21

method of wear debris is divided into feature extraction, classification and density estimation research areas as shown in Fig. 1.

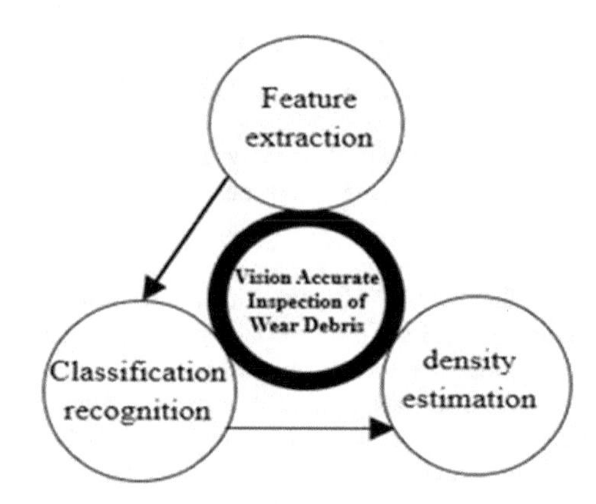

Fig. 1. The frame of wear debris vision inspection method

2 Research Status

2.1 Wear Debris Image Feature Extraction

Wear debris image features is a prerequisite for vision inspection of wear debris. The research on image feature extraction technology is imperative. Hu [6] uses two kinds of filtering method to improve image quality for further target segmentation and extraction. The local texture feature has clear pixel, edge and corner. Those can obtain a certain geometric invariance. As a result, it can get more meaningful expression at a small cost. The different types parameters of wear debris may contain the same or similar features information, namely have information redundancy. Xing [7] and Tang [8] combine with Fourier Transform Wavelet analysis, fractal theory and other modern mathematical analysis theory and applying to automatic analysis technology of wear debris. Based on extracting of grain shape and size, scholars effort to mine features information of wear debris and put forward the new features of wear debris, such as the fractal dimensions of the grain edge and surface texture fractal features [9–12].

Wear debris color information contains composition and abrading section. With the development of image analysis technology of wear debris, scholars gradually use color image processing method to implement grain feature extraction [13, 14]. Such as Wu [15] extracts shape, color, texture and fractal features of wear debris and forms a set of characteristic parameters description system of wear debris.

2.2 Wear Debris Recognition

Wear debris accurate inspection belongs to image object recognition category. Accurate recognition is the core content of wear debris inspection [16]. Yeung [17] composes of image acquisition system with microscope and camera, and then develops computer aided expert analysis system of large particle of aircraft engine filter. Stachowiak [18, 19] devote to research on grain classification and feature recognition and use wear test samples to establish feature database for wear debris automatic recognition and classification. Wang [20] apply morphology theory to grain feature extraction and use color and geometric features to develop a set of wear debris classification and identification parameters. The results shows that process the typical wear debris image have good results. According to the shortcoming of less 2D shape information of wear debris, Yuan [21, 22] proposes a new method of 3D reconstructing wear debris and implement grain recognition. Li [23] adopts D-S method of information fusion evidence to recognition wear debris atlas and typical aero engine wear debris. The simulation results show recognition correct rate reach more than 90%. Peng [24] realizes wear debris recognition analysis and decision by grey system theory and can improve wear debris recognition efficiency. However the analysis process is complicated. Laghari [25] use neural network to automatic identification wear debris. Gu [26] extract their morphologic characters from wear debris samples and are taken as the input of SVM wear particle classifier including roundness, slightness, scatter and concavity. Then judge the sliding, cutting, normal, fatigue erosion wear are taken as the output of SVM wear debris classifier. Finally the paper realizes recognition and classification of wear debris.

Information redundancy can not only reduce the efficiency of information description and wear debris recognition but also have different results for some feature of grain and even the opposite impression. Those unfavorable factors greatly reduce the accuracy of grain recognition [27]. Therefore, it is necessary to optimize the parameters, solve the uncertainty problem of multiple parameters and improve the accuracy of wear debris recognition.

2.3 Estimation of Wear Debris

Wear loss is the accumulation of long-term wear process. Single grain represents some transient wear state. The commonality of multiple grains represents current wear state, so the number estimate of grains is especially important.

Considering that there is a certain error in the particle and actual distribution of wear debris, the density of abrasive particles is defined as the number of abrasive particles in the area. Obviously, density estimation of wear debris is defined as the number of grain in the measurement area. The density obvious is an important basis to judge the quality of the wear debris. Wear debris image segmentation is the first step of counting method for wear.

Wear Debris/Background Segmentation. Wear debris image is obtained with optical microscope using transmitted illumination and indirect illumination. Due to uneven illumination, small depth of field and reflection, the actual point light source may lead to contour blur and distributing uneven background brightness of wear debris Image. Therefore, Chen [28] uses adaptive fuzzy threshold segmentation method and 2D maximum entropy based on genetic algorithm to solve the problem of omit over light or over shading part of traditional wear debris/background image segmentation method. Chen [29] suggests using RGB or HSV color space and Fuzzy Clustering technique to segment the foreground and background regions of wear debris image as much as possible. The Reference [30] research results show that the color space can obtain more accurate segmentation results than the gray threshold method. Fu [31] segment wear debris/background image by 2D entropy and restore RGB color with its position. According to regional difference, the automatic threshold method is used to segment wear debris/background image in Jiang [32]. Xing [7] implement multi-threshold method of wear debris/background image segmentation through wavelet transform for each histogram of each color channel.

Wear Debris Segmentation. Many factors lead to adjacent to each other or some overlap of the wear debris, furthermore single wear debris morphology (shape, color and texture) is random and complex. The wear debris segmentation becomes a difficult problem. Wang [20] respectively uses automatic threshold method segment wear debris/background and wear debris/wear debris in RGB component image. Li [33] adopts corrosion, expansion and Laplace operator to extract single wear debris morphology. Hu [34] uses image enhancement, corrosion, adaptive threshold method, hole filling and image segmentation into a wear debris segmentation method. Wu, et al. [35, 36] pointed out how to effective segment and identify for different types and different scales wear debris, and then judge the wear state and wear degree.

Quantities of wear debris are closely related to wear state and speed of friction surface material. Niu [37] sequentially uses angle average method, angle ratio method and nearest neighbor method to evaluate wears particle distribution based on the wear debris image analysis. But the result is influenced significantly by random sampling point. Su [38] uses the Tyson polygon to define quantity of in measurement area and proposes evaluation method on distribution uniformity of wear debris. This method has the update design idea and certain feasibility. However, the wear debris is simplified into particle so that the results lack of intuition.

3 Key Problems

A Vision inspection of wear debris is undoubtedly a hot topic that has attracted the attention of the academic community. Though researchers have conducted lucubrated and meticulous work from different perspectives and achieve gratifying achievement. Restricted of the above various constraints, the existing methods all have certain limitations and there still exist the following urgent key problems.

Diversity and Variability of Wear Debris Image Features. Smaller grain is inspect under less observation conditions by means of vision inspection technology, and affected by field work environment, sensors and other uncertain factors. The large number of types of abrasive particles, their color, shape, texture, smaller shape and complex morphology result in a large amount of data for detection parameters, and the same kind of particle image shows the characteristics of diversity and variability is difficult to achieve the correct characterization and cognition in different stages and working state. There are many features and parameters that describe abrasive particles. In order to improve the accuracy and speed of abrasive grain identification, further research is needed. Although the dense feature extraction has a large amount of feature data and large redundancy, but feature encoding method by looking for more powerful can obtain more distinguishing and more robust feature expression. This step is crucial to the performance of wear debris detection.

Uncertainty and Complexity of Wear Debris Recognition. The conventional specimen calibration method is time-consuming and labor-intensive, and has the disadvantages of large data redundancy, which results in the complexity of the abrasive particle type identification process. It is imperative to make breakthroughs in the theoretical method of specimen image features calibration and achieve cross-scale matching of abrasive image features under different usage conditions. In the actual abrasive images collected, the local features of densely extracted images often have a certain degree of ambiguity. That is, the local characteristics of an abrasive particle may differ little from multiple visual words. At this time, it is necessary to adopt proper methods to integrate feature sets. Otherwise, other visual words with high similarity will be ignored, and at the same time, the high cost of using the feature set to express the abrasive image is avoided. The next step is to use classifiers to identify them. When the number of calibration samples is limited, SVM can obtain better classification results. The key to this is the need for in-depth comparative study of the kernel function that is optimally adapted to the processing of abrasive particles. Selecting different kernel functions is directly related to the accuracy of abrasive grain identification.

Robustness and Accuracy of Wear Debris Density Estimates. The conventional specimen calibration method is time-consuming and labor-intensive, and has the disadvantages of large data redundancy, which results in the complexity of the abrasive particle type identification process. It is imperative to make breakthroughs in the theoretical method of specimen image features calibration and achieve cross-scale matching of abrasive image features under different usage conditions. In the actual abrasive images collected, the local features of densely extracted images often have a certain degree of ambiguity. That is, the local characteristics of an abrasive particle may differ little from multiple visual words. At this time, it is necessary to adopt proper methods to integrate feature sets. Otherwise, other visual words with high similarity will be ignored, and at the same time, the high cost of using the feature set to express the abrasive image is avoided. The next step is to use classifiers to identify them. When the number of calibration samples is limited, SVM can obtain better classification results. The key to this is the need for in-depth comparative study of the kernel function that is optimally adapted to the processing of abrasive particles. Selecting different kernel functions is directly related to the accuracy of abrasive grain identification.

4 Prospect of Analysis

The paper discusses and looks forward to its key issues, and aims to provide new ideas for the rapid and accurate inspection of wear debris in the future.

Analyze the Expression that Best Embody the Essential Feature of Wear Debris Images. Characteristic is the basis of wear debris vision inspection. Therefore, based on the working environment parameters, the correlation between the features of the wear debris image features is excavated; an accurate description method and its math expressions were established. The math representation of the characteristics of wear debris was used to establish the basic features of wear debris image representation and cognitive theory. Analysis can best embody the characteristics of typical wear debris nature of the image feature is one of the key issues need to break, is directly related to the accuracy and validity of the results of wear debris vision inspection.

Simplified representation or compression vision inspection image features can reduce the redundant information. However, despite the large number of local features higher redundancy, the wear debris may be included in the information richer. Therefore, in recent years, the field of object classification has used more feature-intensive extraction methods, and a number of local feature descriptions have been extracted from images in fixed steps and scales. An important point in deep learning theory is learning from image pixels to task-related feature descriptions [39]. Liu [40] extend gray value of the pixel to multi-dimensional vector, proposed dense SIFT Flow technology. SIFT Flow technology can significantly improve the image feature description ability, rotation and affine invariance [41]. Unfortunately, this method has not yet been applied in the field of wear debris vision inspection.

Establishing Multi-scale Wear Debris Image BoW Model. The traditional feature matching algorithm is suitable for multivariate fusion and matching problems between image feature pairs and sample. However, when there are different states, features and distributions of wear debris, for example, can seriously affect the detection, leading to a drop in detection results. On the other hand, the transformation of image features poses a challenge to computational efficiency and robustness of wear debris recognition. Therefore, how to obtain more precision and efficiency wear debris matching methods based on the multi-scale of the wear debris image feature is one of the key issues to be solved.

The most representative is the highest level of international vision algorithm competition PASCAL VOC as the main line for object recognition algorithm is analyzed in an effort to seek new methods of wear debris image expression and interpretation, extended wear debris vision inspection capability [42]. In recent years, object classification methods mainly focus on studying characteristics expression, including BoW model (Bag of Words, BoW) and the deep learning model [43]. For the representation of information, BoW model is an effective tool in the field of text classification. Siciv [44] introduces it into the field of computer vision in 2004. After that, it quickly attracted the attention of many scholars. Related research mainly focuses on feature coding, feature aggregation and classifier design.

Using the BoW model for effective expression can get better performance than the regional feature detection. Feature encoding abstracts and reduces redundancy will improve robustness and differentiation. The spatial characteristics of convergence are the character of the feature set can be obtained after encoding a compact integration operation as the image character expression. Not only can obtain certain feature invariance, but also avoid the cost of using the feature set for image expression. After a fixed dimensional vector is used to express features and describe them to obtain visual vocabulary features, a classifier needs to be established to complete image classification [45, 46]. Common classifiers include SVM, K-nearest neighbors, neural networks, and random forests. Among them, SVM not only has stronger generalization ability, but also has a large advantage in solving small problems. It has been widely used in classification and identification problems in many fields. With the increase of the visual word and the image expression, the scale of data processing increase, and the design and research of the new classifier has received attention and applications.

Analysis of Texture Features of Wear Debris. Wear debris density estimation may be affected by operating conditions, resulting in very different detection results. Especially in medium and high density, it is difficult to detect wear debris in the image due to adverse factors such as morphology, adhesion, and occlusion, resulting in poor accuracy of the wear debris density. At this time, it is wise to use wear debris texture features to estimate the wear debris density. Because the extraction of texture features directly affects the quality of subsequent processing, it is the key to estimating the wear debris density how to properly extract wear debris texture characteristics under the conditions of live lighting, rotation, background, and wear debris life.

At present, the wear debris density estimation is still in the exploration stage of research. The constraints of the above make the existing wear debris vision inspection methods have certain limitations. In view of the fact that there are few researches on the estimation of wear debris density, the closest to this study are the population density estimation. The population density estimation method based on micro-statistics and macro-analysis [47, 48] The micro-statistics method uses detection or trajectory clustering to obtain population density through population statistics. This method has better performance for estimating low to moderate density populations, but it is severely degraded for high density populations due to the occlusion issue. The method based on macro analysis is to treat the population as a single entity and use techniques such as regression fitting or texture classification to obtain population density. It is not easily affected by occlusion and cluttered backgrounds, and can be better estimated all density levels. Assume that the images of low-density populations present coarse texture features, high-density population images present finer texture features, and different descriptors are used to obtain classification to achieve population density. This kind of method is easy to be interfered by the background noise in the low density situation, has weaker ability to distinguish the high density people's texture, and does not consider the influence of the near-far effect.

5 Conclusion

In order to meet the requirements for development of wear debris inspection, the paper is based on research of wear debris vision inspection, and aims to excavate the key problems of vision accurate inspection theory of wear debris. Combining with the requirement of wear debris and the actual working condition, the existing wear debris vision inspection method cannot effectively solve diversity and time-varying of images characteristics, uncertainty and complexity of type identification, multi-constraint and accuracy of wear debris density estimation. In this paper, we focus on diversity of wear debris image features, rapid classification of wear debris under owed conditions and density estimation of wear debris under multi-constraint and so on, trying to open up new innovation and breakthroughs in the research of wear debris inspection technology.

Acknowledgment. This work is supported by Program for Innovative Research Team (in Science and Technology) in University of Henan Province (15A520056), Supported by Youth Foundation of Henan University of Technology (2016QNJH29), and Research Key Scientific Research Projects of the Higher Education Institutions of Henan Province under Grant (18A430011).

References

1. Chen, W.P., Gao, C.H., Ren, Z.Y.: Advances and trends on abrasive particle characterization. Chin. J. Constr. Mach. **13**(4), 283–289 (2015)
2. Wang, H.W., Chen, G., Lin, T., et al.: Knowledge rules extraction and application of micro debris image recognition. Lubr. Eng. **40**(1), 86–91 (2015)
3. Roylance, B.J., Albidewi, I.A., Laghari, M.S.: Computer-aided vision engineering (CAVE) quantification of wear particle morphology. Lubr. Eng. **50**(2), 111–116 (1994)
4. Wang, J.Q.: Research on ferrograph image segmentation and wear particle identification. The Ph.D. dissertation of Nanjing University of Aeronautics and Astronautics (2014)
5. Bi, J.: Study on evaluation method of wear particle segmentation ferrograph image. The Master dissertation of Nanjing University of Aeronautics and Astronautics (2016)
6. Hu, X., Huang, P., Zheng, S.: Object extraction from an image of wear particles on a complex background. Pattern Recognit. Image Anal. **16**(4), 644–650 (2006)
7. Xing, F., Fan, Y.J.: Segmentation method of color image of ferrography based on multi-threshold wavelet transform. Lubr. Eng. **33**(7), 69–72 (2008)
8. Tang, C.J.: Study on Evaluation method of wear particle segmentation ferrograph image. The Master dissertation of Nanjing University of Aeronautics and Astronautics (2015)
9. Liu, J.H.: Research on analyse and recognition of particle image on digital image processing. The Master dissertation of Beijing Jiaotong University (2007)
10. Surapol, R.: Wear particle analysis-utilization of quantitative computer image analysis: a review. Tribol. Int. **38**(10), 871–878 (2005)
11. Myshkin, N.K., Grigoriev, A.Y.: Morphology: texture, shape, and color of friction surfaces and wear debris in tribe diagnostics problems. J. Frict. Wear **29**(3), 192–199 (2008)
12. Laghari, M.H., Ahmed, F.: Wear particle profile analysis. In: The 2009 International Conference on Signal Processing Systems, Singapore, pp. 546–550 (2009)
13. Pan, Bingsuo, Fang, Xiaohong: Mingyuan Niu.: Image segmentation of protruded diamonds on impregnated bits by fuzzy clustering algorithm. Diam. Abras. Eng. **172**(4), 62–67 (2009)
14. Jiang, L., Chen, G.: A quantitative analysis method in ferrography based on color image processing. In: The 1st International Conference on Modeling and Simulation, Nanjing, pp. 512–515 (2008)
15. Wu, Z.F.: The research of engine wear faults diagnosis based on debris analysis and data fusion. The PhD dissertation of Nanjing University of Aeronautics and Astronautics (2002)
16. Lv, Z.H.: Digital Detection Method of Wear Particle Image. Science press, Beijing (2010)
17. Yeung, K.K., Mckenzie, A.J., Liew, D.: Development of computer-aided image analysis for filter debris analysis. Lubr. Eng. **50**(4), 293–299 (1994)
18. Podsiadlo, G.P., Stachowiak, G.W.: Development of advanced quantitative analysis methods for wear particle characterization and classification to aid tribological system diagnosis. Tribol. Int. **38**(10), 887–897 (2005)

19. Stachowiak, G.P., Stachowiak, G.W., Podsiadlo, P.: Automated classification of wear particles based on their surface texture and shape features. Tribol. Int. **41**(1), 34–43 (2008)
20. Wang, H.G., Chen, G.M.: Theory and Technology of Ferrograph Image Analysis. Science Press, Beijing (2015)
21. Yuan, Chengqing, Yan, Xinping, Peng, Zhongxiao: Three-dimensional surface characterization of wear debris. Tribol. **7**(3), 294–296 (2007)
22. Yuan, C.H., Wang, Z.F., Zhou, Z.H., et al.: Research on mapping model between wear debris and worn surfaces in sliding bearings. J. Wuhan Univ. Technol. **31**(12), 123–126 (2009)
23. Li, Y.J., Zuo, H.H., Wu, Z.H., et al.: Wear particles identification based on dempster-shafer evidential reasoning. J. Aerosp. Power **18**(1), 114–118 (2003)
24. Peng, Z., Kirk, T.B.: Wear particle classification in a fuzzy grey system. Wear **225**(4), 1238–1247 (1999)
25. Laghari, M.S.: Recognition of texture types of wear particles. Neural Comput. Appl. **12**(1), 18–25 (2003)
26. Gu, D.Q., Zhou, L.X., Wang, J.: Ferrography wear particle pattern recognition based on support vector machine. J. Mech. Eng. **17**(13), 1391–1394 (2006)
27. Zhong, X.H., Fei, Y.W., Li, H.Q., et al.: The research of optimization on ferrographic debris feature parameters. Lubr. Eng. **170**(4), 108–110 (2005)
28. Chen, G., Zuo, H.H.: The image adaptive thresholding by index of fuzziness. Acta Autom. Sin. **29**(5), 791–796 (2003)
29. Chen, G.M., Xie, Y.B., Jiang, L.Z.: Application study of color feature extraction on ferrographic image classifying and particle recognition. China Mech. Eng. **17**(15), 1576–1579 (2006)
30. Yu, S.Q., Dai, X.J.: Wear particle image segmentation method based on the recognition of background color. Tribol. **27**(5), 467–471 (2007)
31. Fu, J.P., Liao, Z.Q., Zhang, P.L., et al.: The segmenting method of ferrographic wear particle image based on two-dimension entropy thresold value. Comput. Eng. Appl. **41**(18), 204–206 (2005)
32. Jiang, L., Chen, G., Long, F.: Auto-threshold confirming segmentation for wear particles in ferrographic image. In: The 2008 International Symposium on Computational Intelligence and Design, Wuhan, pp. 61–64 (2008)
33. Li, F., Xu, C., Ren, G.Q., et al.: Image segmentation of ferrography wear particles based on mathematical morphology. J. Nanjing Univ. Sci. Technol. **29**(1), 70–72 (2005)
34. Hu, X., Huang, P., Zheng, S.: On the pretreatment process for the object extraction in color image of wear debris. Int. J. Imaging Syst. Technol. **17**(5), 277–284 (2007)
35. Wu, T.H., Wang, J.Q., Wu, J.Y., et al.: Wear characterization by an on-line ferrograph image. J. Eng. Tribol. **13**(4), 23–34 (2011)
36. Wu, T.H., Mao, J.H., Wang, J.T., et al.: A new on-line visual ferrograph. Tribol. Trans. **52**(5), 623–631 (2009)
37. Niu, M.Y., Pan, B.S., Tian, Y.C.: Quantitative measurement of the uniformity of diamond distribution by acute angle ratiomethod. Diam. Abras. Eng. **173**(5), 28–40 (2009)
38. Su, L.L., Huang, H., Xu, X.P.: Quantitave measurement of grit distribution of diamond abrasive tools. Chin. J. Mech. Eng. **25**(10), 1290–1294 (2012)
39. Lecun, Y., Bengio, Y., Hinton, G.E.: Deep learning. Nature **521**(7553), 436–444 (2015)
40. Liu, C., Yuen, J., Torralba, A.: SIFT flow: dense correspondence across scenes and its applications. IEEE Trans. Pattern Anal. Mach. Intell. **33**(5), 978–994 (2011)
41. Xu, S., Liu, X., Li, C., et al.: An image registration algorithm based on dense local self-similarity feature flow. J. Optoelectron. Laser **24**(8), 1619–1628 (2013)

42. Everingham, M., Eslami, S.M.A., Gool, L.V., et al.: The pascal visual object classes challenge: a retrospective. Int. J. Comput. Vis. **111**(1), 98–136 (2015)
43. Wu, L., Hoi, S.C.H., Yu, N.: Semantics-preserving bag-of-words models and applications. IEEE Trans. Image Process. **19**(7), 1908–1920 (2010)
44. Sivic, J., Zisserman, A.: Video Google: a text retrieval approach to object matching in videos. In: The 9th International Conference on Computer Vision (ICCV'2003), pp. 1470–1477 (2003)
45. Lazebnik, S., Schmid, C., Ponce, J.: Beyond bags of features: Spatial pyramid matching for recognizing natural scene categories. In: The 2006 IEEE Computer Society Conference: Computer Vision and Pattern Recognition, pp. 2169–2178 (2006)
46. Yang, J.C., Yu, K., Gong, Y.H., et al.: Linear spatial pyramid matching using sparse coding for image classification. In: Proceedings of The 12th International Conference on Computer Vision (ICCV'2009), pp. 1704–1801 (2009)
47. Qin, X.H., Wang, X.F., Zhou, X., et al.: Counting people in various crowed density scenes using support vector regression. J. Comput. Aided Des. Comput. Graph. **18**(4), 392–398 (2013)
48. Muhammad, S., Khan, S.D., Michael, B.: Texture-based feature mining for crowd density estimation: a study. In: 2016 International Conference on Image and Vision Computing New Zealand (IVCNZ), pp. 404–410 (2016)

Case Study of Magic Formula Based on Value Investment in Chinese A-shares Market

Min Luo(✉)

School of Management, Shanghai University, Shanghai 200444, China
luominrosemary@gmail.com

Abstract. This paper provides new empirical evidence that the magical formula, put forward by Joel Greenblatt, is effective and applicable in Chinese A-shares market. Referring to other value investing researches and pioneering trading experiences, some new revised strategies based on magical formula are also used to back-test historical returns in Chinese A-shares market, and we compare return of them with that of magical formula, the common value investing strategy and CSI300. Using experimental analysis, we finally find that holding period, capitalization of stocks, selection and weight of financial index can make obvious influences to strategy return. Additionally, the alpha of magical formula is found to exist but small. Our findings highlight the usefulness of value investing in Chinese A-shares market, which has long been regarded as non-standard as well as chaotic, and offer some available options of investment strategy for value investors.

Keywords: Value investing · Magical formula · A-shares market

1 Introduction

Value investing is a classical investing theory that derives from ideas which Benjamin Graham summarized from his stock trading experience. The text *Security Analysis* [1] written by Graham developed and epitomized those ideas. From his perspectives, value investing is built on the basis of sufficient fundamental analysis, since investors should utilize financial analysis to estimate the intrinsic value and consequently find out stocks underpriced. For instance, such a stock can gain may be traded at discounts to book value with a low price-earnings ratio or a low price-book value ratio. In this case, investors can buy this stock in a low price and sell it in a high price ending in a considerable return from bid-ask spread. Except intrinsic value, there is another key point of value investing: margin of safety, which is a discount of market price to intrinsic value. Using margin of safety is a defense mechanism which can protect investors from loss caused by unwise trading decisions and downturns. From the random walk model [2], the changes of stock price are random, and Graham claimed stock price always fluctuates from the intrinsic value, therefore, the fair value or market price is difficult

N. Xiong et al. (Eds.): ISCSC 2018, AISC 877, pp. 177–194, 2019.
https://doi.org/10.1007/978-3-030-02116-0_22

to estimate accurately. Hence, an appropriate margin of safety gives investors enough room for errors in investing. What's more, Cronqvista, Siegelb and Yua [3] also applied behavioral finance theory to value investing research. Over the past few decades, as a great number of successes of value investing strategies has exposed to investors and academics in some west developed countries, value investing turned to be accepted widely and was introduced to China.

Up to now, a great number of prior studies have demonstrated the applicability and superiority of value investing. Fama and French [4] confirmed that book-to-market strategy can help investors gain a superior return and compensate risks to some extent. Doukas, Kim and Pantzalis [5] did an experimental analysis from 1976 to 1997 and found superiority of value stocks didn't come from systematic errors made by investors when they are predicting future trend. What's more, Piotroski [6] also affirmed the function of financial statement analysis when investors use historical financial information to make investment. Chahine [7] confirmed the positive effect of earnings-per-share in undervalued value stocks and value investing strategy is applicable in Euro-markets. Asness, Moskwitz and Pedersen [8] provided evidence that value strategy can help investors gain return premia. Bartov and Kim [9] found that book-to-market can help investors achieve higher returns without evidence of risk increased. Woltering, Weis, Schindler, Sebastianb [10] and Perez [11] examined the risk premium of value stocks within a global investment strategy framework. Besides, Stafford [12] represented an economically large improvement in risk – and liquidity-adjusted returns over direct allocations to private equity funds. Ray [13] innovatively put the artificial intelligence to the value investing. In the area of investor structure, Betermier, Calvet, Sodini [14] discussed the differences between the value investors and growth investors. Also, Johnson [15] found some behavioral obstacles to successful value investing. However, value investing is also denounced by many researchers such as Porta et al. [16] who suggested that the superior return comes from expectational errors about future earnings prospects made by investors.

Finding an effective stock investment strategy is one of the hardest challenges that both individual and institutional investors should face to. Some prior studies have already done a lot of researches in various investment theories, such as technical analysis, value investing and quantitative investing. In practical stock trading, these investment theories are also used to supplement each other for a better return. For example, Ko, Lin, Su [17] verified that the sophisticated investors can do better when they combined the technical analysis and value investing in Taiwan stock market. However, this stock price predicting method is a multistrategy based on both value investing and technical analysis, the academics who focus on unalloyed single value investing are far fewer, especially in mainland of China. In this paper, we try to test whether if a pure value investing strategy can work in Chinese A-shares market. Here, a famous strategy proposed by Joel Greenblatt is investigated [18].

This paper is organized as follows. In Sect. 2, we will review basic concepts about magical formula and we will introduce method of back-test. Especially considering that different financial indicators may cause varying degree impacts to stock return, we also did further researches by extending the empirical research based on magical formula. Section 3 will apply magical formula and some new strategies to Chinese A-shares market, and simulate calculation of stock returns. Therefore, we can analyze magical formula in detail to show the impacts from other influence factors. The last section will give some concluding remarks.

2 Method and Data

2.1 The Magical Formula

As discussed in the last section, value investing is built on fundamental analysis since intrinsic value of a stock is determined by fundamentals of its corresponding listed company. Fundamentals of listed company can be reflected in many aspects, such as in financial reports and operating situations. Generally, if a listed company has good fundamental situations, the market price of their stocks would rise in the holding periods to come. Especially, Carpenter, Lu and Whitelaw [19] has proved that Chinese stock prices connected to fundamentals of enterprises strongly. Hence, here comes a question: how to analyze the fundamentals? This is where financial indexes come to play roles in, and as we will do in this paper. Financial indexes, also known as accounting ratios, are many standard ratios used to evaluate the financial condition of a company. For instance, Ball, Gerakos, Linnainmaa and Nikolaev [20] claim that operating profitability can predict returns as far as ten years ahead. Especially, market price of stocks can be reflected in certain financial indexes, for example, price-earnings ratio. Therefore, financial indexes have been widely used to analyze the fundamentals of stocks since they are dynamical reactions of corporation financial status. Price-earnings ratio, return on equity and price-book value ratio are most common used.

Similarly, magical formula also takes advantages of two certain financial indicators, return on capital and earning yield, to select stocks. Return on capital is the ratio of EBIT (Earnings before interest and taxes) to tangible capital (net floating capital plus net tangible asset), which is used as profitability indicator. Although many value investors user return on equity (ROE) or return on asset (ROA) in practical stock trading when measuring the profitability of listed enterprises, magical formula prefers EBIT. Greenblatt thinks that EBIT can not only help investors to compare operation ability of different companies, but also avoid distortion caused by difference in tax rates and debt levels. Further, tangible capital is used to measure the net asset which companies consume for normal production and operation activities.

Besides, earnings yield is the ratio of EBIT to EV (Enterprise value). In practical trading, value investors tend to choose PE (price-earnings ratio) as valuation index to select stocks, however, Greenblatt use earnings yield. He holds belief that using earnings yield can take both market capitalization and liabilities into account and in this case investors can compare returns of various companies with different liability and tax rates.

Using return on capital and earnings yield, magical formula can help investors find out penny stocks with high quality in stock market. In details, return on capital is used to find out companies whose profits are higher than costs. Besides, earnings yield is used for searching companies whose profits are higher than acquisition price. Greenblatt tested his magical formula in the U.S. stock market. The average return of magical formula in the U.S. market was 30.8%, which was much higher than 12.3% for average return of market, calculated by SP 500.

Some foreign academics have tested Greenblatt's magical formula in their own countries, for example, Persson and Selander [21] made a back test from 1999 to 2008 in Nordic stock market. Magical formula was also proved to be effective since it achieved a profitable return of 21.25%, while the MSCI Nordic market only achieved an average return of 9.28% during that ten years.

2.2 Method

We try to imitate Greenblatt's magical formula in Chinese A-shares market as objectively and comprehensively as possible, thus, we test all A-shares traded in market for the period starting from January 2nd 2004 to December 31st 2016. The back-test we will make can be divided to three steps as following.

Step 1. Rank all stocks in Chinese A-shares market.

At the beginning, we rank stocks according to return on capital and rank stocks from highest to lowest. The company with the biggest return on capital should be marked by sequence number 1. Similarly, the company with the smallest return on capital should be marked by the last sequence number. We rank these stocks by earnings yield afterwards in the same way. The company who has the biggest earnings yield will get a new sequence number 1, and who has the smallest earnings yield should be marked by the last sequence number. Then, we plus these two sequence numbers and generate a series of new synthetical sequence numbers for every stock. Each new sequence number can indicate a stock, and we can use new sequence numbers to rank stocks once again from smallest to largest. This rank method is aiming to find out companies who have outperformance in both two aspects rather than in terms of return on capital only, or in terms of earnings yield only. In this back test case, we choose 10 stocks to create a portfolio annually.

Step 2. Simulate stock trading.

The basic strategy is buying stocks at the first trading date of a year and selling them in the last trading date of that year. The portfolio size is 10 stocks and holding period of each portfolio is one year. Furthermore, we renew stocks in portfolio at the first trading date every year.

Besides, for a better comparison, we also include the return of the portfolio with the consecutive holding period of two as well as three years respectively in this back-test case.The concrete content of holding period is showed in next subsection. But when calculating them, the closing price $p'_{n,y}$ of the stock n in year y should be adjusted to the closing price in the last trading date of year $y+1$ or $y+2$.

Step 3. Define the portfolio return. We use average return of 10 stocks

$$RP_{n,y} = \frac{\sum R_{n,y}}{10} \tag{1}$$

as the portfolio return. In addition, the annualized portfolio return could be calculated as following formula:

$$RP''_{n,y} = \sqrt[m]{\prod (1 + PR_{n,y})} - 1 \tag{2}$$

where m is 12/*holding period*. If the holding period is one year, the m is 12 and if the holding period is two years, the m is 6. Return of a single stock is

$$R_{n,y} = INT(\frac{100}{p_{n,y}}) \cdot \frac{p'_{n,y} - p_{n,y}}{100} \tag{3}$$

We assume that investors allocate 100000 Chinese yuan to each single stock. Because in Chinese A-shares market, the size of stocks traded in market must be in multiple of 100, so the INT( ) is rounding function to calculate size of stocks which can be bought in maximum. $p'_{n,y}$ presents the closing price of stock n in the year y and $p_{n,y}$ presents the opening price of stock n in the year y. Therefore, $p'_{n,y} - p_{n,y}$ presents the bid ask spread.

Step 4. Make a comparison.

We build a new portfolio by selecting stocks with low PE and high ROE, and we call this new selection method the common value investing strategy. We should exclude the stocks whose asset-liability ratio is bigger than 60% in advance in order to avoid risks as much as possible. Then, we rank stocks by PE in a descending order and by ROE in an ascending order. Besides, we also define the return of Chinese A-shares market using CSI300 index:

$$RI_{n,y} = \frac{I'_{n,y} - I_{n,y}}{I_{n,y}} \tag{4}$$

$I_{n,y}$ denotes the opening index of CSI300 in the first trading date of year y and $I'_{n,y}$ denotes the closing index in the last trading date of year y.

We call the portfolios selected by Magical formula *portfolio A* and that selected by the common value investing strategy *portfolio B* respectively. Additionally, the portfolios whose returns measured by CSI300 index are called *portfolio C*. We believe that the *portfolio A* has the best market performance since magical formula is supposed to be a better strategy.

2.3 Potential Issues

The Influences of Holding Period. From Rousseau and Rensburg [22], the return of value investing rewarded from investor's holding period. Since that, portfolio return could be higher when its holding time is longer. Thus, to test the influence of holding period, we also take returns of magical formula with two years holding period and three years holding period into consideration respectively. For example, if we want to generate a portfolio with longer holding period in 2004, we should create portfolio at the beginning of 2004 and reallocate it at the beginning of 2006 or 2007. In this way, we can create 6 portfolios whose holding periods are two years and 4 portfolios whose holding periods are three years. The former group of portfolios called $portfolio\ A1$ and the latter group of portfolios called $portfolio\ A2$. Additionally, we also generate $portfolio\ B1$ and $portfolio\ B2$ for the common value investing strategy and $portfolio\ C1$ and $portfolio\ C2$ for CSI300 in the same way. From the discussions above, we hold the belief that $portfolio\ A2$ can achieve higher return than $portfolio\ A$ and $portfolio\ A1$.

Market Capitalization. Banz [23] found that there is a relationship named size effect between the return and the market capitalization. Fama and French [24] find that value and momentum returns decrease with size. He proved that smaller companies have had higher risk adjusted returns than larger firms due to size effect. Hence, we suggest that the price of a stock may raise or decrease more easily if its market capitalization is small. Conversely, a large cap stock could wait a longer time to show price tendency obviously, because large cap stocks needs far more stronger buying (or selling) forces to push their market prices upward (or downward) due to its big market capitalization. In Chinese A-shares market, a big cap stock is defined as a stock with market capitalization more than 2 billion Chinese yuan, otherwise, it is defined as a small cap stock. Hence, in this paper, we also research the influence from the market capitalization. To achieve this aim, we therefore create two portfolios using large cap stocks and small cap stocks respectively. The former portfolio is called $portfolio\ D$ and the latter one is called $portfolio\ E$.

Depreciation and Amortization. Both EBIT (Earnings Before Interest and Tax) and EBITDA (Earnings Before Interest, Taxes, Depreciation and Amortization) are indicators of a company's profitability, however, the financial fundamentals they describe are different. In details, EBIT = Net Income + Interest + Taxes, and EBITDA = Net income + Interest + Taxes + Depreciation + Amortization. EBIT has been widely used to evaluate the value of enterprises, since it can effectively measure the operating profits of listed companies. In magical formula, Greenblatt assumes that profits are not charged with any cash expenses, which means that depreciation and amortization of assets are not necessary to be considered. However, there may be a bias when we accept Greenblatt's assumption and use EBIT to select stocks, since we focus solely on operating profits and ignore the depreciation and amortization of assets. Hence, we use EBITDA to make a new strategy, which can take variables ignored into consideration, and therefore create a new portfolio called $portfolio\ F$. In the $portfolio\ F$ selection process, the return on capital is the ratio of EBITDA to tangible capital and earnings yield is the ratio of EBITda to EV. We believe *portfolio F* can beat *portfolio A* which created by EBIT.

Price Earning Ratio and Price Book Value Ratio. The magical formula uses earnings yield to measure the purchasing cost: the smaller the earnings yield is, the cheaper the purchase price is. Given that the distortions caused by different debt levels, Greenblatt choose earnings yield to replace PE (stock price/Earnings Per Share), a more common indicator of purchasing cost used by both scholars and investors. However, we still consider PE is effective since it can take the dynamical stock price into consideration, and therefore it can evaluate a stock more sensitively. For instance, Wu [25] claims that PE ratio plays a crucial role in investment and it can reflect company's future growth and risk. Also, considering the influences of debt, we also use price book value PB (stock price/total assets-total debts) and combine it as well as PE to propose a new strategy. The effectiveness of PB has been confirmed by Fairfield [26]. Summarized from the stock trading practices in Chinese A-shares market, we select stocks satisfying the criteria PE*PB<30, PE*PB<50, PE*PB<100 respectively, and then rank stocks by PE*PB in ascending order instead of using earnings yield. In this way, we can find out cheap stocks from all A-shares. The portfolio we create by return on capital and PE*PB is called *portfolio G*. To subdivide, the portfolio with PE*PB<30 is named *portfolio G*1, the portfolio with PE*PB<50 is named *portfolio G*2 and the portfolio with PE*PB<100 named *portfolio G*3. We reckon that these three portfolios are all better than *portfolio A*.

Weight of Index. As discussed in the last section, magical formula is easy to implement since it requires less for professional knowledge and skills of investors. Greenblatt has told investors certain fixed parameters that magical formula required, for example, the holding period is always set as one year. However, some affecting factors may be ignored when using fixed parameters, which could result in a miss of potential stock return.

In magical formula, there are two important parameters which influence return most: holding period and weights of return on capital as well as earnings yield. Hence, we try to analyze effect from weight of these two stock indicators to portfolio return. In magical formula, the weights of return on capital and earnings yield are equal. That's to say, those two indexes are attached same importance to, nevertheless, returns of portfolio with different indicator weights are definitely different and a certain weight couple of indicators which can create the most profitable portfolio may exist elsewhere. Therefore, we calculate portfolio returns based on different weight couples in order to make a profit comparison. The concrete steps are as following:

Step 1. We generate 101 groups of weight couples of return on capital and earnings yield. The first couple is 0 and 1, the second couple is 0.01 and 0.99,$\cdots$, the last couple is 1 and 0. The variation of each indicator weight is 0.01 and the sum of two weights in a couple is 1.

Step 2. Calculate portfolio returns. Using weighted earnings yield and return on capital to create 101 groups of portfolio annually. We use the formula (3) mentioned above and calculate all returns of 101 portfolio annually.

Step 3. We compare 101 groups of returns every year and identify their corresponding weight couples. Finally, we analyze the return series and find out optimal weight couple.

Given that the weight couple (0.5, 0.5) can summarize stock fundamentals more comprehensively and balanced, we reason that the return of portfolio with weight couple (0.5, 0.5) could be the highest, and return of others could approximately diminish with the changes of return on capital and earnings yield.

Profits Exceeding Normal Returns. Is magical formula for real or only a pure luck? The simple back-tests proposed above are not enough to answer this question. Here, we try to introduce EMH theory (Efficient Market Hypothesis) [27] proposed by Eugene Fama to explain this confusion. Fama defined three different degrees of the efficient markets: weak form, semi-strong form and strong form. In the weak form, EMH theory claims that stock prices reflect only historical price movements and volume data, which can not affect current or future stock information. Therefore, fundamental analysis can identify undervalued stocks and get profits exceeding normal returns while technical analysis is invalid. As for semi-strong form, stock prices only reflect all public information, for example, the quarterly reports revealed by listed companies regularly. In this case, neither technical analysis nor fundamental analysis can get abnormal returns, however, insider traders can. Finally, in strong form, prices fully reflect both public and private information, therefore, even insider information cannot offer advantages to investors. Strong form implies that no abnormal return can be earned, regardless of whether strategies you try.

Obviously, Chinese A-shares market is not a strong form efficient market. Rather, especially considering the immaturity of market supervision, Chinese A-shares market is more likely a weak form efficient market, which has been confirmed by some scholars [28,29]. In other words, the randomness of A-shares prices can make technical analysis invalid. There is no need to predict the evolution of stock price, conversely, the only thing we ought to do is to research the fundamentals and to identify the undervalued stocks. Hence, here we try to test whether magical formula can obtain return advantages, or abnormal return, in Chinese A-shares market. In this paper, we use alpha to measure the profits exceeding normal returns and we believe that the alpha of magical formula is positive. Generally, alpha is the intercept term of the regression equation fitted by market return and portfolio return [30]. Since we do this research on Chinese market, we need a reasonable benchmark index to measure the market return of A-shares. Considering the arguments that the distortion of risk-free rate may exist, we finally choose the market premium factor, return of the market portfolio minus return of the risk free rate, of Fama-French's three factor model [31] to measure the market return.

2.4 Data

The data is downloaded from Wind datasets. We use daily stock price of all Chinese A-shares in the Shanghai Stock Exchange and Shenzhen Stock Exchange from 2004 to 2016. The market premium factor can also be accessed directly in Wind datasets. Especially, to avoid the look-ahead bias, which caused by using data not available yet when stock selecting event occurred, we use third quarterly financial reports rather than annual reports.

Further, we adjust the statistics as following in advance:

Remove the stocks whose corresponding listed companies having loss;

Remove the special treated stocks;

Remove the financial stocks (Bankingstocksforinstance).

The above adjustments are proposed by Greenblatt in *The little book that beats the market*.

2.5 Hypothesis

As discussed above about different factors which could impose potential influences to portfolio return, we therefore get a series of hypothesis. They are:

Hypothesis 1: The magical formula is more effective than the common value investing strategy, while both of these two value investing strategies can get better market performances than CSI300, i.e. *portfolio A* achieves higher return than both *portfolio B* and *portfolio C*, also, *portfolio B* achieves higher return than *portfolio C*.

Hypothesis 2: The return of magical formula become larger as the holding period length become longer, i.e. *portfolio A*2 achieves higher return than both *portfolio A* and *portfolio A*1, also, *portfolio A*1 achieves higher return than *portfolio A*.

Hypothesis 3: The portfolio created by small cap stocks achieves higher return than that created by big cap stocks, i.e. *portfolio D* achieves higher return than *portfolio E*.

Hypothesis 4: EBITDA is more effective than EBIT when applying them to magical formula, i.e. *portfolio F* achieves higher return than *portfolio A*.

Hypothesis 5: PE*PB is more effective than earning yield earnings yield in magical formula, i.e.*portfolio G*1, *portfolio G*2 and *portfolio G*3 are better than *portfolio A*.

Hypothesis 6: Returns of portfolios with weight couples (0.5, 0.5) could be the highest, and the return of others could about successively diminish with the changes of return on capital and earnings yield.

Hypothesis 7: Using magical formula can get profits exceeding normal returns, i.e. the alpha is positive.

3 Empirical Results and Analysis

3.1 Return of Magical Formula

After back testing data from 2004 to 2016, we can rank all the A-shares according to their return on capital and earnings yield to generate portfolios. Annual returns of *portfolio A*, *portfolio B* and *portfolio C* can be seen in the following Tables 1 and 2.

Table 1. The returns comparison among different strategies from 2004 to 2016

Portfolio	Annualized	Variance	Mean	Maximum	Minmum
A	15.45%	0.7518	36.11%	201.45%	−52.15%
A1	32.80%	7.0939	110.18%	646.04%	−46.38%
A2	47.32%	0.7919	67.79%	177.09%	−37.05%
B	10.87%	0.8082	31.22%	283.69%	−64.60%
B1	28.85%	3.1370	80.03%	462.81%	−28.61%
B2	71.44%	1.2323	95.76%	245.16%	−23.11%
C	5.84%	0.3087	17.63%	130.43%	−65.39%
C1	15.42%	1.9677	53.93%	353.17%	−37.72%
C2	24.00%	0.2262	31.61%	78.72%	−30.76%

Table 2. The returns comparison among different strategies from 2004 to 2016

Portfolio	Annualized	Variance	Mean	Max	Min
A	15.45%	0.7518	36.11%	201.45%	−52.15%
A1	32.80%	7.0939	110.18%	646.04%	−46.38%
A2	47.32%	0.7919	67.79%	177.09%	−37.05%
B	10.87%	0.8082	31.22%	283.69%	−64.60%
B1	28.85%	3.1370	80.03%	462.81%	−28.61%
B2	71.44%	1.2323	95.76%	245.16%	−23.11%
C	5.84%	0.3087	17.63%	130.43%	−65.39%
C1	15.42%	1.9677	53.93%	353.17%	−37.72%
C2	24.00%	0.2262	31.61%	78.72%	−30.76%

The annualized returns of value investing are proved to be obviously high: 15.45% for magical formula and 10.87% for the common value investing strategy. Besides, the annualized return of CSI300 is only 5.84%. In terms of risks, magical formula still outperforms others since it has the smallest variance. Otherwise, magical formula also has bigger maximum return and smaller minimum return.

For a better understanding, we also use the following line chart Fig. 1 to demonstrate the changing variation in portfolio annual returns. The line chart describes that magical formula is better than the common value investing strategy and CSI300 in most of time. Rather, the magical formula has some inevitable failures such as in 2008. Overall, the magical formula is confirmed to be applicative and effective in Chinese A-shares stock market. Hence, we cannot reject *Hypothesis* 1.

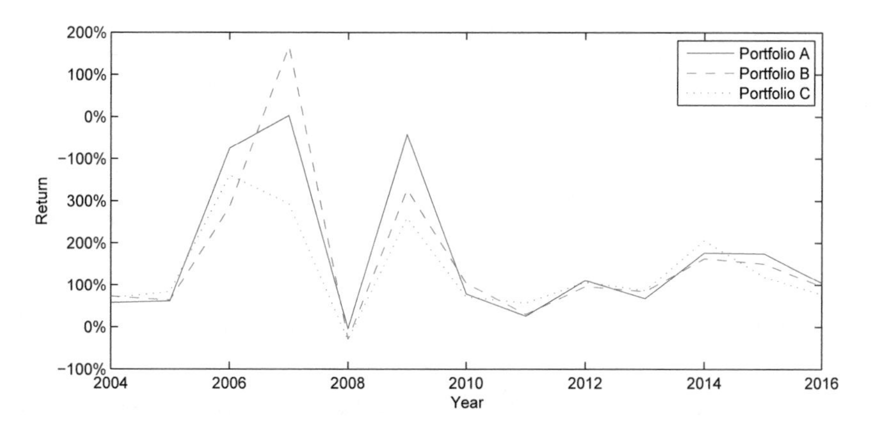

Fig. 1. The aggregated returns of portfolio A, portfolio B and portfolio C from 2004 to 2016

3.2 The Influences of Holding Period

As discussed, we also calculate returns of portfolios with two and three years holding period respectively. Annualized returns and variances are showed in Table 1. More details about time varying returns can be seen in Figs. 2 and 3.

Generally, the portfolios created by magical formula substantially achieve higher return than the common value investing strategy and CSI300. When holding period is three years, magical formula still outperforms CSI300, whereas it is in vulnerable side compared with the common value investing strategy. In terms of Table 1, it is clear that portfolios with three years holding period earn much higher returns than that with one year or two years holding period, also, we find it is interesting that portfolios with two years holding period always face higher risks. To conclude, strategies with three years holding period is more profitable and truly magical formula rewards from time since its return increases as holding period extends. We therefore accept *Hypothesis* 2.

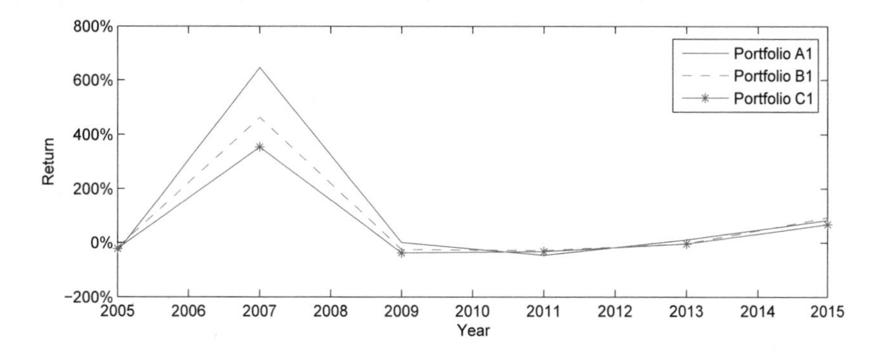

Fig. 2. The aggregated returns of portfolio A1, portfolio B1 and portfolio C1 from 2004 to 2016

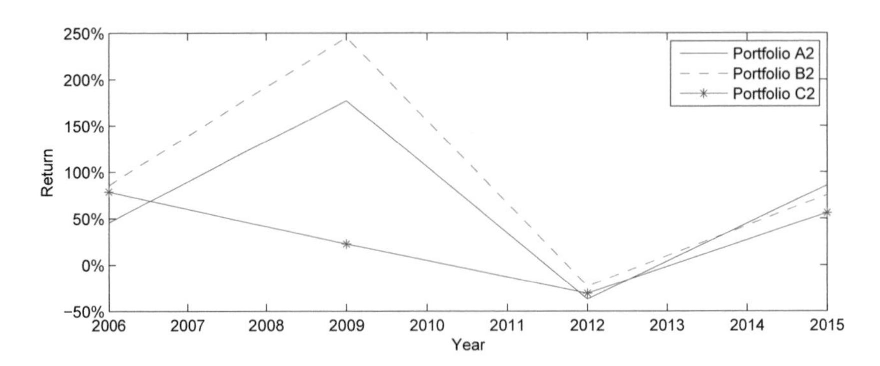

Fig. 3. The aggregated returns of portfolio A2, portfolio B2 and portfolio C2 from 2004 to 2016

3.3 The Influences of Market Capitalization

In the Fig. 4, we can conclude that $portfolio\ D$ achieves higher return than $portfolio\ E$ in most tested years, which is what we expected above. To illustrate the comparison, we make more descriptives for those new strategies and show them in the following Table 3. As seen, both two portfolios achieve impressive annualized return, 15.73% for $portfolioE$ and 49.61% for $portfolio\ D$. Returns of $portfolio\ E$ is significantly higher compared to $portfolio\ D$, especially in 2016, therefore we accept the $Hypothesis$ 3. As seen in Table, on the other hand, the variance, which reflects risk, also shows distinctions: $portfolio\ D$ faces higher risk than $portfolio\ E$. It is understandable, because most of the listed companies who issue big cap stocks have stronger risk resistances capacity, owing to their sufficient capitalization and mature management systems. Furthermore, in Chinese A-shares market, a great number of listed companies are state-owned enterprises and they can sometimes get preferential policy supports from governments. Conversely, most of the listed companies who issue small cap stocks are

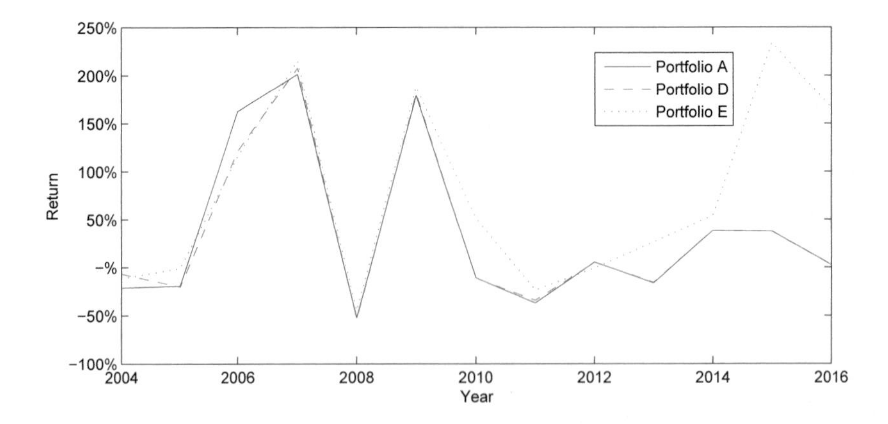

Fig. 4. The aggregated returns of portfolio D and portfolio E 2004 to 2016

private enterprises, or even are entrepreneurial companies, they therefore have more risk exposures.

Table 3. Descriptives for Portfolio A, Portfolio D, Portfolio E, Portfolio F and Portfolio G.

Portfolio	Annualized	Variance	Mean	Max	Min
A	15.45%	0.7518	36.11%	201.45%	−52.15%
D	15.74%	0.6893	34.74%	207.92%	−53.21%
E	49.61%	0.9508	74.54%	233.82%	−44.67%
F	20.37%	0.7030	39.22%	209.19%	−50.02%
G1	29.56%	0.8529	48.58%	283.18%	−23.11%
G2	27.23%	0.9297	47.57%	297.76%	−22.85%
G3	19.64%	1.0467	44.34%	302.95%	−58.35%

3.4 The Influences of Depreciation and Amortization

As seen in the Fig. 5, EBITDA is confirmed to be more effective than EBIT. The annualized return of *portfolio F* is 20.37%, which is roughly 5% higher than the *portfolio A*. The return advantages are kept in most of tested years. As for risks they face, *portfolio F* has a variance of 0.70 compared to 0.75 of *portfolioA*, indicating that the risk of *portfolio F* is smaller than the other portfolio. In summary, we therefore accept *Hypothesis* 4 and conclude than EBITDA is a better stock indicator in Chinese A-shares market.

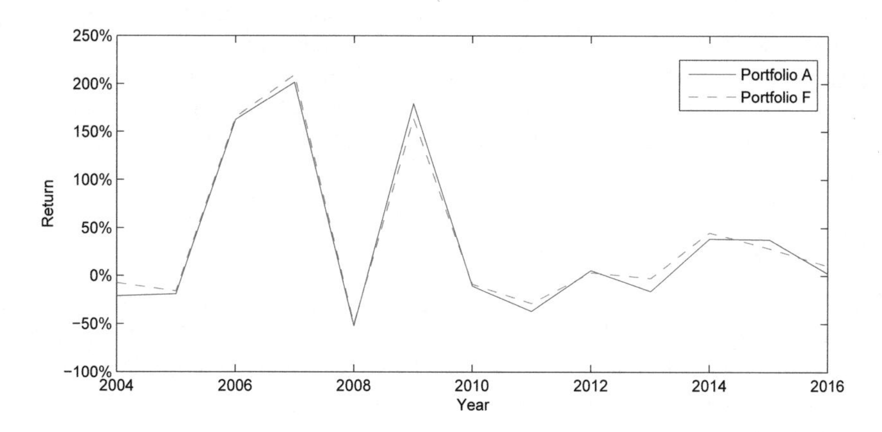

Fig. 5. The aggregated returns of portfolio D and portfolio E 2004 to 2016

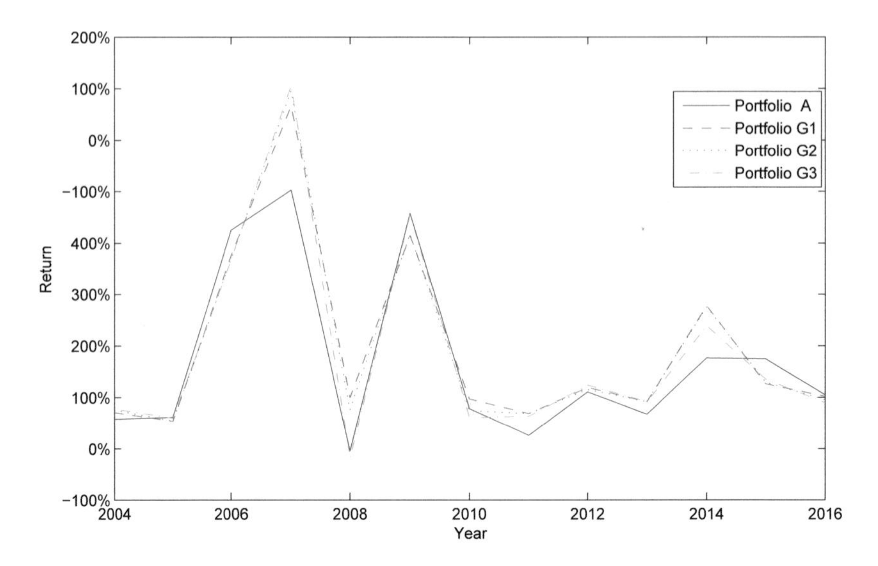

Fig. 6. The aggregated returns of Portfolio A, Portfolio G1, Portfolio G2 and Portfolio G3

3.5 The Effectiveness of Price Earning Ratio and Price Book Value Ratio

We show the returns of $portfolio\ G1$ (29.55%), $portfolio\ G2$ (27.23%), and $portfolio\ G3$ (19.64%) respectively in Fig. 6. As we expected, the new portfolios achieve higher returns than $portfolio\ A$, concluding that PE*PB can identify the cheap stocks more successfully. However, we cannot ignore the flaw of risk. Although our new portfolios have better performances in return, they also face higher risks at the same time. One thing we find amazing is that the relationship between PE*PB and portfolio return (or portfolio risk) is regularly: the bigger the PE*PB is, the higher the portfolio return (or portfolio risk) is. Finally, we accept the $Hypothesis$ 5.

3.6 The Influences of Index Weights

The Fig. 7 shows the returns based on various weights couples (Given that it is not necessary to show all results for 13 years, we only select 5 years randomly, 2004, 2007, 2008, 2011 and 2015, to generate this line chart). From the Fig. 7, it is clearly that there is no obvious trend of return variation. We previously reckoned that returns of portfolios with weight couples (0.5, 0.5) could be the highest, and returns of other weight couples could roughly diminish consequently. Hence, the $Hypothesis$ 6 is failing to be proved. Basically, the portfolio returns seen in Fig. 7 keep stable and even achieve the highest value at terminal vertex, for example, in 2007. That is, portfolio returns do not change with weights of financial indicators regularly. It is surprised that weights of return on capital and

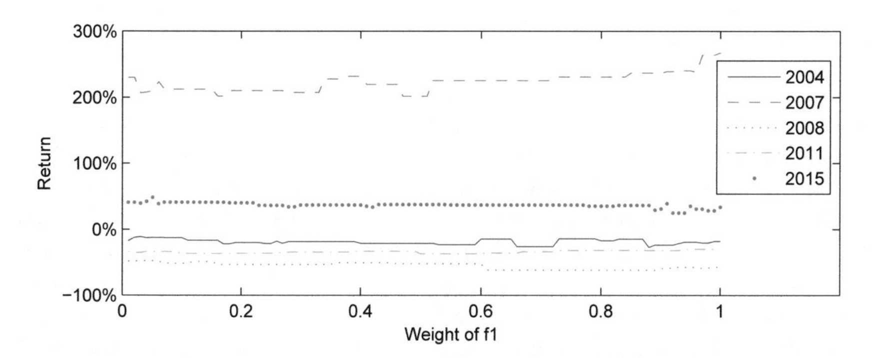

Fig. 7. Returns based on various weights couples for 13 years weighted

earnings yield impacts less on return. However, it can be a beneficial discovery for investors because even when financial information is missing, investors can still use return on capital or earnings yield solely to select stocks profitably. Overall, the effectiveness of magical formula in A-shares market is out of debate, because even the lowest return in series generated by 101 returns is even far higher than return of the common value investing strategy and CSI300.

3.7 The Profits Exceeding Normal Returns

Using the market premium factor as explanatory variable and return of *portfolio A* as dependent variable, the scatter diagram and regression equation can be seen in the Fig. 8. Clearly, the intercept is positive but small, only 0.1621 for alpha. Still, we undoubtedly cannot reject the *Hypothesis* 7. That is to say, magical formula is effective in Chinese A-shares market and it can guide investors get profits exceeding normal return, although not that big.

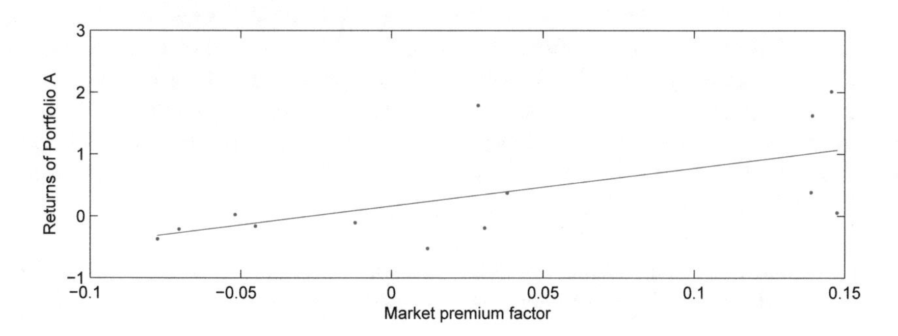

Fig. 8. Regression figure using the return of Portfolio A as dependable variable and market premium factor as explanatory variable

3.8 Discussions

As discussed above, magical formula has an absolute annualized return advantage compared with both the common value investing strategy and CSI300. Furthermore, it is extremely reliable in long-term investment. The highest annual return was 201.45% in 2007, and the lowest return was −52.15% in 2011. Although magical formula does perform superior in most of tested years, it still faces investment failures and unavoidable loses in some years, which investors should pay attention to.

What's more, we also make some changes in Greenblatt's stock selecting indicators to obtain a higher investment return. We find that the portfolio of small cap stocks performs better, achieving a annualized return 49.61%, while the annualized return of portfolio generated by big cap stocks is only 15.73%. However, the risk of the former portfolio is higher because small cap stocks always face more risks than big cap stocks. EBITDA can be a better financial indicator and it can replace EBIT when using magical formula to select stocks, because EBITDA can help investors achieve 5% higher return and face less risk. More interestingly, PE*PB is found to be an effective indicator to identify the cheap stocks, and 30 is an appropriate upper bound compared with 50 and 100, since the portfolio with $PE * PB < 30$ performs superior both in terms of return and risk. When considering weights of two basic stock selection indicators, return on capital and earnings yield, we made a modified empirical research suggesting that the weight impacts less on return. Thus, stocks selection can still be realized in cases when financial data is inadequate. Finally, the profits exceeding abnormal return, also known as alpha, of magical formula is positive but a bit small.

4 Conclusion

In this paper, we successfully examine the superiority of magical formula in Chinese A-shares market. In summary, magical formula is effective and profitable since it can beat the common value investing strategy and CSI300, only using simple fundamental analysis. Especially, longer holding period can make magical formula even more profitable. Moreover, our new strategies proposed in this paper perform even better. On the whole, the significance of magical formula in Chinese A-shares market are beyond all question and we suggest that more investors can use this simple and easy implemented strategy in real stock trading, which is beneficial for both nation and individuals indeed.

Acknowledgment. The author would like to thank the anonymous reviewers for their very insightful and inspiring comments that helped to improve the paper.

References

1. Graham, B.: Security Analysis: The Classic, 1934th edn. McGraw-Hill, McGraw-Hill (2008)
2. Fama, E.F.: Random walks in stock market prices. Financ. Anal. J. **21**(5), 55–59 (1965)
3. Cronqvista, H., Siegelb, S., Yua, F.: Value versus growth investing: why do different investors have different styles. J. Financ. Econ. **117**(2), 333–349 (2015)
4. Fama, E.F., French, K.R.: The cross-section of expected stock returns. J. Financ. **47**(2), 427–465 (1992)
5. Doukas, J.A., Kim, C.F., Pantzalis, C.: A test of the error-expectations explanation of the value/glamour stock returns performance: evidence from analysts' forecasts. J. Financ. **57**(5), 2143–2166 (2002)
6. Joseph, D.: Value investing: the use of historical financial statement information to separate winners from losers. J. Account. Res. **38**, 31–41 (2000)
7. Chahine, S.: Value versus growth stocks and earnings growth in style investing strategies in Euro-markets. J. Asset Manag. **9**(5), 347–358 (2008)
8. Asness, C.S., Moskwitz, T.J., Pedersen, L.H.: Value and momentum everywhere. J. Financ. **68**(3), 929–985 (2013)
9. Bartov, E., Kim, M.: Risk, mispricing, and value investing. Rev. Quant. Financ. Account. **23**(4), 353–376 (2004)
10. Woltering, R., Weis, C., Schindler, F., Sebastianb, S.: Capturing the value premium – global evidence from a fair value-based investment strategy. J. Bank. Financ. **86**, 53–69 (2018)
11. Perez, A.: Value investing and size effect in the South Korean stock market. Int. J. Financ. Stud. **6**(1), 31–41 (2018)
12. Stafford, E.: Replicating Private Equity with Value Investing, Homemade Leverage, and Hold-to-Maturity Accounting (2017). https://ssrn.com/abstract=2720479
13. Ray, K.: Artificial intelligence and value investing. J. Invest. **27**(1), 21–30 (2018)
14. Betermier, S., Calvet, L.E., Sodini, P.: Who are the value and growth investors? J. Financ. **72**(1), 5–46 (2016)
15. Johnson, R.R.: Behavioral obstacles to successful value investing. Fac. Publ. Financ. Serv. **18**(2), 745–752 (2017)
16. Porta, R.L., Lakonishok, J., Shleifer, A., Vishny, R.: Good news for value stocks: further evidence on market effciency. J. Financ. **52**(2), 859–874 (1997)
17. Ko, K., Lin, S., Su, H., Chang, H.: Value investing and technical analysis in Taiwan stock market. Pac.–Basin Financ. J. **26**, 14–36 (2014)
18. Greenblatt, J.: The Little Book That Beats the Market: Your Safe Haven in Good Times or Bad. Wiley, Hoboken (2010). Updated Edition
19. Persson, V., Selander, N.: Back testing The Magic Formula in the Nordic region, Master Thesis. Stockholm School of Economics (2009)
20. Ball, R., Gerakos, J., Linnainmaa, J.Y., Nikolaev, V.: Deflating profitability. J. Financ. Econ. **117**(2), 225–248 (2015)
21. Zhang, Y., Zhou, Y.: Are China's stock markets weak efficient? J. Financ. Res. **3**, 34–40 (2001)
22. Rousseau, R., Rensburg, P.: Time and the payoff to value investing. J. Asset Manag. **4**(5), 318–325 (2003)
23. Banz, R.W.: The relationship between return and market value of common stocks. J.Financ. Econ. **9**(1), 3–18 (1981)

24. Fama, E.F., French, K.: Size, value, and momentum in international stock returns. J. Financ. Econ. **105**(3), 457–472 (2012)
25. Wu, W.: The P/E ratio and profitability. J. Bus. Econ. Res. **12**(1), 67–76 (2014)
26. Fairfield, P.M.: P/E, P/B and the present value of future dividends. Financ. Anal. J. **50**(4), 23–31 (1994)
27. Fama, E.F.: Efficient capital markets: a review of theory and empirical work. Finance **25**(2), 383–417 (1969)
28. Lim, T.C., Huang, W., Yun, J.L.X., Zhao, D.: Has stock market efficiency improved? evidence from China. J. Financ. Econ. **1**(1), 1–9 (2013)
29. Dai, X., Yang, J., Zhang, Q.: Testing Chinese stock market's weak-form efficiency: a root unit approach. Syst. Eng. **23**(11), 23–28 (2005)
30. Sharpe, W.: Capital asset prices: a theory of market equilibrium under conditions of risk. J. Financ. **19**(3), 425–442 (1964)
31. Fama, E.F., French, K.: Common risk factors in the returns on stock and bonds. J. Financ. Econ. **33**(1), 3–56 (1993)

Design and Implementation of Key Modules for B2B Cross-Border E-Commerce Platform

Guodong Zhang, Qingshu Yuan, and Shuchang Xu(✉)

College of Information Science and Engineering, Hangzhou Normal University, Hangzhou, Zhejiang, China
philous@163.com

Abstract. Although most existing B2B (Business-to-Business) e-commerce platforms have relatively complete functions and processes, these platforms lack flexible and comprehensive price management and management of sales teams for wholesaler users. In order to solve these problems, a B2B cross-border e-commerce platform based on spring MVC (Model View Controller) framework is proposed. This platform provides wholesaler users for different customers according to different business needs to develop different product quotation template, manage their own sales team and other functions, which frees up the sales team that allows them to spend more time developing new business.

Keywords: B2B · Cross-border e-commerce · E-commerce platform
Spring MVC · Quotation template

1 Introduction

"2017 China cross-border e-commerce outlet (B2B) development report" shows that China's cross-border e-commerce market continues to grow, the industrial chain continue to improve, and regional advantages are gradually brought into play [1]. Which, B2B electronic business platform showed a growing market and business opportunities. At present, the typical B2B platforms in China include Dunhuang Network [2], Easy Ordering Network [3], Alibaba [4], China Manufacturing Network [5], etc., as well as various online B2B websites with vertical industry segments. The positioning of these platforms is shown in Table 1.

For the B-side user's cross-border e-commerce business, the existing common platform has some minor deficiencies:

- Lack of flexible and comprehensive quote features.
  For merchants, cross-border e-commerce faces customers in different countries, and these customers have different pricing capabilities for the same goods. In addition, businesses also need to provide different pricing for new and old customers. Therefore, a flexible quotation mechanism is needed to meet the differentiated pricing of customers in different regions and different categories, so as to maximize the profits of merchants.

N. Xiong et al. (Eds.): ISCSC 2018, AISC 877, pp. 195–203, 2019.
https://doi.org/10.1007/978-3-030-02116-0_23

Table 1. Positioning of some domestic B2B platforms

Platform	Positioning
Easy ordering network	An omni-channel mobile ordering platform that manages order-centric invoicing process
Alibaba	A B-side buyers and sellers platform which build online display platforms and sales channels for businesses and buyers
China manufacturing network	The online platform of China's product information, provides e-commerce services for Chinese products globally, and focuses on B2B e-commerce platforms made in China
Dunhuang network	The first domestic website for B2B online trading for SMEs, the world's leading online trading platform
Conghui network	Focus on the B2B service platform for electronic product supply chain, electronic and related industry segmentation platforms

- There is a lack of hierarchical management functions for wholesaler users.
  For merchants, there may be a hierarchical sales force. Sales personnel in different roles in the merchant team have different functions, rights, and performance distinctions. However, the existing e-commerce platforms do not have the hierarchical management function for the entire sales team.

In view of the above two issues, this paper proposes a platform for cross-border e-commerce B-end merchants, focusing on the two key modules that can be used for hierarchical management of businesses and flexible pricing. As shown in Fig. 1.

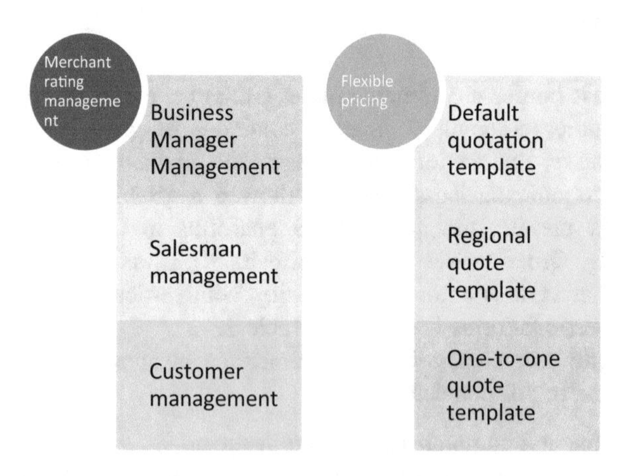

Fig. 1. The article mainly designed content

In the whole system, hierarchical management business is embodied in the business management for business and business manager role user services distribution, including sales account upgrade, distribution business to business manager, etc. For example, a merchant may transfer a salesman of a business manager to another business manager, and promote a salesperson with excellent sales performance as a business manager. As shown in Fig. 2.

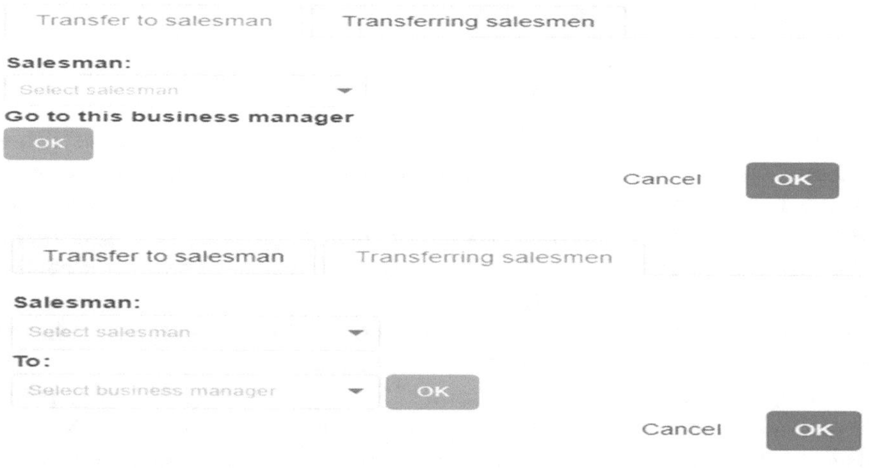

Fig. 2. Business hierarchy management business allocation function diagram

In addition, the flexible price management mechanism provides merchants with one-on-one quotation templates for different regions and new and old customers. For example, for a newly-registered user, the merchant can manage the display of the customer's page by setting, not displaying the price of the product for a newly registered customer but displaying the product information. As shown in Fig. 3.

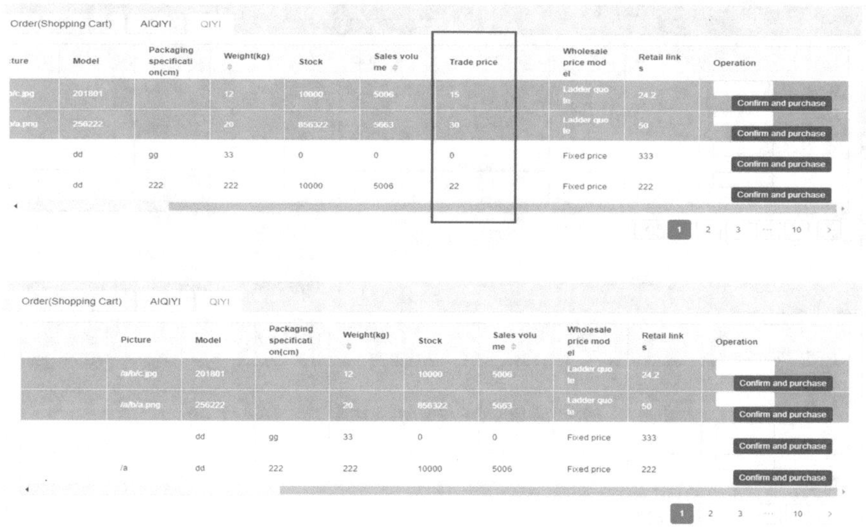

Fig. 3. Businesses display control content for customer pages

2 Architecture Design

The entire system platform is implemented using three-tier architecture - client layer, application layer, and data layer [6].

The client layer is responsible for the interaction between the platform and the user. The front-end adopts the latest version of VueJS of the Model-View-View Model framework and combines the UI (User Interface) framework iView to provide users with page displays [7]. Jumps between pages are implemented via VueRouter. VueJS's state management mode Vuex adopts the state of all components of the centralized storage management application and ensures that the state changes in a predictable way with the corresponding rules [8]. The data exchange between the front end and the back end is realized through Axios and RESTful architecture.

The application layer uses the Spring MVC framework. Spring MVC intercepts and forwards the request to the corresponding controller, and then returns the JSON data by configuring Spring MVC. The spring framework of the back-end system mainly implements dependency injection. The application layer uses the shiro framework to perform transaction management and isolates operations such as logs, access control, and database access to improve the aggregation of services [9].

The data layer uses MyBatis to persist the data [10], Redis as a second-level cache database. Each time the interface is queried, it is first determined whether there is any cache in Redis, and if so, it is read. If not, the database is queried and saved in Redis. The next time it is queried, it will be read directly from the cache [11]. The overall framework of the system is shown in Fig. 4.

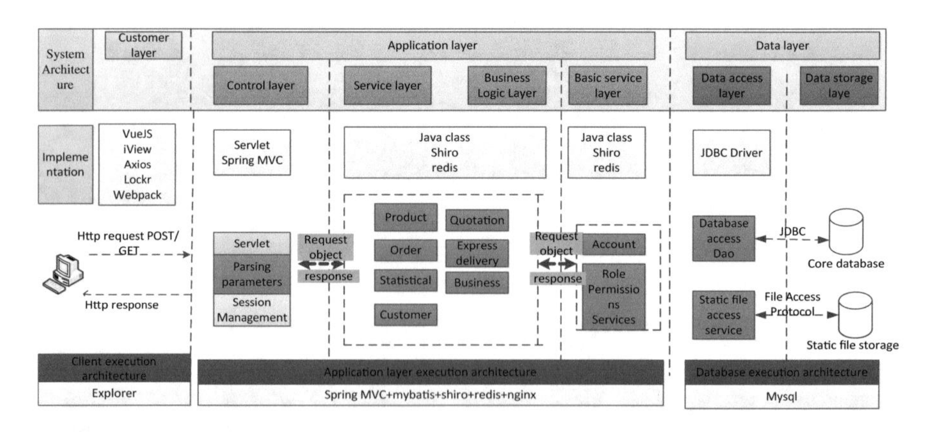

Fig. 4. Cross-border e-commerce platform architecture

3 The Core Module Design

The modules related to flexible pricing and merchant-level management functions, such as account management, pricing management, and customer management, are completely different from existing platforms.

- Account Management
  The entire cross-border e-commerce B2B platform mainly includes three layers of platforms, businesses, and users. The functional permissions of each layer of roles are different. For merchants, they also have the role of merchant administrator, merchant business manager, and merchant salesperson. These three roles with different permissions design. As shown in Fig. 5.

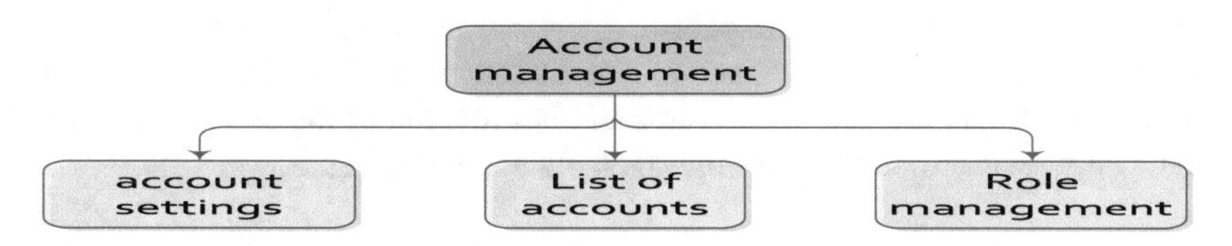

Fig. 5. Account management module diagram

- Pricing Management
  The pricing management module is the core module of the entire platform. His main function is to help merchants develop a unique product quotation template for each customer; liberate the salesperson, so that the salesperson have more time to develop new customers. This module allows users in different roles in the merchant to set different pricing templates, thereby providing merchants with a flexible quotation method. As shown in Fig. 6.

Fig. 6. Quote management module diagram

- Customer Management
  The customer list module shows all the users of the business, businesses can perform some operations on these customers, such as show/hide price. The basic setting module includes setting default quotation templates, express delivery template default settings, customer list display item settings, and order list display item settings for newly registered customers. As shown in Fig. 7.

Fig. 7. Customer management module diagram

4 Core Module Implementation

Merchant role management, as well as pricing modules and customer management are the core modules of the entire system. The following details the implementation of these major modules.

- Merchant Role Management Module
  Businesses have three types of roles: Merchant Manager, Merchant Business Manager, and Merchant Salesperson. Only the merchant administrator has the operation authority of the three sub-modules of account setting, account list and role management, and can see the user account information of the two roles defined by his own salesman and business manager. As shown in Fig. 8.

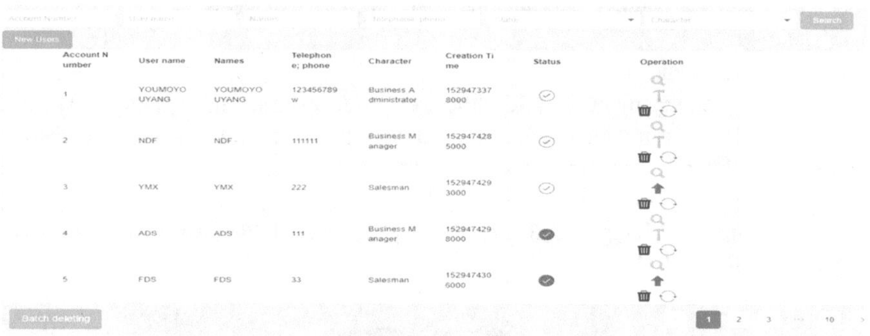

Fig. 8. Merchant administrator account management module

Merchant administrators can add or delete accounts. It also provides the ability to view detailed information, reset passwords, and upgrade accounts, enable/disable accounts and business assignments. The upgrade account is mainly for the salesman of the merchant, and the merchant administrator can upgrade the original salesman's account to the role that the business manager or other merchant has defined for better performance or other reasons. The service allocation function includes the distribution of the salesperson and the distribution to the customer. The merchant administrator can assign the salesperson and the customer to the business manager, and can also assign the customer to the salesperson. It is worth mentioning that after the clerk or business manager leaves the post, the merchant administrator can transfer all customers under the account, including customer data, to other business managers or clerk accounts, which greatly reduces the loss of customers due to staff turnover.

- Pricing Management Module
  The pricing management module is a feature of the entire platform, and its role is mainly two: First, to help businesses develop a product quotation form that is closest to the actual transaction status for each customer, and improve the transaction success rate; The second is to help merchants manage customer information,

retain old customers, liberate salesmen, and allow sales staff to focus more on new business development. Its workflow is shown in Fig. 9.

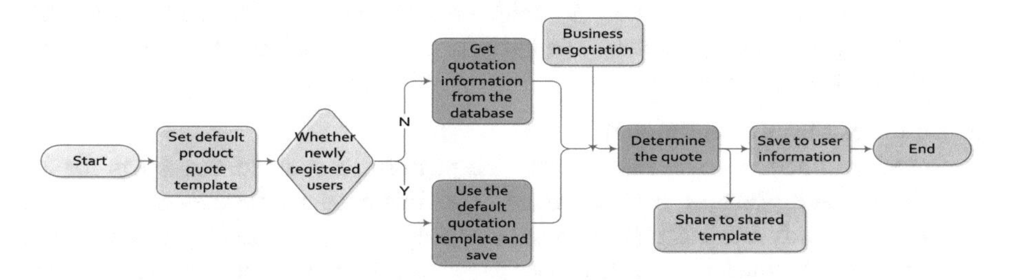

Fig. 9. Quote template flow chart

First, the merchant sets a default product quotation template for the newly registered client on this platform. When the new registered user logs in, the platform will display the price of the product according to the merchant's settings (the platform provides merchants with product information and product price concealment functions. Only if the merchant has opened the product price display function for this customer to display the price), once the customer registers, the quotation template will accompany the customer as its basic attribute. In addition, quotation templates can also be shared among merchants at all levels of sales.

When designing the database table, considering that the quotation template should always be consistent with the merchant's product table, once a new product is added or removed, the quotation template must be changed accordingly. Because the product information is also stored in the quotation template, the platform will add an update operation to the quotation template table when the merchant performs the operation of adding and deleting the product. The quotation template information table design is shown in Table 2.

The "QuotationInfo_id" field in Table 2 indicates the quotation template information table id; the type is int, which is not an empty self-increment primary key. The "QuotationInfo_name", "QuotationInfo_description", and "QuotationInfo_source" fields indicate the name of the quotation template, the description of the quotation

Table 2. Quote template information table

Filed	Type	Is null	Primary key
QuotationInfo_id	Int	No	PK, auto_increment
QuotationInfo_name	Varchar	No	
QuotationInfo_description	Varchar	No	
QuotationInfo_createtime	Timestamp	No	
QuotationInfo_modifytime	Timestamp	No	
QuotationInfo_source	Varchar	No	
QuotationInfo_status	Tinyint	No	

template, and the source of the quotation template. “QuotationInfo_createtime” and “QuotationInfo_modifytime” indicate the creation time of the quotation template and the latest modification time respectively. “QuotationInfo_status” indicates the status of the quotation template. Its value of 0, 1, 2:0 indicates that the template is a basic data table, stores the initial information of all the products of the merchant, 1 indicates that the template is an ordinary quotation template, and the value 2 indicates that Templates are shared quote templates.

- Customer Management Module
  In the customer management module, the platform lists all its customers for the business and manages the customers. The implemented functional operations include: modify the customer’s quotation template, freight template, whether to display the product on the customer page, display the product price, view the customer’s order list, view the customer information, assign the salesperson for the customer, reset the customer account password, and delete the customer account number. As shown in Fig. 10.

Fig. 10. Customer management module

5 Conclusions

In some application scenarios of cross-border e-commerce services, there are still some functions that need to be explored. Based on the SSM framework, the functions of flexible pricing and multi-level management are designed and implemented.

I would like to express my sincere thanks to those who helped me during the project. Thanks to http://www.puzzlewholesale.com/ for their support and my tutor Xu for guiding the project and this paper.

References

1. Reporter Liu Shubo: Cross-Border E-Commerce on Longjiang "Online Silk Road". Harbin Daily, 2018-06-17(002)
2. Dunhuang Net: http://seller.dhgate.com/.2018-04-22
3. Easy Ordering Network: https://sso.dinghuo123.com/.2018-04-22
4. Alibaba: https://www.1688.com/.2018-04-22
5. China Manufacturing Network: https://www.made-in-china.com/.2018-04-22
6. Fang, W.: Java EE Architecture Design and Development Practice. Tsinghua University Press, Beijing (2017)
7. Zhang, Y.: Vue js Definitive Guide. Electronic Industry Press, Beijing (2016)
8. Chen, L.: Vue.js Front-End Development Quick Start and Professional Application. People's Posts and Telecommunications Press, Beijing (2017)
9. Bretet, A.: Spring MVC. Publishing House of Electronics Industry, Beijing (2017)
10. Wu, Y.: Design and Implementation of Stock Market Competition System Based on Spring MVC and MyBatis Framework. Nanjing University (2016)
11. Yang, K.: Java EE Internet Lightweight Framework Integration Development SSM Framework Spring MVC+Spring+MyBatis and Redis Implementation. Publishing House of Electronics Industry, Beijing (2017)

An Analysis of Barriers to Embedding Moral Principles in a Robot

Ping Yan(✉)

Xi'an Shiyou University, Xi'an, China
pyan@xsyu.edu.cn

Abstract. One of the difficulties to embed moral principles in a robot is the barriers due to moral principles themselves. The barriers are analysed from four specific topics. The first is about the difference between value judgments and fact judgments. The following analyses focus on the specific moral barriers resulted respectively from deontology, consequentialism, and virtue ethics. These analyses show that it is much harder than what we thought to make a moral robot. If we forget the reductionism, the method used commonly in moral philosophy, we should find a new direction to solve the problem. It is highly possible that human beings can find a better way to improve their own moral cultivation while they try to make moral machines.

Keywords: Robot · Barrier · Moral principle

1 Introduction

The paper is about how to make moral robots. With the development of Artificial intelligent, the studies of ethics of robotics are getting more and more important. There are two kinds of the problems: one is how human uses robots morally; the other is how to make robots behave morally, or how to create moral robots. This paper focuses on the latter problem.

There are various difficulties for creating moral robots, and the barriers mentioned here are limited to what result from moral principles themselves. To make robots behave morally, the easy way seems to embed in robots proper programs from moral principles. There are various difficulties for this way, and four topics are discussed to analyse the barriers to embedding moral principles in the robots. The first is about barriers result from the difference between fact judgments and value judgments. The studies of the difference between these two kinds of judgments make it distinct that value judgments cannot be translated into simple principles to be embedded in robots. Other three analyses focus on the specific moral barriers resulted respectively from three different moral systems: deontology, consequentialism, and virtue ethics.

Can we find a way to step over the barriers to embedding the moral principles in robots? If we forget the reductionism, the method used commonly in moral philosophy, we should find a new direction to solve the problem.

N. Xiong et al. (Eds.): ISCSC 2018, AISC 877, pp. 204–212, 2019.
https://doi.org/10.1007/978-3-030-02116-0_24

2 Barriers Result from the Difference Between Two Kinds of Judgments

There are different kinds of judgments. One classification method is to sort the judgments according to whether there are subjective moral contents in the judgments. In this way, fact judgments and value judgments are different. Fact judgments describe the basic facts, so this kind of sentence is called descriptive. "He is a teacher." for example, is a fact judgment. Value judgments provide certain value to the world, so this kind of sentence is called prescriptive. "He is an excellent teacher." for example, is a value judgment. Some time fact judgments depend on value judgments, and vice versa, so there is no clear distinction between these two kinds of judgments. But in most cases we can find that they have distinctly different characters. Fact judgments are usually dependent on objective neutral facts, and value judgments are usually dependent on morally subjective worth, or aesthetic criteria.

Many philosophers found that value judgments or moral principles cannot be translated into simple judgments and cannot be expressed by plain language symbols. Based on this, it is clear that these moral principles are difficult to be embedded in robots. Four philosophers are mentioned below to explain this kind of barriers.

English philosopher David Hume gave a clear statement about the effects on the system of morality from the difference between fact judgments and value judgments. "In every system of morality, which I have hitherto met with, I have always remarked, that the author proceeds for some time in the ordinary ways of reasoning, and establishes the being of a God, or makes observations concerning human affairs; when all of a sudden I am surprised to find, that instead of the usual copulations of propositions, is, and is not, I meet with no proposition that is not connected with an ought, or an ought not. This change is imperceptible; but is, however, of the last consequence. For as this ought, or ought not, expresses some new relation or affirmation, it is necessary that it should be observed and explained; and at the same time that a reason should be given, for what seems altogether inconceivable, how this new relation can be a deduction from others, which are entirely different from it. But as authors do not commonly use this precaution, I shall presume to recommend it to the readers; and am persuaded, that this small attention would subvert all the vulgar systems of morality, and let us see, that the distinction of vice and virtue is not founded merely on the relations of objects, nor is perceived by reason." [1] Today, even knowing the difference between these two kinds of statements, we still do not know how to put our value judgments on an objective foundation, let along express them in programs.

Danish philosopher S. Kierkegaard expressed strong concern about the difference between the two kinds of judgments, and emphasized the limit of rational abstract from fact judgments. He criticized Hegel's famous idea: "the Rational is the Real and the Real is the Rational." Kierkegaard argued that this idea confused the distinction between facts and beliefs, and he emphasized that existence, the facts, are different from thought, especially different from the moral judgments and religion beliefs. In other words, language, the abstract logic, divides us from our life experience. His opinion about religion can help us to have a further understanding about the difference between facts and value. He argued that facts judgments are evident, but our belief are usually can not based on these clear evidences. To the contrary, our beliefs are dependent on doubt, dependent on our having no clear evidences for our faith. Doubt even creates a new world: doubt is conquered by faith, and at the same time faith brings doubt into the world.

German philosopher F. Nietzsche struggled with the difference between facts and value in his key academic period. He argued that language "tyrannize over us as a condition of life." [2] It will be helpful for understanding this to review what happened during his writing the last posthumous work *will to power*. In the beginning, he tried to overthrow the basic western moral principles with the great ambition which he called a revaluation of all values. He used "will to power" as the basic principle for human life, and believed that people seek power as general goal in different way. What happened later left us with a deep concern about the general nature of moral system. The later experts on Nietzsche found two Nietzsche's notebooks he used to write his great book, *The Will to Power*. The evidence from the notebooks shows that before Nietzsche's mind shut down, he intended to stop the work from being known to the public, and he went so far as to write a shopping list at the line spacings in his manuscripts. At last, Nietzsche opposed any attempt to reduce the moral acts from one basic principle including his "will to power". Nietzsche found it impossible to give some axioms and then from these axioms deduce theorems and a system of moral philosophy. This seems to break down our dream to find an axiom-deduction moral system. At the same time, we recognized that it is not an easy way to find a logic system from moral principles and then put it in a smart machine.

Another German philosopher L. Wittgenstein gave a similar argument. Wittgenstein belonged to analyst tradition, and his predecessors tried to use logic to solve the language problem, and tried to show that an extension of the basic principles of logic can give us all the fundamental notions of mathematics. Wittgenstein had a bigger ambition to use logic explain the whole world. It is said the early Wittgenstein got the inspiration from a traffic court where people investigated an automobile accident by using dolls as representatives of the real people who involved in the accident, and using toys as representatives of the real cars and trucks they are involved in the accident too. These dolls and toys were put into a system exactly represented the actual space relation in the accident. With the help of these toys and dolls, people successfully explained what happened in the real accident. Wittgenstein asked why this was possible. His answer was that between the space system of the people, the car and truck in the real accident and the space system of the toys and dolls there was the same logic relation. In his famous work *Tractatus Logico-Philosophicus,* he gave a brave conclusion about what the world is (Wittgenstein gave number mark for the all proposition in his work.): The facts in logical space are the world (1.13). Logic pervades the world: the limits of the world are also its limits (5.61). Propositions represent the non-existence and existence of states of affairs (4.1). What should be emphasized here is that Wittgenstein found that from these logic statements we can not give the world value, the meaning of our life. He said: All propositions are of equal value (6.4). The sense of the world lies outside the world (6.41). Ethics cannot be put into words (6.421). The solution to the enigma of life in space and time lies outside space and time (6.4312). What he said in the preface of his work can help us understand this better. He said that the book will, therefore, draw a limit to thinking, or rather—not to thinking but to the expression of thoughts; for, in order to draw a limit to thinking we should have to be able to think both sides of this limit (we should therefore have to be able to think what cannot be thought).

In the latter thought of Wittgenstein, we can also find the opinion about language limits for value. Most people believe that language meaning is to name, but Wittgenstein found that it is difficult to give an invariable name for many objects. Giving name means to find a common feature for a class, and it is some time easy to do this. For example, it is easy to find a common feature and then give a name for triangle (three-straight-lines-sided closed figure). Some time, it is difficult to find a common feature and give a name for a class. We can only find some overlap similarities in our language use. These situations can be found more frequently in moral language because it is more difficult to find common feature in moral expression.

3 Kant's Solution to the Problem How to Give Our Life Value

Deontology was originally from a Greek, *deon,* which means duty or obligation, and today it means the moral system based on duty or obligation. Deontology, consequentialism and virtue ethics are three main branches of moral philosophy. Immanuel Kant is the most important representative of deontology, and from what Kant discovered, we can find distinct barriers to embedding the moral principles in robots. The other two theories, consequentialism and virtue ethics, will be discussed respectively in parts IV and part V.

Kant's great contributions are based on the work to solve the problems Hume proposed. As mentioned before, Hume concerned about the difference between facts judgments and value judgments. Besides this, his empiricism and skepticism are neither acceptable nor refutable. Hume argued that there are two kinds of propositions: analytic propositions and synthetic propositions. The proposition "All bachelors are unmarried." is analytic, and the proposition "The cat is on the mat." is synthetic. Analytic propositions are a priori true by definition, necessarily true, and tautological, i.e. they can not provide us with new information. Synthetic propositions are posteriori, not necessarily true, and can provide us with new information. We cannot find necessary true propositions which can give us new information at the same time. From his empiricist point, Hume argued that we can not find the necessary foundation for the truth of our knowledge. He believed that we do not know if there is external world, and do not know if there is causality, and we even do not know if there are real ourselves. Furthermore, we can not find foundation for our moral principles.

Kant thought that only Hume's skepticism is refused, philosophy, science, moral value and religion can be respectable again. Kant believed that we should have synthetic a priori propositions, i.e. the statements which are meaningful and necessarily true. He divided our mind into three different kinds of faculties: perception, understanding and reason. In our perception, with the help of time and space, a priori perception forms, we can have necessary meaningful information. With the help of categories of the faculty of understanding, we can get more extensive knowledge which is meaningful and necessary. Kant said, "For how should our faculty of knowledge be awakened into action did not objects affecting our senses partly of themselves produce representations, partly arouse the activity of our understanding to compare these representations, and, by combining or separating them, work up the raw material of the sensible impressions into that knowledge of objects which is entitled experience?" [3] From these analyses, Kant provided us with meaningful and necessary true perception and understanding knowledge.

After the reestablishment of the facts judgments, how to give moral judgments necessary foundation is the further problem for Kant. He agreed that beyond the phenomenal world, there is a noumenal world where the God, the eternity, and moral principles exist, but we have no access to the noumenal world. We can not have the knowledge of phenomenal world with the help of time, space and causality used in facts judgments. But if we have no foundation for our moral judgments, for the existence of eternity, we will lose the meaning of our life, let alone have moral programs used in smart machines. Great Kant found a way to solve the problem. He distinguishes our knowledge from beliefs and based our beliefs on moral necessity. How to find moral necessity? He created a new name: categorical imperatives which are intrinsically good and valid, and every one must obey them in all time. These categorical imperatives are different from hypothetic imperatives which are based on worldly desires and needs.

The next problem is how to prove these categorical imperatives are intrinsically good and valid. On Kant's tombstone there is a paragraph from his work *Critique of Practical Reason*: "Two things fill the mind with ever new and increasing admiration and reverence, the more often and more steadily one reflects on them: the starry heavens above me and the moral law within me. I do not need to search for them and merely conjecture them as though they were veiled in obscurity or in the transcendent region beyond my horizon; I see them before me and connect them immediately with the consciousness of my existence." [4] Even though we can not find foundation from facts judgments, but we can have a strong feeling of the sublime and beauty which lead us to the noumenon world, the transcendental existence.

Can we find more specific moral principles? Kant gave us some imperatives after which we should follow in our moral life. The start point is freedom. Kant argued that we human beings must be free and reasonable if we want live a meaningfully moral life. Ought implies can. The reasonable human beings should recognize that we must respect the freedom of all if we want be free. From this start point, Kant gave us the following principles. One is called formula of universal law: Act only according to that maxim by which you can at the same time will that it should become a universal law. We are duty-bound not to lie, and if we all lied, there were no words could be truth. We are duty-bound not to steal, and if we all stole, there were no properties left as private properties, and there were nothing left which could be stolen. We are duty-bound not to kill, and if we all killed, there were no one left to comply with the law. Kant's line of reason is clear: when contradictions follow after a moral command, actions on that command are impossible in the real world; if there are no contradictions, the command is permissible. If we agree with the opinion that all rational beings should recognize that we must respect the freedom of all, we should agree with the following demand that all rational beings must be treated never as a mere means but as the supreme limiting condition in the use of all means, i.e., as an end at the same time.

It seems easy to put these commands mentioned above into the mind of smart machines, but there are a lot of difficulties if we analyse them carefully. A frequently used example to falsify the possibility is as follow: if a murderer asks you where your friend is, you are duty-bound not tell him the truth, rather than not to lie. Someone suggests that we can substitute the misleading truth for the outright lie, and in the situation mentioned above, you should say you do not know. It is a misleading truth because you really do not know where your friend is just now. Is this a good way to solve the difficulty to put a moral principle in smart machines? We can say this solution is just to substitute a new problem for the old one. There are the same difficulties for machines to distinguish the difference between outright lies and misleading truths. Even in ancient Plato's famous work *The Republic*, there is a similar example: what is justice? The first answer in *The Republic* is to give to each what is owed to him. It seems a clear and perfect definition, and probably we can put it in a smart machine's mind. In the work, Socrates immediately falsified the definition: We say that justice is to give to each what is appropriate to him, but "when he is out of his mind, it is, under no circumstances, to be given to him?" "Should one also give to one's enemies whatever is owed to them?" [5].

Following this line of thinking, so many difficulties are found for smart machines to recognise the right moral principles. Can we find another way to give the machine a clear and simple formula to calculate values?

4 The Way to Calculate the Value from Utilitarianism

Utilitarianism argues that the best act is what can maximize utility. It can be also called consequentialism which put the standard of right and wrong on the consequence of acts. J. Bentham, the founder of modern utilitarianism, gave a clear description about the opinion: "Nature has placed mankind under the governance of two sovereign masters, pain and pleasure. It is for them alone to point out what we ought to do… By the principle of utility is meant that principle which approves or disapproves of every action whatsoever according to the tendency it appears to have to augment or diminish the happiness of the party whose interest is in question: or, what is the same thing in other words to promote or to oppose that happiness. I say of every action whatsoever, and therefore not only of every action of a private individual, but of every measure of government." [6]

The utilitarianism is not necessary egoistic, and we can be motivated by the pleasure of ourselves and others. This idea is reflected in Bentham's most famous maxim: It is the greatest happiness of the greatest number that is the measure of right and wrong. How to decide which happiness is the greatest happiness? Bentham's method seems to be clear and simple: "one person, one vote principle" can easily decide what the greatest happiness is. Bentham found a set of standard for the measure of happiness and pain. This set of standard includes several categories: intensity, duration, certainty, proximity (how soon the happiness can be experienced), fecundity (how many more other happiness can follow this), purity, and extent. To prevent individuals from making mistakes in measurement, the laws are necessary for the whole society to create maximum pleasure. Following Bentham's method, can we give a smart machine clear and right moral demands? If we examine the principle carefully, it is evident that the principle fails to respect individual right, and it is impossible to calculate all values according to Bentham's principle.

Mill is the follower of Bentham, and Mill tried to resolve the problem resulted from Bentham's principle. He criticized the pure quantity measurement of consequences: "It is quite compatible with the principle of utility to recognize the fact, that some kinds of pleasure are more desirable and more valuable than others. It would be absurd that while, in estimating all other things, quality is considered as well as quantity, the estimation of pleasures should be supposed to depend on quantity alone" [7].

Mill's solution faces new problem: What is the best quality? One of the critics is "the doctrine worthy only of swine". If a pig has a right to chose, wallowing in the mud, instead of studying philosophy, would be the best. Even Aristotle expressed the same attitude in his *Nicomachean Ethics*. He said that identifying the good with pleasure is to prefer a life suitable for beasts. Mill argued some pleasures are evidently superior to others: "Few human creatures would consent to be changed into any of the lower animals, for a promise of the fullest allowance of a beast's pleasures; no intelligent human being would consent to be a fool, no instructed person would be an ignoramus, no person of feeling and conscience would be selfish and base, even though they should be persuaded that the fool, the dunce, or the rascal is better satisfied with his lot than they are with theirs… A being of higher faculties requires more to make him happy, is capable probably of more acute suffering, and certainly accessible to it at more points, than one of an inferior type; but in spite of these liabilities, he can never really wish to sink into what he feels to be a lower grade of existence… It is better to be a human being dissatisfied than a pig satisfied; better to be Socrates dissatisfied than a fool satisfied. And if the fool, or the pig, is of a different opinion, it is because they only know their own side of the question…" [8]. A lot of new problems followed Mill's answer. One of them is that Mill's answer can not satisfy the demand to solve the conflicts among different cultures. Different cultures can not be divided in different grades because there are no single common recognised standards. If we just give a smart machine what Mill said, it still loses its head. Another counterview is that Mill ignored the justice and duty. Even though Mill gave some respects to quality of utilities, there are a lot of examples to prove that measurement of consequence can not give enough respect for human's dignity, honor and other similar things. If we pay more attention to this kind of thing, we will return to Kant, and face the same difficulties which Kant already met.

Utilitarians haggle over every ounce to measure the happiness, but found a lot of difficulties to put its principles in smart machines.

5 The New Way from Virtue Ethics

Virtue ethics is different from other two main normative ethics: deontology and consequentialism. Deontology focus on duty of acts, and consequentialism regards the results of acts as the standard of right and wrong. In some way, we can use Kant's categorical imperatives, or utilitarian maximization principle to regulate our acts. Virtue ethics is contrasted to these two, and concerns about human beings themselves rather than their acts. This moral theory provides a method to help people live a positive and self-examination life. Morality comes as the intrinsic nature of human.

Virtue ethics has a long history, and a simple review will help us have a good understanding. Some ancient ethics did not believe that moral principles are requirements outside people; but instead, ethics is about the perfection of human. A rational being has no choice but to cultivate himself, and this self-examination goes through his life. Aristotle argued that moral principles are intelligible only if they have the meaning of the purpose (Aristotle: *telos*) for human life. The purpose of a knife is sharpness; the purpose of a racehorse is speed; and similarly, the purpose of human is virtue, or at least one of the most important purposes. He said: "The Good of man is the active exercise of his soul's faculties in conformity with excellence or virtue, or if there be several human excellences or virtues, in conformity with the best and most perfect among them. Moreover, to be happy takes a complete lifetime; for one swallow does not make a spring." [9] Based on this, Aristotle listed some specific moral demands: wisdom, prudence, justice, fortitude, courage, liberality, magnificence, magnanimity, and modesty in the face of shame or shamelessness temperance. Following Aristotle, Scholasticist Thomas Aquinas emphasized even more acts (*actus* in Latin), and stressed their close relation with being (*esse* in Latin): "There is no essence without existence and no existence without essence." [10] From what Thomas said, we can find the famous statement "existence precedes essence". The essence or the meaning of our life results from our existence, from our hard work, hard choices.

There are a lot of objections to the virtue ethics, and one of them is that this ethics focus on what make people good rather than what are permitted in society and how to decide what is virtue. It is still a hard work to find a simple set of moral principles for smart machines.

6 Conclusion

Among three different morality schools, duty theory and virtue ethics can not offer us a simple system of moral principles; utilitarianism seems give a clear formula for us to calculate what is the maximum value, but it is difficult to use it in practice. Resulting from the difference between value judgments and fact judgments there are still a lot of barriers to embedding ethic procedures in robots. These analyses show that it is much harder than what we thought to make moral robots.

Can we find a way to step over the barriers to embedding the moral principles in robots? If we change our method to understand moral principles, probably we can find a new direction to step over the barriers. Today the method most commonly used is reductionism. This method believes that the surfaces which confront us are determined by some basic elements, and we should tries to reduce and explain appearances by these basic elements. This method is successfully used in modern science, and most of us forget its limit. In moral philosophy, many people try to find the basic elements which can determine the specific moral principles, but they failed. Hume, Kierkegaard, Nietzsche, and Wittgenstein noticed the limit of the method, but their arguments failed to be used in other specific researches. Wittgenstein argued: "But what are the simple constituent parts of which reality is composed?—What are the simple constituent parts of a chair?—The bits of wood of which it is made? or the molecules, or the atoms? —'Simple' means: not composite. And here the point is: in what sense 'composite'? It makes no sense at all to speak absolutely of the 'simple parts of a chair.'" [11] According to reductionism, changes must result from something that changes; therefore, moral principles must be someone's principles. This thought and method lead to worry about human's value, the consequence, and the duty, and further the barriers mentioned. If we forget the reductionism, we should find a new direction to solve the problem. Phenomenology is different from the traditional reductionism, and emphasizes the analyses of the phenomena that just happen in acts. The father of phenomenology Edmund Husserl argued: "I am aware of a world, spread out in space endlessly, and in time becoming and become, without end. I am aware of it, that means, first of all, I discover it immediately, intuitively, I experience it. Through sight, touch, hearing, etc., corporeal things are for me simply there, 'present' whether or not I pay them special attention." [12] Follow this thought, our immediate concern is not whose value, whose consequence, and whose duty, but the value, consequence, and duty themselves.

It is highly possible that human beings can find a better way to improve their own moral cultivation while they try to make moral machines.

References

1. Hume, D.: Treatise of Human Nature, pp. 469–470. Wikisource (2003)
2. Nietzsche, F.: The Will to Power, trans. Walter Kaufmann and R.J. Hollingdale, p. 535. Vintage Books, New York (1968)
3. Kant, I.: Critique of Pure Reason, trans. N. Smith, p. 41. Macmillan and Co., Limited, London (1929)
4. Kant, I.: Critique of Practical Reason, trans. M. Gregor, p. 129. Cambridge University Press, Cambridge (2015)
5. Plato: The Republic, trans. C. Reeve, p. 6. Hackett Publishing Company, Inc., Cambridge (2004)
6. Bentham, J.: An Introduction to the Principles of Morals and Legislation, p. 14. Liberty Fund Inc., Indianapolis (2002)
7. Mill, J.: Utilitarianism, p. 56. Oxford University Press, Oxford (1998)
8. Mill, J.: Utilitarianism, pp. 56-57. Oxford University Press, Oxford (1998)
9. Aristotle: Nicomachean Ethics. I.7.1098a
10. Copleston, F.: A History of Philosophy, vol. 2, p. 53. Image Books/Doubleday and Co., New York (1962)
11. Wittgenstein, L.: Philosophical Investigations, trans. G. E. M. Anscombe, p. 21. Basil Blackwell Ltd., Oxford (1986)
12. Husserl, E.: Ideas: General Introduction to Pure Phenomenology, trans. W. R. Boyce Gibson, p. 91. Collier-Macmillan, London (1969)

Multi-outlet Routing Policy Analysis and Improvement of Campus Network Based on User Group

Haifeng Lu[1(✉)] and Xiangyuan Zhu[2]

[1] Information Center, Zhaoqing University, Zhaoqing 526061, China
haifeng.lu@qq.com
[2] School of Computer Science, Zhaoqing University, Zhaoqing 526061, China

Abstract. With the continuous increase of campus network resources and users, increasing outlet bandwidth and the introducing multiple ISPs have become two key concerns for the development of colleges and universities informatization. Therefore, Zhaoqing University has upgraded and transformed the campus network's outlets. Based on the existing CERNET (China Education and Research Network) and CT (China Telecom) networks, the university brought CMCC (China Mobile Communication Group Co., Ltd) networks and use user group-based policy routing to solved the problem of lacking outlet bandwidth while bringing a multi-ISP healthy competition mechanism. The actual application results prove that this transformation scheme utilized the outlet bandwidth resources efficiently and reasonably and achieved the expected results.

Keywords: Address Translation · Static routing · Policy routing
Minimalist export

1 Introduction

With the increasing popularity of colleges and universities informatization, the campus network resources are becoming more and more abundant, and the number of users continues to grow. As a result, the requirements for campus network infrastructure are also increasing. In recent years, colleges and universities have added ISP outlet routes on the basis of CERNET to increase the speed of campus network's access to the Internet. However, the original routing policy fails to reasonably allocate the network resources of various users, making users' Internet experience unsatisfactory and affecting the normal operation of the campus network [1–3]. In this paper, I took Zhaoqing University as an example to develop a low-input and high-efficiency campus network transformation scheme.

N. Xiong et al. (Eds.): ISCSC 2018, AISC 877, pp. 213–221, 2019.
https://doi.org/10.1007/978-3-030-02116-0_25

2 Campus Network Outlet Transformation Scheme

2.1 Campus Network Status

In Zhaoqing University, there are 25,000 teachers and students, up to 12,000 wired and wireless network terminals (online peak) at the same time, over 100 external websites and over 200 information systems of various types at present. Therefore, higher requirements on the performance and stability of the campus network infrastructure have been raised.

The original campus network outlets included three links, namely, 100M CERNET, 1G CT and 4G CT for student, in the peak period of using Internet, the excessive bandwidth consumed by applications such as video, download, or games that students often use, resulting in slow speed of network and high network packet loss rate, seriously affects the normal operation of the campus network. At the same time, some flow policies are used unreasonably. Each type of user can only use a certain one of the outlet lines, and if the users want to change other outlet routes, a great deal of configuration and testing are required [4].

In order to enable the campus network to truly become a public platform for teaching, research, learning and living, to meet the needs of various network applications, as well as to provide safe, stable and fast services, in this transformation, one outlet route is considered to be added to relieve the pressure on current bandwidth use, and some dual-line hot standby for important services can be done [5]. In addition, the international flow on CERNET will be drained to CT and CMCC outlets which reduce the international flow costs of CERNET routes. User-group based policy routing is used to optimize and adjust the outlet route, improving the quality of network access [6].

2.2 Transformation Purpose

(1) Campus users access the resources on CERNET through the CERNET route;
(2) Office users and teacher users of faulty access non-CERNET resources through CT 1G route;
(3) CT student users access non-CERNET resources through CT 10G outlet and CMCC student users access non-CERNET resources through CMCC 20G outlet;

2.3 Transformation Scheme

The original campus network outlet topology is shown in Fig. 1. It consists of a core switch and an outlet router.

Because the original outlet router NPE60 does not support the route selection of the outlet routes based on the operators' package purchased by students, and its number of ports is not enough to expand to a larger bandwidth, the bottleneck in performance is reached, and the device is unstable due to the long service life.

In this transformation, the outlet router was changed to Ruijie RSR7708-X. This device can support linkage routing selection with the radius system, multi-operator management, log capture, support minimal outlets and Simplistic-Campus multiple authentication methods that will be deployed later. Teacher users of faulty and student

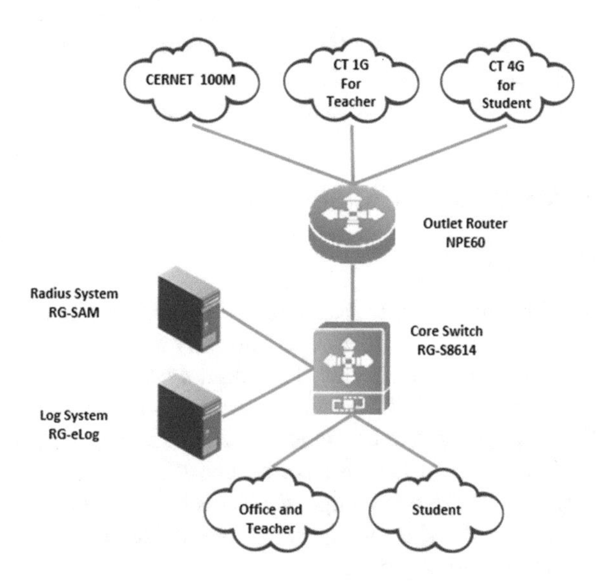

Fig. 1. Campus network outlet structure before transformation

users need to be authenticated before accessing the campus network. After succeed in the authentication, the page will automatically jump to the operator radius server for secondary web authentication or SAM-ISP-Server authentication. During the use process, users can also apply for different outlet routes according to their own needs (Fig. 2).

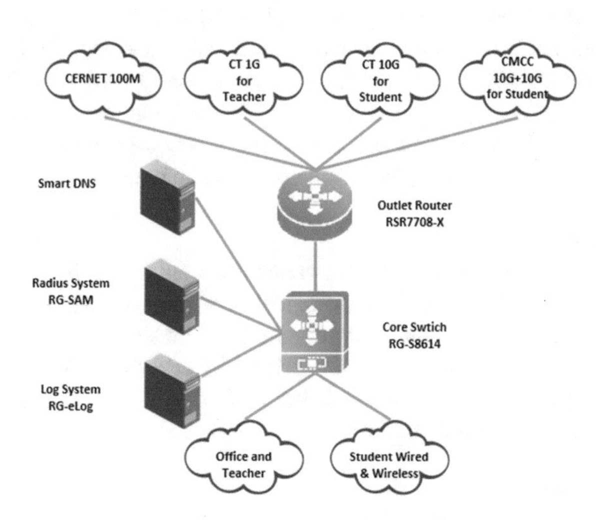

Fig. 2. Campus network outlet structure after transformation

The structure divides the entire outlet into four parts, among which 1G bandwidth is allocated to the server in CMCC outlet. Through the smart DNS, return to the server outlet address of different routes according to the source address of different ISPs [7].

This will reduce the burden on faculty CT 1G, and avoid the instability that may occur when extranet users access cross ISPs. Currently, dual-line access is deployed for high-security services such as school homepages, department homepages and VPNs.

3 Technical Analysis and Application

3.1 Technical Analysis

The core switch and the outlet router are connected by 10 Gigabit fiber and can also be extended to dual 10 Gigabit in the future. Students only need to pass the authentication of the school when they access CERNET. They do not need to be authenticated by the ISP. When accessing non-CERNET resources, they need to pass the authentication of the selected ISP. We assign private IP address to each terminal, no matter which outlet, there is a separate address pool for NAT conversion [8].

Because student flow is relatively large, the CT 1G outlet is dedicated to Office and Teachers to meet the needs of teaching and research. Students can choose different ISP's outlets according to their preferences.

3.2 Technical Application

3.2.1 Address Translation Technology

Network Address Translation (NAT) is a technology that translates internal network address into external network address through certain rules. This technology can save valuable public network address, hides internal networks and enhance security [9]. NAT is mainly classified into two types: NAT and PAT. NAT is a one-to-one translation of internal network address to external network address. PAT (Port Address Translation) is the translation of internal network address to external network address and can be used to determine the uniqueness of its multiple internal hosts by multiple port numbers [10, 11]. In this scheme, the PAT technology is adopted for students Internet access. For the external part of the server port, static NAT technology is used.

3.2.2 Static Route

Static route refers to route information manually configured by users or network administrators. When the network topological structure or link status change, the network administrator needs to manually modify relevant static route information in the route table [12]. Static route information is private by default and will not be passed to other routers. The static route in this scheme is used to specify the campus resource address and CERNET resource address.

3.2.3 Policy Route

Policy route is a more flexible data package forwarding mechanism than static route [13]. The router will determine how to process the data packages that need to be routed through the route map. The route map determines the next-hop forwarding router of a data package.

Policy route includes destination address routing, source address routing, and MAC address routing [14]. Its advantage is that it is very flexible and can achieve various complex requirements. Its disadvantage is that it consumes more CPU and memory resources. When configuring, users need to carefully think about the priority relationship between different policies, otherwise it is not easy to troubleshoot [15]. In this scheme, a policy for forwarding according to different user groups is deployed in policy routing, and there are also policies for forwarding to different routes according to different hard demands, such as sending a DNS request to the external network.

4 Specific Configuration

According to the analysis of campus network structure and application of our school, we should follow the following steps to configure:

(1) Configure the campus static route of the core switch. The detail route points to each resource IP in the campus and the default route points to the IP of the outlet router.
(2) Configure the static route of the outlet router. The detail route points to the free address segment of CERNET and the default route points to the IP at the opposite end of the CT 1G route.
(3) Configure the Policy route of outlet router:

a. The IP address to two CMCC 10G outlets (mandatory)
b. Student CMCC users
c. Student CT users
d. The IP address to CERNET 100M (mandatory)
e. Default address to CT 1G

4.1 Static Routes of Core Switch

```
ip route 0.0.0.0 0.0.0.0 10.10.9.1              // Default routers point to outlet routers
ip route 172.16.0.0 0.0.16.255 10.20.9.2        // Server IP section 1
ip route 10.0.1.0 0.0.0.255 10.20.9.2      // Server IP section 2
```

4.2 Outlet Router

4.2.1 Static Route

```
ip route 0.0.0.0 0.0.0.0 10.10.9.2              // The default route of the core switch
ip route 202.192.0.0 255.240.0.0 210.38.191.1 // The resources of the CERNET are directed point to
CERNET outlet
```

4.2.2 Policy Route

a. The IP address to two CMCC 10G outlets (mandatory)

```
route-map student permit 8
 match ip address 138
 set ip default next-hop 192.168.36.61
 set ip default next-hop 192.168.36.62
exit
ip access-list extended 138 permit ip host 10.0.1.20 any   // Radius server
```

b. Student CMCC users

```
route-map student permit 9
 match ip address 135
 set ip default next-hop 192.168.36.62
 set ip default next-hop 192.168.36.61
exit
ip access-list extended 135 permit ip user-group STU-CMCC any // Student CMCC user
```

c. Student CT Users

```
route-map student permit 11
 match ip address 189
 set ip default next-hop 10.100.1.1
exit
ip access-list extended 189 permit ip user-group STU-CT any // Student CT users
```

d. Default address to CT 1G

```
route-map student permit 102
 match ip address 133
 set ip default next-hop 10.101.1.1
exit
ip access-list extended 133
10   permit ip user-group Teacher any          // Office and Teacher
11   permit ip any any                         // default
exit
```

4.2.3 Static NAT Mapping

```
//CT 1G out mirror of reverse proxy server
ip nat inside source static 10.0.1.77 159.234.15.21 match GigabitEthernet 1/2/2 permit-inside
//CMCC 10G outlet image of reverse proxy server
ip nat inside source static 10.0.1.77 120.19.24.26 match TenGigabitEthernet 3/2/0 permit-inside
```

```
//DNS server mapping on the CT 1G outlet
ip nat inside source static tcp 10.0.1.88 53 159.234.15.15 53 permit-inside
```

5 Optimization Result

This transformation scheme has been running for more than six months since its completion in October 2017. At present, the number of peak users is 8000 and outlets are 31.1G. There are no major failures. The effects are as follows:

(1) The performance utilization of the core switch is reduced, and the forwarding work is done to increase the number of users that can be carried. As shown in Fig. 3, from the end of October 2017, the CPU utilization remains at about 5%, even in the end of May 2018, users increased by about 3000, and CPU utilization increased, but it was still lower than the same period in 2017.

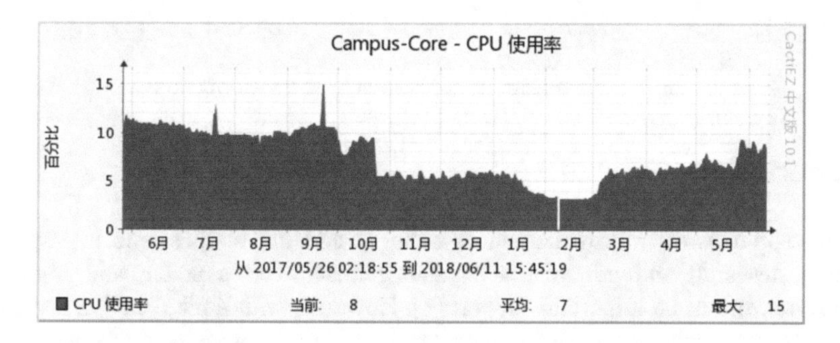

Fig. 3. Core switch CPU utilization before and after transformation

(2) As shown in Fig. 4, the increase in user network speed does not affect the faculty office network and the network does not drop or buffer. Besides, the video is smooth, students have a high degree of self-selection, and all of above is in line with relevant government laws.

(3) Reduce the cost of operation and maintenance, simplify the wiring, and add ISPs subsequently simply by adding interfaces and user groups.

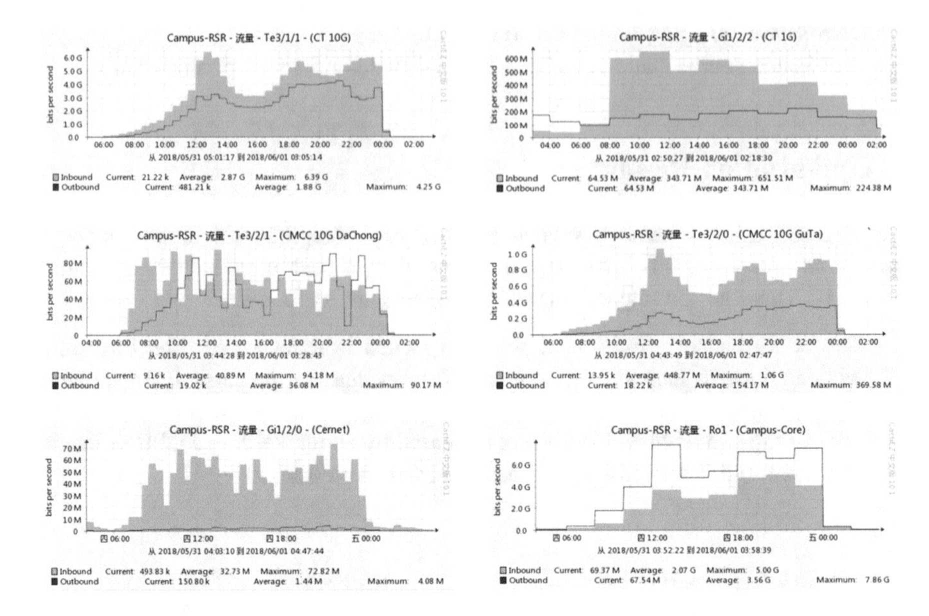

Fig. 4. Flow chart of each outlet after transformation

6 Conclusion

The development of informatization of colleges and universities, and all aspects of teaching, research and administrative management need a stable and high-speed campus network as an important support [1]. However, with limited capital investment in schools, how to integrate existing network resources, improve the performance of the network, and improve the efficiency, security, flexibility, and expandability of campus network is a focus of all constructors of informatization of colleges and universities and it is also a long-term research direction.

References

1. Jian, S.: Analysis and application of campus network routing policy. Exp. Technol. Manag. **11**, 122–125 (2013)
2. Song D., Ma, F.: Application of policy route in the campus. Inf. Technol. Informatiz. **6**, 94–95
3. Zeng, Q., Chen, L., Liang, S., Du, Y.: Application of the new model of campus network topology structure: taking Jiangxi University of Engineering as an example. J. Yichun Univ. **9**, 43–46
4. Xing, L., Fang, Y., Jiang, L., Li, G.: Integrated meteorological WAN design based on BFD in Hubei province. J. Wuhan Univ. Technol. (Inf. Manag. Eng.) **5**, 635–642 (2017)
5. Guo, Y., Zhan, W., Qin, W., Zhao, B.: Application of two-wire hot standby in Taizhou meteorological bureau. Meteorol. Hydrol. Mar. Instrum. **32**(4), 55–58 (2015)
6. Chunxu, D.: Analysis of advantages and networking modes of cooperation between campus network exit and multi-operators. Educ. Mod. **11**, 96–97 (2018)

7. Shan, Q., Yan, P., Nan, F.: Discussion on the optimization of college DNS servers in multi-outlet environment. Chin. J. ICT Educ. **7**, 89–93 (2017)
8. Shengqin, B., Hongbo, W., Xiaolong, C.: Realization of NAT experiment teaching on packet tracer software. Exp. Sci. Technol. **16**(2), (2018)
9. Yong, L.: Design and realization of NAT simulation experiment. Exp. Technol. Manag. **4**, 132–135 (2018)
10. Sun, Y.: Application of network address translation technology (NAT). Electron. Technol. Softw. Eng. **10**, 17 (2015)
11. Gao, T., Chen, J.: Design and realization of NAT module based on distributed architecture. Study Opt. Commun. **5**, 25–28 (2017)
12. Yong, L., Xinling, G.: Design and realization of integrated routing experiment based on packet tracer. Res. Explor. Lab. **9**, 111–114 (2015)
13. Zhang, H., Yuliang, X.: Realization of campus network exit policy route independently selected by users. J. Commun. s2 (2013)
14. Wang L., Tang, Y.: Research on the construction of large campus network based on ROS. Inf. Sci. **7**, 88–95 (2017)
15. Jia, J., Yan, Z., Guanggang, G., Jin, J.: Research on BGP route leakage. Chin. J. Netw. Inf. Secur. **8**, 54–61 (2016)

Ranking the Emotion Factors Sensitive to English Test Performance

Mei Dong(✉) and Ying Xue

Xi'an Shiyou University, Xi'an 710065, People's Republic of China
2548121615@qq.com

Abstract. Since the performance of some high-stake English tests often goes hand in hand with the future career and happiness of test takers, emotions aroused by them are enormous, which has caught much attention within the past few decades. However, due to statistical limitations, so far, few research has been done as to ranking the emotion factors based on their sensitivity to English test performance. Therefore, in this paper, a comparison study concerning the ranking of seven extensively investigated emotion factors has been conducted with gray relational analysis, a statistical method that has no requirements of both sample size and distribution. Results show that motivation-to-avoid-failure, a branch of achievement motivation, is the most sensitive factor to English test performance (comprehensive gray correlation = 0.874, $n = 20$) with performance-avoidance goal right following it (comprehensive gray correlation = 0.709, $n = 20$).

Keywords: Gray correlation analysis · Emotion factors
English test performance

1 Introduction

Over 25 million people take part in large-scale English proficiency tests, such as TOEFL, IELTS, TOEIC and College English Test (CET), for career or academic promotion in over 90 countries around the world each year [1]. As the performance of these test takers usually goes hand in hand with their future success and happiness, emotions aroused by them are enormous, which has caught much attention from researchers within the past few decades [4].

To date, several emotion factors have been identified to be associated with the performance of test takers [4–8], which include test anxiety [2], self-esteem [3], English test anxiety [4], trait anxiety [4], state anxiety [5], achievement motivation [6, 7] and achievement goal [8], to name just a few. However, no research has been carried out yet as for the ranking of the emotion factors that is sensitive to English test performance. And this may lead to the following severe consequences. First, confusion would be aroused in choosing input variables when researchers try to set up a neural network model between emotion factors and English test performance for a better interpretation of test scores. Second, some debilitating emotion factors that are highly sensitive to English test performance will continue to be neglected in educational and testing contexts. As a result, test-takers' negative emotions cannot be soothed in a

N. Xiong et al. (Eds.): ISCSC 2018, AISC 877, pp. 222–231, 2019.
https://doi.org/10.1007/978-3-030-02116-0_26

timely manner. What's more, no effective precaution measure would be taken due to the ignorance of the potential harms caused by these emotion factors. Third, the performance of test-takers with emotion problems would be seriously underestimated, which greatly undermines the principle of equality of all existing large-scale English proficiency tests and thus arouse doubts about the validity of those tests. Forth, the future happiness and career development of a quite a lot of the test takers who are less blessed emotionally are being ruined simply due to their poor performance in English tests. Thus, a comparison study of emotion factors that are sensitive to English test performance must be started immediately.

However, the reality is that conventional statistical methods cannot deal with this study for reasons listed as follows. Luckily, Gray relational analysis may help crack the hard nuts that cannot be dealt with using conventional statistical methods because it has no requirements of both sample size and sample distribution [25]. What's more, it needs only a little calculation, therefore is very time- and energy-efficient. Moreover, it is particularly developed for those systems whose information remain partly known and partly unknown [25]. And this is perfectly appropriate for the "emotion factor - English test performance" system.

Hence, in this paper, we are going to compare seven extensively investigated emotion factors [2–8] based on their sensitivity to English test performance with the method of gray relational analysis. The seven emotion factors selected are test anxiety, English test anxiety, self-esteem, achievement motivation, state anxiety, trait anxiety and achievement goal.

2 Ranking the Selected Emotional Factors with Gray Correlation Analysis

2.1 Research Questions

In the first place, research questions of this study are carefully devised and presented as follows.

1. Among the seven emotion factors (including their branches), which are the top three factors sensitive to English test performance?
2. Is English test anxiety more sensitive to English test performance than the other kinds of anxieties involved in this study?
3. Among the four branches of English test anxiety, English listening test anxiety, English speaking test anxiety, English reading test anxiety and English writing test anxiety, which is most sensitive to English test performance?

2.2 Research Samples

20 Grade 2 undergraduate have taken part of this research, 10 of whom are male students. Two of them major in business administration, 3 in oil engineering, 4 in chemicals, and 1 in English. Their ages range from 19 to 21.

2.3 Research Tools

This study has employed six questionnaires that are widely adopted in the academic world to collect its data. Information about these questionnaires are demonstrated as follows:

1. State-trait anxiety inventory: It is a widely adopted inventory for measuring state and trait anxiety and was developed in 1983 by Charles Spielberger and his other four colleagues [28, 29]. The inventory can be applied in clinical and research settings for diagnosing anxiety as well as distinguishing it from depressive syndromes. Form Y is adopted in this study, which includes two sub-scales, state anxiety inventory and trait anxiety inventory, each of which is of 20 items. The state anxiety inventory include such items as: “I am tense”; “I am worried”; “I feel calm”; “I feel secure.” The trait anxiety inventory include items such as “I worry too much over something that really doesn’t matter.”; “I am content.” or “I am a steady person.” All items are rated on a 4-point scale from “Almost Never” to “Almost Always”. Higher scores indicate greater anxiety. The internal consistency coefficients for the scale range from 0.86 to 0.95, while test-retest reliability coefficients range from 0.65 to 0.75 over a 2-month interval [29].
2. English test anxiety scale: It is a scale for measuring the debilitating effect of test anxiety against the background of English proficiency tests. It was developed in 2013 [4] and is used mainly for identifying anxious test takers during English proficiency tests. The scale is consisted of four sub-scales, English listening test anxiety scale, English speaking test anxiety scale, English reading test anxiety scale and English writing test anxiety scale. It involves items such as “I am sure to fail in this English proficiency test.”, “During the oral part of an English proficiency test, I just stuck there and couldn’t recall the vocabulary I used to be very familiar with.”, “My heart beat faster and faster during the listening part of an English proficiency test.”, “I always miss some important information during the listening part of an English proficiency test.” As it can be seen, the scale is retrospecting in nature. English test anxiety scale is proved to have a close correlation with Westside test anxiety scale ($\alpha = 0.57$, $n = 229$, $p < 0.001$). Its test-retest reliability coefficients are 0.89, 0.91, 0.92 and 0.91 respectively [4].
3. Westside test anxiety scale: It is a reliable ten-item screening instrument meant to identify students with anxiety impairments under test situations [30]. The scale combines six items assessing impairment, four on worry and dread. The Westside scale is of high face validity, in that it includes some highly relevant cognitive and impairment factors but omits the marginally relevant over-arousal factor [32]. It is right with this scale that Driscall has found a significantly negative correlation between test anxiety and test scores ($\alpha = 0.40$, $p < 0.01$) [32]. Then, Driscall repeated such a test twice and got similar results ($\alpha = 0.49, 0.40$, $p < 0.01$) [32]. It must be mentioned that those results are proved to be consistent with that of Hembree and Seipp [2, 9].
4. Achievement motivation scale: It is a reliable measurement originally developed by T. Gjesme, and R. Nygard, R, Norwegian psychologists from Oslo university based in Norway in 1970. The scale is designed for measuring two factors simultaneously,

which are motivation-to-avoid-failure and motivation-to-approach-success. Its test-retest reliability coefficients are 0.83 and 0.84 respectively. The scale was translated into Chinese in 1988 and slightly revised in 1992 for measuring achievement motivation of Chinese middle school students and college students [34]. The scale adopted in this study is the revised Chinese version, which has been widely applied in China.

5. Self-esteem scale: The scale was carefully designed by Rosenberg in 1965 and then was translated into Chinese by Ji Yifu and Yu Xin. Meanwhile, they also made some trivial revisions to cater to their Chinese samples in 1993 [35]. It was a popular tool for measuring self-esteem in China with its test-retest coefficient being 0.79. Right with this scale, Tian and Guo [36] finds that self-esteem is significantly associated with state anxiety ($r = -0.39$, $n = 216$, $p < 0.01$), which in turn verifies the validity of the scale to some extent.
6. Achievement goal inventory: In 2001, Elliot and McGregor worked together and developed the achievement goal inventory. It is actually composed of four sub-scales, scale of grasp-approach achievement goal, scale of grasp-avoidance achievement goal, scale of performance-approach achievement goal and scale of performance-avoidance achievement goal. It is a pity that no validation study concerning the achievement goal inventory has been found.

Besides, what also needs to be mentioned here is that in this study, the CET-4 scores of the 20 sample students are used to represent their English test performance.

2.4 Data Collecting and Processing

1. All sample students were arranged in one classroom. They were asked to listen to the directions on how to fill in the questionnaires of this study. The directions were played through a cassette recorder.
2. The sample students started to complete the questionnaires and all of them finished within 45 min. All together, 19 valid questionnaires were collected.
3. Data of the six questionnaires were recorded in Excel form for later use.
4. All of the recorded data were changed into dimensionless data with a formula presented as follows:

Suppose $X_i = (x_i(1), x_i(2), \cdots, x_i(n))$ is the behavior sequence of factor X and D_1 is the sequence operator, then,

$$X_i D_1 = (x_i(1)d_1, x_i(2)d_1, \cdots, x_i(n)d_1),$$

In which $x_i(k)d_1 = x_i(k)/x_i(1)$, $k = 1, 2, \cdots, n$
The results are presented in Tables 1, 2 and 3 together with CET-4 scores.

5. The data of the factors which are negatively related to test performance are conversed to positively related ones by using 1 to divide them and the results are presented in Tables 4 and 5.

Table 1. The dimensionless data sequences of this study (part 1)

CET-4 scores	State anxiety	Trait anxiety	General test anxiety	Self-esteem
0.991	0.952	0.957	1.146	0.863
1.045	1.452	1.165	2.308	0.694
1.023	1.198	1.259	2.084	0.969
1.074	1.098	1.132	1.849	1
1.074	1.148	1.129	1.773	0.687
0.971	1.234	1.132	2.226	0.687
0.963	1.147	1.128	1.149	0.546
0.843	1.443	1.037	1.769	0.968
1.452	1.054	1.176	1.776	0.794
0.932	1.192	0.869	1.543	0.758
1.043	1.147	1.168	1.916	0.856
1.092	1.047	1.088	2.463	0.929
1.083	0.948	0.827	1.765	0.587
0.785	1.548	1.257	1.847	0.929
1.112	1.135	0.776	1.376	0.342
1	1.123	0.869	1.618	0.687
0.978	0.893	0.612	1.225	0.856
0.974	1	1	2.231	0.967
1.10	1.154	1.043	0.765	0.587

Table 2. The dimensionless data sequences of this study (part 2)

English listening test anxiety	English speaking test anxiety	English reading test anxiety	English writing test anxiety	Motivation-to-approach-success
1.049	1.237	1.037	1.125	1.017
1.867	1.675	2.447	1.876	0.69
1.743	1.734	2.037	1.657	1.024
1.557	1.512	1.967	1.435	0.967
1.389	1.398	1.634	1.224	0.887
1.563	1.617	1.765	1.723	1.286
0.984	0.887	1.312	0.914	0.647
1.724	1.643	1.723	1.254	1
1.498	1.598	1.487	0.932	1.128
1.611	1.443	1.814	1.943	1.049
1.537	1.443	1.768	1.816	0.969
1.724	1.536	2.089	1.645	1.109
1.372	1.398	1.324	1.579	0.835
1.501	1.235	1.923	1.509	1.024
1.213	1.109	1.176	1.718	0.698

(*continued*)

Table 2. (*continued*)

English listening test anxiety	English speaking test anxiety	English reading test anxiety	English writing test anxiety	Motivation-to-approach-success
1.192	1.201	1.587	1.378	0.759
0.924	0.934	1.364	1.443	0.857
1.473	1.547	1.876	1.968	1.023
0.609	0.626	0.721	0.658	1.458
1.049	1.237	1.037	1.125	1.017
1.867	1.675	2.447	1.876	0.69
1.743	1.734	2.037	1.657	1.024
1.557	1.512	1.967	1.435	0.967
1.389	1.398	1.634	1.224	0.887
1.563	1.617	1.765	1.723	1.286
0.984	0.887	1.312	0.914	0.647
1.724	1.643	1.723	1.254	1
1.498	1.598	1.487	0.932	1.128
1.611	1.443	1.814	1.943	1.049
1.537	1.443	1.768	1.816	0.969

Table 3. The dimensionless data sequences of this study (part 3)

Motivation-to-avoid-failure	Grasp-approach	Grasp-avoidance	Performance-approach	Performance-avoidance
0.709	0.674	0.831	0.628	0.429
0.809	0.327	0.327	0.268	0.432
1.287	0.827	1	0.728	1.576
1.142	0.745	1	0.634	1.287
0.857	0.923	0.674	0.734	1.147
1.476	0.334	0.334	0.269	0.427
0.812	0.567	0.501	0.457	0.426
1.243	0.668	0.826	0.536	1.721
1.142	0.665	0.671	0.543	1.147
1.048	1.084	1	0.265	1.287
1.186	1	1	0.543	1.146
1.332	1	0.826	0.816	1.289
0.947	0.834	0.834	0.823	1.154
1.243	0.256	0.835	0.634	1.576
0.954	0.834	0.332	0.537	0.567
1	0.746	0.667	0.538	0.716
1	0.918	1	0.543	0.723
1.243	0.487	1.327	0.628	1
0.947	1	1	1.089	1.147

Table 4. The conversed data sequences of trait anxiety, test anxiety and the four branches of english test anxiety

Trait anxiety	Test anxiety	English listening test anxiety	English speaking test anxiety	English reading test anxiety	English writing test anxiety
1.042	0.874	0.954	0.812	0.964	0.891
0.853	0.432	0.532	0.598	0.412	0.526
0.787	0.479	0.569	0.598	0.487	0.599
0.879	0.543	0.637	0.658	0.512	0.687
0.876	0.557	0.718	0.704	0.623	0.832
0.875	0.477	0.688	0.619	0.555	0.576
0.881	0.867	1.023	1.123	0.758	1.102
0.976	0.557	0.576	0.612	0.579	0.76
0.867	0.547	0.667	0.628	0.665	1.059
1.151	0.654	0.623	0.687	0.543	0.519
0.852	0.524	0.648	0.687	0.564	0.546
0.923	0.409	0.576	0.648	0.479	0.613
1.213	0.557	0.729	0.708	0.764	0.643
0.793	0.546	0.674	0.834	0.523	0.657
1.276	0.724	0.831	0.894	0.848	0.579
1.152	0.618	0.839	0.828	0.624	0.721
1.643	0.812	1.087	1.078	0.737	0.686
1	0.452	0.675	0.647	0.531	0.511
0.964	1.302	1.643	1.588	1.389	1.521

Table 5. The conversed data sequences of motivation-to-avoid-failure, state anxiety and performance-avoidance goal

Motivation-to-avoid-failure	State anxiety	Performance-avoidance goal
1.412	1.047	1.198
1.232	0.687	3.029
0.779	0.823	1
0.879	0.912	1
1.157	0.869	1.486
0.663	0.828	3.032
1.231	0.869	1.98
0.812	0.711	1.201
0.876	0.947	1.489
0.946	0.829	1
0.843	0.867	1
0.745	0.798	1.23

(*continued*)

Table 5. (*continued*)

Motivation-to-avoid-failure	State anxiety	Performance-avoidance goal
1.053	1.054	1.298
0.813	0.647	1.289
1.054	0.914	0.032
1	0.909	1.487
1	1.108	1
0.813	1	0.747
1.05	0.869	1

2.5 Results

All the above data sequences are input into GM, a software specifically developed for handling calculation related to gray theory. Then the gray correlation coefficient of each factor is presented in Table 6.

Table 6. The results of the gray relational analysis

Factor	Ranking	Comprehensive gray correlation
Motivation-to-avoid-failure	1	0.874
Performance avoidance goal	2	0.709
Trait anxiety	3	0.640
Motivation-to-approach success	4	0.607
State anxiety	5	0.591
Grasp-avoidance goal	6	0.580
English listening test anxiety	7	0.540
Grasp-approach goal	8	0.538
English writing test anxiety	9	0.537
English speaking test anxiety	10	0.531
English reading test anxiety	11	0.530
Self-esteem	12	0.529
Test anxiety	13	0.528
Performance-approach goal	14	0.527

3 Discussion

The results of this research can be summarized as follows: It is confirmed that motivation-to-avoid failure, performance-avoidance goal and trait anxiety are the top three emotion factors sensitive to English test performance with motivation-to-approach success right following them, which ranks No. 4. Both state anxiety and English listening test anxiety are very much sensitive to English test performance. Finally, although several studies have investigated the correlations among the seven emotion factors and test performance [2–19], the present one is the first to report the ranking of the seven well-investigated emotion factors with gray relational analysis, a method more appropriate for investigating gray system such as the "emotion factor-English test performance" system, than conventional statistical methods including correlation analysis, regression analysis and principal component analysis.

Before concluding, we should reiterated that there are two procedural decisions that constraint the interpretation of the present rankings. First, although the sample students are from various parts of China, they are actually selected from the same university. Therefore, repeated study shall choose samples from multiple universities. Second, only seven emotion factors are ranked. There possibly exist emotion factors that are highly sensitive to English test performance which have not been involved. Thus, it is suggested that an exploratory study be conducted before doing gray relational analysis.

References

1. 360doc. http://www.360doc.com/
2. Seipp, B.: Anxiety and academic performance: a meta-analysis of findings. Anxiety Res. **4**, 27–41 (1991)
3. Liu, M.: Research on the relationship among self-esteem, academic performance and interpersonal attribution style. Sci. Psychol. **3** (1998)
4. Dong, M.: English Anxiety Study. China Petroleum Engineering Press, Beijing (2013)
5. Zeidner, M.: Test anxiety: the state of the art. Plenum, New York (1998)
6. Atkinson, J.W.: An Introduction to Motivation. Van Nostrand, New Jercy (1964)
7. Arkinson, J.W.: Motivation for achievement. In: Blass, T. (ed.) Personality Variables in Social Behavior, pp. 47–67. Lawrence Erlbaum, New York (1977)
8. Elliot, A.J.: A conceptual history of the achievement goal construct. In: Elliot, A., Dweck, C. (eds.) Handbook of Competence and Motivation, pp. 52–72. Guilford Press, New York (2005)
9. Hembree, R.: Correlates, causes, effects and treatment of test anxiety. Rev. Educ. Res. **58**, 47–77 (1988)
10. Gao, X.L.: A study of the relationship among test anxiety, achievement motivation and CET-4 test performance. J. Southwest Jiaotong Univ. **11**, 20–24 (2010)
11. Elliot, A.J.: Approach and avoidance motivation and achievement goals. Educ. Psychol. **34**, 169–189 (2001)
12. Linnennbrink, E.A., Pintrich, P.A.: Multiple pathways to learning and achievement: the role of goal orientation in fostering adaptive motivation, affect, and cognition. In: Sansone, C., Harackiewicz, J.M. (eds.) Intrinsic and Extrinsic Motivation: The Search for Optimal Motivation and Performance, pp. 82–89. Academic Press, New York (2000)
13. Yu, A., Yu, L., Zhou, D., Zhu, K., Wang, G.: Correlation analysis among anxiety, depression, self-esteem and academic performance of middle school students. Shandong Arch. Psychiatry **4**, 222–242 (2005)
14. Sun, Y.L.: The impact of negative body awareness of juveniles on their academic performance, unpublished master thesis. Jiangxi Normal University
15. Gu, X.Y.: The correlation analysis between the psychological state and academic performance of middle school students. Career Health **15**, 21–41 (1999)
16. Liu, M.: A research on the cause-effect relationship among high school students' self-esteem, academic performance and interpersonal relationship. J. Sci. Psychol. **3**, 281–282 (1998)

17. Endler, N.S., Kantor, L., Parker, J.D.A.: State-trait coping, state-trait anxiety and academic performance. Pers. Individ. Differ. **16**, 663–670 (1994)
18. Somuncuoglu, Y., Yildirim, A.: Relationship between achievement goal orientations and use of learning strategies. J. Educ. Res. **92**, 267–277 (1999)
19. Ayumi, T., Yoshiho, M., Takuhiro, O.: Achievement goals, attitudes toward help seeking and help seeking behavior in the classroom. Learn. Individ. Differ. **13**, 23–35 (2002)

The Optimization Model for River Rafting Management

Na Wang(✉), Xinshe Qi, Yuyi Li, and Xin Wang

College of Information and Communication,
National University of Defense Technology, Xi'an 710106, China
syesun@hotmail.com

Abstract. In this paper, we concerned two problems. Firstly, based on the hypothesis during the rafting, the river carrying capacity model is established. Secondly, we choose four 6-nights-motorized of no encounter to enter the river, then get rid of one 6-nights-motorized trip and add some other duration trips in, until the restrict of varying duration and propulsion are satisfied. According to our launching schedule, composite rate could reach 100%.

Keywords: River carrying capacity · Launching schedule · Optimal mix of trip

1 Introduction

The Big Long River trips all start at First Launch and exit the river at Final Exit, 225 miles downstream. Passengers take either oar-powered rubber rafts, which travel on average 4 mph or motorized boats, which travel on average 8 mph. Trips range from 6 to 18 nights. The campsites on the Big Long River distributed fairly uniformly throughout the river corridor. Trips travel down the Big Long River each year during a six-month period due to the weather. And no two sets of campers can occupy the same site at the same time [1]. They want to determine how they might schedule an optimal mix of trips, of varying duration (measured in nights on the river) and propulsion (motor or oar) that will utilize the campsites in the best way possible (Fig. 1).

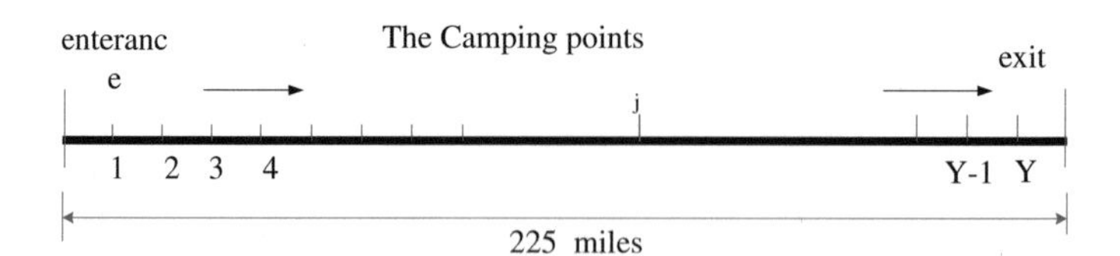

Fig. 1. The big long river with Y-campsites uniformity distributed.

N. Xiong et al. (Eds.): ISCSC 2018, AISC 877, pp. 232–237, 2019.
https://doi.org/10.1007/978-3-030-02116-0_27

2 The Model of Carrying Capacity

In order to avoiding conflict resulted by too many trips during the Big Long River's rafting season, we must consider the river carrying capacity firstly. Then according to the river carrying capacity, we can determine the max number boat trips could be added to the Big Long River's rafting season.

Because the river trips all start at First Launch and exit the river at Final Exit, 225 miles downstream, and the trips range from 6 to 18 nights of camping on the river, start to finish. Passengers take either oar-powered rubber rafts, which travel on average 4 mph or motorized boats, which travel on average 8 mph. In order to finish the trip within the prescribed time, we must limit the speed of drifting.

2.1 Confining to Daily Rafting Distance

Further analyzing of n-night trips. One n-nights trip daily average rafting distance and time are calculated as following:

$$\overline{S} = \frac{S}{n+1} \tag{1}$$

$$\bar{t} = \frac{\overline{S}}{v} \tag{2}$$

According to above formulate, when $S = 225$ miles and passengers is take either oar-powered rubber rafts, which travel on average 4 mph or motorized boats, which travel on average 8 mph. The trips range from 6 to 18 nights of camping on the river, we can obtain the time of using oar-powered rubber and motorized boats rafts as Table 1.

Table 1. Time of using oar-powered rubber and motorized boats rafts.

n (night)		6	7	8	9	10	11	12	13	14	15	16	17	18
$\bar{t}$(h)	Oar-powered	9.4	7.0	6.2	5.6	5.0	4.6	4.2	4.0	3.6	3.4	3.2	3.0	2.9
	Motorized boat	4.7	3.5	3.1	2.8	2.5	2.3	2.1	2.0	1.8	1.7	1.6	1.5	1.4

The 6-nights trip is the fastest trip, in this way, the visitors spend 6 nights at camp site and rafting 7 daytime on the river. If this way is oar-powered, the visitor must rafting 9.4 h. So the 6-nights trip is must be motor-powered.

According to the materials, we can not only experience the enjoyment of the drafting, but also appreciate the beautiful scenery. During the Big Long River's trip, we hike most every day from 1–5 h generally. The paper confirmed that we drifting most every day 4.5 h. In view of the above Table 1, we know that the visitors who drift 6–11 days with the motor-powered, while visitors who drift 11–18 days with oar-powered.

The 6-nights trip is the fastest trip, in this way, the visitors spend 6 nights at camp site and rafting 7 daytime on the river. If this way is oar-powered, the visitor must raft 9.4 h. So the 6-nights trip is must be motor-powered.

The slowest trip is 18-night trip. The rafting time is too short for motorized boats, so we must consider that it is oar-powered rubber rafts [2]. If n = 5, then average rafting distance: $\overline{S}_5 = \frac{S}{(n+1)} = \frac{225}{(5+1)} = 37.5$, and $\overline{t_5} = \frac{\overline{S_5}}{v_M} = \frac{37.5}{8} = 4.7$.

In order to ensure that the minimum drifting time is six nights, we must travel certain distance which is inferior to 37.5 miles every day, otherwise, this drifting will be early termination. If n = 18, then average rafting distance: $\overline{S_{18}} = \frac{S}{(n+1)} = \frac{225}{18+1} = 11.84$, and $t_o = \frac{\overline{S_{18}}}{v_O} = \frac{11.84}{4} = 2.96 \approx 3.0$.

Similarly, to ensure the maximum drifting time is 18 nights, we must travel certain distance which is more than driving distance 11.84 miles every day, and otherwise, 18 nights can't complete the tourism Conclusion:

(1) Hiking is also a major focus of our trips. So we should keep a certain amount of time and energy after rafting. And then we must limit the time of rafting $3 \le T_j \le 4.5$ each day.
(2) In order to take fully advantage of the two powered ways. We analysis to identify the powered ways of different kinds of trips. We adopt: motorized boats for 6–11 nights' trips; and oar-powered rubber rafts for 12–18 nights' trips.
(3) The distance of every trip must be limited each day: $11.84 < S_{O,j}\ or\ S_{M,j} < 37.5$.

2.2 *N*-Nights Trip Scheme Model

Due to the campsites on the Big Long River, distributed fairly uniformly throughout the river corridor, and we named them from the 1st, 2nd, ..., y^{th}. Each journey is expressed as an array $\mathbf{M}(k_1, k_2, \cdots, k_n)$. For example: A seven-night-trip can be expressed as (2, 6, 10, 14, 19, 25, 31). That is to say, we will stop at the campsites that the number is 2, 6, 10, 14, 19, 25, 31 from the first to the seventh night. We confirm the campsites of one day by the lengths of the rafting:

$$k_1 = \left[\frac{S_{i1}}{d}\right], k_2 = k_1 + \left[\frac{S_{i2}}{d}\right], \cdots, k_n = k_{n-1} + \left[\frac{S_{in}}{d}\right] \quad (3)$$

s.t.

$$\sum_{j=1}^{n+1} S_{i,j} > 225 \quad (4)$$

$$11.84 < \left|S_{i,j+1} - S_{i,j}\right| < 37.5 \quad (5)$$

$$1 \le j \le n,\ 1 \le i \le 2 \quad (6)$$

According to the following case, we provide the camping way of different times [3]. Motorized boats: $11.84 < S_{M,j} < 37.5$, according to the modal above we can get the

n-nights trip schedule. The calculation of the case put the 6-night motorized trip as an example. By the previous argument, we can see that the most rafting distance of the oar is four multiplied by four point five, be eighteen miles, $S_{o\max} = 4 * 4.5 = 18$ miles; and the fewer rafting distance of the motor is three multiplied by eight, be twenty-four miles $S_{m\min} = 3 * 8 = 24$ miles. In order to avoid that there are two sets of campers can occupy the same site at the same time. So we can consider that $|S_{m\min} - S_{m\max}| \leq d_{\min}$. Therefore $d_{\min} = 6$, $Y_{\max} = \left|\frac{225}{d_{\min}}\right| = 36$. $24 \leq S_m \leq 36, 6 \leq S_o \leq 18$. According to the formula, we can get 168 kinds of 6-night trips scheme via setting up and running a procedure with MATLAB.

2.3 Determining an Optimal Mix of Trips Schedule

According to the requirement of the problem, management require an optimal mix of trips varying duration (measured in nights on the river) and propulsion (motor or oar) that will utilize the campsites in the best way possible. If we put the slower into the river first, the faster can certainly catch up with the slower after a few hours. We assume the position of the campsites random. They must contact in the campsites. If we put the faster into the river first, the slower can't catch up with the faster in a certain time range. The distance from the slower and the faster will be longer. The campsites will be in idle. So we must limit the distance of the boats each day. In other words, the routes of the boats must be fixed. Obviously, the shorter of the trips schedule, the more trips can be accommodated. When we admit only the 6-night-trip launched, we can achieve the max number of trips (Table 2).

Table 2. The different route of the 6-nights trips.

3	9	15	21	27	33
4	8	14	20	26	32
5	10	16	22	28	34
6	11	17	23	29	35

Then we remove one of the 6-night's trips and add a 7-nights trip. In order to content different time and propulsion trips, it can be achieved via the following scheme keeping the max number of trips [4].

Step 1: There are four 6-night-trips of different operation modes. They put into the river in different time (Get minimal contact on the river). Because of this reason, we can utilize the campsites in the fastest speed possible.

Step 2: In order to improve the utilization rate and make the 6-nights-trips' operation modes accord with our regulations. We remove one of the 6-night's trips and add a 7-nights trip.

Step 3: Extend the rest of the 6-nights trips with above method. According to the requests of the problem, we must add oar-powered rubber rafts into the river.

Step 4: Draw the relevant data and prove the campsites have no repeat use.

Step 5: Determine the departure time of the drifting ships. We must get minimal contact on the river. We obtain the result by using the MATLAB software (Table 3).

Table 3. Campsites number.

n	Campsites number													
6	6	11	16	20	26	32								
7	4	10	14	19	24	29	34							
8	3	8	13	17	22	27	31	35						
14	2	5	7	9	12	15	18	21	23	25	28	30	33	36

3 The Optimal Mix Model of Trips

Considering that shorter the trip duration is the more group there are. If all trips are 6-nights-motorized, the largest number of boats would be permitted to enter the river for rafting. We first choose four 6-nights-motorized of no encounter to enter the river, then get rid of one 6-nights-motorized trip and adding some other duration trips in, until the restrict of varying duration (measured in nights on the river) and propulsion (motor or oar) are satisfied.

Table 4. Optimal mix of trips schedule.

n-night trip	Propulsion	The first night's campsite	Departure time
6-night	Motor	6	8:00
7-night	Motor	4	10:00
8-night	Motor	3	14:00
15-night	Oar	1	Random
n-night trip	Propulsion	The first night's campsite	Departure time
6-night	Motor	6	8:00
7-night	Motor	4	10:00
9-night	Motor	1	14:00
14-night	Oar	2	Random
n-night trip	Propulsion	The first night's campsite	Departure time
6-night	Motor	6	8:00
8-night	Motor	1	14:00
8-night	Motor	3	10:00
14-night	Oar	2	Random
n-night trip	Propulsion	The first night's campsite	Departure time
7-night	Motor	1	14:00
7-night	Motor	4	8:00
8-night	Motor	3	10:00
14-night	Oar	2	Random

Note: When the current three motors that determined by the way of the operation, according to this problem, we must set up an oar which drives into the river. So we could list the first three groups of the remaining campsites, and compare the distance between two points, in order to make it satisfy: $|S_{i+1} - S_i| \leq S_{\max}$. Then, comparison and scheduling again, we must ensure the campsites reached the highest efficiency.

We find the campsite that the first campsite is empty. So it can be added into the 6-nights-trip, 7-nights-trip, 8-nights-trip and 14-nights-trip. Then we can get the 7-nights-trip, 8-nights-trip, 9-nights-trip and 15-nights-trip. In this method, the campsites can be used totally. We obtain six kinds of trips: 6-night-trip, 7-night-trip, 8-night-trip, 10-night-trip, 14-night-trip, 15-night-trip. We obtain four optimal arrangement plans as the Table 4 shown.

4 Conclusion

This paper has described the procedure of seeking the optimum scheme. According to the requirements of the problem, we get all the available reasonable results via setting up and running a procedure. And we find the best combination which can make the river accommodate more passengers and the utilization rate reached one hundred percent. At last, we ascertain the max number of the campsites is 36 and the number of the travel times is 720.

References

1. Underhill, A.H., Xaba, A.B., Borkan, R.E.: The wilderness use simulation model applied to Colorado River boating in Grand Canyon National Park, USA. Environ. Manag. **10**(3), 367–374 (1986)
2. Jalbert, L.: Update of river research at Grand Canyon. Color. River Sound. **5**(8), 1–3 (1998)
3. Schechter, M., Lucas, R.C.: Simulation of Recreational Use for Park and Wilderness Management. RFF Press, Washington, USA (2011)
4. Gimblett, R., Roberts, C.A., Daniel, T.C., Ratliff, M., Meitner, M.J., Cherry, S., Stallman, D., Bogle, R., Allred, R., Kilbourne, D., Bieri, J.: An intelligent agent based model for simulating and evaluating river trip scenarios along the Colorado River in Grand Canyon National Park. In: Gimblett, H.R. (ed.) Integrating GIS and Agent Based Modeling Techniques for Understanding Social and Ecological Processes, pp. 245–275 (2000)

Research on Smart Library Big Service Application in Big Data Environment

Jialing Zhao[1(✉)], Wenwei Cai[2], and Xiangyuan Zhu[2]

[1] Information Center, Zhaoqing University, Zhaoqing 526061, China
jialing.zhao@qq.com

[2] School of Computer Science, Zhaoqing University, Zhaoqing 526061, China
109924221@qq.com

Abstract. In the big data environment, the Smart Library with the development of information technology, changes in user needs continue to transform and reform, form a new form. This paper introduces the concept and characteristics of big service. According to the research status of the smart library, it puts forward the application of large service in the University Library and its architecture, and looks forward to the development trend of the smart Library in Colleges and universities.

Keywords: Big service · Internet of Services · Smart library

1 Introduction

With the rapid development of Internet, Internet of things (IoT), cloud computing and big data, the interconnection and interworking between people and people, things and things, as well as people and things reflect the technological revolution and innovation increasingly. The trend of computing service is increasingly evident, and "everything as a service (EaaS) [1]" is undergoing tremendous changes from connotation to environment. The integration of the physical world and the digital world makes up the cyber-physical system (CPS) [2], and the online to offline (O2O) service system in and out of the Internet world constitutes the service network world, which generates the "Internet of Services (IoS) [3]". The further development of the IoS gives rise to the "big service".

The big service has been applied to various industries of the modern society, such as smart city, smart community and smart campus. Among them, most smart services involve complex cross-network, cross-field and cross-world services, which need the support of big service systems or platforms.

University libraries, which provide important support for teaching and scientific research in universities, have highly educated readers with more professional and profound information demand in professional fields, stronger needs of personalized service, higher acceptance of new technology, higher popularity of intelligent equipment, and more urgent demand for smart library services. This paper discusses the application of big service in smart libraries.

N. Xiong et al. (Eds.): ISCSC 2018, AISC 877, pp. 238–245, 2019.
https://doi.org/10.1007/978-3-030-02116-0_28

2 Concept and Features of Big Service

The big service is a kind of complex service model (or complex service network) arising out of the integration and synergy of cross-world (real world and digital world), cross-field, cross-region and cross-network heterogeneous services. The big service, made up of large-scale smart services based on the rules of big data and various intelligent services, supports the solution for the physical information system through the association of big data in business process and business applications, and creates values.

The service system based on the big service forms the big service system. It is a complex service ecosystem or complex service network constituted by the big service under the background of big data. The service system is oriented towards large-scale personalized user requirements, achieves the cross-network, cross-field and cross-community service integration in real world and virtual world, establishes the best service solution rapidly, and creates values for various parties.

The big service has the following 7 major features [4].

Massive: Under the environment of cloud computing and big data, the online cross-field and cross-network large-scale open services constitute the foundation of big service

Complicated: The big service is made up of complex services (virtual service, physical service, artificial service, etc.) from different fields

Cross: The big service is a complex service network covering various cross-field, cross-network and cross-world services. The service solution is usually made up of cross-field services

Convergence: The cross-field service in big service needs to be aggregated to form an integrated service solution in order to meet the user requirements. The service communities in big service are aggregated by services of smaller granularity in various fields.

Customized: The big service system establishes service composition, binds service resources, and completes service tasks as required according to the needs of the large-scale personalized users. The composite services can be adjusted according to the dynamic changes of the client demand.

Credit: The credibility of the big service is very important. The key to establishing the cross-field credit system of the big service is to set up the credit standard and credit chain for such service

Value-Added: Big service is value oriented. In the big service environment, the service value is defined, decomposed, transformed, and delivered to each phase, process, entity as well as activity in the service life cycle.

3 Application of Big Service in Smart Library

3.1 Development of Big Service

The development of the Internet technology enables all the Internet users to participate in, exchange, and share various kinds of Internet information resources and services, and realizes the business interaction and transformation between the virtual world of the Internet and the real world. The focus of the Internet has also shifted from the network, data, information, content, etc. to applications and services. Cloud computing technology provides users with massive resources (mass data, cloud computing resources, etc.) and various cloud services in a variety of application models (such as anytime access, use on demand, infinite extension, and pay per use). The focus of cloud computing also transfers from massive resources, virtualization, and cloud computing resources to services.

With the development of the Internet and cloud computing, big data presents the characteristics of mass data scale, immediate processing, diversified data types as well as huge data value [5]. Big data technology are gradually shifting attention to behavior data from information data processing, with bigger focus on complicated intelligent services, business processing rules, business relations, etc., all of which require an association with a complex service system to eventually form IoS [6]/big service.

The ever-changing information technology makes the interaction between teachers and students in university campus and the ubiquitous personalized learning anytime and anywhere. While new "smart campus" with intelligent teaching management, integrated education resources and technical service is emerging gradually, and "smart library" is replacing "digital library" to become the mainstream of university library.

3.2 Development of Smart Library

The concept of smart library was first put forward by Finnish scholar Aittola in 2003. He believes that smart library is a mobile library without space limitation that can be perceived by the readers and is composed of intelligent technology and intelligent equipment [7]. In the past five years, more and more attention has been paid to the research of smart library in our country and the theoretical research has been increasingly deepened and detailed. By December 2017, there have been 1037 literatures on the theme of "smart library" in the CNKI platform. Every year there is an upward trend in the number as shown in Fig. 1.

Among them, Dong Yan, a Chinese scholar, believes that the smart library = library + IoT + cloud computing + intelligent equipment, which realizes intelligent services and management through IoT [8]. Scholar Shiwei Wang believes that smart library is a library based on digital, network and intelligent information technologies and focusing on collaborative management, ubiquitous benefit service and green sustainable development of the future library [9].

By studying the relevant literatures, the author thinks that smart library cannot be realized by simple resource digitization, equipment intellectualization and service humanization, but is based on the advanced information technology to realize the cross-network and cross-field service integration, achieve the business management, reader

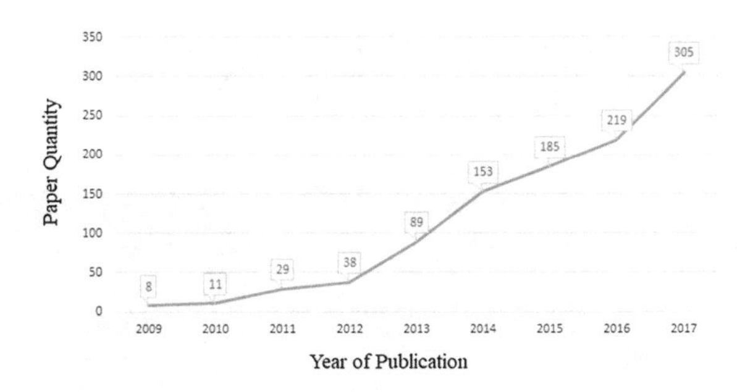

Fig. 1. Trend of paper quantity

service as well as the excavation and dissemination of knowledge in and out of the library, and maximize the utilization of space, personnel and resources. The emergence of big service meets the development needs of smart library.

After more than a decade of digitized campus construction, information resources and application systems of various functional departments within the campus have been enriched and improved gradually. However, only a small amount of public data can be presented to the users in the digital portals and most of the important data and information still scatter in various systems. It has not truly integrated the "isolated information" or realized the co-construction or sharing of resources.

Similar problems exist in the construction of university libraries. Numerous advanced information technologies are applied to library construction, such as wireless network, intelligent equipment, virtualization technology, and sound business management system. In practice, however, various resources in different divisions are not fully integrated, service systems in different areas fail to achieve the collaborative management, online services fail to synchronize with offline services, and users have not really experienced the ubiquitous smart service anywhere. None of those is close to the goal of the smart library construction.

3.3 Big Service for Smart Library

Big service is a new concept of service derived from the basis of "big data + IoS", referring to mass complex aggregation services that emerge after the cross-network and cross-field aggregation in the Internet virtual world and real world [10]. Big service is applied in the smart library by using the laws of big data to produce the large-scale distributed intelligent services, constitute a complex service ecosystem or IoS, and solve the problems of the business processing and business application associated with big data in the library.

As shown in Fig. 2, the smart library service system covers a wide range of services in many fields, including data center, reader service, resource service, inter-library service, internal library services, telecommunication service, media and social networking. For routine and unexpected events such as reader consultation, borrowing and returning, reading and sharing, equipment service, electronic resource download, inter-library loan, it is necessary to aggregate the services in the above-mentioned areas to dynamically form an integrated, personalized, context-aware complex service network. And these complex services can be seen as "big service" in the field of smart library.

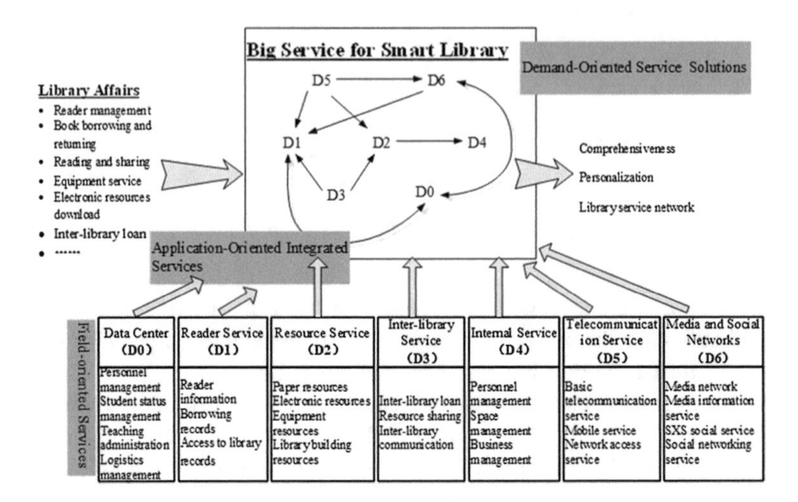

Fig. 2. Big service for smart library

The application of big service brings new positioning and transformation in library functions and user services. The smart library breaks down the tangible wall of a library, connects to the campus data center, and realizes the interconnection within the campus. It creates a network of reading, communication, and promotion for librarians and readers. Every reader is a knowledge provider, a learner and a communicator. Thus, a new type of social network for readers emerges that integrates "library and library", "library and reader", and "reader and reader". The development of the smart library can effectively upgrade the service concept, reorganize the business process, and enrich the service models. It is fully intelligentized and sublimated into a more humanized and smarter high-level library.

4 Architecture of Big Service for Smart Library

The architecture of the "big service for smart library" mainly includes the local service layer, industry service layer, service solution layer, and the upper and lower external layers [11], as shown in Fig. 3

Local Service Layer: This is the basic layer of big service, including many local basic services developed by numerous organizations, institutions, companies, communities and individuals. Most of these services are closely related to specific services and their resources, some of which are virtualized physical services or human services.

Industry Service Layer: The major services at this level are field-oriented composite services. These services consist of various basic services. A certain range of service

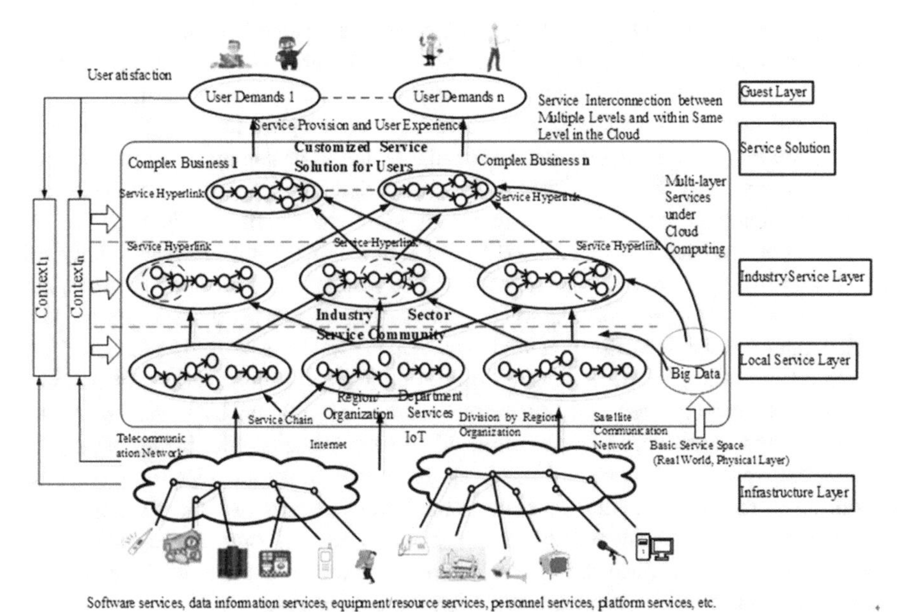

Fig. 3. Architecture of big service for smart library

communities or IoS are formed by composite services according to each industry or their characteristics. These service communities can be composed of services from various regions or networks and include service links. This is the core layer of big service.

Service Solution Layer: Services at this level are complex solutions to meet users' demands. Service solutions are often aggregated across fields, networks, and worlds to satisfy the needs of large-scale and personalized users and maximize their values.

Upper and Lower External Layers: It includes infrastructure layer and customer layer. The infrastructure layer consists of cloud computing platforms, physical services, personnel services, IoT, infrastructure, etc. The customer layer focuses on analysis and presentation of user needs, user interface, etc. Service demand engineering is an important factor in such layers.

5 Development Trend of Big Service for Smart Library

The ultimate goal of the big service is to achieve the integration of realization and virtualization of the service applications, by networking and aggregation of service business and service computing across time and space. Based on the continuous development of key technologies such as IoT, cloud computing, big data, and service computing, the big service for smart library will present a series trends of development.

Service Intelligentization: Personalized demands are designed for massive users. To ensure customers' satisfaction, personalized services are offered to every user by analyzing the characteristics and application environment of various user groups in a complicated way of integrating and coordinating services [12]. During the operational process, the big service system adopt dynamic evolution or deploy dynamic resources or services rapidly by the way of model-driven and dynamic mapping according to user needs, application scenarios and changes in their characteristics, so that the best operating performance can be ensured and effective cooperation between supply side and demand side in virtual and real space can be reached. The service resources, data resources and knowledge are classified, distributed and managed intelligently. With service intelligentization, the potential user needs can be analyzed, proceeded and filtered to provide users with detailed and comprehensive information. In addition, differentiated, personalized and humanized services are offered by providing recommendation and customization of customized service. For example, in the big service system, the smart library may actively push the proper books and electronic resources according to the class schedule in educational administration system; if a teacher has declared for the establishment of his or her scientific research project, the smart library will capture, filter and push the related resources to the user dynamically and intelligently after obtaining such information from the data center.

Diversified Service: In order to help various users use big service and resources conveniently, the big service/IoS will provide the users with efficient and rapid construction methods and convenient human-machine interaction mechanisms. The big service, IoS and IoT will be integrated closely, so that people can access different types of services via various mobile devices any time anywhere. It enables users to enjoy the convenient "tailored" service of the smart library by getting rid of the constraints of time and space. There is no need to go back and forth in different departments for service anymore. It will get rid of the constraints of time and space, offer round-the-clock personalized services to meet the needs of getting resources for massive users, and improve service quality. For example, users can inquire the library resources and other related businesses any time anywhere, download, read or borrow from another place, which will save time and improve the efficiency.

Resource Virtualization: It integrates the physical resources of the real world and information resources of the digital world to the utmost extent by the way of cloud service, and realizes the discovery, integration, collaboration, management, and utilization of resources through virtualization technologies to offer cross-time-and-space services [13]. In the past, the library simply converted paper resources into digital resources. But in the future, it will converge, integrate and manage the physical resources uniformly and deeply, including equipment, collections, and network resources through virtualization technology. It will allocate, convert and deliver the physical resources properly according to user needs, so that they can enjoy efficient and reliable cloud services.

Application Territorialization: With the help of experts in various fields and more detailed applications, the service resources, contents, and businesses of the Internet will provide underlying, granular services vertically and form various innovative service models to meet more subtle needs of users.

6 Conclusion

The big service is becoming a new scientific and technological concept that is symbiotic with the future Internet, cloud computing, smart earth, and big data after Internet and IoT. It is a good interpretation of large complicated service network system in the big data environment. The application of big service in the library has brought new opportunities and challenges to the services of the smart library. In the big service environment, the library realizes the smart interconnection between the physical world and the digital world. The depth and breadth of its connection, as well as the diversity of connected objects and the continuity of the connections make the smart library rich and varied.

References

1. Ghosh, A.: Cloud Computing: Everything as a Service (XaaS or EaaS). https://thecustomizewindows.com/2011/08/cloud-computing-everything-as-a-service-xaas-or-eaas/
2. Lee, E.A.: Cyber physical systems: design challenges. In: Proceedings of the 2008 11th IEEE Int'l Symposium On Object Oriented Real-Time Distributed Computing (ISORC), pp. 363–369 (2008) https://doi.org/10.1109/isorc.2008.25
3. European Future Internet Portal: http://www.futureinternet.eu
4. Xu, X., Wang, Z., Feng, Z.: The big service, applications and their impacts in big data environment. Commun. CCF **2**(2), 24–30 (2017)
5. Dobre, C., Xhafa, F.: Intelligent services for big data science. Futur. Gener. Comput. Syst. **37**, 267–281 (2014)
6. Xu, X., Wang, Z.: Internet of Services in future internet environment. Commun. CCF **7**(6), 8–12 (2011)
7. Aittola, M., Ryhanen, T., Ojala, T.: Smart Library: Location-Aware Mobile Library Service [EB/OL]. http://pdfs.semanticscholar.org/35ec/f598c237068dfa64504eae70a74c713feee4.pdf
8. Dong, Y.: Smart library based on internet of things. J. Libr. Sci. **7**, 8–10 (2010)
9. Wang, S.: On three main features of smart library. J. Libr. Sci. China **6**, 22–28 (2012)
10. Xu, X.: Towards big service in the big data decade. In: Keynote Speed in NICST 2013 (International France-China Workshop on New and smart Information Communication Science and Technology to support Sustainable Development), Clermont-Ferrand, France, 18–20 September 2013
11. Xu, X., Sheng, Q.Z., Zhang, L.-J., Fan, Y., Dustdar, S.: From big data to big service. Computer **48**(7) (2015)
12. Rasch, K., Li, F., Sehic, S., et al.: Context-driven personalized service discovery in pervasive environments. World Wide Web Internet Web Inf. Syst. **14**(4), 295–319 (2011)
13. Zhang, Q., Cheng, L., Boutaba, R.: Cloud computing: state-of-the-art and research challenges. J. Internet Serv. Appl. **1**(1), 7–18 (2010)

Fault Diagnosis of Roller Bearing Based on Volterra and Flexible Polyhedron Search Algorithm

Yongqi Chen(✉), Yang Chen, and Qinge Dai

College of Science and Technology,
Ningbo University, Ningbo 315211, People's Republic of China
chenyongqi@nbu.edu.cn

Abstract. In this paper, the basic theory of Voltera series is briefly introduced. The kernel of Volterra is identified by the polyhedron search algorithm. Different from the traditional diagnostic method based on signal processing, this method applies the Volterra series based on the system model to the fault diagnosis of the roller bearing system. The change of the Volterra series kernel is used to distinguish the different fault state of the roller bearing. The experimental results verify the feasibility and effectiveness of the method.

Keywords: Volterra · Flexible polyhedron search algorithm · Fault diagnosis

1 Introduction

The nonlinear fault diagnosis method based on Volterra series is a typical nonparametric model estimation method. It uses the input and output signals of the system to establish the system model, and determines whether the system is in a fault state by judging the change of the Volterra kernel. Volterra series is often applied to nonlinear system analysis [1–3]. Some method are proposed to obtain Volterra kernel [4]. Li put forward the idea of fault diagnosis based on Volterra functional series theory [5]. In this paper, the nonlinear fault diagnosis method based on Volterra series is applied to the roller bearing system, and the flexible polyhedron search algorithm is used to establish the Volterra series model of the roller bearing system. By studying the change of Volterra series kernel, one can judge the change of system fault state. Simulation and experiment prove the feasibility and effectiveness of the method.

N. Xiong et al. (Eds.): ISCSC 2018, AISC 877, pp. 246–251, 2019.
https://doi.org/10.1007/978-3-030-02116-0_29

2 Nonlinear Volterra Model of Time Series

The phase space reconstruction of the chaotic time series {x (n)} by using the delayed coordinate method, and the point in the phase space can be expressed as:

$$\mathbf{x}(n) = \{x(n), x(n-\tau), \cdots, x(n-(m-1)\tau\}$$

where, m is embedded dimension, τ is time delay. $\mathbf{x}(n)$ is the system input. $x(n+1)$ is the system output. The input and output relation can be expressed as the following Volterra series form:

$$x(n+1) = h_0 + \sum_{n=1}^{p} \sum_{i_1=0}^{m-1} \cdots \sum_{i_k=0}^{m-1} [h_k(i_1, \cdots i_k) \times \prod_{j=1}^{k} x(n-j\tau)] \tag{1}$$

where $h_k(i_1, \cdots, i_k)$ is the kth order Volterra time domain kernel. The three order Volterra time domain is as follows:

$$\begin{aligned} x(n+1) = & \sum_{i=0}^{m-1} h_1(i)(n-i\tau) + \sum_{i=0}^{m-1} \sum_{j=i}^{m-1} G(i,j) h_2(i,j) x(n-i\tau) x(n-j\tau) \\ & + \sum_{i=0}^{m-1} \sum_{j=i}^{m-1} \sum_{k=j}^{m-1} L(i,j,k)[h_3(i,j,k) \\ & \times ux(n-i\tau)x(n-j\tau)x(n-k\tau)] + e(n) \end{aligned}$$

$$G(i,j) = \begin{cases} 1, if\ i = j \\ 2, if\ i \neq j \end{cases}$$

$$L(i,j,k) = \begin{cases} 1 & if\ i = j = k \\ 6 & if\ i \neq j andf\ j \neq k\ andf\ i \neq k \\ 3 & else \end{cases} \tag{2}$$

where $e(n)$ is the truncation error. In order to solve the three order Volterra time domain, the following cost function can be constructed:

$$\min J(n) = \sum_{k=0}^{n} \gamma^{n-k} [x(k+1) - \mathbf{H}^T(n)\mathbf{X}(k)]^2$$

$$\begin{aligned} \mathbf{X}(n) = & [x(n), \cdots, x(n-(m-1)\tau), x^2(n), \\ & 2x(n)x(n-\tau), \cdots, x^2(n-(m-1)\tau), x^3(n), \\ & 3x^2(n)x(n-\tau), \cdots, x^3(n-(m-1)\tau)]^T \end{aligned}$$

$$\begin{aligned} \mathbf{H} = [h_1(0), \cdots, h_1(N-1), h_2(0,0), h_2(0,1), \cdots, h_2(N-1, N-1), \\ h_3(0,0,0), h_3(0,0,1), \cdots, h_3(N-1, N-1, N-1)]^T \end{aligned} \tag{3}$$

where $0 \leq \gamma \leq 1$ is the forgetting factor.

3 The Resolve of Time Domain Kernel Bases on the Flexible Polyhedron Search Algorithm

In the Volterra model, the appropriate kernels are selected by the flexible polyhedron search algorithm. The polyhedron is constituted by m vertexes in n-dimensional space. The fitness value for each vertex is calculated. The worst vertex can be eliminated through the compare of every fitness value. For this reason, the new polyhedron moves closer to the feasible region. Each vertex in the iterative process is constantly moving to the point of optimal fitness value until this algorithm meets the convergence criteria. As what has been mentioned above, volterra kernels become the vertexes In function regression, the fitness function is the sum square error between the real output data of the system and the output data of the volterra model in the same input. The process obtaining the parameters by the flexible polyhedron search algorithm is as follows:

Step 1: generate the initial vertexes of the polyhedron;
Step 2: calculate the fitness value of each vertex, determine the greatest value vertex x_{best}^{k}, the least value vertex x_{worst}^{k} and the second greatest value vertex x_{better}^{k} of fitness function for all vertex;
Step 3: calculate the centroid of this polyhedron $x_{cen}^{k} = \frac{1}{m-1}(\sum_{i=1}^{m} x_{i}^{k} - x_{worst}^{k})$;
Step 4: calculate the reflection point of this polyhedron $x_{refect}^{k} = x_{cen}^{k} + \alpha(x_{cen}^{k} - x_{worst}^{k})$. α is the reflection coefficient and $\alpha > 1$;
Step 5: If the fitness value of the reflection point is better than the greatest value vertex x_{best}^{k}, the expansion point can be calculated as follows. If not, algorithm must be transferred to step 6;

(1) Calculate the expansion point $x_{enl\arg e}^{k} = x_{cen}^{k} + \lambda(x_{fefect}^{k} - x_{cen}^{k})$, $\lambda > 1$;
(2) the least value vertex x_{worst}^{k} is modified as follows:

$$x_{worst}^{k} = \begin{cases} x_{refect}^{k}, & x_{refect}^{k} \ is\ better\ than\ x_{enlarge}^{k} \\ x_{enlarge}^{k}, & x_{refect}^{k} \ is\ worse\ than\ x_{enlarge}^{k} \end{cases}$$

(3) $k = k + 1$ then go to step 9

Step 6: If the fitness value of the reflection point is only better than the second greatest value vertex x_{better}^{k}, the least vertex is modified as $x_{worst}^{k} = x_{refect}^{k}$. Then, go to step 9. If the fitness value of the reflection point is only better than the least value vertex x_{worst}^{k}, the reduction point can be set as $x_{compress}^{k} = x_{cen}^{k} + \eta(x_{fefect}^{k} - x_{cen}^{k}), 0 < \eta < 1$. If the fitness value of the reflection point is worse than the least vertex, the reduction point is calculated using the following compression point $x_{compress}^{k} = x_{cen}^{k} + \eta(x_{worst}^{k} - x_{cen}^{k}), 0 < \eta < 1$.
Step 7: If the fitness value of the reduction point is better than the least value vertex x_{worst}^{k}, the least value vertex can be changed $x_{worst}^{k} = x_{comress}^{k}$. Set $k = k + 1$ and go to

the step 9; if the reduction point is worse than the least vertex, the polyhedron must be compressed;

Step 8: The greatest vertex is maintained and the other vertexes is closed towards the greatest vertex. The formula is $x_i^k \leftarrow x_{best}^k + 0.5(x_i^k - x_{best}^k)$. Then, go to step 9;

Step 9: If $\sum_{i=1}^{m} \left\| x_i^k - x_{best}^k \right\| < (m-1)\varepsilon_1$. It is indicating that this polyhedron has been adjusted to the optimal state. An optimum kernel parameters of volterra is obtained as $x_{best}^k = (c_{best}, \sigma_{best}^2)$. This algorithm runs over, or else go to step 2.

4 Fault Diagnosis Experiment for the Rolling Bearing

In this paper, the rolling bearing fault data is used to verify the effectiveness of the voltera model. Bearing state consists of four categories: normal, inner race fault, outer race fault and ball fault. Status of fault damage bearings is single damage. All the vibration data of rolling bearing analyzed in this paper comes from Case Western Reserve University (CWRU) bearing data center.

Three fault types, such as inner race faults, outer race faults, ball faults, are used in this experiment. Some fault vibration signals are listed in Fig. 1.

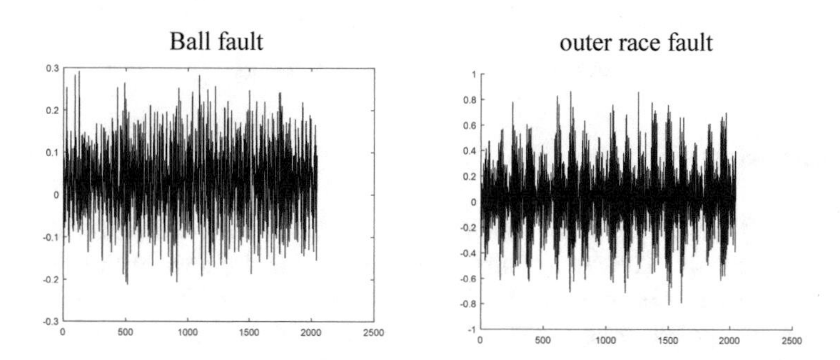

Fig. 1. Waveforms of roller bearing fault vibration signals

According to the theory of Volterra series, the vibration signals measured at drive end are taken as input, and vibration signals at fan end are taken as outputs. The length of the first order time domain kernel vector is 3, the length of the two order time domain kernel is 9, and the length of the three order time domain kernel vector is 27 (Fig. 2).

As can be seen from Fig. 3, the waveforms of the time domain kernel in ball fault state and outer race fault state are distinctly different. The maximal value of the time domain kernel of the ball fault is far greater than the outer race rotor, The peak value is more obvious for ball fault. The cusp of the kernel of ball fault is more than that of the outer race fault. Thus, the nonlinearity of the roller bearing fault is clearly reflected in the higher order kernel of Volterra.

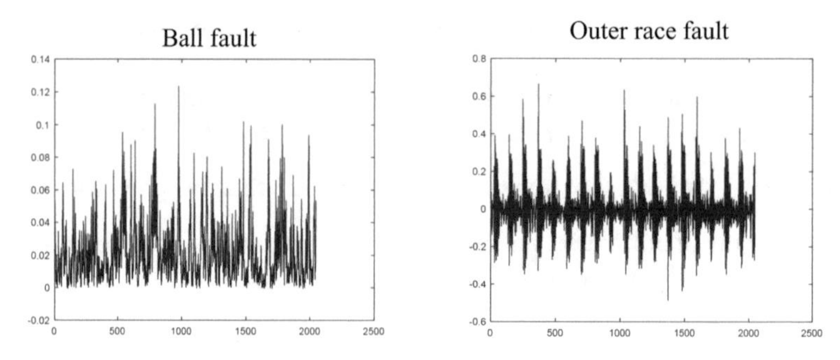

Fig. 2. Volterra simulation output

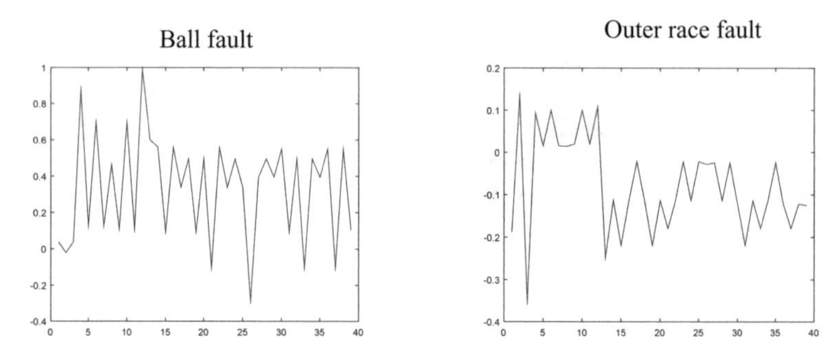

Fig. 3. Volterra time domain kernel

5 Conclusion

In this paper, The Volterra time domain kernel is used to solve the roller bearing fault classification. Firstly, The method first uses the input and output of the synchronous sampling of the roller bearing system. Then, Volterra series of the system is identified by the flexible polyhedron search algorithm. Finally, Volterra kernel is used to determine the system fault state.

Acknowledgements. This research was supported by Zhejiang Provincial Natural Science Foundation of China under Grant No. LY16E050001, Public Projects of Zhejiang Province (LGF18F030003), Ningbo Natural Science Foundation (2017A610138).

References

1. Li, Z., Jiang, J., Chen, J., Wu, G., Li, X.: Volterra series identification method based on quantum particle swarm optimization. J. Vib. Shock **32**(3), 60–74 (2013)
2. Jiang, J., Tan, G.: Fault diagnosis method based on Volterra kernel identification for rotor crack. J. Vib. Shock **38**(23), 138–140 (2010)
3. Schetzen, M.: The Volterra and Wiener Theories of Nonlinear Systems. Krieger, Melbourne (2006)

4. Wray, J., Green, G.G.R.: Calculation of the Volterra kernels of nonlinear dynamic systems using an artificial neuronal network. Biol. Cybern. **71**(3), 187–195 (1994)
5. Evans, C., Rees, D., Jones, L., et al.: Periodic signals for measuring nonlinear Volterra kernels. IEEE Trans. Instrum. Meas. **45**(2), 362–371(1996)

Classification Based on Association with Atomic Association—Prediction of Tibetan Medicine in Common Plateau Diseases

Xuexi Wang[1], Lei Zhang[2,3(✉)], Lu Wang[1], Xiaolan Zhu[1], and Shiying Wang[1]

[1] State Key Laboratory of Plateau Ecology and Agriculture, Department of Computer Technology and Applications, Qinghai University, Xining 810016, China

[2] College of Computer Science, Sichuan University, Chengdu 610065, China
zhanglei@scu.edu.cn

[3] Information Management Center, Sichuan University, Chengdu 610065, China

Abstract. In order to make up for the deficiencies in the research of drug combinations in the study of the prescriptions of Tibetan medicines, this paper applies the Drug-atomic combination association rules to the study of Tibetan medicine prescriptions based on the intrinsic link between drugs. Based on the data of 223 chronic atrophic gastritis cases in Tibetan hospitals in Qinghai Province, using the correlation analysis and association rule algorithm, the method of using commonly used Tibetan medicine pairs for atomic combination to research was proposed for the first time; in the process of extracting common drug pairs, base on the method that limit the length of frequent item sets generated to improve the efficiency of the traditional Equivalence Class Clustering and bottom up Lattice Traversal algorithm (ECLAT algorithm). In this paper, the Classification Based on Association algorithm (CBA algorithm) based on the atomic combination of drugs is creatively applied to the prediction of Tibetan medical drugs. Base on the pulse, urine, and tongue diagnosis data of chronic atrophic gastritis with Tibetan characteristics to predict the diagnostic medicine of doctors. Used the class association rules obtained before were applied to chronic atrophic gastritis in Tibetan medical treatment data, and a 5-fold cross-validation of the rule set was performed. The accuracy rate was 76%. This achieved the preliminary goal of predicting Tibetan medicines based on Tibetan medical vein, urine, and tongue diagnostic data.

Keywords: Association rules · Classification Based on Association
Drug prediction model · Tibetan medicine prescription

This paper is partially supported by The National Natural Science Foundation of China (No.61563044, 71702119, 61762074, 81460768); National Natural Science Foundation of Qinghai Province (2017-ZJ-902) Open Research Fund Program of State key Laboratory of Hydro science and Engineering (No.sklhse-2017-A-05);.

N. Xiong et al. (Eds.): ISCSC 2018, AISC 877, pp. 252–262, 2019.
https://doi.org/10.1007/978-3-030-02116-0_30

1 Introduction

In the actual diagnosis and treatment, the Tibetan doctors' prescribe for patients are judged based on personal experience combined with typical cases, which makes the drug combinations of some diseases greatly affected by personal experience. A variety of drug combinations results often appear for a disease. The complex drug combinations make each patient's drug use too specific, it is difficult to form a standardized diagnosis and treatment system, is not conducive to the inheritance and promotion of Tibetan medical drugs, this study through the Tibetan drug prescription analysis, to contribute to the standardization of Tibetan medicine. Some scholars have carried out relevant researches on this issue. Based on the drug theory of Tibetan medicine, Ren [1] studied the Chinese Tibetan medicine prescriptions in the literature, and started with a simple Chinese Tibetan medicine prescription to provide research ideas for the study of Tibetan medicine prescriptions. Zuo et al. [2] and Yang et al. [3] applied the association rule Apriori algorithm to the study of the formulation of the pterocephalus hookeri heck and myrobalan. The prescriptions of these two drugs are commonly used and the drug combinations for the different diseases are analyzed separately. Guo et al. [4] studied cerebral infarction based on the algorithm of association rules, using time series analysis the prescription regularity according to the characteristic of Tibetan medicine that take medication in line with the time. Cai et al. [5] propose to use the association rules and classification methods of data mining to find the prescription regularity of Tibetan medicine, and combined the Tibetan medicine with the nature and parts of the disease, to provide a way to research Tibetan medicine.

The use of Tibetan medicine follows the principle of mutualism, and different combinations of drugs will be produced for different conditions. Simple research on single drug cannot fully reflect the interrelationship between drugs. However, summing up and drawing lessons from previous studies, most scholars have neglected the consideration of drug combinations. Therefore, starting from the two drug combinations (drug pairs), the research on the rules of the prescriptions and the new research direction are be tried.

Chronic atrophic gastritis is a common disease in the plateau, its high incidence, repeated lingering illness, when accompanied by intestinal metaplasia, dysplasia increased risk of cancer, clinical treatment is so tricky [6], as one of gastric precancerous lesions are attention [7], while Tibetan medicine for the treatment of chronic atrophic gastritis have a good effect. Therefore, based on the data of chronic atrophic gastritis, this paper proposed a method based on the study of drugs atom combinations, finally obtained 10 pairs of commonly used drugs. Choosing pulse-taking, urinalysis and tongue diagnosis of Tibetan medicine and using correlation classification algorithm to classify the commonly used effective medicines, we get 9 articles of strong correlation rules as the classification basis.

The data of this study was taken from the Tibetan Hospital of Qinghai Province. The hospital has a high status in Tibetan medical treatment. Because the cultural background and regional environment of the Tibetan medicine is complex, there is no standardized guiding principle for Tibetan medicine. Tibetan medical data also differs to different extents with different regions, schools, and teachers. Based on this data background, 223 chronic atrophic gastritis data were studied in this study.

2 Research Contents

In this paper, R is used as a research tool to study the regularity of Tibetan medicine in chronic atrophic gastritis based on the association rules of drug atom combinations. After data preprocessing, drug combinations analysis, and generate class association rules, the "symptoms==>drugs pairs" model are finally established. The illustration of the research framework is shown in Fig. 1.

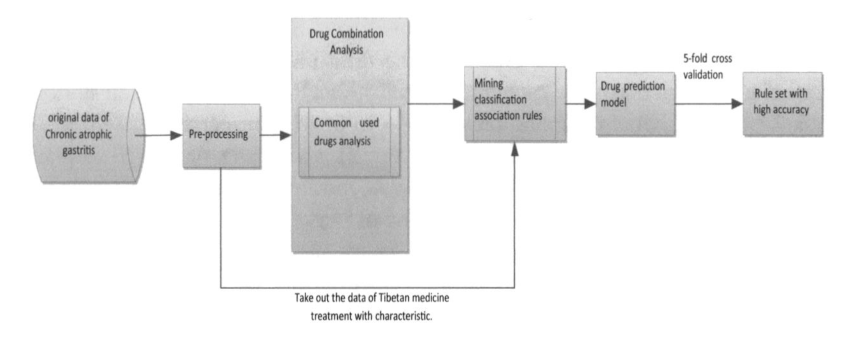

Fig. 1. Illustration of the research framework

This paper preprocesses data from 223 cases of chronic atrophic gastritis obtained from Tibetan hospitals, removes noise from the original data, and only retains the data which use of drugs during treatment. It is helpful for the study of the prescription of patients' medications. And through the method of correlation analysis, the commonly used method of drug attribute reduction based on the frequency of drugs used was verified, and the use of drugs with low frequency was eliminated. After the preprocessed, the length of frequent item sets and the threshold of minimum support are limited to improve the efficiency of ECLAT algorithm. In this way to obtain the common drug pairs for chronic atrophic gastritis; the commonly used drug pairs are linked together as atomic combinations as the foundation for studying a variety of drugs atom combinations and further establishing a more comprehensive "symptoms ==> drugs" rule base. Based on the associative classification CBA algorithm to discover the association rules of "symptoms==>drugs" and form rule sets, establish a Tibetan medicine model for the diagnosis of pulse-taking, urine diagnosis, and tongue diagnosis of Tibetan characteristics. After 5 folds cross-validation, the correctly classified instances of this prediction model reached 76%.

3 Pre-processing

To avoid the original data led to noise interference on the process of data mining and affect the accuracy of the final result, the data needs to be pre-processed before the research begins.

In all the samples, because of different conditions, the drugs used are also different, and patients with milder conditions do not need to use drugs. In order to solve this situation, data pre-processed methods such as data cleaning, data transformation, data reduction and others are carried out on the raw data, finally obtaining the 115 case of patient medication data and the Tibetan medicine 89 flavors.

Not all of the attributes in the data are helpful for the study. In order to reduce the redundant attributes, the attribute selection of the data is necessary. There are many ways to choose attributes, and most of the studies on data mining of Tibetan medicine often use the methods of subtracting the less frequently used drug attributes. However, this method is very effective for the field of data mining of traditional Chinese medicine that is already very mature. But for Tibetan medicine, because of the diagnosis and treatment and the system of medication is different from traditional Chinese medicine, and whether the laws of traditional Chinese medicine prescriptions can be applied still need to be verified by scientific methods. This paper uses the method of correlation analysis to explore the feasibility of this method of attribute selection. Use Web network diagram (see Fig. 2.) to intuitively show the relevance of chronic atrophic gastritis drugs.

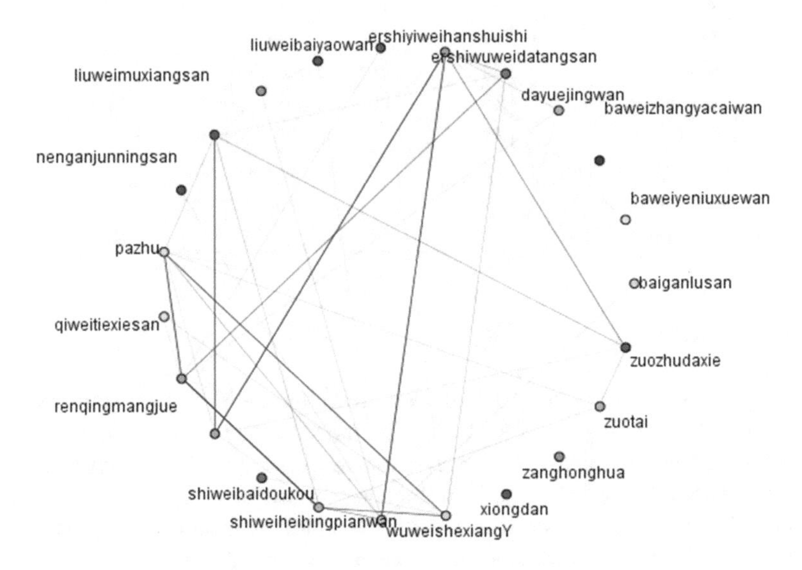

Fig. 2. Correlation analysis web network diagram

After discussions with professional Tibetan medicine personnel, because there is no accurate translation system for Tibetan medicine for the time being, Chinese Pinyin is currently used as a substitute for the English of Tibetan medicine. Therefore, this article also uses pinyin to translate drugs names.

The thickness of the line in the web network diagram in Fig. 2 represents the degree of association between two drugs. The greater the degree of association is, the thicker the line between the two drugs is, and otherwise the thinner the line is. The other drugs that do not appear in the web graph are scattered too high and the correlation is too weak was excluded [8]. And finally found a correlation between about 23 flavors of the drugs. Compare the 23 Tibetan medicines obtained from correlation analysis with the result obtained from attribute reduction methods screened by frequency, that the drugs extracted by correlation analysis were similar to those using more than 10 times drugs. After verification, combined with the data and correlation analysis results, we decided to screen out the chronic atrophic gastritis drugs use frequency of more than 10 times drugs for the prescription regularity analysis. The drugs and their frequency are shown in Table 1.

Table 1. Drugs used more than 10 times.

No.	Drugs name	Frequency	No.	Drugs name	Frequency
1	renqingmangjue	68	13	qiweitiexuesan	19
2	wuweishexiang	60	14	wuweijinsewan	17
3	ershiyiweihanshuishi	58	15	baweizhangyacaiwan	16
4	liuweimuxiangsan	40	16	baweiyeniuxiewan	15
5	zuotai	38	17	liuweihanshuiwan	14
6	shiweiheibingpianwan	34	18	xiongdan	13
7	pazhu	34	19	shiweibaidoukou	13
8	shiliujianweiwan	34	20	liuweibaiyaowan	13
9	zuozhudaxie	33	21	zhituojiebaiwan	13
10	ershiwuweidatangsan	24	22	shisanweiqinglansan	13
11	liuweianxiaowan	20	23	nenganjunningsan	12
12	dayuejingwan	20	24	baiganlusan	12

4 Drug Combination Analysis

The drug combinations can be two, three or more drugs combinations. At the beginning of the study, the combination of two drugs is first selected for research, new research methods are attempted, and a systematic research system is formed prepare for the study of multiple drug combinations in the later period.

Due to the low efficiency of using Apriori algorithm in generating frequent itemsets and it will generate many redundant frequent itemsets, this paper adopts ECLAT algorithm to mine frequent item sets when studying common drug pairs. It is a depth-first algorithm proposed by Zaki [9] in 2000. This algorithm uses a different structure of transaction database than traditional algorithms. It reduces the number of database scans and improves algorithm efficiency in the process of mining frequent itemsets [10]. In this experiment, when carrying out common drug pair excavation, compare the results of setting minsupp from 0.1 to 0.5, and it is found that setting the minimum support threshold to 0.2 is optimal, so in the final experiment, min sup = 0.2. During the generated frequent items the limitation of the set length is that the algorithm terminates when the length of frequent is 2, avoiding the redundant operation of the post order to delay the algorithm runtime.

Definition 1: Tidset: For item X, the set of identifiers for all transactions with item X is called Tidset [10] for item X.

Based on the improved ECLAT algorithm, the process of common drug pairs mining is as Fig. 3:

Input: TDB: Tibetan medicine's transaction database,

Minsup: minimum support threshold:,

Minlen: minimum frequent item set length ,

Maxlen: maximum frequent item set length .

Output: L: all Tibetan items that meet the conditions are frequent item sets。

1. The first scan TDB, transform the horizontal format of Tibetan medicine data set into vertical format.

2. The second scan of TDB, obtain frequent 1-itemsets of Tibetan medicine. denoted by L1, $L = L \bigcup L_1$;

3. ECLAT(L1):

4.{

5. { for all $X_i \in L_1$ do{

6. $X_j \in L_1$, j<i do{

7. Generate a new candidate set by cross operation R= $X_i \bigcup X_j$,

8. Tidset (R) = $Tidset(X_i) \bigcap Tidset(X_j)$;

9. if($|Tidset(R)| \geq$ min sup)

10. { $L = L \bigcup R$;

11. $Ti = Ti \bigcup R$; //It is empty at the beginning

12. if ($Ti \neq \phi$)

13. {if ($\min len \leq len(T_i) \leq \max len$)

14. transfer ECLAT(Ti); }

15. else exit;

16. }

17. }

18. }

Fig. 3. The process of common drug pairs mining

Table 2. Commonly used drug combinations obtained by ECLAT algorithm

No.	Drug combinations	Support
1	ershiyiweihanshuishi-wuweishexiang	0.32
2	ershiyiweihanshuishi-renqingmangjue	0.31
3	liuweimuxiangsan-wuweishexiang	0.29
4	renqingmangjue-wuweishexiang	0.28
5	pazhu-shiweiheibingpianwan	0.26
6	shiliujianweiwan-zuotai	0.24
7	wuweishexiang-zuotai	0.24
8	ershiyiweihanshuishi-zuotai	0.22
9	ershiyiweihanshuishi-liuweimuxiangsan	0.22
10	renqingmangjue-zuotai	0.21

Commonly used drug pairs obtained based on improved ECLAT algorithm, show in Table 2.

Visualize the above common drug pairs (see Fig. 4).

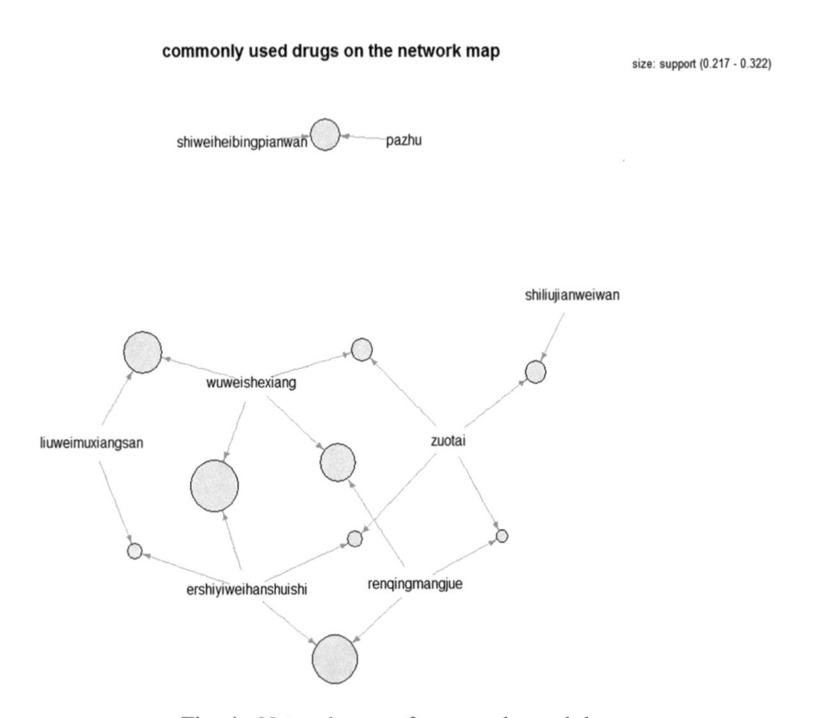

Fig. 4. Network map of commonly used drugs

The size of the dots in Fig. 4 represents the degree of support between drugs. If the support between drugs is high, the dot between drugs will be big. For example, the maximum support for “ershiyiweihanshuishi - wuweishexiang” is 0.32, so the point between ershiyiweihanshuishi and wuweishexiangis biggest.

5 Classification Prediction Model

Based on the previous experimental results, a drug-to-atom combination was used as a classification attribute to establish a model for the prediction of Tibetan medicine using chronic atrophic gastritis. Due to the fact that veins diagnosis, urine diagnosis, and tongue diagnosis are different from other medical methods in Tibetan medical treatment. For example, the urine diagnosis in Tibetan medicine is based on three major factors: “rLung, nKhris-pa, Badkan”. And it is believed that urine is dominated by the three major factors, once the three major imbalances, it wills performance in urine [8]. Therefore, in this paper, the use of drugs for chronic atrophic gastritis in Tibetan medicine pulse diagnosis, urine diagnosis, and tongue diagnosis are predicted. In the actual situation, there are cases where one sample uses multiple drug pairs at the same time. It is difficult to classify using traditional algorithms such as decision trees, naive Bayes, etc. Therefore, this paper uses association rule classification algorithm to build a classification prediction model. Association rule classification is to mine rules such as condset->c, where condset is a set of items and c is a class label, which is called a class association rule [13]. The CBA algorithm used in this paper is an association rule classification algorithm proposed by Liu Bing et al [15] at the International Conference on Database Knowledge Discovery in New York. This algorithm is a method for constructing a classification model after using Apriori algorithm to extract rules [14, 15]. This article only uses the CBA algorithm to achieve the purpose of symptom classification ==> drug classification. The detailed association classification method implementation process will be reflected in the follow-up articles.

Based on Apriori Algorithm to Mining the Class Association Rules:

(1) Generate frequent itemsets. Scan diagnostic drug data of pulse, urine, and tongue and generate frequent itemsets as described in Chapter 4.
(2) Perform further processing on the generated frequent item-sets, and add conditional restrictions on the generation of rules. The left-set of rules is the diagnostic data of pulse, urine, and tongue. The right-set of the rules is the classification attribute that is the common drug pairs. In addition, the three factors, namely the support degree of the rule, the degree of confidence, and the length of the left part of the rule, were used to select the classification association rules. The class association rules that are prioritized according to the degree of confidence (show in Table 3).

In order to make the visualization more clear, to visualize the first 7 rules in Table 3, establish the graph of “symptom ==> drug pairs” (shown in Fig. 5).

Table 3. Class association rule

No	Class association rules	Support	Confidence
1	{tongue red}==>{ershiyiweihanshuishi wuweishexiang}	0.16	0.61
2	{pulse condition is slow, pulse condition is weak, tongue coating thin}==>{pazhu shiweiheibingpianwan}	0.15	0.60
3	{yellow urine, tongue coating thick}==>{wuweishexiang zuotai}	0.16	0.59
4	{tongue red}==>{ershiyiweihanshuishi renqingmangjue}	0.15	0.57
5	{yellow urine, tongue coating thick}==>{renqingmangjue zuotai}	0.15	0.54
6	{ tongue coating thick}==>{shilijianweiwan zuotai}	0.16	0.54
7	{yellow urine, pulse condition is taut}==>{wuweishexiang zuotai}	0.16	0.52
8	{Urine sediment is thin, pulse condition is taut}==> {ershiyiweihanshuishi renqingmangjue}	0.24	0.51
9	{white urine, yellow urine, tongue color yellow}==> {ershiyiweihanshuishi wuweishexiang}	0.21	0.50

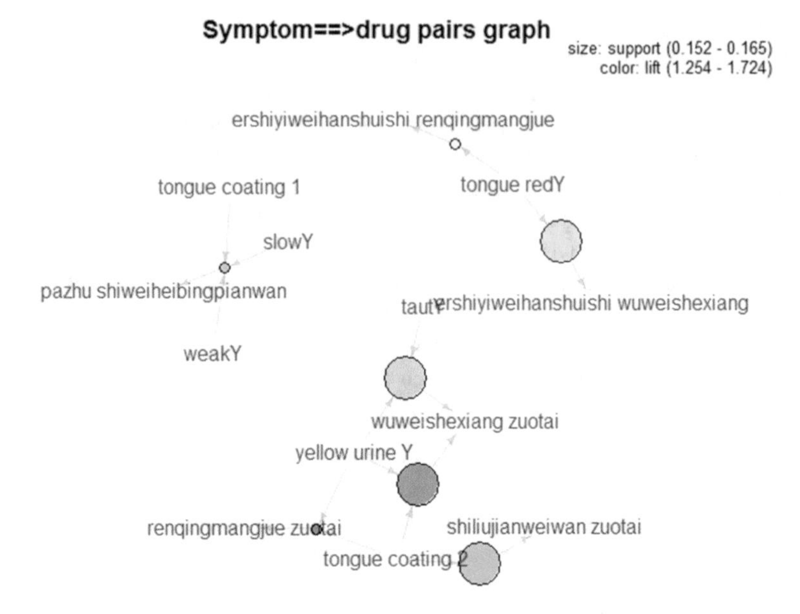

Fig. 5. Symptom==>drug pairs graph

In Fig. 5 "tongue coating 1" indicates thin tongue coating in diagnostic data, "tongue coating 2" indicates thick tongue coating in diagnostic data, "slow Y" indicates slow pulse, and "weak Y" indicates weak pulse, "taut Y" indicates that the pulse is taut and "yellow urine Y" indicates that the urine is yellow. The size of the point in the graph represents the size of the support; the depth of the color of the point represents the degree of lift.

Based on this rules set, a classification model is built and the classification results are verified. Because the sample set used to build the model is small, the method of dividing train and test set will affect the training of the model. Therefore, this paper adopts a 5-fold cross validation method to test the prediction model. The Correctly Classified Instances of the final rules set is 76%. Prove that the classification model has practical application value.

6 Conclusion

A study was conducted on Tibetan medicine prescriptions for chronic atrophic gastritis in Tibetan medicine. After the data pre-processing stage, to clean original data set and reduce attributes, to remove redundancy and noise in the data, and the commonly used attribute screening methods in the drug formulation are verified. The R language was used to analyze the cleaned clinical medical data, and the formulas of prescription were explored by implementing ECLAT algorithm. Based on the drug pairs of Tibetan medicine, the CBA algorithm was used to establish the classification prediction model, and the medication rule for the characteristic diagnosis and treatment of Tibetan medicine was found.

ECLAT algorithm is efficiently used to mine common drug pairs. After screening and analyzing the results, the commonly used drug pairs in the treatment of chronic atrophic gastritis by Tibetan medicine are obtained, which can be used as a reference for Tibetan medicine doctors when selecting drugs.

Using the CBA algorithm, after classification association rules mining and sorting, a classification prediction model was established to establish a connection between the pulse diagnosis, urine diagnosis, tongue diagnosis and medication data for the realization of chronic atrophic gastritis medication diagnosis and prediction model. And the feasibility of the research method is verified, it provide assistance for the Tibetan medical treatment.

For discovering the prescription features of Tibetan medicines, a combination of Tibetan medicines was studied. Based on the advantages of previous studies, new ideas and methods for the study of Tibetan medicine prescriptions were proposed starting from simple drug pairs. To establish the symptoms of chronic atrophic gastritis ==>drug pairs predictive model, through a 5 folds cross-validation, the model correctly classified instances of 76%, based on the pulse, urine and tongue diagnostic data this model can effectively predict the used of Tibetan medicine. In the subsequent research, this method will be optimized, and drug combinations of 3 and 4 flavour Tibetan drugs will be studied to achieve a more systematic basis for the use of Tibetan medicine.

References

1. Ren, X., Wang, M., Mao, M., Wu, H., Zhu, R., Guo, H., et al.: Discussion on the prescription of Tibetan medicine prescriptions based on the theory of Tibetan medicinal properties. World Sci. Technol. Mod. Tradit. Chin. Med. **18**(5), 894–897 (2016)
2. Zuo, F., Wei, Z., Tang, C., Wang, W., Tong, D., Meng, X., et al.: Data mining based study on prescriptions of Pterocephalus hookeri. Chin. J. Chin. Mater. Med. **42**(16), 3213–3218 (2017)
3. Yang, W., Ze, W.Y., Zhang, Y., Yue, Z., Nie, J.: Analysis of the formula containing myrobalan of Tibetan medicine based on data mining technology. Chin. J. Mater. Med. **42**(6), 1207–1212 (2017)
4. Guo, H., Ren, X., Mao, M., Wang, M., Nima, Z., Dunzhu, et al.: Study on the rule of drug use in the treatment of cerebral infarction in Tibetan medicine based on association rules. World Sci. Technol. Mod. Tradit. Chin. Med. **18**(4), 594–599 (2016)
5. Cai, R.N.J., Ren, J., Dorje, C., Lausanne, D., Li, X.: Application thinking of data mining technology in the study of compatibility laws of Tibetan medicine prescriptions. China J. Chin. Mater. Med. **37**(16), 2366–2367 (2012)
6. Wang, P.: Prof. Tang Xudong's experience in data mining for the treatment of chronic atrophic gastritis. Mod. Chin. Med. Clin. **20**(1), 25–30 (2013)
7. Hu, Y., Zang, J., Jia, Y., Jiu, X., Yang, C., Qi, X.: Clinical observation on the treatment of chronic atrophic gastritis by Tibetan medicine. Chin. J. Med. **17**(5), 17–19 (2011)
8. Wang, M.: Exploration and analysis of the symptoms and medication laws of Tibetan medicine in cerebral infarction based on data mining. Doctoral dissertation, Beijing University of Chinese Medicine (2017)

9. Zaki, M.J.: Scalable algorithms for association mining. IEEE Trans. knowl. data Eng. **12**(3), 372–390 (2000)
10. Geng, X.: Research and application of ECLAT algorithm in association rules. Doctoral dissertation, Chongqing University (2009)
11. Liu, H., Guo, R., Jiang, H.: Research and improvement of Apriori algorithm for mining association rules. Comput. Appl. Softw. **26**(1), 146–149 (2009)
12. Han, J., Kamber, M.: Data mining concept and techniques (2001)
13. Qin, D.: Research on association classification algorithm based on frequent closed itemsets. Doctoral dissertation, Chongqing University (2009)
14. Wang, W.: Research on improved and unbalanced data classification algorithms for association classification. Doctoral dissertation, Longnan Normal University (2016)
15. Liu, B., Hsu, W., Ma, Y.: Integrating classification and association rule mining. In: Proceedings of the 4th International Conference on Knowledge Discovery and Data Mining, pp. 80–86 (1998)

Review of Tibetan Medical Data Mining

Lei Zhang[1,2], Xiaolan Zhu[1(✉)], Shiying Wang[1], and Lu Wang[1]

[1] State Key Laboratory of Plateau Ecology and Agriculture, Department of Computer Technology and Applications, Qinghai University, Xining 810016, China
zxlscu@126.com

[2] College of Computer Science, Information Management Center, Sichuan University, Chengdu 610065, China

Abstract. Firstly, this review analyzed the prospect and the main problems of data mining in the Tibetan medicine field. Secondly, it focused on the elaboration of the current research status of data mining technology applied in Tibetan areas, and carried on detailed analysis of the research in clustering, association rules, classification technology in Tibetan field respectively. Finally, this paper put forward a novel decision support framework of Tibetan medicine treatment innovatively based on the above problem, which aims to realize the personalized treatment of Tibetan medicine and provide effective support for the scientific treatment of common diseases of Tibetan plateau further.

Keywords: Tibetan medicine · Data mining · Knowledge base
Decision support

1 Introduction

The Tibetan medicine, which has a long history, belongs to the traditional Chinese medicine and pharmacy. It has systematic theory, unique clinical curative effect and characteristic of drug use. It is the first choice to seek medicine for medicine for the farmers and herdsmen in Northwest China.

In recent years, the extensive application of data mining technology in the field of traditional Chinese medicine has also promoted its development in Tibetan medicine. Through the analysis of the clinical data of massive Tibetan medicine and the classical literature of Tibetan medicine, the internal relations among the symptoms, drugs, individual characteristics and symptomatic types of the patients were explored, and the compatibility of Tibetan medicine was found. Decision-making support of the diagnosis of Tibetan medicine was provided, which will be of great significance to the research in the field of Tibetan medicine.

2 The Problems Facing of Data Mining in Tibetan Medicine

The research of data mining in Tibetan medicine was still in its initial stage. Research status of data mining in Tibetan Medicine was analyzed [1–4] at home and abroad, and found that the main reason was that its information technology coverage was narrow.

N. Xiong et al. (Eds.): ISCSC 2018, AISC 877, pp. 263–273, 2019.
https://doi.org/10.1007/978-3-030-02116-0_31

At present, there still exist some problems of applying to data mining in Tibetan medicine. Firstly, because of the complexity of the cultural background and regional environment, the clinical research foundation is weak and the clinical characteristic technology is nonstandard. There are some absent domains, which were unified standard lexicon, name, term and diagnosis processes of Tibetan Medicine. The shortcomings will result in the difficulty of communication and decline the trust of research. Thirdly, Tibetan medicine has no clinical guiding principle of Tibetan medicine that based on the theory of Tibetan medicine. The subjective factors of the doctor are strong, and there are different degrees of difference in the diagnosis and treatment of the disease. Fourth, the information technology of Tibetan medicine area is relatively backward, the way of clinical diagnosis and treatment record is unscientific, and the correct statistical method is lack. The collected data sets are relatively rough, the sample content is small, and there is a high loss rate. At the same time, the accuracy of various data mining methods to assist the diagnosis of Tibetan medicine is not consistent, and it is not clear which method is better for a certain disease, which is related to sample collection, feature extraction, regional difference and so on.

Data mining technology has been widely applied in traditional medicine, such as traditional Chinese medicine. However, Tibetan medicine and traditional Chinese medicine have many differences in clinical diagnosis and treatment. On one hand, the utility of Tibetan medicine is totally different from the understanding of traditional Chinese Medicine. What's more, there is no guiding principle for Tibetan medicine clinical research based on Tibetan medicine theory. On the other hand, there are many differences between Tibetan medicine and its traditional Chinese medicine in drug theory and efficacy. Furthermore, the data mining technology used on traditional Chinese medicine could not be directly applied to the study of Tibetan medicine because of great differences in the theory and the efficacy of the drug property and treatment theory.

Therefore, based on Tibetan medicine theory and the large scale of clinical diagnosis and treatment data, data mining technology can be used to establish clinical standard and store implicit medical diagnosis rules and patterns for common diseases of the plateau and make a positive contribution for the standardization of Tibetan medicine clinical diagnosis, treatment and pharmacy, which will provide decision supports for Tibetan doctors to diagnosis disease and improve the diagnostic accuracy.

3 Research Status of Data Mining in Tibetan Medicine

The purpose of data mining is to obtain hidden knowledge from large amounts of data [5]. According to data mining tasks, these knowledge patterns can be classified into descriptive and predictive. The former is the description of the intrinsic characteristics of data, and the latter is the induction modeling of current data in order to make predictions for the future. At present, the research of data mining in the field of traditional Chinese medicine has been carried out in depth and systematically. In clustering analysis, the literature [6] proposed a clustering method of mixed attribute data, which provided a new way for the study of sub classification of cancer. The literature [7] aims to explore the distribution rules and methods of the syndrome of type

2 diabetes, after the formulation of the syndrome questionnaire of diabetes mellitus (diabetic nephropathy), 180 cases of type 2 diabetes patients were analyzed and the results showed that yin deficiency syndrome and Yang Deficiency Syndrome, Qi deficiency syndrome and yin deficiency syndrome generally appeared simultaneously in the deficiency syndrome of type 2 diabetes. In addition, the syndrome of heat syndrome and blood stasis syndrome is also seen synchronously, which proves that the method of cluster analysis for traditional Chinese medicine syndrome research is feasible. In the literature [8], the high yield disease and the key source of the patients were divided by using clustering method and the correlation analysis to discover the complications of diabetes and the correlation of drug use in uremia. Literature [9] analyzed the association relation between "Theory-Methods-Formulas-Drugs" was discussed in view of the application of association rules in the diagnosis of liver disease in traditional Chinese medicine. To a certain extent, it provides reference for the clinical treatment of traditional Chinese medicine, the teaching of traditional Chinese medicine and the development of Chinese traditional medicine. In terms of classification and prediction, Literature [10] used decision tree algorithm to predict coronary heart disease, and used division and hierarchical clustering algorithm to classify hospitalized population. The literature [11] described the method of medical data mining based on computational intelligence, and introduced the application of artificial neural network, fuzzy logic, genetic algorithm, rough set theory and support vector machine in medical data mining.

3.1 Research Status of Data Mining Framework for Tibetan Medicine

Because the research of data mining in the area of Tibetan medicine is still in its infancy, a systematic framework of Tibetan medical data mining research hasn't been put forward. That is, how to combine data mining technology with Tibetan medicine to build a perfect medical treatment decision system. Aiming at high altitude polycythemia (HAPC), literature [12] summarized the characteristics of the diagnosis and treatment of Tibetan Medicine from the aspects of the etiology, pathogenesis, clinical symptoms, diagnostic criteria, treatment plans, preventive measures and common complications of the Tibetan medicine for diagnosis and treatment of HAPC. On the basis of inheriting the theory of Tibetan medicine, it is necessary to combine the original thought of Tibetan medicine with modern science and technology. As a result, the collection of Tibetan medicine for the treatment of HAPC ancient books were strengthened and the standard of clinical treatment and the scheme of drug use are standardized. But the corresponding knowledge base of HAPC is not proposed. The member of this research team Nanjia [13] firstly put forward the study of the nature and location of disease by combining drug factors with the premise of not departing from the theoretical framework of Tibetan medicine. Compatibility rules and ideas of Tibetan medicine prescriptions can be found to use data mining technology. But it is based on theoretical level and there is no further experimental verification. Data mining method was used to compare the compatibility composition, chemical composition, pharmacology and clinical information of the seventy flavor pearl pills and twenty-five pearl pills by literature [14], which provided a way of thinking for the study of the material basis and mechanism of the Tibetan medicine. But there is no in-depth study.

In view of Tibetan medicine treatment of viral hepatitis, the literature [15] provided a standardized way, but there is no clear way to form Tibetan medicine subject thesaurus.

3.2 Clustering Analysis

Clustering is based on the principle of maximizing inter class similarity and minimizing the similarity between classes. The essence of the clustering is to divide the data into several categories according to the intrinsic relevance. The literature [16] established a method of clustering data mining based on the collection and collation of ancient books of Tibetan medicine, and established the original protection of Tibetan medicine ancient books according to the characteristics of ancient books. However, there is no standardized and uniform terminology criterion for the study of Tibetan medicine ancient books. So there is a lot of irregularities and errors in the study, it is difficult to establish a more accurate clustering model. The literature [17] conducted searches and collation of existing Tibetan acupuncture works and modern magazines, and statistically analyzed acupuncture points contained in Tibetan acupuncture, and applied clustering algorithms and association rule mining research to find much Tibetans treated diseases (symptoms), that up to 489 species. However, this study only analyzes the literature, lacks the support of modern Tibetan medicine clinical verification, and cannot guarantee the authenticity and effectiveness of clinical guidance.

For the study of clustering in Tibetan medicine, the author believes that it can be discussed in depth from the following three aspects: First, the clinical diagnosis and treatment data of Tibetan medicine can be used to find the common characteristics of similar disease groups by using cluster analysis methods when it is not clear how to classify them to achieve the initial classification of the disease. The second is to analyze the drug factors using clustering methods such as K-means and their improved algorithms [18–20], fuzzy C-means [21, 22], DBSCAN [23, 24], and hierarchical clustering [25]. Extracted the traits of drugs with the same efficacy were and the characteristics for discover the Tibetan medicine prescriptions used in the clinical diagnosis and treatment of Tibetan medicine and find the implicit compatibility rules. The third is to cluster the clinical diagnosis and treatment data of Tibetan medicine, to obtain the phenotypic characteristics of the disease, to extract risk factors that affect the disease, to reveal the hidden patterns in the massive cases, and to provide decision support for the diagnosis and treatment of the healers.

3.3 Association Rules

Association rule algorithm is used to find the association between item sets and determine the relationship between different attributes in the data set. In the Tibetan medical treatment research, it can be used to explore the relationship between the prescription and rules of Tibetan medicine prescriptions and the relationship between the characteristics of the prescription and the symptoms, to find out the norms and theoretical basis for its use, and to provide reference for the clinical diagnosis and treatment of Tibetan medicine. Literature [26] discussed the rules and theoretical basis for the compatibility of traditional Tibetan medicine prescriptions for treating liver diseases in the “Four-Volume Medical Code” by arranging all the prescriptions for the treatment of heat liver disease in the "Four Medical Classics" and using association rules. The important role of Tibetan medicine "nature, efficacy, and taste" theory for routine clinical use and individualized treatment has a strong guiding significance for the development of new drugs for Tibetan medicine. But the study has not yet deepened, especially for the treatment of therapeutic liver diseases. The prescription knowledge base of the formula does not propose a specific method. Literature [27] based on the use of drugs in the medical records of patients with Stagnancy disease (cerebral infarction) who were treated in Tibet Tibetan Autonomous Hospital and met the inclusion and exclusion criteria, using simple association rules, sequence association, and other association rules to mine Methods: Analyze the frequency of drug use in medical records, high-frequency drug combinations and sequence association rules and combination rules between drugs, explore drug use patterns of Tibetan medicine, and screen out drug combinations with strong correlations. The dosing principle of the drug combination for the treatment of Stagnancy disease has a good guiding role, but the study did not dig deeper into the complex association rules that include individual characteristics and disease influencing factors. Literature [28] used association rules Apriori algorithm, improved mutual information method and other data mining methods to analyze the frequency of commonly used drugs, combinations of drugs, association rules and core drug combinations in Hankezi Tibetan medicine. The study summed up the drug use law of Hankezi Tibetan medicine, defined the characteristics of Hankezi in Tibetan medicine treatment, the clinical prescription and the characteristics of commonly used drug combinations, and provided important reference for clinical application, but also did not establish a knowledge base of prescription compatibility.

For the study of association rules in Tibetan medicine, the author thinks that it can be discussed in depth from the following two aspects: First, the use of association rules classic algorithms such as Apriori algorithm to explore more relevant drug combinations and the relationship of "drug-symptoms" and "drug-treatment" "symptoms-symptoms", and "symptoms-treatment" found the Tibetan medicine prescription clinical characteristics and implied laws in the treatment of Tibetan medicine; secondly, the complex correlation of medical data and the hidden value of data itself. Fully use of a large amount of clinical data, use association rule algorithms to mine the complex relationships among treatment, symptoms, and drugs in clinical diagnostic data, and explore the complex correlations between individual characteristics, medication characteristics, disease characteristics, and symptom characteristics, it means that to discover the clinical diagnosis and treatment data of Tibetan medicine of "individual characteristics-drug-symptom-syndrome-type" mining precious knowledge and experience of Tibetan medicine in treating diseases. This is of great value in seeking the rules of diagnosis and treatment of Tibetan medicine experts and forming a personalized diagnosis and treatment plan.

3.4 Classification

Classification is to construct a classifier by analyzing the characteristics of the data set and the classifier maps each input unknown class instance to a given class. In the Tibetan medical treatment field, the classification is based on the predictors of the disease, and classify the discrete variables attribute factors of the data type to realize the decision support of clinical diagnosis and treatment. The literature [29] suggested adopting data mining technology, especially classification technology, to develop the theory of drug properties of Tibetan medicine, elucidating the laws of medicinal properties of Tibetan medicine, and achieving the purpose of scientifically understanding the theory of medicinal properties of Tibetan medicine. The study only provided a theoretical method and did not conduct a deeper practical exploration. Literature [30] applied the artificial neural network method to Tibetan medicine to treat hemorrhagic disease in clinical research, making it possible to independently identify and judge hemorrhagic disease, and conducted a pioneering exploration of classification methods in the application of Tibetan medicine diagnosis and treatment. The study did not analyze the relationship between the characteristics of hemorrhagic disease and medication, and no comparison was made for the prediction effect of multiple classification methods. The preliminary research results of this research team [31], through the clustering and correlation analysis of clinical data of common diseases of the high altitude (atrophic gastritis), formed the standard knowledge base of common diseases of the highland and proposed a new distance discrimination KNN algorithm based on the gray box method, combined with individual patient characteristics and typical symptoms of common diseases in the plateau, to achieve the Tibetan Plateau medical diagnosis and treatment prediction model, the accuracy rate is 80.1%. However, due to the small amount of sample data, the accuracy of the prediction model needs to be further verified.

For the classification in Tibetan medicine, the author believes that the classic classification algorithms of data mining in medical research can be used - C4.5, KNN, Logistic Regression (Logical Regression), BP neural Network [32, 33], Classifier Integration [34, 35] and other methods to train the massive Tibetan medicine clinical data and classic literature, to find out the characteristics of the disease and its relevance, establish a corresponding Tibetan disease prediction model, to provide decision supporting with diagnosis and treatment, improve the accuracy rate of clinical diagnosis of common diseases in the highland, realizing the intelligence of clinical diagnosis and treatment, and then providing decision support for the future Tibetan medicine research work.

4 The Frame of Tibetan Medicine Decision Support Systems

Based on the characteristics of the Tibetan medical data and the research status of data mining in the medical treatment of Tibetan Medicine, the primary problem of building a Tibetan medical diagnosis and treatment system was constructing a knowledge base of the clinical diagnosis and treatment of Tibetan Medicine. As shown in Fig. 1, the disease bank, which symptom library and Medicine store focus on. From the perspective of Tibetan Medicine theory, the connotative and complex association rules between “individual characteristic - symptom - Medicine - syndrome” were analyzed. Meanwhile, the knowledge base of clinical diagnosis and treatment of common Tibetan Medicine in plateau was realized.

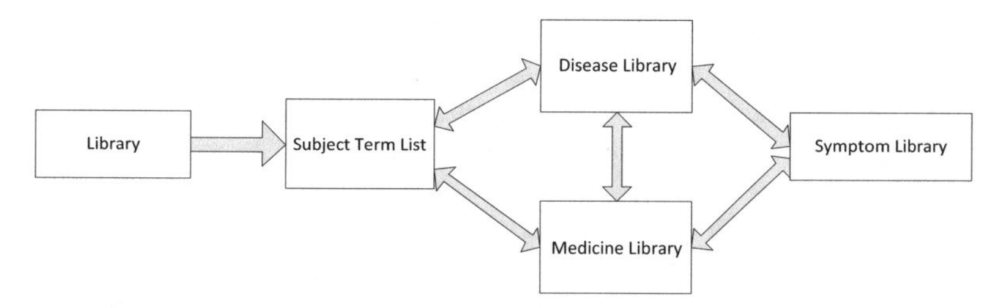

Fig. 1. The under framework of Tibetan Medicine knowledge base system

According to the above, the knowledge of the disease library and the Medicine library come from the classical literature and clinical data of the Tibetan Medicine. The clinical manifestation of disease could be linked to the symptom library through the disease bank, then the Medicine treatment in the disease bank could be linked to Medicine library and the Medicine library will describe the effect and the main disease. From the library, the symptom characteristics of the clinical diagnosis and treatment were recorded in the symptom library and could be linked to the symptom library through the Medicine library. Different medication will reveal the different treatments, and then the symptom library could link to the Medicine library.

Secondly, on the basis of the Tibetan Medicine "Four Medical Classics" and Combine the research results of Qinghai Tibetan Hospital and the Tibetan Medicine College of Qinghai University, the description items of the knowledge base were classified. Finally, the disease bank, the medicine store and the symptomatic library were formed. The three important standard libraries and their relationships were shown in Fig. 2.

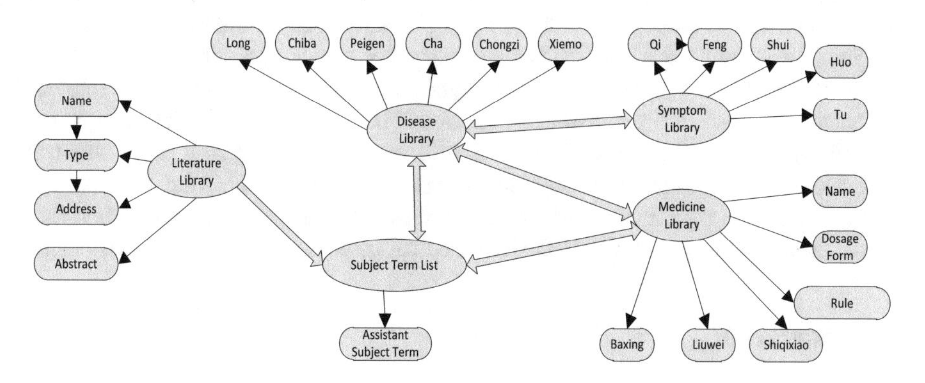

Fig. 2. Thinned image of Tibetan Medicine knowledge base

The characteristics of Medicine in Tibetan Medicine generally included Medince Name, Dosage Form, Compatibility Rule, Baxing, Liuwei and Shiqixiao. The characteristics of disease included the nature, location, season, regional climate characteristics, physical condition of patients, diet habits, age and sex and other factors, but from the text, it could only reflect the nature and location of disease. The nature of the disease could be divided into seven categories. They were Long, Chiba, Peigen, Cha, Quse, Chongzi, Xiemo and so on. The location of the disease could be divided into different levels, such as Shang, Zhong, Xia, Nei, Wai, Zhoong, the viscera, feature and different angles. The symptom bank has five symptoms which were Qi, Feng, Shui, Huo and Tu.

Finally, on the basis of the standard framework for the knowledge base system of medical diagnosis and treatment proposed by Houli [36], the author put forward the framework of the clinical decision support systems for the medical treatment of Tibetan Medicine, which was shown in Fig. 3.

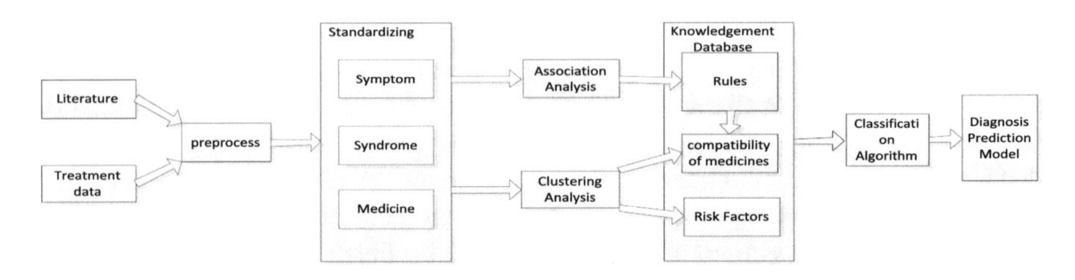

Fig. 3. System framework of Tibetan Medicine Decision Support systems

Firstly, According to the preprocessing of the incompleteness, inconsistent and redundant clinical data in Tibetan Medicine, we summarized the characteristic theme words of Medicine and diseases to standardize the characteristics of Medicine and disease in the clinical diagnosis and treatment of Tibetan Medicine. Secondly, through clustering analysis, the characteristics of Tibetan Medicine prescription were found and the risk factors which affect the diseases were extracted. Then, through the association rules, we build the association between the characteristics of Medicine, disease and symptoms and found the connotative and complex association rules between "Individual Characteristic-Symptom-Medicine-Syndrome". Meanwhile, the knowledge base of clinical diagnosis and treatment of common Tibetan Medicine in plateau was realized. Finally, through the classification algorithm and according to the age structure, gender differences, altitude, related laboratory examination data, the frequency of Tibetan Medicine selection, symptom characteristics, complications and other characteristics, the final model of the Tibetan medical disease prediction was established.

5 Summarize

To sum up, the research of data mining in the diagnosis and treatment of Tibetan medicine was still in its initial stage, and it is a broad cross-disciplinary study with broad application prospects. With the in-depth research of data mining technology and its wide application in medical field, the field of Tibetan medicine with unique system will provide more research space for data mining. In the next step of research, the author's research team will be based on the characteristics of Tibetan medical data and key problems in the diagnosis and treatment system of Tibetan medicine. By standardization of pharmaceutical factors and disease factors, to explore the rule of "Characteristic-Symptom-Medicine-Syndrome", construct the knowledge base of clinical diagnosis and treatment of Tibetan medicine, realize standardization of clinical diagnosis and treatment, clinical diagnosis and evidence-based treatment, and standardized treatment system. Ultimately, it provides decision-making support for clinical diagnosis and treatment of Tibetan to improve the intelligent level of traditional Tibetan medicine clinical diagnosis and treatment.

Acknowledgments. This paper is partially supported by The National Natural Science Foundation of China (No. 61563044, 71702119, 61762074, 81460768); National Natural Science Foundation of Qinghai Province (2017-ZJ-902); Open Research Fund Program of State key Laboratory of Hydroscience and Engineering (No. sklhse-2017-A-05).

References

1. Luo, H., Zhong, G.: Status survey of clinical literature on Tibetan medicine in China. Chin. Tibet. (04), 156–161 (2013)
2. Zhang, M.: The pertinence and applicability of informatization technology in the research of Tibetan medicine. Sci. Technol. Inf. (28) (2012)
3. Ren, X.: Reflections on the clinical research methods of Tibetan medicine. Chin. J. Tradit. Chin. Med. **20**(01), 8–11 (2013)
4. Dawa, T.: The significance and thoughts of Clinical research on Tibetan medicine. World Med. Technol. TCM Mod. **15**(05), 1019–1022 (2013)
5. Hong, S., Zhuang, Y., Li, K.: Data Mining Technology and Engineering Practice. Mechanical Industry Press, Beijing (2014)
6. Bai, T.: Research on the Clustering Method of Biomedical Data. Computer Science and Technology of Jilin University, Jilin (2012)
7. Zhou, D., Zhao, J., Mou, X., Liu, W., Zhou, D., Liu, Y., Ma, G., Hu, Y., Shou, C., Chen, J., Wang, S.: Based on the "disease" in clustering analysis of syndromes in type 2 diabetes. Chin. J. Tradit. Chin. Med. **27**(12), 3121–3124 (2012)
8. Zuo, Y.: Application of Data Mining in Medical Data Analysis. National University of Defense Technology, Changsha, Hunan (2007)
9. Ou, F., Wang, Z.: Application of data mining technology based on association rules in TCM diagnosis. J. Henan Inst. Eng. (Nat. Sci. Ed.) **02**, 53–58 (2011)
10. Li, W.: Research on Data Mining Based on Medical Information Data Warehouse. Medical Examination Department of Chongqing Medical University, Chongqing (2009)
11. Zhu, L., Wu, B., Cao, C.: Techniques, methods and applications of medical data mining. J. Biomed. Eng. (03), 559–562 (2003)
12. Huang, Y.: Research Development on Tibetan Medicine Prevention and Treatment of High Altitude Polycythemia. Modernization of Traditional Chinese Medicine and Materia Medica-World Science and Technology (2015)
13. CaiRang, N., RenZeng, D., DuoJie, C., LuoSang, D., Li, X.: The application of data mining technology in the study of the compatibility rule of Tibetan medicine prescription. Chin. J. Tradit. Chin. Med. **37**(16), 236–2367 (2012)
14. Xun, W., Sun, W., Wang, Z., Liu, G., Liang, Y.: The comparison and analysis of the modern research progress of RNSP and 25-flavor pearl pill based on literature data mining. J. China Natl. Folk. Med. **26**(04), 59–63 (2017)

15. Sun, X., Chen, J., Li, X.: Clinical and experimental study on the treatment of viral hepatitis by Tibetan medicine. J. Chin. Natl. Med. **8**(02), 17–19 (2002)
16. Nie, J., Zhang, Y., Deng, D., JiangYong, S.: Collection and sorting of the classical literature of Tibetan medicine and the model research of data mining. Chin. Folk. Med. (04), 1–2 (2015)
17. Ou, Y.: A Comparative Study of TIBETAN BLood and Moxibustion Therapy Based on Data Mining. Beijing University of Chinese Medicine, Beijing (2011)
18. Zhou, S., Xu, Z., Tang, X.: A new k-means algorithm was used to determine the optimal cluster number. Comput. Eng. Appl. (16), 27–31 (2010)
19. Ying, W., Yao, C.L.: Application of improved K-Means clustering algorithm in transit data collection. In: 20103rd International Conference on Biomedical Engineering and Informatics (BMET) (2010)
20. Zhu, Y., Li, Y., Cui, M., Yang, X.: A clustering algorithm for improving k-means algorithm CARDBK. Comput. Sci. **42**(03), 201–205 (2015)
21. Zanaty, E.A.: Determining the number of clusters for kernelized fuzzy C-means algorithms for automatic medical image segmentation. Egypt. Inform. J. **13**(1), 39–58 (2012)
22. Qi, M., Zhang, H.: Improved fuzzy c-means clustering algorithm. Comput. Eng. Appl. (20), 133–135 (2009)
23. Wang, G., Wang, G.: Improved fast DBSCAN algorithm. Comput. Appl. (09), 2505–2508 (2009)
24. Kumar, K.M., Reddy, A.R.M.: A fast DBSCAN clustering algorithm by accelerating neighbor searching using Groups method. Pattern Recogn. **58**, 39–48 (2016)
25. Wang, L.: Study on clustering method based on hierarchical clustering. Hebei University (2010)
26. CaiRang, N., RenZeng, D.: Based on data mining, the four medical book classics of Tibetan medicine were used to treat the compatibility of hot liver disease prescription. China Exp. Formul. J. (10), 216–219 (2015)
27. Guo, H., Ren, X., Mao, M., Wang, M., NiMa, C., Dun, Z., Dawa, T.: Study on the law of drug use in Tibetan long lag brucella patterns (cerebral infarction) based on association rules. World Sci. Technol. Mod. Chin. Med. (04), 594–599 (2016)
28. Yang, W., Ze, W., Yong, Z., Zhang, Y., Geng, Z., Nie, J.: Analysis of the rules of the medicine of the Tibetan medicine based on data mining (06), 1207–1212 (2017)
29. Ren, X., Mao, M., Guo, H., Wang, M., Wu, H.: The theory of Tibetan medicine and its implications for the theoretical study of modern Tibetan medicine. World Sci. Technol. Mod. Chin. Med. (09), 1911–1916 (2015)
30. Tong, L., De, L., PengMao, Z., Yang, F., ReZeng, C., SuoNan, D.: Based on data mining platform of Tibetan medicine and lung disease system. Chin. J. Natl. Med. (07), 60–62 (2011)
31. Wang, S., Zhang, L., Wang, L., Cairang, N., Zhu, X., Li, H., Wang, X.: Research on syndromic Prediction Model of Tibetan Medicine Diagnosis and Treatment Based on Data Mining. In: 2016 12th International Conference on Signal-Image Technology & Internet-Based Systems (SITis2016), pp. 497–502 (2016)
32. Tan, W., Wang, X., Xi, J., Wang, J.: Research on the diagnosis method of diseases based on BP neural network model. Comput. Eng. Des. (03), 1070–1073 (2011)
33. Huang, L.: Improvement and application of BP neural network algorithm. Chongqing Normal University (2008)
34. Li, Y., Liu, Z., Zhang, H.: An overview of the integrated classification algorithm for unbalanced data. Comput. Appl. Res. **05**, 1287–1291 (2014)

35. Guo, H., Yuan, J., Zhang, F., Wu, C., Fan, M.: A new method of combined classifier learning. Comput. Sci. **41**(07), 283–289 (2014)
36. Hou, L., Huang, L., Li, J.: Discussion on the construction of ontology based clinical medicine knowledge base system. In: The 24th National Conference on Computer Information Management, no. 11, pp. 42–47 (2010)

Establishment of the Database of Men's Suits and Design of Customized Systems

Hongxing Li(✉), Suying Chen(✉), and Wen Yin(✉)

Qingdao University, Qingdao, China
195829978@qq.com, chensyhk@163.com

Abstract. The paper takes men's suits as breakthrough point. According to the existing size and popular fabric in the market, the paper establishes fabric database and size database and then establishes suit style database by modularizing every style element of men's suit. On the basis of this, the paper designs men's suit customization systems which can realize consumers' personalized suit customization through online customized systems.

Keywords: Men's suits · Database · Customization

1 Preface

Custom-made clothing is a garment that is tailored and made according to the size of the individual [1]. The development of custom-made clothing runs through ancient and modern times. For traditional customization, the choice of fabrics and styles is less. And it is unable to meet needs of modern men for personalized clothing.

The paper takes men's suits as breakthrough point and aims at the important parts of customization including size, fabric and style. The paper establishes corresponding database namely size database, fabric database, style database. Which makes it more intuitive and convenient for consumers to make personalized customization.

2 Establishment of Size Database

There are two types of size customization for men's suits: the choice of existing size and measurement of one's size. The former provides customers with ready-made size selection and description of body size specifications. Referring to the GBl335-97 standard men's 5.4 series specification [2], system options are shown in Table 1. The latter provides a graphic description of size measurement method and a custom size input box. Customizable dimensions include height, cervical point height, arm length, waistline height, chest circumference, neck circumference, shoulder width, waist circumference and hip circumference. It gives specific measurement of each item. For example, type classification is illustrated in chart form (as shown in Table 2, Fig. 1). Classification code of body type is expressed by "Y","A","B","C", in which "Y" type is lean body with broad shoulders and thin waist, "A" type is a normal type, "B" type is abdomen slightly protruding, and "C" type is obese.

N. Xiong et al. (Eds.): ISCSC 2018, AISC 877, pp. 274–282, 2019.
https://doi.org/10.1007/978-3-030-02116-0_32

Table 1. System optional size

A series of men's 5.4				
165/84Y	170/88Y	175/92Y	180/96Y	185/100Y
165/84A	170/88A	175/92A	180/96A	185/100A
165/84B	170/88B	175/92B	180/96B	185/100B
165/84C	170/88C	175/92C	180/96C	185/100C

Table 2. Adaptation range of somatotype classification code

Somatotype classification code	Y	A	B	C
Thoracolumbar difference	17–22	12–16	7–11	2–6

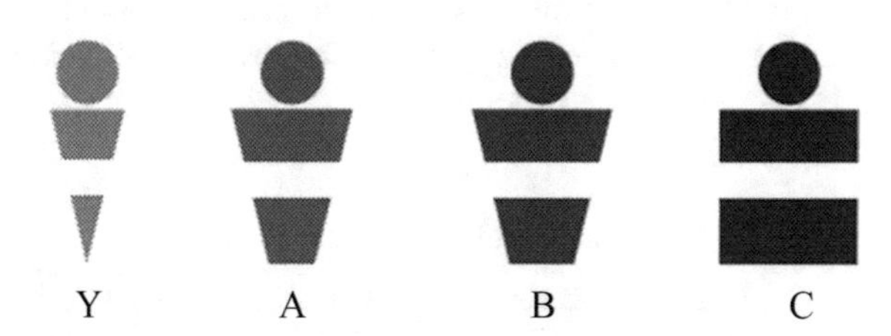

Fig. 1. Classification of body types

3 Establishment of Size Database Fabric Database

According to the situation of fabric markets, enterprises should carry out extensive market research regularly, and collect, sort out, analyze information of domestic and foreign markets in time. According to orthodox or casual style features of suits, enterprises can absorb individual style elements, select dozens or even hundreds of fabrics for customers to choose through multi-directional comparison. And finally, they set up a database of fabrics that can be used in suits. Enterprises should revise database every six months to keep the fabric database in sync with market rhythm [3].

Table 3. Part fabric display

Pure color fabric			...
Striped fabric			...
Jacquard fabric			...
Lattice fabric			...

In the process of filling fabric library, we use real fabric photos as basic material [4]. This paper explains it in details including names, colors, compositions, gram weights and performance of each kind of fabrics. So that it can facilitate choice of customers and realize customization of fabrics. Take pure wool fabrics as an example, some fabrics are shown in Table 3.

4 Establishment of Style Database

In order to facilitate customers to choose each style element, this thesis carries on the modularization processing for men's suit style. The paper takes men's suit jacket as the research object and modularizes its style so we can encode each module (as shown in Table 4). There are six modules in men's suit jacket: profile, front closure style, collar style, cuff style, pocket style, and clothing style. The article uses Adobe Illustrator to editor each module diagram of men's suits.

Table 4. Module code of men's suit

Module	Profile	Front closure	Breast pocket	Garment pocket	Collar	Cuff	Back vent	Hem
Code	*"H"/"Y"/"V"*	*"a"*	*"b"*	*"c"*	*"d"*	*"e"*	*"f"*	*"g"*

4.1 Profile

The profile of men's suit is represented by *"H"*, *"Y"*, *"V"*, as shown in Table 5.

Table 5. Classification of body profile

Profile of a suits	Type *"H"*	Type *"Y"*	Type *"V"*

4.2 Front Closure

Front closure has two forms single-breasted and double-breasted. Single-breasted has one button, two buttons, three buttons and four buttons in four forms. Double-breasted has four buttons and six buttons in two forms. In this paper front closure is represented by "*a*", single-breasted and double-breasted forms are coded as shown in Table 6. For example, one-button of single-breasted suits is represented by "*a*01".

Table 6. Classification and coding of suit's closure

Front closure ctyle "*a*"	Single-breasted	One button	Two buttons	Three buttons	Four buttons
		01	02	03	04

	Double-breasted	Four buttons	Six buttons	Four buttons	Six buttons
		05	06	07	08

4.3 Pocket

Suit pocket has two types breast pocket and body pocket. Breast pocket is denoted by *"b"*. Body pocket is denoted by *"c"*. They are coded separately and these details are shown in Table 7.

Table 7. Classification and coding of pocket style

Suit pocket style	Breast pocket "*b*"	Patch pocket	Box pocket		
		01	02		
	Body pocket "*c*"	Double wire inset pocket	Double wire oblique inset pocket	Waist pouch	Patch pocket with a cover
		01	02	03	04
		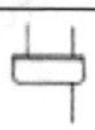			

4.4 Suit Collar Type

Suit collar include notched lapel, peaked lapel, shawl collar, etc. Suit collar is represented by *"d"* as shown in Table 8.

4.5 Cuff Style

Cuff is represented by *"e"* and it has four forms one button, two buttons, three buttons and four buttons, as shown in Table 9.

Table 8. Classification and coding of collar style

Suit collar style "d"	Notched lapel	Peaked lapel	Half peaked lapel	Shawl collar
	01	02	03	04

Table 9. Classification and coding of cuff style

Cuff style "e"	One button	Two buttons	Three buttons	Four buttons
	01	02	03	04

4.6 Body Style

Back vent (as shown in Table 10) and hem (as shown in Table 11) are expressed respectively in *"f"* and *"g"*. Back vent has three types hook vent, side vent, no vent. Hem includes ordinary round type, large fillet type, small fillet type, large square type, small square type. The choice of back vent and hem should match the style of other parts.

Table 10. Classification and coding of back vent style

Back vent style "f"	Hook vent	Side vent	No vent
	01	02	03
	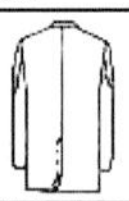	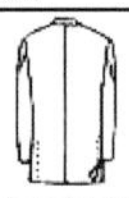	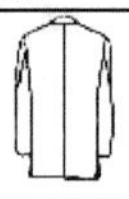

Table 11. Classification and coding of hem style

Hem style "g"	Ordinary round	Large fillet	Small fillet	Large square	Small square
	01	02	03	04	05

5 Design of Customized System

Online customized system adopts modularization idea, each module is independent but interrelated. The idea of modularization brings great convenience to the design of interface and perfection and maintenance of the system. It enhances the interaction performance between systems and users [5]. According to actual situations, user interface can be adjusted to make it quick and intuitive, easy to operate.

Custom system is divided into two subsystems, user interface and back-stage management system (Fig. 2). Users enter the men's suit custom website from system login interface. They can login, consult customer service, input data, custom suit, and so on. Back-stage management system is that administrator manages users' registration information, order information and so on. At the same time, they can add, delete and modify styles and fabrics.

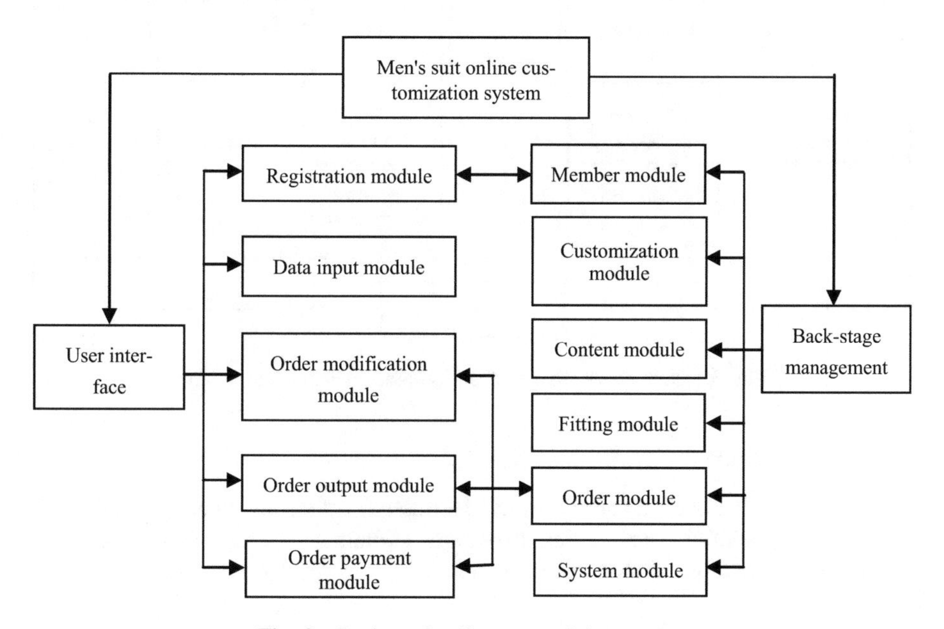

Fig. 2. Design of online customizing system

There are six parts in back-stage management system. They are system module, member module, custom module, fitting module, content module, order module (Fig. 3). The paper takes back-stage management as the starting point to explain interface design of men's suit customized system.

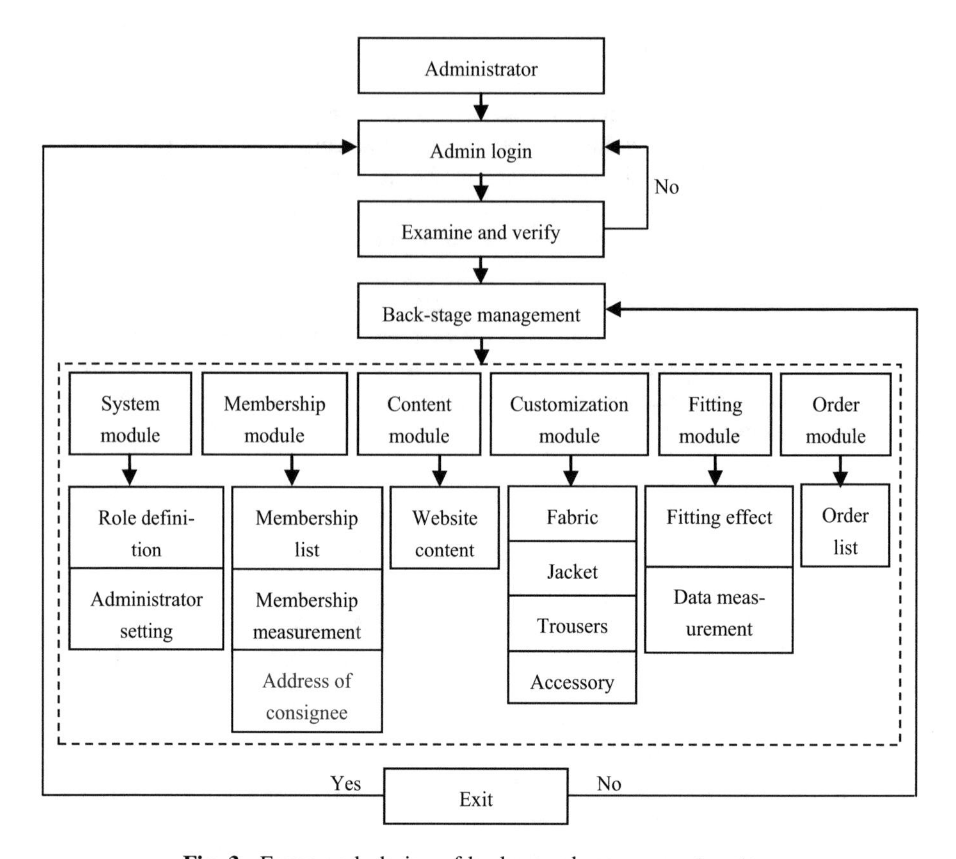

Fig. 3. Framework design of background management system

5.1 System Module

System module is a website administrator permission setting module. It includes role definition and administrator setting two parts. Administrator can login back-stage system and operate on all modules.

5.2 Membership Module

Membership module is a module that manages information of all registered users. It mainly includes membership list, receiving address and member measurement.

5.3 Customization Module

Customization module is one of the most important operating modules in the background system. Products of the system are mainly divided into five parts: fabric, jacket, trousers, vest and accessory product. All products update in the website are implemented through customization module.

5.4 Content Module

Content module is a module that edits content linked to a navigation button in the top frame. The custom website uses "T" structure. The top frame of the page is designed as website logo and menu. The bottom frame is company address. The left frame is navigation button. The right frame is display content. The left frame and the right frame are linked to each other.

5.5 Order Module

After submitting an order through the website, user can view relevant information in order management. It includes order number, order status, payment method, commodity price, freight, etc. Administrator can view orders and manage orders in backstage system.

5.6 Fitting Module

Virtual fitting is a 3D garment virtual fitting technology based on virtual reality and digital simulation technology. It refers to the virtual mannequin instead of users trying on various styles of virtual clothing. Users can intuitively see try-on effect of men's suit.

5.7 Application

In the actual production, the application of customization system can make clothing customization more convenient. The choice of fabrics and styles is rich and diverse. It can meet the needs of most customers. Garment enterprises can change modules or update databases according to their own conditions.

6 Conclusion

With the change of people's life style and the improvement of aesthetic taste, clothing customization is gradually developed. Men's suit, as the most popular and social costumes for men, has also been liberalized and personalized in recent years. The paper splits elements of men's suit and builds an optional database. Which enables consumers to choose fabrics and styles according to their own preferences. They can match different elements into a man's suit with individuation, when they customize a man's suit. This paper designs the online ordering system to realize the online customization of suits. This is of great significance to the transformation and upgrading and sustainable development of enterprises.

References

1. General Business Department of China Light Industry Federation. Compilation of Standards for China's Light Industry: fur and Tannery. China Standard Press, Beijing (2006)
2. Ruipu, L.: Design Principle and Application of Garment Pattern. Menswear. China Textile Publishing House, Beijing (2008)
3. Wu, R.: Design and operation of individual customization pattern for suit. Sci. Technol. Innov. Rep. **1**, 222 (2009)
4. Li, B.: Study on Wedding Dress Design System Based on Personality Customization. Tianjin University of Technology (2015)
5. Wan, L.: Research on Online Customization of Individual Body-Fitted Blouses Based on MTM. Donghua University (2012)

Experimental Study on Deep Learning Oriented to Learning Engagement Recognition

Xiaoming Cao[1], Yonghe Zhang[1](✉), Meng Pan[1], and Hongyang Zhou[2]

[1] Normal College Shenzhen University, Shenzhen, China
yhzhang@szu.edu.cn
[2] Shenzhen Middle School, Shenzhen, China

Abstract. Learning engagement recognition is an important scenario for the application of artificial intelligence technology in education. Its accuracy is influenced by many elements such as data sets and algorithm model. This paper selects five typical algorithm models of convolutional neural network to conduct deep learning experiments on raw data sets, and to evaluate their accuracy and efficiency. Furthermore, the balanced data set and enhanced data set are further introduced to conduct experiment to compare the influence of different data sets on accuracy and efficiency. Through the experiment, we found that: High quality data sets are characterized by large amount of samples, balanced number of each grade sample, and clear images of data images for each level. The fewer the classification level, the easier the model classification is, and the higher the classification accuracy is. According to the performance of the classification model, the Xception model has the best comprehensive performance for the data set of this experiment. It shows that Xception model is better at identifying and classifying images of students' learning engagement.

Keywords: Deep learning · Learning engagement recognition
Artificial intelligence · Multimodal fusion · Attention

1 Introduction

Learning engagement is a necessary condition for learning and has an important impact on the quality of education. The goal of learning engagement recognition is to quantify the participation of students, thus to help teachers understand the learning states of students and make timely intervention, and inspire students reflect on their learning and promote their further participation in the learning process. This is the basis for studying the internal motivations of students' learning effects and the decision of smart teaching. Therefore, real-time and convenient collection, diagnosis and analysis of students' learning engagement that displayed in visual form provides a basis for teachers to adjust teaching strategies. It is the entry point of AI application in the current field of education. It is also a common problems and general needs that prevail in the current smart campus.

The main method of Learning engagement recognition is to classify Learning engagement by deep learning technology on AI, identify the state images of students in

N. Xiong et al. (Eds.): ISCSC 2018, AISC 877, pp. 283–295, 2019.
https://doi.org/10.1007/978-3-030-02116-0_33

class and determine the Learning engagement level of students at a certain time. The accuracy of learning engagement recognition is related not only to data sets, but also to specific algorithm models. Especially when the data set has been determined, there is a big difference in the accuracy and operation efficiency of the algorithm model. The main goal of this experimental is to verify the difference between the accuracy and the execution efficiency under different algorithm models, and then find the optimal model in this situation. At present, the deep learning method based on convolutional neural network (CNN) has been widely studied and applied in image recognition [1]. So this experiment studies on 5 typical image classification network models under convolution neural network are trained by Learning engagement recognition classification training, and then the model classification accuracy and training time cost are collected to compare and analyze the performance of the model. Our research is an exploration of deep learning in the field of education, hoping to provide a reference for researchers in this direction.

2 Technology and Methods

In 1962, Hubel and Wiesel proposed the concept of receptive field by studying the system of visual cortex in cats. They further discovered the hierarchical processing mechanism of information in the visual cortex pathway [2]. In the middle of 80s, a neural perceptron (Neocognitron) based on the concept of receptive field is proposed by Fukushima and Miyake [3]. This can be regarded as the first realization of convolutional neural network and also the first artificial neural network based on local connectivity and hierarchical structure between neurons. Since then, researchers began experimenting with a multilayer perceptron (actually contains only one hidden layer node of the shallow layer neural network model) to replace the manual feature extraction. At the same time, a simple stochastic gradient descent method is used to train the model. In 1986, Rumelhar et al. released the famous back propagation algorithm [4], which is later proved to be very effective [5]. In 1990, Lecun et al. studied the problem of handwritten digit recognition, and a convolutional neural network model trained by gradient back propagation algorithm was firstly proposed [6]. This method shows better performance than other methods in MNIST handwritten numeric data set. In recent years, convolutional neural networks have made breakthroughs in many fields including speech recognition, face recognition, general object recognition, motion analysis, Natural Language Processing and even brain wave analysis. This has laid the foundation for our CNN based learning engagement recognition research [7].

2.1 The Network Structure of CNN

Convolutional Neural Network (CNN) is a multilayer neural network. Its basic structure has the Input Layer, Convolutional Layer, Pooling Layer, Fully Connected Layer and Output Layer. The specific network structure is shown in Fig. 1. CNN mimics the visual information processing process of simple cells and complex cells in the visual nerve. It uses the convolution operation simulation of simple cell edge processing information in different directions in the neighborhood, the output of simple cells with similar cells accumulated in simulation of complex operation of the pool [8]. CNN contains multiple coiling and pooling layers. The first layer convolution directly processes the pixel values of the two-dimensional input image, and outputs the results to the pool layer for fusion. The result of fusion can be regarded as some feature images extracted from input images. New feature images are passed down layer by layer. The convolution and pooling of each layer are the feature extraction of input images. Finally, fuzzy classification or recognition of the full link layer and output layer is carried out.

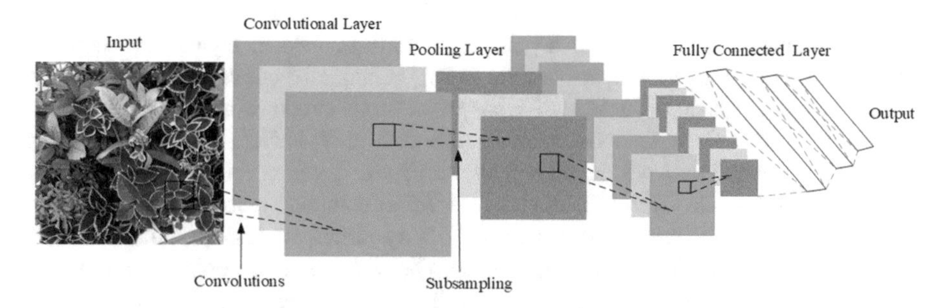

Fig. 1. Diagram of CNN network structure.

The Convolutional Layer usually contains a number of Feature Map. Each characteristic plane is made up of neurons arranged in rectangles. The neurons of the same feature plane share weights, and the weights shared here are convolution kernels. Convolution kernels are usually initialized in the form of random decimal matrices. In the process of network training, convolution kernel will learn to get a reasonable weight. The direct benefit of shared weight (convolution kernel) in the process is to reduce the connection between different layers of the network and the risk of overfitting. Usually, the convolutional layer is a feature extraction layer. By using convolution operation, original signal may enhanced and the noise may be reduced [9].

The Pooling Layer is also known as the lower sampling layer. Pooling operations usually have two forms: mean pooling and max pooling. It uses data processed by convolution as input. Pooling can be regarded as a special convolution process. It compresses the results to reduce the size of data and the number of parameters, which, in return, improves computational efficiency and controls fitting. That is to say, the pooling is used for the quadratic feature extractions, which can be regarded as a fuzzy filter.

The Fully Connected Layer is at the end of the convolution neural network. It transforms the two-dimensional feature map of the convolution output into a one-dimensional vector to improve the purity of the feature extraction. It is also more convenient for the final classifier or regression [10].

2.2 Typical Network Model of Image Classification

CNN has excellent research results in image classification and recognition. In the development process, many excellent network models have emerged. According to the comparison of the performance of different network models published by Canziani et al. in their paper in 2016 [11], and excellent papers of network model in the ImageNet Large Scale Visual Recognition Challenge (ILSVRC). We decided to use the 5 network models, including VGG16, VGG19, Inception V3, ResNet50 and Xception, as students' Learning engagement recognition experimental model.

VGG16 and VGG19

As the name suggests, the network structure of VGG16 has 16 layers (weight layer), which contains 13 convolutional layers and 3 fully connected layers. The network structure of VGG19 has 19 layers (weight layer), contains 16 convolutional layers and 3 fully connected layers. These two types of network structure belong to VGGNet. They are two models with good classification skill in VGGNet. VGGNet is a deep convolutional neural network developed by the Visual Geometry Group of University of Oxford and Google Group Geometry DeepMind company researchers. It was formally proposed by the team in the paper written by Simonyan and Zisserman in 2014 [12]. VGGNet mainly discusses the relationship between network depth (layer number) and model performance. Compared with other network structures, VGGNet has significantly reduced the error rate and won the 2nd place in the ILSVRC 2014 competition classification and the 1st place in the Localization project.

By repeatedly stacking 3×3's small convolution kernel and 2×2's maximum pool level, VGGNet extends the depth of the convolutional neural network to the 19 layers. There are 6 different configurations of VGGNet, named A-E (VGG16 is D, VGG19 is E). The depth is from 11 layers (8 convolutional layers and 3 full connected layers) to 19 layers (16 convolution kernel layers and 3 fully connected layers). The depth of each volume layer is from the 64 in the beginning to 512 in the end (every time passes a max-pooling, depth doubles). Although each level of network has deepened from A to E, the amount of network parameters does not increase a lot. This is because the parameters are mainly consumed in the last 3 fully connected layers. The convolution part of the front is very deep, but the parameter consumption is not large. However, the time consuming part of training is still convolution because of its large amount of computation.

Inception v3

The Inception V3 model is a network structure in Inception. Inception microarchitecture appeared in Szegedy's et al. paper in 2014 [13]. It is a sparse representation with high efficiency. It first appeared in the ILSVRC 2014 competition (the same year as VGGNet) and won the first place with a obvious advantage. The Inception in that competition is usually called Inception V1, also known as GoogleNet. Its biggest feature is to control the amount of computation and parameter, and, at the same time, get excellent classification performance. In the subsequent published paper [14], improvements were made to Inception v1 to form a new network structure, named Inception vN (N is the version number set by Google). Among them, the Inception V3 network structure is a follow-up paper from Szegedy et al. It is the third Inception network structure that has been officially launched.

The Inception V3 model has a good classification effect. The main transformation of the researchers is to decompose a large two-dimensional convolution into two smaller one dimensional convolutions. For example, the 3 × 3 convolution is disassembled into 3 × 1 convolution and 1 × 3 convolution. This saves a lot of parameters, speeds up the calculations and reduces overfitting. A layer of nonlinear extended model expression ability is also added. It is pointed out in the paper [14] that the result of splitting the asymmetric convolution structure is more obvious than that of the symmetry of the same small convolution kernel. It can deal with more and more abundant spatial features and increase feature diversity.

ResNet50

The depth of the ResNet50 model has 50 layers (weight layer). It is a type of network structure in the ResNet series. ResNet (Residual Neural Neural Network) is also known as the deep residual network. It is proposed by He et al. from Microsoft Research Institute [15]. They successfully trained neural network of 152 layers deep by using Residual Unit. They won the championship in ILSVRC in the 2015 game, with 3.57% top-5 error rate. This shows its outstanding effect. The emergence of ResNet makes it possible to train hundreds, or even thousands of layers of neural networks. The results of training are also commendable. It can be regarded as the most pioneering work in the field of computer vision (or deep learning) in recent years.

The traditional convolution layer or full link layer will lose information or get loss when its information is transmitted. To some extent, ResNet solves this problem, by transmitting the input information directly to the output. In order to protect the integrity of information, the whole network only needs to learn the difference between input and output. This simplifies learning goals and difficulties. Moreover, the training error of ResNet network will decrease with the increase of the number of layers. The performance on the test set will also be improved.

Xception

Xception was released by Google in April 2017. It was proposed by Fran OIS Chollet [16], the author and maintainer of Keras library. Xception is not a model compression technique, but a CNN architecture design strategy with few parameters. Its aim is to improve performance. It is an improvement to Inception v3, a type of Extreme Inception. It brings the principle of Inception to the extreme, hence its name Xception. It mainly refers to the use of (non use) deep separable convolution (depth wise separable convolution) to replace the convolution operation in the original Inception v3. The performance of Xception on ImageNet data set is slightly better than that of Inception v3. This advantage is more obvious in classifying datasets including 350 million pictures and 17000 categories. Most importantly, it has the same number of model parameters as Inception, which means it may have faster computation efficiency.

Xception network structure is divided into Entry flow, Middle flow and Exit flow, totaling 36 weight layers. Among them, Entry flow contains 8 conv, Middle flow contains 3 × 8 = 24 conv, and Exit flow contains 4 conv. These 36 coiling layers are used as the feature extraction basis of Xception network structure, and are divided into 14 modules. Except for the last module, there are linear residual connections between the other modules. After the convolution is completed, we can directly use the logical regression layer and add a complete connection layer before the logical regression layer. In one word, the Xception structure is a linear stacking layer with residual join.

3 Experimental Design

3.1 The Purpose of the Experiment

This experiment uses several typical image classification models mentioned above to classify open datasets in Kamath's paper [17]. We compared the classification performance of these models about learning engagement by recording the classification accuracy and training time cost of each model. At the same time, it also reflects the application of depth learning in the field of education.

3.2 Experimental Tools

Experimental Instrument and Environment
This experiment was carried out in the laboratory of Shenzhen University Normal College. It was tested with the Alienware desktop computer, with the memory 16 GB. The processor is Intel(R) Core(TM) i7-8700 K CPU @ 3.70 GHz, and the graphics card is NVIDIA GeForce GTX 1080 Ti. Experimental software Environment: 64-bit Win10 operating system; Anaconda3-5.1.0; PyCharm-community-2018.1.3. Among them, the deep learning modeling environment is the Keras library with TensorFlow as the background of calculation.

Experimental Dataset
The data set used in the experiment is an open dataset in Kamath's paper [17]. The data set was subject's learning videos of 23 courses under the online learning environment (only part of the subject's face and upper body was photographed). The subjects were undergraduates and graduate students for 18–24-year-old. By the research analysis, Aditya Kamath decided to use the subject image obtained from the video frame, instead of the entire video clip. So, the captured video is sampled at a fixed rate and extracted frames to get the image. Crowdsourcing was used to evaluate and mark the learning engagement of the subjects in the image. The degree of engagement was divided into three levels: 1 for non-participation, 2 for general participation, and 3 for active participation. Finally, a majority voting aggregate technique was used to determine the final tag of the image, that is, the level of engagement. At this point, the experimental dataset of this experiment was generated. The experimental dataset has a total of 4408 images, including 400 images of engagement level 1, 2245 images of level 2, and 1763 images of level 3.

3.3 Experimental Process

The experimental process is divided into four stages: pre-preparation, experiment, data analysis, and summary.

Pre-preparation

We first downloaded and organized the experimental data sets that need to be used in the experiment. Then, we installed and configured the software, Anaconda and PyCharm, of the computer used in the experiment. Next, we configured Keras deep learning modal environment with TensorFlow as the calculation background. At the same time, we improved the preparation of several typical image classification models.

Experiment

We first used the Xception model to train and study the experimental data set, and set batchsize = 128. And then we set the epochs to 200, 300, 400 for training and observed the average accuracy rate generated after training the model-two-dimensional map of training times. Since the Xception model has a tendency to change when the epochs is 200 or 300, we decided to set the epochs to 400.

(a) Experiment 1: Using Xception, Inception v3, VGG16, VGG19 and ResNet50 models to train and study the experimental data sets. The parameters were set to batchsize = 128, epochs = 400.
(b) Experiment 2: Because the paper the experimental data set belongs to mentioned: the images of the Level 2 participation should be removed, only classify the participation of level 1 and Level 2, accuracy will be improved. So then use the above 5 classification models to eliminate the level 2 data sets for training and learning, parameter set to batchsize = 128, epochs = 400.
(c) Experiment 3: In the process of Experiment 1 and Experiment 2, we found out some problems of the original data sets, such as unbalanced quantity levels, small data set and single images. These problems lead to the unideal outcome of classification. So, we try to change this situation by code on the basis of the data sets we own now. We used the Class weight method to balance the sample to form a balanced data set (the original dataset is called the Raw dataset). And then, we trained the data set with the above 5 models, and also learned from the balanced data set that excludes level 2.
(d) Experiment 4: On the basis of the balanced dataset, we used the Keras Image Data Generator to rotate, offset and flip the data set image to generate new images. This supplement the dataset data volume and the varied image state to form an enhanced data set. Then, the data set is trained by the above 5 models, and the enhanced data set is trained to eliminate level 2.

Data Analysis

The experimental data analysis means that after the implementation of the above four experiments, data such as the accuracy of learning engagement degree and training time cost of the 5 models collected are statistically collated and presented in the form of graphs. Then, the data are analyzed comprehensively to explain the experimental results.

Summary

The summary is the elaboration, analysis and the summary to the question which exists in the experiment process, to analyze the error which may influence the experiment result.

4 Experimental Results and Analysis

4.1 The Raw Dataset

By implementing Experiment 1 and Experiment 2, the classification effect data (average accuracy and training time cost) of different models under different levels of the original data set (excluding participation level 2 data images) are collected. Concrete data is shown in Table 1.

Table 1. Average accuracy rate and training time cost of different models under the Raw dataset.

Models	Number of levels			
	Level 1, 2, 3		Level 1, 3	
	Average accuracy	Training time cost (min)	Average accuracy	Training time cost (min)
Xception	55.39%	68.3	90.53%	40.5
Inception v3	62.86%	70.5	94.69%	43.7
VGG16	40.94%	60.3	56.44%	37.7
VGG19	40.94%	68.9	56.44%	42.1
ResNet50	50.34%	228.2	90.53%	134.4

According to the Table 1 data, we see that the classification accuracy of each model for the three participation levels (non-participation, general participation, and active participation) is low, but once the level 2 data images are removed, only two types of participation (not participating and actively participating) data images are used for training and learning, and the participation classification accuracy rate will be greatly improved. This phenomenon is normal. In general, the fewer the classification targets, the easier the classification and the higher the accuracy are. Therefore, after removing the level 2 data images, the classification accuracy has improved. The second reason is that level 2 is for general participation. When this level is used as the classification of the images, it is easy to confuse with level 3, which may result in unclear data image. This cannot well represent the level 2 participation, that is, the level 2 data image itself may have problems, which may lead to unsatisfactory training results.

For the case that the classification accuracy of different models is low under the three levels of participation, we found that, besides the data set itself, there is a certain problem: the imbalance of the number of participation level samples may also lead to the accuracy of model classification. So we used the Class_Weight method to equalize the sample, and then train again to see the classification effect.

4.2 The Balanced Dataset

By implementing experiment 3, the classification effect data (average accuracy and training time cost) of different models in the balanced dataset (excluding participation level 2 data images) were collected, as shown in Table 2.

Table 2. Average accuracy rate and training time cost of different models under the balanced dataset.

Models	Number of levels			
	Level 1, 2, 3		Level 1, 3	
	Average accuracy	Training time cost (min)	Average accuracy	Training time cost (min)
Xception	56.38%	69.2	93.18%	40.3
Inception v3	67.34%	71.7	92.05%	44.5
VGG16	40.94%	61.3	73.11%	35.5
VGG19	33.33%	67.7	43.56%	41.3
ResNet50	50.34%	228.2	90.53%	134.4

It is known from Table 2 that the accuracy of classification of data images with different models under the balanced data sets is much higher than that of the classification accuracy before culling. Moreover, the classification accuracy rate of different models under the balanced data set has a certain degree of improvement compared with Table 1, especially for the classification accuracy of 3 kinds of participation degree. This shows that the uneven number of samples has certain influence on the accuracy rate of the model classification.

Although the accuracy of the model classification is improved after the number of samples is balanced, it is still not ideal. So considering the number of the whole dataset and whether the single data image has an effect on the classification accuracy of the model. Based on the balanced data set, we can use the Keras Picture Generator (image data generator) to rotate, offset and flip the dataset image to generate new images, to supplement the dataset's data volume, and to diversify the image state. Then, we trained to see the classification effect again.

4.3 The Enhanced Dataset

By implementing experiment 4, the classification effect data (average accuracy and training time cost) of different models in the enhanced dataset (excluding participation level 2 data images) were collected, as shown in Table 3.

It is shown from Table 3 that the classification accuracy of different models under the enhanced dataset is also much higher than the classification accuracy before the elimination of the level 2 data. Compared with the data of Tables 3 and 2, it can be seen that the classification accuracy of the model is increased under the enhanced data set, and the classification accuracy of the Xception, Inception V3 and RESNET50 models for the 3 types of participation degree is still considerable. This shows that the

Table 3. Average accuracy rate and training time cost of different models under the enhanced dataset.

Models	Number of levels			
	Level 1, 2, 3		Level 1, 3	
	Average accuracy	Training time cost (min)	Average accuracy	Training time cost (min)
Xception	73.15%	64.3	98.48%	37.6
Inception v3	71.59%	67.3	95.08%	42.3
VGG16	40.94%	57.9	71.59%	36.1
VGG19	40.94%	65.7	85.61%	41.3
ResNet50	73.83%	275.9	97.35%	166.3

number of rich datasets and images will improve the accuracy of model classification to some extent, that is to say, the number of datasets and the state of data images have some influence on the accuracy of model classification.

4.4 Overall Analysis

To facilitate the intuitive understanding and comparison of the classification effects of different models under different datasets, the data for Tables 1, 2 and 3 are integrated together, as shown in Table 4.

According to the comprehensive analysis of Table 4 data, we can see that:

Table 4. The average accuracy rate of different models under different dataset.

Models	Datasets					
	Raw dataset		Balanced dataset		Enhanced dataset	
	Level 1, 2, 3	Level 1, 3	Level 1, 2, 3	Level 1, 3	Level 1, 2, 3	Level 1, 3
Xception	55.39%	90.53%	56.38%	93.18%	73.15%	98.48%
Inception v3	62.86%	94.69%	67.34%	92.05%	71.59%	95.08%
VGG16	40.94%	56.44%	40.94%	73.11%	40.94%	71.59%
VGG19	40.94%	56.44%	33.33%	43.56%	40.94%	85.61%
ResNet50	50.34%	90.53%	61.52%	92.05%	73.83%	97.35%

- In any data set, the classification accuracy of the 5 models has been greatly improved after removing the level 2 data images.
- Under the enhanced data set, the classification accuracy of 5 models is obviously better than other data sets. It can be said that the enhanced data set is an ideal data set in this experiment, and it is a data set formed by improving the original data set through code methods.
- The specification of learning engagement classification of VGG16 and VGG19 under 3 data sets is not high, or even low. Also, under the raw, balanced and

enhanced these 3 datasets, the accuracy rate is still basically unchanged. However, this does not mean that these two types of classification, VGG16 and VGG19, are not good. They are just only less effective in the data sets in this experiment, and not be good at classifying facial movements.

- Because the accuracy of model classification under the enhanced dataset is quite impressive, based on the Table 5 data, comprehensive comparison of the average accuracy and training time cost, we got the conclusion that the Xception model is the best performance classification model in this experiment, followed by the inception V3 model, then the ResNet50 model.

5 Research Conclusions and Suggestions

5.1 Research Conclusion

Based on the data analysis, we can see that there are some problems in the original data sets, which leads to the lowest accuracy of each model. After dealing with the original data set by some code methods, the data set is trained again and again, and the accuracy of the model classification is basically improved. This means that the high-qualitied data set is the premise and key of model training learning. It has a great influence on the classification effect of the model. Generally, the less the classification targets, the easier the classification and the higher the accuracy. The classification accuracy of the learning engagement of 2 levels (non participation and active participation) is far higher than the classification accuracy of 3 levels (non participation, general participation, active participation). However, this does not mean that there are many classification targets, and the classification accuracy will be low. If the data set used for training and learning is large and clear, and the appropriate classification model is selected, the classification accuracy will not be bad.

Deep learning is mostly used in industrial and commercial fields, and rarely used in education. This experiment shows the use of deep learning in students' learning engagement recognition. Except for objective perspectives, the classification accuracy of learning engagement is ideal. At the same time, we can also use the best model that is experimentally analyzed for the follow-up projects on students' learning engagement recognition. This reflects that the deep learning can be applied in the field of education. We can believe that deep learning will be applied more in the field of education in the future.

5.2 Research Suggestions

The data set of this experiment comes from the open dataset in Aditya Kamath paper [12]. However, there are some problems in the original data (for example, insufficient sample data, inaccurate classification images, and uneven amount of sample data et al.). Although the code method is used to adjust, the classification accuracy of the model is still not high. Therefore, we suggest that a high-qualitied dataset should be established as the primary factor for future research. After all, high-qualitied data set is the premise and key of model training study. We should pay attention to and overcome the problems mentioned above. Meanwhile, in the process of collecting similar experimental data sets, we should avoid the Hawthorne effect to ensure the quality of data sets.

Because the 5 models used in this experiment are the convolutional neural network (CNN), and the shared weight (convolution kernel) on its convolutional layer is usually initialized in the form of a random fractional matrix. This may lead to the same classification model of each round of training after learning the accuracy rate is not the same, even with a big difference. So the optimal performance classification we get from the study is only for reference. It cannot be applied to all situations. At the same time, if the time and equipment conditions allowed, we suggest that people can try more training on learning model, and select the best collection of classification accuracy of training model.

6 Summary and Prospect

From the training data set, the data set in large quantities, and the number of samples of each grade and balance, image data in each level which is clear and diverse are the quality of the good data set. These are the precondition and key of the excellent calculation model. So, the collection and processing of data sets is an important and worthwhile part of deep learning. From the number of classification level, the fewer the classification level, the easier the model classification, and the higher the classification accuracy rate. But this does not mean that when the number of classification levels is large, the classification effect is poor, and the classification accuracy is low. When the data set of training learning is of high quality, the classification model is appropriate, and the classification effect is not bad. Viewing from the performance of the classification model, the Xception model is the best in the 5 models used in this experiment. When its classification accuracy is almost the same as that of the ResNet50 model with the highest classification accuracy, its training time cost is only about 1/4 of the ResNet50 model. It can be seen that the Xception model is better at identifying and classifying the images of students' learning engagement, which provides a reference model for the following researchers.

Deep learning has become a hot research direction in the field of machine intelligence. More and more researchers focus on the research and application on deep learning. As a popular deep learning framework, CNN has shown more and more obvious advantages in image recognition and classification. Facial expression and emotional state recognition and classification belong to image recognition. At the same time, CNN has the characteristics of simplified network model, automatic training parameters, which make it has potential advantage and good application prospect in facial expression recognition and emotion classification. If facial expression and emotion recognition are widely applied to the field of education, we may get application feedback from learning engagement recognition. In the aspect of data volume expansion, the method based on crowdsourcing will greatly promote the construction of learning engagement feature data set. Thus lay the foundation for more refined recognition. On the other hand, the development of deep learning technology will further enhance the accuracy and efficiency of learning engagement recognition. Although AI application still has a long way to go in the field of education, we believe that in the near future, deep learning will do better in the field of education such as learning engagement recognition.

Acknowledgments. The work described in this paper was fully supported by a grant from the Humanities and Social Sciences Foundation of the Ministry of Education in China, China (Project No.: 13YJC880001) and Guangdong Provincial Education Department Foundation (Project No: 2017JKDY43).

References

1. Xu, K.: Application of convolution neural network in image recognition. Zhejiang University (2012)
2. Hubel, D.H., Wiesel, T.N.: Receptive fields, binocular interaction and functional architecture in the cat's visual cortex. J. Physiol. **160**(1), 106–154 (1962)
3. Fukushima, K., Miyake, S.: Neocognitron: a new algorithm for pattern recognition tolerant of deformations and shifts in position. Pattern Recogn. **15**(6), 455–469 (1982)
4. Rumelhart, D.E., Hinton, G.E., Williams, R.J.: Learning representations by back-propagating errors. Nature **323**(6088), 533–536 (1986)
5. Ruck, D.W., Rogers, S.K., Kabrisky, M.: Feature Selection Using a Multilayer Perceptron. Neural Netw. Comput. **2**, 40–48 (1993)
6. LeCun, Y., Boser, B., Denker, J., Henderson, D., Howard, R.: Handwritten digit recognition with a backpropogation network. World Comput. Sci. Inf. Technol. J. **2**, 299–304 (1990)
7. Zhou, F., Jin, L., Dong, J.: Review of convolution neural networks. J. Comput. Sci. **40**(6), 1229–1251 (2017)
8. Liu, R., Meng, X.S.: Based on deep learning, multimedia picture emotion analysis. In: Audio-Visual Education Research (1), pp. 68–74 (2018)
9. Zhang, Z.: CNN deep learning model for expression feature extraction method. In: Modern Computer (3), pp. 41–44 (2016)
10. Chen, Y., Guo, X., Tao, H.: Overview of image recognition models based on deep learning. In: Electronic World (4), pp. 65–66 (2018)
11. Canziani, A., Paszke, A., Culurciello, E.: An analysis of deep neural network models for practical applications (2016)
12. Simonyan, K., Zisserman, A.: Very deep convolutional networks for large-scale image recognition. In: Computer Science (2014)
13. Szegedy, C., Liu, W., Jia, Y., et al.: Going deeper with convolutions, pp. 1–9 (2014)
14. Szegedy, C., Vanhoucke, V., Ioffe, S., et al.: Rethinking the inception architecture for computer vision. In: Computer Science, pp. 2818–2826 (2015)
15. He, K., Zhang, X., Ren, S., et al.: Deep residual learning for image recognition, pp. 770–778 (2015)
16. Chollet, F.: Xception: deep learning with depthwise separable convolutions. In: IEEE Computer Society, pp. 1800–1807 (2017)
17. Kamath, A., Biswas, A., Balasubramanian, V.: A crowdsourced approach to student engagement recognition in e-learning environments. In: Applications of Computer Vision, pp. 1–9 (2016)

Application Research About Aircraft Readiness Rate Evaluation Based on Improved Neural Network

Zhengwu Wang(✉) and Shenjian Yao

Aviation Maintenance NCO Academy of Air Force Engineering University, Xinyang 464000, Henan, China
wangzhengwu000@sina.com

Abstract. The paper gave the idea, the method and the process about the aircraft readiness rate evaluation based on improved neural network. It constructed the tunable activation function neural network to evaluate the aircraft readiness rate and replaced conventional transfer function with mutative function to extend the evaluation capability, and used the improved algorithm to increase the evaluation precision. The application indicated the evaluation effect was perfect and exact.

Keywords: Aircraft readiness rate · Evaluation · Improved neural network Application research

1 Foreword

The aircraft readiness rate was the ratio that equaled the date of the aircraft in good condition divided by all the date of the aircraft under certain stated environmental conditions. It was one important index to evaluate the aircraft reliability and safety and the maintenance supporting capacity. Many factors and situations would influence the aircraft readiness rate, just like: the aircraft reached its given lifetime, the overhauling in factory, the engine changing process, the periodic maintenance, the daily maintaining and repairing, the lack of aviation repairing materiel, and others couldn't flying situations. Too many factors influenced the aircraft readiness rate, so it was difficult to evaluate the results. The neural network was widely used in the evaluation process because of its peculiarities and functions. With the neural network method maturated gradually [4, 5], we could use it to evaluate the aircraft readiness rate, the application indicated the effect was perfect and exact.

2 The Aircraft Readiness Rate Evaluation Process

2.1 The Evaluation Thought

We selected some appropriate training samples and used its parameters to train the improved neural network, here the parameters included the date because of the parked aircraft maintenance, the overhauling in factory, the changing engine process, the

N. Xiong et al. (Eds.): ISCSC 2018, AISC 877, pp. 296–300, 2019.
https://doi.org/10.1007/978-3-030-02116-0_34

periodic maintenance, the daily maintaining and repairing, the lack of aviation repairing materiel, and so on. Then used its nonlinear changing, storage and mapping capability to reflect the non-training samples, and got the purpose to evaluate the aircraft readiness rate.

2.2 The Evaluation Construction [1–3]

Using the neural network to evaluate the aircraft readiness rate, it needed us to concern the reflection about the unknown samples, it depended on two factors, one was the neural network construction, another was the training samples. So, we could set up the appropriate neural network construction during the evaluation process, it included the number of the layer and nodes. Here we only adopted one middle layer and took the advantage of the base function to enlarge the neural network storage and response capability, the construction just like Fig. 1: x_i (i = 1, …, n) were the known parameters and represented the input signals value, y_j (j = 1, …, r) were outcomes was disposed by the nerve cell activation function upon the input signals, o was the output, it represented the evaluation value, $(\omega_{ij})_{n\times r}$, c_j were the connection weights; the nerve cell activation function were composed by a series of base function $(\phi_0(x_i),\ldots,\phi_h(x_i))$, just like a series of base function operated on the inputted signals and executed non-linear changing, the transmission function were $Y_j = f(\phi_i(x_i))$, here we took $\phi_0(x) = \frac{e^x - e^{-x}}{e^x + e^{-x}}$, $\phi_h(x) = \phi_0^{(h)}(x)$ to denote the response activation function, and used the tan-sigmoid function to replace the sigmoid function, to enlarge the response values.

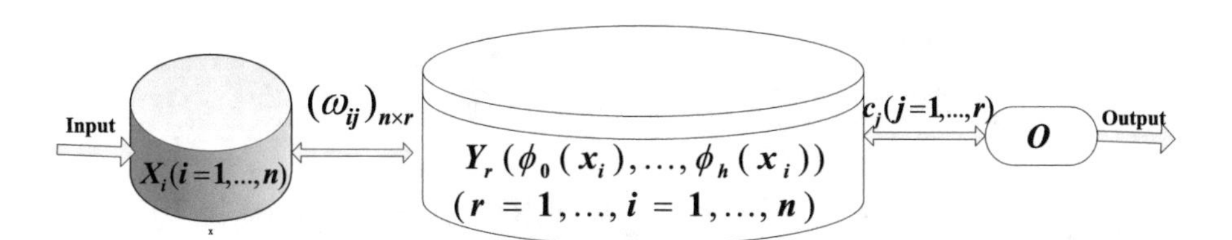

Fig. 1. The neural network configuration

2.3 The Evaluation Process

2.3.1 Confirm the Parameters

Set up the equation

$$\min_{\omega_{ij},C_j} \varepsilon^{(k)} = \sum_{\alpha=1}^{p} (o^{(\alpha,k)} - \overset{\leftrightarrow}{o}^{(\alpha)})^2, \tag{1}$$

We controlled the parameters changing through the errors between the actual outputs and the ideal outputs. Under the process of finding the equation optimal solution, to decide the neural network parameters values, $\left\{\overset{\leftrightarrow}{o}^{(\alpha)}, o^{(\alpha)}\right\}$ were the actual outputs and ideal outputs of the training sample, $\alpha = 1 \sim p$, denoted the sample number, k was the number of training time. Using the following process to decide the parameters, we denoted like follows for convenience:

$$x_j^* = \sum_{i=1}^{n} \omega_{ij} \cdot x_i,\ y_j = \sum_{i=0}^{h} a_i \cdot \phi_i(x_j^*) \tag{2}$$

$$O^* = \sum_{j=1}^{r} c_j \cdot y_j,\ o = \sum_{i=0}^{h} b_i \cdot \phi_i(O^*) \tag{3}$$

a_i, b_i were the basic function coefficients, we adopted the fast speed descend method, let weights change along the function error negative grads direction, and ensure the error convergence to the smallest.

$$\Delta\omega_{ij}^{(k+1)} = -\frac{\partial\varepsilon^{(k)}}{\partial\omega_{ij}^{(k)}},\quad \Delta C_j^{(k+1)} = -\frac{\partial\varepsilon^{(k)}}{\partial C_j^{(k)}}, \tag{4}$$

To quicken the changing speed, and avoid the positive values and negative values counteracted each other during the adding process, we took

$$\frac{\partial\varepsilon^{(k)}}{\partial o^{(k)}} = 2 \cdot \sum_{\alpha=1}^{p} \left|(o^{(k,\alpha)} - \vec{o})\right| \tag{5}$$

$$\frac{\partial\varepsilon^{(k)}}{\partial c_j^{(k)}} = 2 \cdot \sum_{\alpha=1}^{p} \left|(o^{(k,\alpha)} - \vec{o})\right| \cdot y_j \cdot \sum_{t=0}^{h} b_t \cdot \phi_t'(O^{*(k)}) \tag{6}$$

$$\frac{\partial\varepsilon^{(k)}}{\partial\omega_{ij}^{(k)}} = (2 \cdot \sum_{\alpha=1}^{p} \left|(o^{(k,\alpha)} - \vec{o})\right| \cdot c_j \cdot \sum_{t=0}^{h} b_t \cdot \phi_t'(O^{*(k)})) \cdot x_i \cdot \sum_{t=0}^{h} a_t \cdot \phi_t'(x_j^{*(k)}) \tag{7}$$

2.3.2 The Training Samples Collection and Pretreatment

We took the parked aircraft maintenance date, just like: the overhauling in factory, the engine changing process, the periodic maintenance, the daily maintaining and repairing, the lack of aviation repairing materiel, and so on as the samples inputs. These values affected the neural network generalization capability. During the pretreatment process, the network representation demanded the samples had some properties: just as the global property, the particularity, the balanced property and the data veracity; the number of the training samples must be appropriate, it could cause excess simulation or the values wave for more or few data. Here let:

$$x_i^{(1)} = \frac{x_i - \bar{x}}{\max x_i - \min x_1} + 1 x_i^{(2)} = \frac{x_i^{(1)}}{\sum_i x_i^{(1)}}, \tag{8}$$

$\bar{x}$ was the mean, changed the values into the interval (0–1), it could decrease the negative influence about the singularity data including too large or too small.

2.3.3 The Iteration Process

Took the B-P algorithm with momentum factors δ_1 and δ_2 to control the constringency speed properly.

$$\omega_{ij}^{(k+1)} = \omega_{ij}^{(k)} + (\Delta\omega_{ij}^{(k)} + \delta_1 \cdot (\omega_{ij}^{(k)} - \omega_{ij}^{(k-1)})) \tag{9}$$

$$c_j^{(k+1)} = c_j^{(k)} + (\Delta c_j^{(k)} + \delta_2 \cdot (c_j^{(k)} - c_j^{(k-1)})) \tag{10}$$

3 The Application Simulation

We utilized the upper methods to evaluate one aviation unit aircraft readiness rate. Constructed six inputs, four middle nodes and one output neural network to evaluate it. Here the inputs represented the samples values, x_1: the aircraft reached its given lifetime, x_2: the overhauling in factory, x_3: the engine changing, x_4: the periodic maintenance, x_5: the faults eliminating, x_6: the lack of aviation repairing materiel. We used 20 groups data to train the neural network, the parts values lies in Table 1. The output o represented the evaluation value, four nerve cell basic function were

Table 1. Parts of the training data

Inputs	Number			
	1	2	…	20
x_1	36	42		25
…				
x_6	12	10		8

$$\phi_0(x) = \frac{e^x - e^{-x}}{e^x + e^{-x}},\ \phi_1(x) = \frac{4}{(e^x + e^{-x})^2} \tag{11}$$

$$\phi_2(x) = \frac{-8(e^x - e^{-x})}{(e^x + e^{-x})^3} \tag{12}$$

$$\phi_3(x) = \frac{16(e^x - e^{-x})^2 - 32}{(e^x + e^{-x})^4}, \tag{13}$$

let the error $\varepsilon = 10^{-5}$, the beginning weights, basic function coefficients and momentum factors be 0.5, the basic function coefficients and momentum factors increase according to the step $\Delta h = 0.001$, to get the optimal values. The simulation results told us the effect was perfect and exact comparatively.

4 The Conclusion

The application indicated that we could use the construction and algorithm improved neural network to evaluate the aircraft readiness rate, the method was feasible, the error was small and the precision was high. However, during the evaluation process, we must eliminate some un-normal data or it could affect the evaluation result, how to dispose the un-normal data, it needed us research more.

References

1. Wu, Y., Zhao, M.: Adjusting study and application about tunable activation function neural network model. Sci. China (series E) **31**(3), 263–272 (2001)
2. Jiang, X., Tang, H.: System analysis for generalization of MFNN. Syst. Eng. Theor. Pract. **20** (8), 36–40 (2000)
3. Shen, Y., Wang, B.: A fast algorithm about tunable activation function neural network. Sci. China (series E) **33**(8), 733–739 (2003)
4. Hu, S.: An Introduction to Neural Network. Defence Science University Press, Changsha (1998)
5. Wang, J., Gao, T., Lu, C.: The application of an improved BP algorithm in forecasting project cost. Chongqing J. Sanxia Univ. (Science) **23**(2), 129–133 (2001)

An Integrated Method of Tactical Internet Effectiveness Evaluation

Hongling Li(✉), Guannan Li, and Xinwang Yang

Department of Information Communication, Army Academy of Armored Forces, Beijing 100072, China
libian0503@163.com, championman@163.com

Abstract. Based on the analysis of the composition and characteristics of Tactical Internet, an integrated method with the single and whole network effectiveness evaluation is proposed. In this paper, the index systems of the single network and the whole network effectiveness evaluation are given out. Furthermore, the pertinency evaluation algorithm is researched.

Keywords: Tactical Internet · Network efficiency · Evaluation algorithm Index system

1 Introduction

As the main means of communication below the Brigade, the Tactical Internet will be widely used in the future. In the past, the research of effectiveness evaluation method mainly focused on the single equipment in the network, but there was not so much research on the effectiveness evaluation rule of the Tactical Internet. With the rapid improvement of the actual combat requirements of the troops, how to evaluate the overall operational effectiveness of the Tactical Internet on the basis of meeting the performance indicators of the single device has become a key problem in the future equipment test and the training of the troops. In order to provide useful reference for improving the effectiveness of equipment and the combat effectiveness of the army This paper proposes a method to evaluate the performance of the Tactical Internet with the combination of single network and whole network.

So far, there are many researches and literatures on Tactical Internet and more in-depth research results have been obtained [1–5]. Based on the analysis of the protocol and criterion of SINCGARS radio stations, according to the characteristic of Ad hoc network, literature [4] analyzed the evaluation of the interference effect. In document [5], the AODV, DSR and DSDV protocols in MANET are analyzed, and the network simulation models are built to evaluate the network effectiveness based on these protocols. Although these documents give more and more valuable reference, they still do not evaluate the whole Tactical Internet effectively.

N. Xiong et al. (Eds.): ISCSC 2018, AISC 877, pp. 301–307, 2019.
https://doi.org/10.1007/978-3-030-02116-0_35

2 The Composition and Characteristics of Tactical Internet

Through the network interconnection protocol (IP protocol), we connect the tactical radio network, the field integrated service digital network and other communication networks, systems and information terminal equipment are connected as one of the mobile tactical communication system [6]. In a word, Tactical Internet is composed of Tactical Radio Internet and field integrated service digital network, in which the Tactical Radio Internet includes short wave, ultra -short wave and high speed data radio [7, 8]. Take the American Tactical Internet as an example, as shown in Fig. 1.

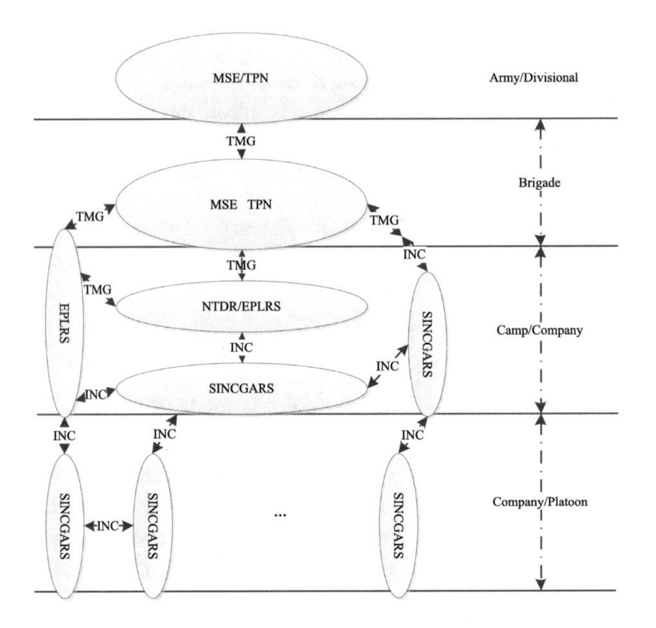

Fig. 1. The American Tactical Internet

From definition about the Tactical Internet and the composition of American Tactical Internet, we can see that the Tactical Radio Internet and the Field Integrated Service Digital Network (the US Army also called it the Field Area Communication Network) belong to different networks. There are great differences between the network architecture of these two kinds of network, for example, the protocol and the form of organization are things similar in form but distinct in kind. The main network of the Tactical Radio Internet is mainly Ad hoc network, the organization form is the access of the tactical sub network, and the Field Integrated Service Digital Network is the backbone network based on the ATM switches. We have to develop a new method to synthesize the effect of single network independently and whole network.

3 An Integrated Method Effectiveness Evaluation Method

To evaluate the effectiveness of Tactical Internet, we should not only set up the evaluation of each sub net in the network, to evaluate the whole network are also needed. The following sections we are going to discuss methods are used to analyze the effectiveness of single network and whole network from index system and evaluation method.

3.1 Construction of Index System

(1) Single network index system
From the application level, whether it is the sub network of ultra short wave radio in the tactical radio internet, the subnetwork of short wave radio, the sub network of high speed data station or the Field Integrated Service Digital Network, the evaluation of the single subnet in the Tactical Internet mainly includes the transmission speed, the transmission reliability and the transmission accuracy which can be used in these aspects. The index system of single network efficiency evaluation is shown in Table 1.

Table 1. Performance index of single network

Network Qos performance	Performance index	Qos parameters
	Transmission speed	IP packet transmission delay (IPTD)
		IP packet delay vibration (IPDV)
	Transmission accuracy	IP packet error rate (IPER)
	Transmission reliability	IP packet loss rate (IPLR)

① IP packet transmission delay (IPTD)
In the network, the IPTD is the end-to-end transmission delay of all the successful data packets which can be calculated in the whole simulation process. For any data packet message, the time interval between the time when the source node is sent out and it when the packet is successfully received by the destination node is the specific end-to-end delay of the data transmission. It includes routing time, packet forwarding time and packet queue waiting time in seconds. The smaller the average end-to-end delay, the faster the transmission speed of the network, the better the performance of sending data packets.

② IP packet delay vibration (IPDV)
The IPDV is defined as the one-way delay difference between two packets in the same data stream. In the application of packet transmission, the merits and demerits of the network retransmission mechanism can be measured according to the variation of IP packet delay, which is related to the link bandwidth and the link utilization rate, and is an important index to evaluate the link access performance.

③④ IP packet error rate (IPER)
The IPER is the ratio of the result of error IP packet transfer to the sum of error IP packet and successful IP packet transmission.

④ IP packet loss rate(IPLR)

The IPLR is the ratio of the lost IP packet results to all IP packets transmission, and is complementary to the packet delivery rate. The greater the value of the packet delivery rate, the less the number of data packets dropped in the transmission process, the better the performance of the routing protocol. The grouping of the parameter statistics refers to the user data packet, not including the Routing control message packet.

(2) Whole network index system

In the whole performance evaluation of Tactical Internet, the operational robustness reflects the general characteristics of the network performance. The network robustness reflects the specific characteristics of the network or we also called the structural characteristics, including the reliability, connectivity, destruction, vulnerability and robustness of the network topology, These underlying factors interaction constraints make it difficult to distinguish effectively, so from the efficiency aspect it was attributed to the network failure rate and the efficiency of network topology recovery. The efficiency of network construction reflects the rationality of link access mechanism, and the efficiency of network recovery is related to the time of network reconfiguration. The index system is shown in Fig. 2.

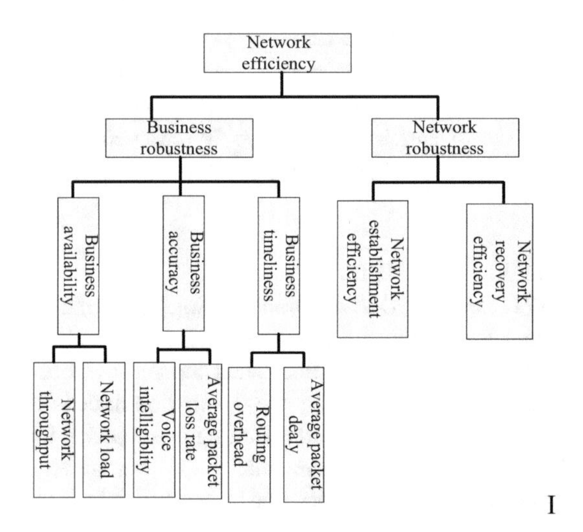

I

Fig. 2. Whole network index system

Each index system is calculated and explained as shown in Table 2.

Table 2. Evaluation element table of network effectiveness index

	Parameters	Remarks
1	Network throughput (bytes/s)	The amount of data that all nodes receive in unit time, which directly reflects the capacity of the network for data services
2	Network load (byte/s)	Packet transmission rate (/s) × packet size (bytes/bits) × number of sources
3	Voice intelligibility	Ratio of obtaining semantic information to listening sounds
4	Average packet loss rate	Taking arithmetic mean
5	Routing overhead	Ratio of routing control packet to data packet, usually normalized routing control overhead
6	Average packet delay	Taking arithmetic mean
7	Network establishment efficiency	Average network establishment time
8	Network recovery efficiency	Average topology reconfiguration time

3.2 Evaluation Algorithm

(1) Single network effectiveness evaluation algorithm
When evaluating the single network efficiency, we need to test the above single network index firstly, and the test method of the index is relatively simple. We take transmission delay test as example, in the case of the transmitter and the receiver synchronize, the transmitter sends 100 groups of IP packets with time marking and group number, and the receiver records time that IP packet arrival, the difference between two time stamps indicates that one-way delay. Other testing methods are no longer described here. On the basis of obtaining the index test data, it is necessary to determine the importance of each node and link. In this case, the node and link weights are determined by the node contraction method.

Figure 3 is a network topology. The node Vi contraction means that all Ki nodes that connected to the node Vi are connected to the node Vi, that is, we can use a new node Vi instead of the Ki + 1 nodes, which are originally associated with the new nodes.

The schematic diagram of the node contraction is shown in Fig. 4.

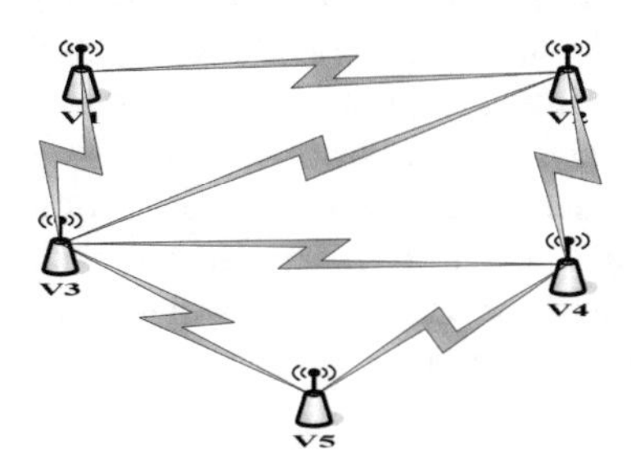

Fig. 3. Network topology

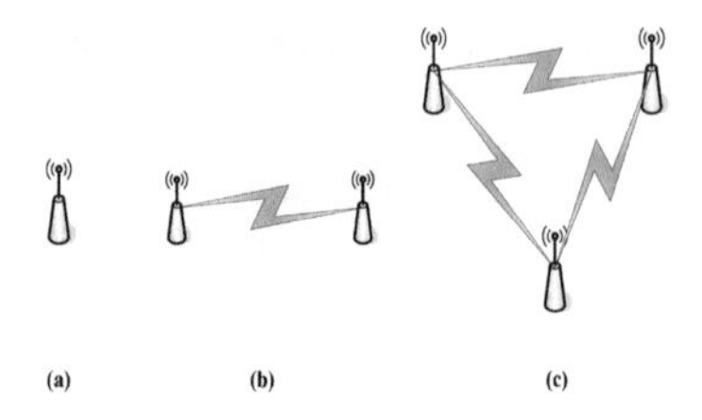

Fig. 4. A schematic diagram of node contraction

Where (a) is the graph of the node V3 after shrinks, (b) is the graph of the node V2 or V4 after shrinks, (c) is the graph of the node V1 or V5 after contraction. According to the graph after node contraction, we can get the result of node importance as shown in Table 3.

Table 3. Significance assessment results of each node

Node	V1	V2	V3	V4	V5
Node weight	0.3333	0.5	1	0.5	0.3333

The calculation of link importance K_x is shown in the following expressions:

$$K_x = (V_i + V_j)/2$$

Among them, V_i and V_j constitute the two nodes of K_x link respectively.

Based on the above test data and the weights of nodes and links, the single network can be evaluated by using the numerical weighting method.

(2) Whole network evaluation

At present, there are many network evaluation algorithms. In general, according to the method of dividing index system as we talked before, the analytic hierarchy process (AHP) can be used. The analytic hierarchy process is more mature, however, it is necessary to point out that, when using this method to evaluate the efficiency of the whole network, although a single index weight is not very high, it would also cause a sudden drop in the performance of the whole network when the value of the index tends to a limit. For example, the signaling station exists in the tactical radio internet, if it is attacked or failed, the success rate of the two parties is nearly 0, and the whole network is in a state of paralysis. Therefore, in the actual application, the performance evaluation of the Tactical Internet must take full account of the limit of the index value. Whatever evaluation algorithm can not be carried out simply according to the mathematical calculation, it must be considered synthetically according to the actual equipment system and organization mode.

4 Summary

In this paper, a method of evaluating the effectiveness of Tactical Internet based on single network and whole network is given, which provides a more effective evaluation method for the equipment construction and army application of Tactical Internet. Of course, there are still a lot of work to be done to evaluate the effectiveness of the Tactical Internet, including the improvement of the network effectiveness index system, such as whether the whole network evaluation has included all the indicators that reflect the performance of the whole network; the optimization of the efficiency evaluation algorithm in the whole network is mainly used in this paper, but as mentioned in this paper The analytic hierarchy process can not reflect the limit of values sometimes. All these work need further in-depth study.

References

1. Hang, Z., He, L.: Study on the evaluation method of Tactical Internet effectiveness based on grid clustering. Sichuan J. Mil. Eng. **32**(9), 111–112 (2011)
2. Shi, R.: Research and performance evaluation of Tactical Internet. Master thesis of Xi'an University of Electronic Science and Technology (2009)
3. Wang, Y.: Research on the method for evaluating the resilience of Tactical Internet based on network topology. Inf. Commun. **6**, 41–42 (2015)
4. Chen, Z., et al.: Evaluation of the effect of Tactical Internet interference. Fire Command Control **41**(4), 126–130 (2016)
5. Wang, S., et al.: Application simulation and evaluation of Tactical mobile self-organizing network. J. Inst. Ordnance Eng. **29**(3), 54–58 (2017)
6. Military Language of the People's Liberation Army of China. Military Science Press, Beijing (2011)
7. Jin, Y.: Research on the architecture of the U.S. military Tactical Internet system. Commun. Technol. **44**(9), 105–107 (2011)
8. Huo, J., Li, H.: Tactical Communication System. National Defense Industry Press, Beijing (2016)

The Research and Implementation of Database Optimization Method Based on K-Nearest Neighbor Algorithm

Hongxing Wu(✉) and Long Zhang

Anhui Construction Engineering Group, Hefei, Anhui, China
hongxingwu@163.com

Abstract. The core technology of relational database in the "Internet plus" era is still widely used, but the performance optimization of relational database inevitably involves a lot of parameter adjustment. This paper put forward a kind of database optimization method based on K-nearest neighbor algorithm, realized the database configuration parameters automatically optimize management. It not only greatly improved the generality of database optimization, but also reduced the cumbersome manual setup, and avoided the disadvantages of empirical setting of the existing technology.

Keywords: K-nearest neighbor algorithm · Database parameters Database optimization

1 Research Background

With the advent of the era of "Internet plus", database technology has experienced a period of rapid development after more than ten years of relative stability. At present, although new databases such as the object-oriented database, the distributed database and multimedia database obtained great progress, but they have not fundamentally got rid of the core technology of the relational database. Therefore, the current commercial database still choose the relational database, and relational database performance optimization inevitably involves a large number of parameter adjustment, in the existing technology database parameter adjustment is still mainly rely on manual operation, the disadvantages of empirical setting is unavoidable.

Relational database performance is not static, but changes with the change of time and environment, it is not only influenced by the server hardware configuration, but also has a direct relationship with the software application environment. Therefore database optimization should be considered the CPU frequency and quantity, frequency and capacity of the memory, disk I/O capability, number of application connections and average business throughput, and so on [1, 2]. Among them, hardware configuration is a static parameter; number of application connections, average business throughput and others are dynamic parameters, which need adjust the parameter value dynamically and timely according to the change of application environment [3]. In view of this situation, it is very necessary and urgent to introduce and improve

N. Xiong et al. (Eds.): ISCSC 2018, AISC 877, pp. 308–315, 2019.
https://doi.org/10.1007/978-3-030-02116-0_36

appropriate algorithms to dynamically adjust database parameters and effectively improve database performance.

2 Algorithm Research

2.1 K-Nearest Neighbor Algorithm

In short, K-nearest neighbor algorithm adopts the method of measuring the distance between different feature values to classify [4]. The basic working principle is to have a training sample set, each sample in the sample set has a label, and the classification of each sample is known in advance. Enter new samples (no label), compare each feature of the new sample with the characteristics of the sample in the sample set, extract the first K classification labels with the most similar (nearest neighbor) features in the sample set, take the label that appears most often as the classification of the new samples. Because the K-nearest neighbor algorithm has high precision, no data input assumptions, suitable numerical model, nominal data and other characteristics, so the database configuration parameters automatically optimize management can be achieved, manual operation is reduced and the disadvantages of setting up by experience in existing technologies can be avoid based on the improved and application of the K - nearest neighbor algorithm.

The general flow of the algorithm:firstly, use any method to collect sample data; secondly, prepare the sample data for calculating the distance between samples, preferably in structured data format; use any method to analyze sample data; and then train and test algorithm, K-nearest neighbor does not need training algorithm, as long as error rate is calculated; finally, the algorithm is used to input sample data, and k-nearest neighbor algorithm is run to identify the classification of input data.

2.2 Sample Distance Calculation of Nearest Neighbor Algorithm

In K-nearest neighbor algorithm, the distance between samples needs to be calculated. There are many ways to calculate the distance, such as European distance, Manhattan distance and Minkowski distance. The most commonly used is the Euclidean distance, which refers to the true distance between two points in m-dimensional space, or the natural length of a vector (i.e. the distance from the point to the origin). The Euclidean distance in two dimensions and three dimensions is the actual distance between two points. The Euclidean distance is relatively intuitive in data representation, that is, the smaller the Euclidean distance, the greater the similarity between two samples, and the greater the Euclidean distance, the smaller the similarity between two samples [5]. In the mean time, Euclidean distance calculation need to make sure that all dimensions on the same level of scale, which is based on absolute values of each dimension characteristic parameters, and database characteristics completely conform to the requirements of the calculation, so when the database parameter optimization the real similarity can be gained by using of Euclidean distance.

3 Implementation Method

Database optimization method based on K-nearest neighbor algorithm is mainly composed of software and hardware configuration parameter acquisition module, the database configuration parameter optimization module and database module, the system architecture is formed by collecting layer, optimization layer and configuration layer. As shown in Fig. 1, the software and hardware configuration acquisition module is used to obtain the database server hardware configuration information and application requirements indicators. The database parameter optimization module is used to generate the database parameter optimization set according to the hardware configuration information and application requirements. The database parameter configuration module is used to configure the database parameter optimization set into the database startup parameter file.

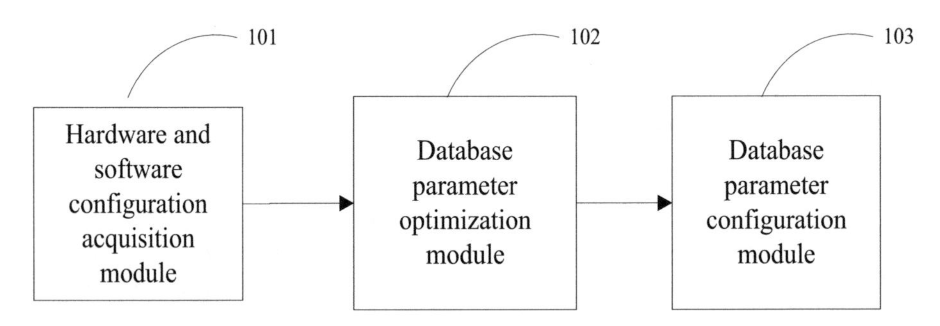

Fig. 1. Database optimization method based on k-nearest neighbor algorithm system architecture schematic diagram

3.1 Acquisition Layer - Software and Hardware Configuration Acquisition

Database server hardware configuration information includes the number of server CPU, main frequency, kernel number, number of threads, memory frequency and capacity, and disk I/O capability. Application requirements metrics include the number of user connections, the number of concurrency, the number of peaks, and the average business throughput [6]. According to the actual hardware configuration information and application requirement index, the virtual environment of database server is constructed.

The collection of software and hardware configuration information of database server can be realized through desktop management interface (DMI) and a series of operating system commands. DMI (Desktop Management Interface) is one of the open technical standards drafted by the industry guidance organization DMTF (Desktop Management Task Force) and is applicable to any platform and operating system. DMI acts as the interface between the management tool and the system layer, establishing a standard manageability system to facilitate the collection of configuration information about the system hardware. In a Linux environment, for example, the command dmidecode, which retrieves information about hardware, is used to decode the information in the DMI database and display it in readable text. As shown in Fig. 2, CPU and memory configuration commands are commonly used in Linux environments.

```
# dmidecode -t 4 |grep "Version:"|sort |uniq
        Version: Intel(R) Xeon(R) CPU E7-4820 v2 @ 2.00GHz

# dmidecode -t 4 |grep "Version:" -c
        4
# dmidecode -t 4 |grep "Core "|sort |uniq
        Core Count: 8
        Core Enabled: 8
# dmidecode -t 17 |more
Handle 0x003A, DMI type 17, 40 bytes
Memory Device
        Array Handle: 0x0000
        Error Information Handle: Not Provided
        Total Width: 72 bits
        Data Width: 64 bits
        Size: 16384 MB
        Form Factor: DIMM
        Set: 25
        Locator: DIMM 9
        Bank Locator: CPU 3 Bank 1
        Type: DDR3
        Type Detail: Synchronous
        Speed: 1600 MHz
        Manufacturer: Samsung
        Serial Number: 408007F7
        Asset Tag: Unknown
        Part Number: M393B2G70DB0-YK0
        Rank: 2
        Configured Clock Speed: 1600 MHz
        Minimum Voltage:  1.5 V
        Maximum Voltage:  1.35 V
        Configured Voltage:  1.35 V
# dmidecode -t 19 |more
# dmidecode 2.12-dmifs
# SMBIOS entry point at 0xbb7fb000
SMBIOS 2.8 present.

Handle 0x0001, DMI type 19, 31 bytes
Memory Array Mapped Address
        Starting Address: 0x00000000000
        Ending Address: 0x01FFFFFFFFF
        Range Size: 128 GB
```

Fig. 2. An example of a Linux system hardware configuration collection

3.2 Optimization Layer - Database Parameter Optimization

Use database server actual configuration index as input values, adopt the K-nearest neighbor algorithm to find out the similarity between it and the sample in the experience knowledge base, such as the actual configuration index (2 CPU, 2 G internal storage,

30 user concurrency) as point A (x1, x2, x3), one sample in experience knowledge base (4 CPU, 4 G internal storage, 80 user concurrency) as point B (y1, y2, y3), the distance of two points is the square root: $(x1 - y1)^2 + (x2 - y2)^2 + (x3 - y3)^2$. To find the distance between it and each sample in the experience knowledge base, use the following formula:

$$dist(X, Y) = \sqrt{\sum_{i=1}^{n} (x_i - y_i)^2}$$

Sort according to similarity, for example, sort according to the actual configuration index and the distance between each sample of experience knowledge base from small to large [7]. Select the first K database parameter values closest to the sample, respectively for the virtual application testing (i.e., the real-time simulation of the user work data application, and the application of limit state) [8], such as take the first k-nearest neighbor sample (4 CPU, 32 G internal storage, 200 user concurrency) corresponding parameter value (the processes = 500, sessions = 800, pga_aggregate_target = 4096 M) to complete the virtual application testing; confirmed whether the database parameter is the optimal value by testing. If the result is "yes", then it automatically summarizes to form the database parameter optimization set, and If the result is "no", the K value is re-adjusted (as shown in Fig. 3).

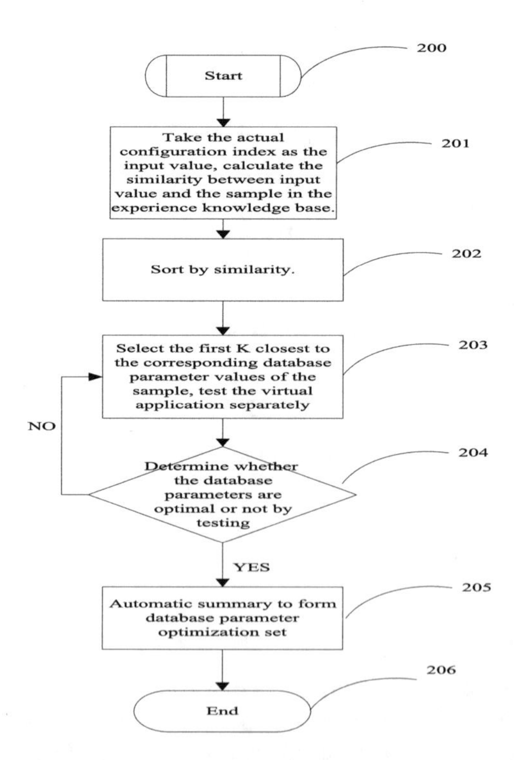

Fig. 3. Flow diagram of database optimization method based on k-nearest neighbor algorithm

3.3 Configuration Layer - Database Parameter Configuration

Through the database interface (including the ODBC and JDBC), configure parameter optimization set to database launch file, perform database service restart detection operation. Its operating mainly includes: after the configuration to take effect, write a successful update prompt message to the system log, automatically save the database launch file; reset the parameter optimization set before the configuration take effect.

Database launch parameters file is a series of parameters and the corresponding value of the operating system file, it loads when the database instance startup, determines the physical structure, memory of the database, the limit of database, a lot of system default values, all kinds of physical properties of the database, the file name and path of specified the database control information. It is an important document for database design and performance tuning. Oracle database launch parameters file, for example, can be divided into two types: (1) the Initialization parameters file (Initialization Parameters Files), is often referred to as PFILE file, stylistic content is pure text format, can be directly modified using a text editor. (2) the Server Parameter file (Server Parameter Files), namely the SPFILE file, the file is in a binary format, cannot be modified directly by using a text editor, only through online modify SQL commands, the dynamic parameters of all the changes can take effect immediately, and static parameters changes must restart the instance. There are also three modes to modify the SPFILE parameter: scope = both (the default mode, which takes effect immediately and permanently), scope = SPFILE (restart takes effect), and scope = memory (which takes effect immediately but fails at the next start). The common commands are as follows:

```
SQL>alter system set parameter = value scope=spfile/both;/* modify parameters */
SQL>show parameters parameter name;/* view parameters */
SQL>select * from v$system_parameter where issys_modifiable= value;
/* use the value of issys_modifiable to determine whether the Oracle parameter is
static or dynamic */
```

Note here is in the scope = spfile or scope = both mode, any parameter changes are not automatically synchronized to the PFILE file, only will be written into the SPFILE file, so it is best to use the create pfile from spfile command to the sync parameters and create PFILE file, to ensure that the next time through the PFILE file, the database can be normally started (specific steps shown in Fig. 4). In practice, you can store all the parameter configuration content (that is, the parameter optimization set) as a script (dbparam.sql) and then execute the script (sqlplus system/passwd-@dbparam.sql).

```
SQL> show parameter processes;

NAME                                 TYPE        VALUE
------------------------------------ ----------- ---------
aq_tm_processes                      integer     1
db_writer_processes                  integer     8
gcs_server_processes                 integer     0
global_txn_processes                 integer     1
job_queue_processes                  integer     1000
log_archive_max_processes            integer     4
processes                            integer     150
SQL> alter system set processes=2000 scope=spfile;
SQL> create pfile from spfile;
SQL> SHUTDOWN IMMEDIATE;
SQL> STARTUP;
SQL> show parameter processes;

NAME                                 TYPE        VALUE
------------------------------------ ----------- ---------
aq_tm_processes                      integer     1
db_writer_processes                  integer     8
gcs_server_processes                 integer     0
global_txn_processes                 integer     1
job_queue_processes                  integer     1000
log_archive_max_processes            integer     4
processes                            integer     2000
SQL> exit
```

Fig. 4. An example of the Oracle database static parameter adjustment proces

4 Conclusions

Database optimization method based on K-nearest neighbor algorithm is first access to the database server hardware configuration information and application requirements, and according to the actual configuration of index information construction the virtualization environment; through the k-nearest neighbor algorithm, selecting multiple parameter sample values closest to the actual configuration metrics from the experience knowledge base, for virtual application testing separately. Then, the optimal value of the tested database parameters is automatically summarized into the database parameter optimization set and configured into the database launch parameter file. Finally, restart the database service to check whether the parameter optimization set configuration data is valid. This optimization method realizes automatic optimization management of database parameter configuration. It not only greatly improves the universality of database optimization, but also reduces the tedious manual operation, and avoided the disadvantages of empirical setting of the existing technology.

References

1. Fan, X., et al.: Performance optimization of relational database in ERP environment. Jiangxi Electr. Power **29**(2), 37–39 (2005)
2. Jiang, L., et al.: The optimization of relational database in ERP environment. CD Softw. Appl. (10), 140 (2014)
3. Zhao, K.: Optimization of database connection pool performance with dynamic parameter adjustment in the big data background. J. Huaibei Vocat. Tech. Coll. **16**(1), 133–135 (2017)
4. Yang, J., et al.: A novel template reduction K-nearest neighbor classification method based on weighted distance. J. Electron. Inf. Technol. **33**(10), 2378–2383 (2011)
5. Yuan, F., et al.: A gradual Chinese text classification technology based on the k-nearest neighbor method. J. South China Univ. Technol. (Natural Science Edition) **32**(s1), 88–91 (2004)
6. Wei, A.: Analysis on Oracle database performance adjustment and optimization methods. Comput. Knowl. Technol. (21), 8–9 (2015)
7. Su, X., et al.: A research on an adaptive flexible database optimization method. Comput. Program. Ski. Maint. (11), 61–64 (2017)
8. Fan, D., et al.: Research on optimize database server performance methods. Electron. Technol. Softw. Eng. (11) (2015)

General Quasi-Laplacian Matrix of Weighted Mixed Pseudograph

Xingguang Chen[1(✉)] and Zhentao Zhu[2]

[1] Jianghan University, Wuhan 430056, People's Republic of China
cxg@nju.edu.cn
[2] Nanjing Institute of Technology, Nanjing 211167, People's Republic of China

Abstract. Many practical problems can be described as a kind graph model having multiple weighted directed edges, undirected edges and weighted vertices simultaneously however, which are lack of sufficient considerations in previous literatures. We discuss some basic definitions for this more general graph and derive some relevant results for its quasi-Laplacian spectrum. Basic non negative results of its eigenvalues are proposed. Furthermore, we argue some dynamic properties between graph spectral layout and its structural variation. Here two types networks evolutionary manner involving insert of new edge or new vertex are considered. The strictness of related conclusions is guaranteed by theoretical analysis. These initial investigation of weighted mixed pseudograph maybe conduct potential theoretical or applied values for a numerous of practical problems.

Keywords: Spectral graph theory · Complex networks
Weighted graph · Weighted mixed graph
General quasi-Laplacian matrix

1 Introduction

In recent years, graph theory has established itself as an important mathematical tool in a wide variety of subjects, ranging from operational research and chemistry to genetics and linguistics, and from electrical engineering and geography to sociology and architecture ([5,8]). The ordinary cases is simple graph, we usually denote such graph $G = (V, E)$ ([6,11]), where V is vertex set of order n and $E \subseteq [V] \times [V]$ is a edge set. If we consider a directed graph (or digraph), we usually denote it $G(V, A)$, where A denotes arc set.Usually, one arc is different from one edge. An edge $\{v, w\}$ is said to join the vertices v and w, and is usually abbreviated to vw.We state vw an undirected edge of $G(V, E)$. The edge vw can equally well be viewed as a directed graph where (v, w) is an arc meanwhile (w, v) is an another arc. If $(v, w) \in E$ and $(w, v) \notin E$, the G has an oriented edge $v \rightarrow w$. The item arc means ordered pairs of vertices in G and edge means unordered pairs of vertices in G. A oriented graph can denote $G(V, A)$, here A is arc set and $A \subseteq [V] \times [V]$. Then we have five types graph: ① simple graph:

N. Xiong et al. (Eds.): ISCSC 2018, AISC 877, pp. 316–323, 2019.
https://doi.org/10.1007/978-3-030-02116-0_37

$G_s(V,E)$; ②multidigraphs: $G_m(V,E)$, ③ oriented graph: $G_o(V,A)$, ④ directed graph $G_d(V,E)$; ⑤ directed pseudograph: $G_{dp}(V,E)$. We have the relation below: $G_s(V,E) \subseteq G_m(V,E)$; $G_o(V,E) \subseteq G_d(V,E) \subseteq G_{dp}(V,E)$. The $G_s(V,E)$ and $G_m(V,E)$ are belong to undirected graph meanwhile $G_o(V,A)$, $G_d(V,E)$ and $G_{dp}(V,E)$ are belong to directed graph. The other graph type is so-called mixed graph which generalize all the five types mentioned above. Bapat *et.al.*[2] has given some properties of mixed graph, which denotes $G=(V,E)$. Here we denote $G_{mix}(V,E,A)$, the item A means *arc* set same as $G(V,A)$.

For directed graph ([1]) and mixed graph ([2,12]), many authors has intensive studied. There is another very important graph which we seldom consider it, which have multi weighted directed edges, undirected edges and weighted vertices simultaneously. Because the weight values of vertices edges or arcs will provides very useful information when we face complex practice problems. Sometimes we have to consider this situation that one graph is a mixed graph which has vertex and edge (arc) weight simultaneously, while it has been studied scarcely. We name this type graph as weighted mixed pseudograph and denote $G(V,E,W)$, where $W \subseteq \mathbb{R}$ is weight set including all vertices and edges, which means the weighted including vertex-weighted, edge-weighted and arc-weighted respectively[1]. ω is a map defined to be: $\omega : E \mapsto W$. The weighted mixed pseudograph cover all these graph types mentioned above, that is $G(V,E,W) \supseteq G_{mix}(V,E,A)$. Below Figs. 1, 2 and 3. illustrates the various type of graphs mentioned before.

(a) simple graph:$G_s(V,E)$ (b) multi-digraph:$G_m(V,E)$

Fig. 1. Simple graph vs Multi-digraph

This paper is organized as follows. In Sect. 2, Some basic notations of weighted mixed pseudograph and some preliminaries of its Laplacian spectrum are discussed. We focus on how the spectrum changes according with the variation of its structure from the dynamic aspects and relevant fundamental results are proposed in Sect. 3. The main conclusions and implications of our presented model are summarized in final section.

2 Preliminaries

Let's consider a weighted mixed pseudograph $G(V,E,W)$. The order of $G(V,E,W)$ is $|V|=n$, and the size of $G(V,E,W)$ is $|E|=m$. W is weight

[1] For simplicity of notations, here E includes all edges(unoriented edges) and arcs(oriented edges).

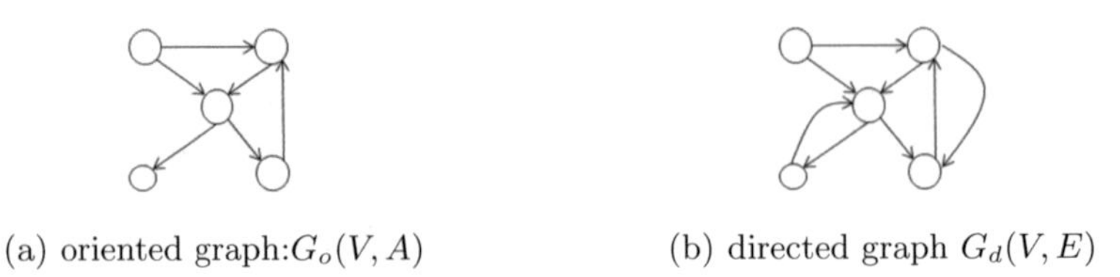

(a) oriented graph:$G_o(V, A)$ (b) directed graph $G_d(V, E)$

Fig. 2. Oriented graph vs Directed graph

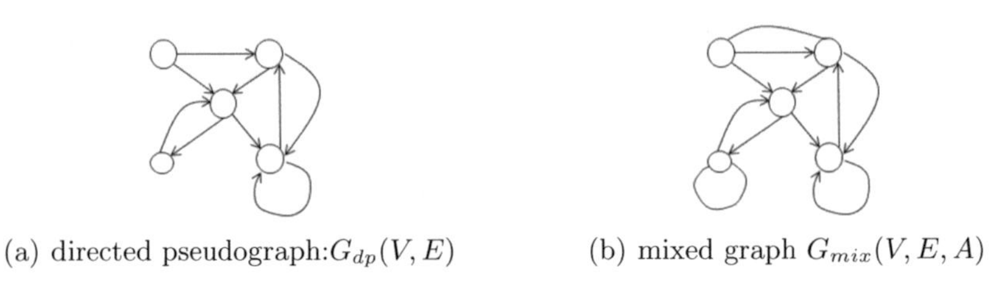

(a) directed pseudograph:$G_{dp}(V, E)$ (b) mixed graph $G_{mix}(V, E, A)$

Fig. 3. Directed pseudograph vs Mixed graph

set of edges, note the elements of W is non negative number. We use init(e) denote initial vertex for a directed edge e, and ter(e) denotes terminal vertex for a directed edge e (The notations reference to Godsil and Royle [7]).

Definition 1. *The quasi-incidence matrix $B(G) = (b_{ij})_{n\times m}$ of graph $G(V, E, W)$ with vertices set $V = \{v_1, \cdots, v_n\}$, and edges set $E = \{e_1, \cdots, e_m\}$ is defined by*

$$b_{ij} := \begin{cases} w_{e_j} & \text{if } ① \, v_i = ter(e_j)(e_j \text{ is oriented edge}), \\ & \quad ② \, e_j \text{ is unoriented edge between } v_i \text{ and } v_j, \\ & \quad ③ \, e_j \text{ is loop of } v_i \\ -w_{e_j} & \text{if } v_i = init(e_j)(e_j \text{ is oriented edge}) \\ 0 & \text{otherwise.} \end{cases}$$

For example, the quasi-incidence matrix of the graph $G(V, E, W)$ of Fig. 4 is $B(G) = (b_{ij})_{n\times(m+1)}$=

$$\begin{array}{c} \\ v_1 \\ v_2 \\ v_3 \\ v_4 \end{array} \begin{array}{c} \begin{matrix} e_1 & e_2 & e_3 & e_4 & e_5 & e_6 & e_7 & e_8 & w_v \end{matrix} \\ \begin{pmatrix} 0.2 & 0 & 0 & 0 & 2.2 & -1 & 0 & 0 & 1.5 \\ -0.2 & 0.5 & 0.2 & 6 & 0 & 0 & 0 & 0 & 3 \\ 0 & -0.5 & -0.2 & 6 & -2.2 & -1 & -5 & 0 & -2 \\ 0 & 0 & 0 & 0 & 0 & 0 & 5 & 0.2 & -3 \end{pmatrix} \end{array}$$

The last column identified with w_v is the weight values of vertices from v_1 to v_4. Note the quasi-incidence matrix $B(G)$ can determine one weighted mixed

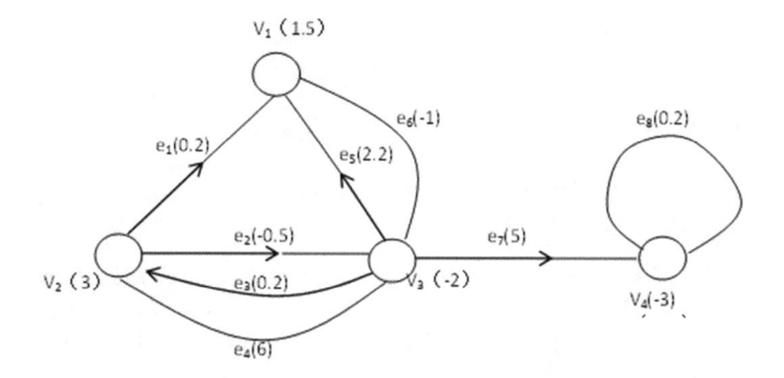

Fig. 4. An example of weighted mixed pseudograph

graph completely and uniquely, i.e., if we know one quasi-incidence matrix, then we can represent the networks structure sketch exactly, and vice versa. Let's scan quasi-incidence matrix's columns, the former m columns (excluding the last column which denotes vertices' weight value) must be two kinds, the one is including two non-zero numbers, and the other is including one non-zero number only. If the column has two non-zero numbers, which means the edge is unoriented edge or oriented edge. And there are two situations in this case, if the two numbers have the same sign, which means the edge is unoriented edge, for example, e_4 and e_6 are unoriented edges in Fig. 4. If the two numbers have opposite sign, which means the edge is oriented edge, for example, e_1,e_2,e_3,e_5 and e_7 are all oriented edges in Fig. 4. On the other hand, if the column includes only one non-zero number, which means the edge is unoriented loop or oriented loop like as e_8 in Fig. 4. It is should be noted that although we can draw unoriented loop or oriented loop in an networks sketch, but these two types loop are equivalent thoroughly if we consider their practical implications, i.e., one object acts on itself, positive weight denotes positive effects, and negative weight denotes negative effects. So, in our new networks model, all loops are tackled as unoriented loop (weighted or non-weighted) uniformly. The other fact should be noted is that if we restore the networks patterns according by our quasi-incidence matrix definition, there may be occur two sketch like as below Fig. 5(a) and (b), but in fact these two types illustrations are identical as well.

(a) $v_1 \rightarrow v_2 : weight = 0.5$ (b) $v_1 \leftarrow v_2 : weight = -0.5$

Fig. 5. Two equivalent oriented relations between two nodes

If we notate $\boldsymbol{p}_i (i \in \{1, \ldots, n\})$ is row vector on $\mathbb{R}^{(m+1)}$ of matrix $B(G) = (b_{ij})_{n \times (m+1)}$. $\boldsymbol{q}_j (j \in \{1, \ldots, m+1\})$ is column vector on $\mathbb{R}^n$ similarly. Then the quasi-incidence matrix of $G(V, E, W)$ can be denoted as $B(G) = \begin{pmatrix} \boldsymbol{p}_1 \\ \vdots \\ \boldsymbol{p}_n \end{pmatrix} = (\boldsymbol{q}_1, \ldots, \boldsymbol{q}_{(m+1)})$. We'll define the quasi-Laplacian matrix of a weighted mixed pseudograph with $B(G)$.

Definition 2. *The quasi-Laplacian matrix of a weighted mixed pseudograph $G(V, E, W)$ is $Q(G) := BB^T \in \mathbb{R}^{n \times n}$. We denote the eigenvalues of quasi-Laplacian matrix of a weighted mixed pseudograph $G(V, E, W)$ is $\{\lambda_i(L(G))\}, i \in \{1, \cdots, n\}$, and call $\{\lambda_i(L(G))\}, i \in \{1, \cdots, n\}$ is the quasi-Laplacian spectrum.*

Merris ([9]) has already given the Laplacian matrix's eigenvalues upper bound for simple graph $G_s(V, E)$. Here we have a simple result for weighted mixed pseudograh $G(V, E, W)$.

Lemma 1. *Let $G(V, E, W)$ be a weighted mixed pseudograph on n vertices with quasi-Laplacian matrix $Q(G)$, then its eigenvalues are all non negative,i.e. $\lambda_i \geq 0. \forall i \in \{1, \cdots, n\}$.*

Proof. Since $Q = BB^T$, then $Q^T = (BB^T)^T = BB^T$, we get Q is real symmetric. For $\forall \boldsymbol{u} \in \mathbb{R}$, $\boldsymbol{u}^T Q \boldsymbol{u} = \boldsymbol{u}^T BB^T \boldsymbol{u} = (B^T \boldsymbol{u})^T B^T \boldsymbol{u}$, let $B^T \boldsymbol{u} = \boldsymbol{v}, \boldsymbol{u}^T Q \boldsymbol{u} = \boldsymbol{v}^T \boldsymbol{v} = (\boldsymbol{v}, \boldsymbol{v}) \geq 0$. Thus Q is positive semidefinite. So its eigenvalues are all non negative. □

Corollary 1. *The Quasi-Laplacian matrix's eigenvalues of a weighted mixed pseudograph satisfy one of these inequalities at least:*

$$|\lambda - \|\boldsymbol{p}_i\|^2| \leq \sum_{1 \leq j \leq n, j \neq i} |(\boldsymbol{p}_i, \boldsymbol{p}_j)|. \tag{1}$$

here, $\lambda \in \{\lambda_i : i \in \{1, \ldots, n\}\}$, $(\boldsymbol{p}_i, \boldsymbol{p}_j)$ is inner product of vector $\boldsymbol{p}_i$ and $\boldsymbol{p}_j$, $\|.\|$ is vector norm and $\|\boldsymbol{\alpha}\| := \sqrt{(\boldsymbol{\alpha}, \boldsymbol{\alpha})}$.

Proof. From **Definitions** 1 and 2, we can get $Q(G) = BB^T = \begin{pmatrix} \boldsymbol{p}_1 \\ \vdots \\ \boldsymbol{p}_n \end{pmatrix} \cdot (\boldsymbol{p}_1^T, \ldots, \boldsymbol{p}_n^T) = \begin{pmatrix} \boldsymbol{p}_1 \boldsymbol{p}_1^T & \cdots & \boldsymbol{p}_1 \boldsymbol{p}_n^T \\ \vdots & \vdots & \vdots \\ \boldsymbol{p}_n \boldsymbol{p}_1^T & \cdots & \boldsymbol{p}_n \boldsymbol{p}_n^T \end{pmatrix} = \begin{pmatrix} \|\boldsymbol{p}_1\|^2 & \cdots & (\boldsymbol{p}_1, \boldsymbol{p}_n) \\ \vdots & \vdots & \vdots \\ (\boldsymbol{p}_n, \boldsymbol{p}_1) & \cdots & \|\boldsymbol{p}_n\|^2 \end{pmatrix}$. Therefore assertion (1) follows from *Gerschgorin Theorem* ([6]). □

3 Updates

Now, we'll focus on the dynamic properties of weighted mixed pseudograph, which will provide some insights between the spectral and its graph structural variation. Brandes *et al.* ([4]) discussed some general principles for dynamic graph layout and derive a dynamic spectral layout approach. But they consider only for simple graph $G_s(V,E)$ and we'll extend to more general situation on weighted mixed pseudograph $G(V,E,W)$. Here real sequence $\{\lambda_i^{(j)} : \lambda_i^{(j)} \in \mathbb{R}\}$ denote eigenvalues of weighted mixed pseudograph $G^{(j)}$.

Proposition 1. *Let $G^{(0)}$ be an original weighted mixed pseudograph, and $G^{(0)} \to G^{(1)}$ by interpolate one edge* [2]*. then $\exists i \in \{1,\ldots,n\}$, satisfy $\lambda_i^{(1)} > \lambda_i^{(0)}$.*

Proof. Use the column vector form of Quasi-Laplacian matrix, we have $Q^{(0)} = B^{(0)}B^{(0)T} = (\boldsymbol{q}_1^{(0)},\ldots,\boldsymbol{q}_{m+1}^{(0)}) \begin{pmatrix} \boldsymbol{q}_1^{(0)T} \\ \vdots \\ \boldsymbol{q}_{m+1}^{(0)\ T} \end{pmatrix} = \sum_{i=1}^{m+1} \boldsymbol{q}_i^{(0)} \boldsymbol{q}_i^{(0)T}$. When we insert one edge in $G^{(0)}$, we'll get $Q^{(1)} = B^{(1)}B^{(1)T} = (\boldsymbol{q}_1^{(1)},\ldots,\boldsymbol{q}_{m+2}^{(1)}) \begin{pmatrix} \boldsymbol{q}_1^{(1)T} \\ \vdots \\ \boldsymbol{q}_{m+2}^{(1)\ T} \end{pmatrix}$. From *Definitions* 1 and 2, we know that $\boldsymbol{q}_i^{(1)} = \boldsymbol{q}_i^{(0)}, (i = 1,\ldots,m)$, $\boldsymbol{q}_{m+2}^{(1)} = \boldsymbol{q}_{m+1}^{(0)}$, so

$$Q^{(1)} = B^{(1)}B^{(1)T} = \sum_{i=1}^{m+2} \boldsymbol{q}_i^{(1)} \boldsymbol{q}_i^{(1)T} =$$

$$\sum_{i=1}^{m} \boldsymbol{q}_i^{(0)} \boldsymbol{q}_i^{(0)T} + \boldsymbol{q}_{m+2}^{(1)} \boldsymbol{q}_{m+2}^{(1)\ T} + \boldsymbol{q}_{m+1}^{(1)} \boldsymbol{q}_{m+1}^{(1)\ T} =$$

$$Q^{(0)} + \boldsymbol{q}_{m+1}^{(1)} \boldsymbol{q}_{m+1}^{(1)\ T}. \tag{2}$$

We can know from linear algebra that $\sum_{i=1}^n \lambda_i^{(0)} = \mathrm{tr}Q^{(0)}$, $\sum_{i=1}^n \lambda_i^{(1)} = \mathrm{tr}Q^{(1)}$, substitute $\boldsymbol{q}_{m+1}^{(1)} = \begin{pmatrix} b_{1,m+1}^{(1)} \\ \vdots \\ b_{n,m+1}^{(1)} \end{pmatrix}$ in formula (2), we can get that $\mathrm{tr}Q^{(1)} = \mathrm{tr}Q^{(0)} + \sum_{i=1}^n {b_{i,m+1}^{(1)}}^2$, since $\exists i = 1,\ldots,n, b_{i,m+1}^{(1)} \neq 0$, then $\mathrm{tr}Q^{(1)} > \mathrm{tr}Q^{(0)}$, that is $\sum_{i=1}^n \lambda_i^{(1)} > \sum_{i=1}^n \lambda_i^{(0)}$ for $i = 1,\ldots,n$. Note $\lambda_i^{(0)}$ and $\lambda_i^{(1)}$ are all non negative, therefore, the claim follows. □

Now let's consider two different situations, $G^{(0)}$ is the initial graph, if add one edge to it and the new graph denote $G^{(1)}$, then, let we add a new isolated vertex to $G^{(0)}$, the new graph denote as $G^{(2)}$, from below proposition, we can distinguish the weight of the two new elements (one new edge and one new vertex) from the sum variation of the quasi-Laplacian spectrum.

Proposition 2. *Assume ① : $G^{(0)} \to G^{(1)}$ by add one edge which has weight ω_e,② : $G^{(0)} \to G^{(2)}$ by add one* ***isolated*** *vertex which has weight ω_v.*

[2] Means the weight value of the edge $\neq 0$.

Then we can claim that if $\sum_{i=1}^{n}\lambda_i^{(1)} - \sum_{i=1}^{n}\lambda_i^{(0)} > \sum_{i=1}^{n+1}\lambda_i^{(2)} - \sum_{i=1}^{n}\lambda_i^{(0)}$, *then the weight satisfy* ① $|\omega_e| > \frac{\sqrt{2}}{2}|\omega_v|$, *if new edge isn't a loop.* ② $|\omega_e| > |\omega_v|$, *if new edge is a loop.*

Proof. For $G^{(0)} \to G^{(2)}$, use the row vector form, we have $Q^{(0)} = B^{(0)}B^{(0)^T} = \begin{pmatrix} \boldsymbol{p}_1^{(0)} \\ \vdots \\ \boldsymbol{p}_n^{(0)} \end{pmatrix} (\boldsymbol{p}_1^{(0)^T}, \ldots, \boldsymbol{p}_n^{(0)^T}) = \begin{pmatrix} \boldsymbol{p}_1^{(0)}\boldsymbol{p}_1^{(0)^T} & \cdots & \cdots \\ & \ddots & \\ \cdots & \cdots & \boldsymbol{p}_n^{(0)}\boldsymbol{p}_n^{0\,T} \end{pmatrix} = \begin{pmatrix} \|\boldsymbol{p}_1^{(0)}\|^2 & \cdots & \cdots \\ & \ddots & \\ \cdots & \cdots & \|\boldsymbol{p}_n^{(0)}\|^2 \end{pmatrix}$,

$$\sum_{i=1}^{n}\lambda_i^{(0)} = \mathrm{tr}Q^{(0)} = \sum_{i=1}^{n}\|\boldsymbol{p}_i^{(0)}\|^2. \tag{3}$$

Similarly, $Q^{(2)} = B^{(2)}B^{(2)^T} = \begin{pmatrix} \|\boldsymbol{p}_1^{(2)}\|^2 & \cdots & \cdots \\ & \ddots & \\ \cdots & \cdots & \|\boldsymbol{p}_{n+1}^{(2)}\|^2 \end{pmatrix}$,

$$\sum_{i=1}^{n+1}\lambda_i^{(2)} = \mathrm{tr}Q^{(2)} = \sum_{i=1}^{n+1}\|\boldsymbol{p}_i^{(2)}\|^2. \tag{4}$$

notice that $\boldsymbol{p}_i^{(0)} = \boldsymbol{p}_i^{(2)}, i = 1, \ldots, n$, $\boldsymbol{p}_{n+1}{}^{(2)} = (b_{n+1,1}{}^{(2)}, \ldots, b_{n+1,m+1}{}^{(2)}) = (0, \ldots, 0, \omega_v)$, substitute in (3) and (4), we can get that $\sum_{i=1}^{n+1}\lambda_i^{(2)} - \sum_{i=1}^{n}\lambda_i^{(0)} = \|\boldsymbol{p}_{n+1}^{(2)}\|^2 = \sum_{i=1}^{m+1}{b_{n+1,i}^{(2)}}^2 = \omega_v^2$. From the proof of *proposition* 1, we can know that $\sum_{i=1}^{n}\lambda_i^{(1)} - \sum_{i=1}^{n}\lambda_i^{(0)} = \sum_{i=1}^{n}{b_{i,m+1}^{(1)}}^2$.

From **Definition** 1, if the new added edge isn't a loop, $\boldsymbol{q}_{m+1}^{(1)} = \begin{pmatrix} b_{1,m+1}^{(1)} \\ \vdots \\ b_{n,m+1}^{(1)} \end{pmatrix} = \begin{pmatrix} \vdots \\ \pm\omega_e \\ \vdots \\ \mp\omega_e \\ \vdots \end{pmatrix}$. It obviously shows that $\sum_{i=1}^{n}{b_{i,m+1}^{(1)}}^2 = 2{\omega_e}^2$. On the other hand, if the new added edge is a loop, then $\boldsymbol{q}_{m+1}^{(1)} = \begin{pmatrix} b_{1,m+1}^{(1)} \\ \vdots \\ b_{n,m+1}^{(1)} \end{pmatrix} = \begin{pmatrix} \vdots \\ \omega_e \\ \vdots \\ 0 \\ \vdots \end{pmatrix}$. In this situation, $\sum_{i=1}^{n}{b_{i,m+1}^{(1)}}^2 = {\omega_e}^2$, then from the condition of *proposition* 2, we have $2{\omega_e}^2 > \omega_v^2$ or $\omega_e^2 > \omega_v^2$, which means $|\omega_e| > \frac{\sqrt{2}}{2}|\omega_v|$(new edge isn't a loop) or $|\omega_e| > |\omega_v|$(new edge is a loop), therefore, the result is proved. □

4 Conclusion

Many complex systems could be demonstrated as networks structures, for many years, relative simple networks models including regular network, random network and, recently, two kinds famous network patterns such as small world [10] and scale-free network [3] are employed to express the wide connections and diversity of real world. In these network models, the edge between nodes is single and, directed and undirected coexisting is not permitted as well. However, some components of nature and artificial systems emerge apparent multiple relations, some relations are undirected and some other relations are directed, while current existing network models are not competent to describe these characteristics thoroughly. To address this nontrivial issues, a novel network model described as weighted mixed pseudograph is proposed. Its static and dynamic properties, involving definition, quasi-Laplacian spectrum, graph spectral layout and its structural variation are discussed, which shed light on the complex characteristics of weighted mixed pseudograph and start up further research for diverse composite networks mentioned in this paper. The future direction of this work will focus on two facets. First is theoretical investigation and the second is about applications regard of this weighted mixed pseudograph model.

Acknowledgments. This work is supported by the National Natural Science Foundation of China (Grant No. 71471084), Open foundation of Wuhan Research Institution of Jianghan University (Grant No. IWHS2016104, and Grant No. jhunwyy 20151204).

References

1. Bang-Jensen, J., Gutin, G.Z.: Digraphs. Theory Algorithms and Applications. Springer, London (2008)
2. Bapat, R.B., Grossman, J.W., Kulkarni, D.M.: Gneralized matrix tree theorem for mixed graphs. Linear Multilinear Algebr. **46**(4), 299–312 (1999)
3. Barabsi, A.L., Albert, R.: Emergence of scaling in random networks. Science **286**(5439), 509–512 (1999)
4. Brandes, U., Fleischer, D., Puppe, T.: Dynamic spectral layout with an application to small worlds. J. Graph Algorithms Appl. **11**(2), 325–343 (2007)
5. Chao, F., Ren, H., Cao, N.: Finding shorter cycles in a weighted graph. Graphs Comb. **32**(1), 65–77 (2016)
6. Diestel, R.: Graph Theory, 3rd edn. Springer, Heidelberg (2006)
7. Godsil, C., Royle, G.: Algebraic Graph Theory. Springer, New York (2001)
8. Hu, J.R., Lin, Q.: Universal weighted graph state generation with the cross phase modulation. Eur. Phys. J. D **70**(5), 1–7 (2016)
9. Merris, R.: A note laplacian graph eigenvalues. Linear Algebr. Appl. **285**(1–3), 33–35 (1998)
10. Watts, D.J., Strogatz, S.H.: Collective dynamics of small-world networks. Nature **393**(6684), 440–442 (1998)
11. Wilson, R.J.: Introduction to Graph Theory, 4th edn. Person Education Limited, London (1996)
12. Zhang, X.D., Li, J.S.: The laplacian spectrum of a mixed graph. Linear Algebr. Appl. **353**(1–3), 11–20 (2002)

The Josephson Current in an s-Wave Superconductor/Noncentrosymmetric Superconductor Junction

Zhuohui Yang(✉)

College of Science, Jinling Institute of Technology, Nanjing 211169, China
zhuohuiyang@jit.edu.cn

Abstract. We study theoretically the Josephson current in an s-wave superconductor/noncentrosymmetric superconductor (SWSC/NCSC) junction. Based on the Bogoliubov-de Gennes equation and the quantum scattering method, we show that the Josephson current periodicity is π not 2π when the NCSC has only a pure p-wave component, whereas when both the s-wave and p-wave components are considered together in the NCSC, the Josephson current periodicity recovers to be the usual one 2π. It is also shown that the coupling strength of the junction as well as the Rashba spin-orbit coupling in the NCSC can significantly affect the periodicity of the Josephson current and its phase dependence. Our findings are useful to discern the order parameter symmetry in the NCSC.

Keywords: Josephson current
Noncentrosymmetric superconductor · Rashba spin-orbit coupling

1 Introduction

The Josephson junction is an active research field in the condensed matter physics, because it has fundamental research merit and also, it is a building block of superconducting electronic devices [1–3]. For instance, the quantum qubit is vital to realize a quantum computer while the superconductor (SC) Josephson junction is one of good candidates for fabricating quantum qubits. Besides, the π state Josephson junction was also considered as the circuit element for the quantum computation [4,5]. The conventional s-wave Josephson junctions have been widely studied [6–12]. In recent years, some noncentrosymmetric materials with superconductivity were found, such as $CePt_3Si$, UIr, Li_2Pd_3B, Li_2Pt_3B and $Cd_2Re_2O_7$ [13–17]. In these materials, the spin-orbit coupling induced by the noncentrosymmetric potential can mix the spin singlet and triplet components in the pairing potential [18–20], which can lead to very peculiar superconductivity. The concrete order parameter symmetry in these noncentrosymmetric superconductors (NCSC) is still an open question, and it is a decisive factor for possible applications in quantum computations. Researchers turned to study the current transport properties of the NCSC junctions [21,22] in order to identify the pairing components in the NCSC. This motivates us to study the characteristic of

N. Xiong et al. (Eds.): ISCSC 2018, AISC 877, pp. 324–331, 2019.
https://doi.org/10.1007/978-3-030-02116-0_38

the Josephson current in an s-wave superconductor (SWSC)/NCSC junction, which was not studied before.

Based on the Bogoliubov-de Gennes equation [23–25] and the quantum scattering method, we study the Josephson current in the SWSC/NCSC heterojunction in this work. It is found that the periodicity of the Josephson current is heavily dependent on the concrete components of the NCSC, and the coupling strength of the junction as well as the RSOC in the NCSC can significantly affect the phase dependence of the Josephson current or even the current periodicity. This can be definitely used to discern the order parameter symmetry of the NCSC material.

2 Model and Formulation

We consider a clean two-dimensional SWSC/NCSC Josephson junction as shown in Fig. 1. The interface of the system is an infinite narrow insulator barrier located at $x = 0$, and it can be described as a delta function, $U(x) = U\delta(x)$. In the left side of the junction, the SWSC contains only the spin singlet state, while in the right side of the junction, the NCSC may contain both the spin singlet state and the spin triplet state due to the Rashba spin-orbit coupling (RSOC), which comes from the noncentrosymmetric potential in the NCSC.

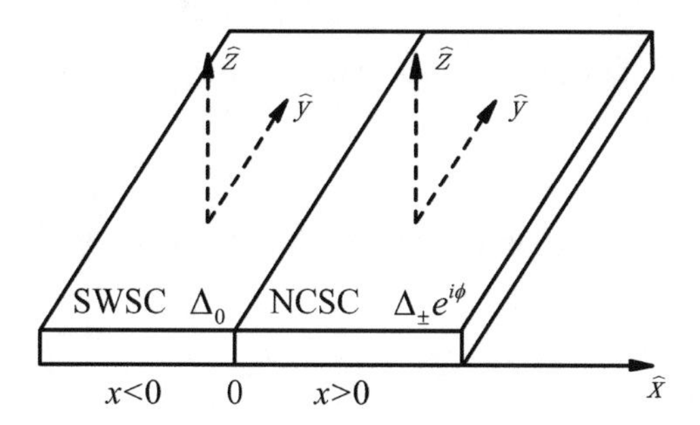

Fig. 1. The schematic diagram of the SWSC/NCSC junction

The Bogoliubov-de Gennes equation describing the junction is given by

$$\begin{pmatrix} H(\mathbf{k}) & \Delta(\mathbf{k}) \\ -\Delta^{\dagger}(-\mathbf{k}) & -H^{*}(-\mathbf{k}) \end{pmatrix} \begin{pmatrix} \hat{u} \\ \hat{v} \end{pmatrix} = E \begin{pmatrix} \hat{u} \\ \hat{v} \end{pmatrix} \tag{1}$$

with $H(\mathbf{k}) = H_0(\mathbf{k}) + H_{RSOC}(\mathbf{k})\theta(x)$ and $H_0(\mathbf{k}) = \varepsilon_{\mathbf{k}} - \mu + U$. μ is the chemical potential of the system. $H_{RSOC} = \frac{\lambda}{\hbar}(\sigma_x p_y - \sigma_y p_x)$ describes the RSOC in the NCSC with λ the RSOC constant. p_x and p_y are the two components of the momentum operator. $\theta(x)$ is the Heaviside step function. The superconducting energy gaps on both sides of the system are expressed as

$$\Delta(\mathbf{k}) = \begin{cases} i\Delta_0 d_0(\mathbf{k})\sigma_y & (x < 0) \\ i[\Delta_s d_0(\mathbf{k}) + \Delta_t \mathbf{d}(\mathbf{k}) \cdot \boldsymbol{\sigma}]\sigma_y e^{i\phi} & (x > 0) \end{cases} \tag{2}$$

where Δ_0 is the spin singlet energy gap in the SWSC, Δ_s and Δ_t are the spin singlet and triplet energy gaps in the NCSC respectively. ϕ is the SC phase difference. As mentioned in [26], the triplet pair state is chosen as $\mathbf{d}(\mathbf{k}) \propto (k_y, -k_x, 0)$, which is parallel to the pseudo magnetic field induced by the RSOC.

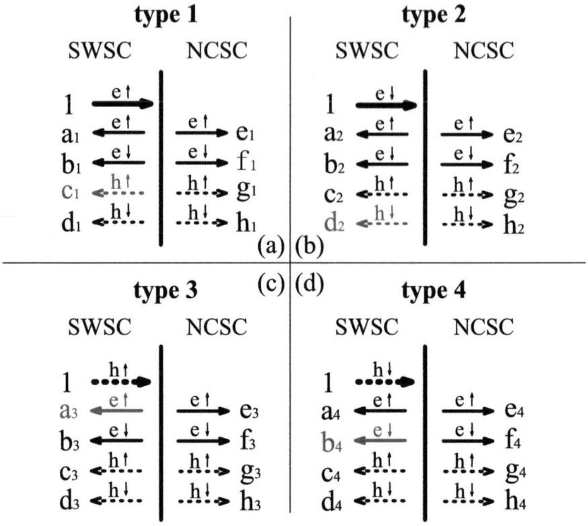

Fig. 2. The schematic diagram of the four quantum scattering processes in the SWSC/NCSC junction

Using the Furusaki-Tsukada formalism [27], we express the radial Josephson current as

$$I_J = \frac{e\Delta_0}{\hbar} \sum_{k_\parallel} \int \frac{dE}{2\pi} \left[\frac{(k_e + k_h)}{2\Omega} \left(\frac{c_1}{k_e} + \frac{d_2}{k_e} - \frac{a_3}{k_h} - \frac{b_4}{k_h} \right) + c.c. \right] f(E). \quad (3)$$

In this equation, k_e and k_h are respectively the x-component wave vectors of the electron-like and hole-like quasiparticles in the NCSC lead. $k_\parallel$ is their transverse component parallel to the interface and $\Omega = \sqrt{E^2 - \Delta_0^2}$ with E being the quasiparticle energy. c_1, d_2, a_3 and b_4 are the Andreev reflection (AR) coefficients in four quantum scattering processes as shown in Fig. 2., c_1, d_2 are the spin-resolved AR coefficients of the electron-like quasiparticles. a_3, b_4 are the spin-resolved AR coefficients of the hole-like quasiparticles.

In order to obtain these four AR coefficients, we need to determine the wave functions of each scattering process in the two SC leads. For example, when the spin up electron-like quasiparticles are injecting from the left SC to the interface, as shown in Fig. 2(a), the wave function in the left side can be expressed as

$$\psi_1(x<0)=e^{ik_\parallel y}\left[\begin{pmatrix}u_0\\0\\0\\v_0\end{pmatrix}e^{ik_e x}+a_1\begin{pmatrix}u_0\\0\\0\\v_0\end{pmatrix}e^{-ik_e x}+b_1\begin{pmatrix}0\\u_0\\-v_0\\0\end{pmatrix}e^{-ik_e x}\right.$$
$$\left.+c_1\begin{pmatrix}v_0\\0\\0\\u_0\end{pmatrix}e^{ik_h x}+d_1\begin{pmatrix}0\\-v_0\\u_0\\0\end{pmatrix}e^{ik_h x}\right] \tag{4}$$

with $u_0=\sqrt{\frac{1}{2}+\frac{\Omega}{2E}}$, $v_0=\sqrt{\frac{1}{2}-\frac{\Omega}{2E}}$. The wave function in the right side can be expressed as

$$\psi_1(x\geq 0)=e^{ik_\parallel y}\left[\frac{e_1}{\sqrt{2}}\begin{pmatrix}u_1e^{i\phi}\\u_1e^{-i\chi(\alpha_1)+i\phi}\\-v_1e^{-i\chi(\alpha_1)}\\v_1\end{pmatrix}e^{iq_{e1}x}+\frac{f_1}{\sqrt{2}}\begin{pmatrix}u_2e^{i\phi}\\-u_2e^{-i\chi(\alpha_2)+i\phi}\\v_2e^{-i\chi(\alpha_2)}\\v_2\end{pmatrix}e^{iq_{e2}x}\right.$$
$$\left.+\frac{g_1}{\sqrt{2}}\begin{pmatrix}v_1e^{i\phi}\\v_1e^{-i\chi(\alpha_{1r})+i\phi}\\-u_1e^{-i\chi(\alpha_{1r})}\\u_1\end{pmatrix}e^{-iq_{h1}x}+\frac{h_1}{\sqrt{2}}\begin{pmatrix}v_2e^{i\phi}\\-v_2e^{-i\chi(\alpha_{2r})+i\phi}\\u_2e^{-i\chi(\alpha_{2r})}\\u_2\end{pmatrix}e^{-iq_{h2}x}\right], \tag{5}$$

where $u_{1(2)}=\sqrt{\frac{1}{2}+\frac{\Omega_{1(2)}}{2E}}$, $v_{1(2)}=\sqrt{\frac{1}{2}-\frac{\Omega_{1(2)}}{2E}}$, with $\Omega_{1(2)}=\sqrt{E^2-|\Delta_s\pm\Delta_t|^2}$. In Eqs. (4) and (5), a_1 and b_1 are the normal reflection coefficients. c_1 and d_1 are the AR coefficients. e_1 and f_1 are the transmission coefficients of the electron-like quasiparticles. g_1 and f_1 are the transmission coefficients of the hole-like quasiparticles. q_{e1}, q_{e2}, q_{h1} and q_{h2} are the x-component wave vectors of the spin-dependent electron-like and hole-like quasiparticles in the NCSC respectively. $\chi(\alpha_i)=\pi/2-\alpha_i$, $\alpha_{ir}=\pi-\alpha_i$, where $\alpha_{i=1(2)}$ is the angle between the quasiparticle wave vectors and the x-axis. We ignore the wave vector difference between the electron-like and hole-like quasiparticles. The wave vectors in the SWSC lead are $k_{e0}=k_{h0}=k_F$, where $k_F=\sqrt{2mE_F/\hbar^2}$ is the Fermi wave vector and E_F is the Fermi energy. The wave vectors in the NCSC can be expressed as $q_{e10}=q_{h10}=\sqrt{k_R^2+k_F^2}-k_R$, $q_{e20}=q_{h20}=\sqrt{k_R^2+k_F^2}+k_R$, where $k_R=m\lambda/\hbar^2$ is the Rashba wave vector. The system maintains the translation symmetry in the interface direction with $k_e=k_h=\sqrt{k_{e0}^2-k_\parallel^2}$, $q_{e1}=q_{h1}=\sqrt{q_{e10}^2-k_\parallel^2}$ and $q_{e2}=q_{h2}=\sqrt{q_{e20}^2-k_\parallel^2}$. Define the dimensionless parameter $Z=2mU_0/\hbar^2k_F$ to represent the insulator barrier strength and $\beta=2k_R/k_F$ to represent the RSOC strength in the NCSC.

The eight scattering coefficients in Eqs. (4) and (5) need to be determined by the boundary conditions [27] at the interface

$$\psi(x)|_{x=+0}=\psi(x)|_{x=-0}, \tag{6}$$

$$v_x\psi(x)|_{x=+0}-v_x\psi(x)|_{x=-0}=Z_0\tau_3\psi(0), \tag{7}$$

with $Z_0 = Zk_F/im$ and

$$\tau_3 = \begin{pmatrix} 1 & 0 & 0 & 0 \\ 0 & 1 & 0 & 0 \\ 0 & 0 & -1 & 0 \\ 0 & 0 & 0 & -1 \end{pmatrix}. \tag{8}$$

In the boundary conditions, the velocity operator along the x-direction is given by

$$v_x = \frac{\partial H}{\hbar \partial k_x} = \begin{pmatrix} \frac{\hbar}{im}\frac{\partial}{\partial x} & \frac{i\lambda}{\hbar}\theta(x) & 0 & 0 \\ -\frac{i\lambda}{\hbar}\theta(x) & \frac{\hbar}{im}\frac{\partial}{\partial x} & 0 & 0 \\ 0 & 0 & -\frac{\hbar}{im}\frac{\partial}{\partial x} & -\frac{i\lambda}{\hbar}\theta(x) \\ 0 & 0 & \frac{i\lambda}{\hbar}\theta(x) & -\frac{\hbar}{im}\frac{\partial}{\partial x} \end{pmatrix}. \tag{9}$$

It is necessary to solve the eight linear equations to obtain the four required AR coefficients c_1, d_2, a_3 and b_4. Then we obtain the Josephson current in this system.

3 Results

We present the numerical results of the Josephson current flowing in the SWSC/NCSC junction in the clean limit at zero temperature $T = 0$K. As a theoretical study, we can fix different values with Δ_s and Δ_t to study different Josephson junctions. If $\Delta_s = 0$, $\Delta_t \neq 0$, the NCSC has only a pure p-wave component and the system is a SWSC/p-wave superconductor (PWSC) junction. If $\Delta_s \neq 0$, $\Delta_t \neq 0$ the NCSC has both the s-wave and p-wave components and the system is a SWSC/NCSC junction. For simplicity, the Fermi energies are set the same in both sides of the system.

We present the Josephson current flowing in the x-direction as a function of the phase difference ϕ with the RSOC strength $\beta \in \{0, 1, 2\}$ and the injection direction angle $\theta = 0.2\pi$. The current is plotted in Fig. 3, Figs. 4 and 5 and the current unit is taken as a dimensionless one $eR_N I_J/\Delta_0$ with R_N being the resistance of the normal junction. In Fig. 3, the SC energy gaps in the right lead of the junction are fixed with $\Delta_s = 0$ and $\Delta_t = \Delta_0$. The right SC has only a pure p-wave component and the system is a SWSC/PWSC junction. The SC energy gaps are fixed with $\Delta_s = 0.5\Delta_0$, $\Delta_t = \Delta_0$ in Fig. 4 and $\Delta_s = \Delta_t = \Delta_0$ in Fig. 5. The right SC has both the s-wave and p-wave components and the system is a SWSC/NCSC junction. In Figs. 3(a), 4(a) and 5(a), the insulator barrier strength is fixed with $Z = 0.01$ and the junction has a strong coupling. In Figs. 3(b), 4(b) and 5(b), the coupling of the junction is weak with $Z = 1$. The black solid line represents the RSOC strength $\beta = 0$. The red dash line represents $\beta = 1$. And the blue dot line represents $\beta = 2$.

We first focus on the non-RSOC case $\beta = 0$ (black solid line). The Josephson current in Fig. 3(a) has a discontinuous jump at $\phi = n\pi$ with $Z = 0.01$ because the Andreev bound states in the energy gap are degenerate at $E = 0$ when

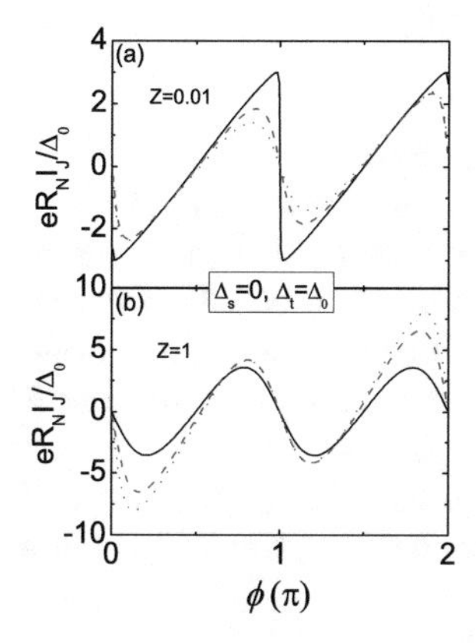

Fig. 3. The dependence of the Josephson current I_J on the phase difference ϕ with $\Delta_s = 0$, $\Delta_t = \Delta_0$.

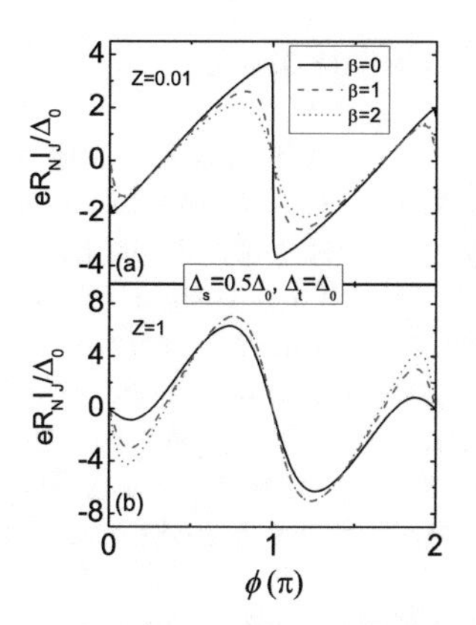

Fig. 4. The dependence of the Josephson current I_J on the phase difference ϕ with $\Delta_s = 0.5\Delta_0$, $\Delta_t = \Delta_0$.

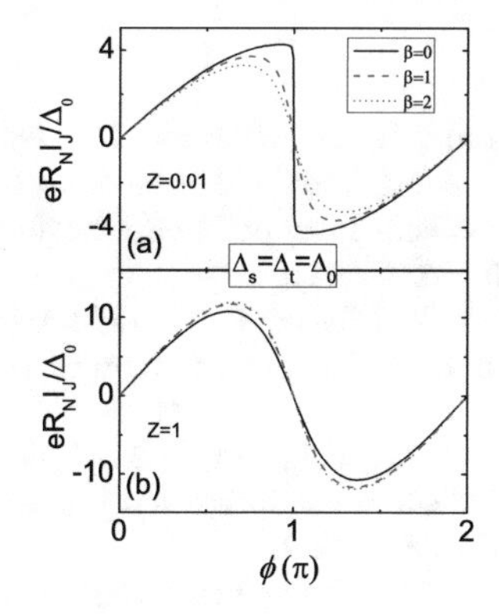

Fig. 5. The dependence of the Josephson current I_J on the phase difference ϕ with $\Delta_s = \Delta_t = \Delta_0$.

$\phi = n\pi$. The Josephson current periodicity is π not 2π. In Fig. 5(a), the discontinuous jump appears at $\phi = (2n + 1)\pi$ and the Josephson current period recovers to be the usual one 2π. With the increase of the barrier strength, the Josephson current is smoothed into sinusoidal curve.

The RSOC is now turned on in the right SC lead. With $Z = 0.01$, the discontinuous jump of the Josephson current no longer exists. Because the RSOC makes the Andreev bound states in the energy gap no longer degenerate. With $Z = 1$, The regular curves show the height deviation between adjacent two cycles and the Josephson current periodicity changes from π to 2π. The Josephson current curves in the SWSC/PWSC junction display a significant antisymmetry. It should be noted that in the NCSC, as long as the s-wave component is not zero, its rigorous period is 2π.

4 Conclusions

In this paper, we have investigated the Josephson current in the SWSC/NCSC junction by considering both the pure p-wave component and mixed s-wave and p-wave components in the NCSC. It is shown that the Josephson current periodicity with a pure PWSC is equal to π. When the NCSC has both the singlet and triplet pair states, the junction turns into a typical SWSC one, and the current periodicity recovers to be 2π. The junction coupling and the RSOC strength were also shown affecting the phase dependence of Josephson current as well as the current periodicity. Our findings are useful in identifying different order parameter symmetries and the pairing components in the NCSC materials.

Acknowledgements. Supported by the Doctoral Scientific Research Startup Foundation of Jinling Institute of Technology (No. jit-b-201202) and Outstanding Young Backbone Teachers Program of Jinling Institute of Technology.

References

1. Golubov, A.A., Kupriyanov, M.Yu., Il'ichev, E.: The current-phase relation in Josephson junctions. Rev. Mod. Phys. **76**, 411–469 (2004)
2. Buzdin, A.I.: Proximity effects in superconductor-ferromagnet heterostructures. Rev. Mod. Phys. **77**, 935–976 (2005)
3. Bergeret, F.S., Volkov, A.F., Efetov, K.B.: Odd triplet superconductivity and related phenomena in superconductor-ferromagnet structures. Rev. Mod. Phys. **77**, 1321–1373 (2005)
4. Ioffe, L.B., Geshkenbein, V.B., Feigel'man, M.V., Fauchère, A.L., Blatter, G.: Environmentally decoupled sds-wave Josephson junctions for quantum computing. Nature (London) **398**, 679–681 (1999)
5. Blatter, G., Geshkenbein, V.B., Ioffe, L.B.: Design aspects of superconducting-phase quantum bits. Phys. Rev. B. **63**, 174511 (2001)
6. Barash, Y.S., Bobkova, I.V.: Interplay of spin-discriminated Andreev bound states forming the $0-\pi$ transition in superconductor - ferromagnet - superconductor junctions. Phys. Rev. B. **65**, 144502 (2002)
7. Volkov, A.F., Anishchanka, A.: Alternative mechanism for the sign-reversal effect in superconductor-ferromagnet-superconductor Josephson junctions. Phys. Rev. B. **71**, 024501 (2005)
8. Ryazanov, V.V., Oboznov, V.A., Rusanov, A.Y., Veretennikov, A.V., Golubov, A.A., Aarts, J.: Coupling of two superconductors through a ferromagnet: evidence for a π junction. Phys. Rev. Lett. **86**, 2427–2430 (2001)
9. Kontos, T., Aprili, M., Lesueur, J., Genêt, F., Stephanidis, B., Boursier, R.: Josephson junction through a thin ferromagnetic layer: negative coupling. Phys. Rev. Lett. **89**, 137007 (2002)
10. Zdravkov, V., Sidorenko, A., Obermeier, G., Gsell, S., Schreck, M., Müller, C., et al.: Reentrant superconductivity in Nb/$Cu_{1-x}Ni_x$ bilayers. Phys. Rev. Lett. **97**, 057004 (2006)
11. Pajović, Z., Božović, M., Radović, Z., Cayssol, J., Buzdin, A.: Josephson coupling through ferromagnetic heterojunctions with noncollinear magnetizations. Phys. Rev. B. **74**, 184509 (2006)

12. Wang, J., Chan, K.S.: Josephson current oscillation in a Rashba ring. J. Phys. Condens. Matter. **21**, 245701 (2009)
13. Bauer, E., Hilscher, G., Michor, H., Paul, C., Scheidt, E.W., Gribanov, A., et al.: Heavy fermion superconductivity and magnetic order in noncentrosymmetric $CePt_3Si$. Phys. Rev. Lett. **92**, 027003 (2004)
14. Akazawa, T., Hidaka, H., Fujiwara, T., Kobayashi, T.C., Yamamoto, E., Haga, Y.: Pressure-induced superconductivity in ferromagnetic UIr without inversion symmetry. J. Phys. Condens. Matter **16**, L29 (2004)
15. Yuan, H.Q., Agterberg, D.F., Hayashi, N., Badica, P., Vandervelde, D., Togano, K., et al.: S-wave spin-triplet order in superconductors without inversion symmetry: Li_2Pd_3B and Li_2Pt_3B. Phys. Rev. Lett. **97**, 017006 (2006)
16. Sergienko, I.A., Keppens, V., McGuire, M., Jin, R., He, J., Curnoe, S.H., et al.: Metallic 'ferroelectricity' in the pyrochlore $Cd_2Re_2O_7$. Phys. Rev. Lett. **92**, 065501 (2004)
17. Yogi, M., Kitaoka, Y., Hashimoto, S., Yasuda, T., Settai, R., Matsuda, T.D., et al.: Evidence for a novel state of superconductivity in noncentrosymmetric $CePt_3Si$: a ^{195}Pt-NMR study. Phys. Rev. Lett. **93**, 027003 (2004)
18. Gor'kov, L.P., Rashba, E.I.: Superconducting 2D system with lifted spin degeneracy: mixed singlet-triplet state. Phys. Rev. Lett. **87**, 037004 (2001)
19. Kimura, N., Ito, K., Saitoh, K., Umeda, Y., Aoki, H., Terashima, T.: Pressure-induced superconductivity in noncentrosymmetric heavy-fermion $CeRhSi_3$. Phys. Rev. Lett. **95**, 247004 (2005)
20. Sugitani, I., Okuda, Y., Shishido, H., Yamada, T., Thamizhavel, A., Yamamoto, E., et al.: Pressure-induced heavy-fermion superconductivity in antiferromagnet $CeIrSi_3$ without inversion symmetry. J. Phys. Soc. Jpn. **75**, 043703 (2006)
21. Zhang, H., Wang, J., Liu, J.F.: Anomalous Josephson effect in noncentrosymmetric superconductors. Appl. Phys. Lett. **108**, 102601 (2016)
22. Liu, J.F., Zhang, H., Wang, J.: Novel $0-\pi$ transitions in Josephson junctions between noncentrosymmetric superconductors. Chin. Phys. B. **25**, 097403 (2016)
23. de Gennes, P.G.: Superconductivity of metals and alloys. Benjamin, New York (1966)
24. de Gennes, P.G., Saint-James, D.: Elementary excitations in the vicinity of a normal metal-superconducting metal contact. Phys. Lett. **4**, 151–152 (1963)
25. de Gennes, P.G.: Boundary effects in superconductors. Rev. Mod. Phys. **36**, 225–237 (1964)
26. Frigeri, P.A., Agterberg, D.F., Koga, A., Sigrist, M.: Superconductivity without inversion symmetry: MnSi versus $CePt_3Si$. Phys. Rev. Lett. **92**, 097001 (2004)
27. Furusaki, A., Tsukada, M.: Dc Josephson effect and Andreev reflection. Solid State Commun. **78**, 299–302 (1991)

Annotation of Type 1 Diabetes Functional Variants Through RegulomeDB

Aysha Karim Kiani[1(✉)], Peter John[2], and Sidrah Jahangir[2]

[1] Department of Biology and Environmental Sciences, Allama Iqbal Open University Islamabad, Islamabad, Pakistan
ayshakiani@gmail.com

[2] Department of Health Care Biotechnology, National University of Sciences and Technology Islamabad, Islamabad, Pakistan

Abstract. Type 1 diabetes is an autoimmune disorder with multiple genes involved in its pathogenesis. To understand the regulatory role of the single nucleotide polymorphisms (SNPs) associated with these genes, functional annotation of SNPs associated with Type 1 Diabetes (T1D) and their proxy SNPs was performed. For this study, 70 target SNPs were selected on the bases of their significant association with Type 1 Diabetes (p-value < 5E−08) in different population taken from five previous studies. Using SNAP web portal SNPs (proxy SNPs) that were in Linkage Disequilibrium (LD) (r2 $\geq$ 0.08) with 70 reported target SNPs were searched. All the target and proxy SNPs were analyzed for potential regulatory functions using RegulomeDB database. RegulomeDB is a useful resource to assess the functional nature of implicated SNPs. Out of the total 1612 SNPs only 187 SNPs exhibited potential regulatory functions by scoring "less than 3" in RegulomeDB whereas 124 out of them scored "1" indicating its highly significant role. From these 187 SNPs with suggested regulatory functions, only 13 were the reported T1D associated genome-wide significant SNPs which are IL27/rs4788084 (score = 1a), CENPW/rs9388489 (score = 1a), HLA/rs9272346 (score = 1b), CD69/rs4763879 (score = 1d), CLEC16A/rs12708716 (score = 1f), ERBB3/rs2292239 (score = 1f), PRKCQ/rs11258747 (score = 1f), HLA/rs9268645 (score = 1f), intergenic/rs2165738 (score = 1f), RAB5B/rs11171710 (score = 1f), CTSH/rs3825932 (score = 1f), IL10/rs3024505 (score = 2b), PTPN22/rs2476601 (score = 2b). Remaining 175 SNPs were in LD with the 70 reported GWAS SNPs. Hence, regulatory role of the Type 1 Diabetes associated SNPs was revealed using RegulomeDB.

Keywords: Type 1 diabetes · RegulomeDB · Functional variants

1 Introduction

Type 1 diabetes is an autoimmune disorder that causes damage of pancreatic β cells liable for producing insulin in pancreas. The recent advances in the area of molecular genetics has helped in the identification and mapping of various genetic variants associated with various phenotypes of T1D. GWAS studies have shown that 88% of the identified variants lie in the non-coding regions of the genome. Large consortia as well

N. Xiong et al. (Eds.): ISCSC 2018, AISC 877, pp. 332–339, 2019.
https://doi.org/10.1007/978-3-030-02116-0_39

as individual researches are generating large amount of functional data in numerous cell types which provides better understanding of genome's noncoding portions [1].

RegulomeDB, collects regulatory information from a variety of sources, including the ENCODE project, and then assigns a score indicating potential functionality of variants based upon these data. To generate these functional projects, large data sets are used which comprise of the information from the manually curated regions that have been shown to play key roles in the regulation of the gene expression. Many GWAS studies have identified many SNPs associated with Type I Diabetes (T1D) risk. A large number of these associated SNPs are in the intronic and intergenic regions. In the past, the functional assessment of such SNPs was not readily possible. For this purpose RegulomeDB is a useful resource to assess the functional nature of implicated SNPs [2].

RegulomeDB provides a scoring system, with categories ranging from 1 to 6. These calculated scores indicate the likelihood of putative functions. Category 1 is further divided into subcategories 1a to 1f. Because lower scores indicate increasing evidence that a variant is located in a functional region, a variant scored as 1a has the highest likelihood of affecting binding and the expression of a target gene. Details of RegulomeDB scores and their respective putative functions are listed in Table 1. Recently,

Table 1. RegulomeDB variant classification scheme [2]

Category	Description
Likely to affect binding and linked to expression of a gene target	
1a	eQTL + TF binding + matched TF motif + matched DNase footprint + DNase peak
1b	eQTL + TF binding + any motif + DNase footprint + DNase peak
1c	eQTL + TF binding + matched TF motif + DNase peak
1d	eQTL + TF binding + any motif + DNase peak
1e	eQTL + TF binding + matched TF motif
1f	eQTL + TF binding/DNase peak
Likely to affect binding	
2a	TF binding + matched TF motif + matched DNase footprint + DNase peak
2b	TF binding + any motif + DNase footprint + DNase peak
2c	TF binding + matched TF motif + DNase peak
Less likely to affect binding	
3a	TF binding + any motif + DNase peak
3b	TF binding + matched TF motif
Minimal binding evidence	
4	TF binding + DNase peak
5	TF binding or DNase peak
6	Motif hit

Rosenthal et al. applied RegulomeDB to GWAS data to demonstrate a possible demonstration for the association between noncoding GWAS SNPs and late-onset Alzheimer's disease. The association of various disease polymorphisms have been performed using RegulomeDB scoring system for various cancers as well [1]. Therefore, selection of putative functional SNPs using RegulomeDB may also be useful in future association studies [2].

2 Methodology

2.1 SNP Selection

A total of 70 already published GWAS SNPs were selected. These selected SNPs were investigated for association with T1D in 5 different studies in different populations [3].

2.2 Linkage Disequilibrium

Following the selection of SNPs, SNP Annotation and Proxy search (SNAP) tool [4] was used to identify the SNPs in linkage disequilibrium (LD) (r2 $\geq$ 0.80) with the SNPs of our interest [5]. The SNAP web portal determines LD based upon LD determined using the CEU populations from the International HapMap (v3) or 1000 Genomes Pilot 1 projects. The web portal than identifies the proxy SNPs through the determined LD. The web tool identified SNPs in LD with the selected SNPs. In order to better assess the associations among SNPs the search was repeated using r2 values of 0.90 and 1.00. The number of SNPs found in LD with the selected SNPs decreased, as the r2 value increased (Table 2).

Table 2. Number of SNPs in linkage disequilibrium for published GWAS SNPs from hapmap3 and 1000 genomes populations at tested r^2 thresholds

	Linkage disequilibrium (r^2) threshold		
	0.80	0.90	1.0
1000 Genomes	1320	816	509
Hap Map 3	292	254	125
Total (overlaps removed)	1612	1070	634

2.3 RegulomeDB

All the putative SNPs and the SNPs in LD with the selected SNPs were analyzed for potential regulatory functions using RegulomeDB data base [6].

3 Results

RegulomeDB is a useful resource to assess the functional nature of SNPs. Here we have utilized RegulomeDB to assess the regulatory function of 70 reported T1D GWAS SNPs from five different studies. Using SNAP webportal (accessed on 14 March 2017) proxy SNPs that were in LD (r2 $\geq$ 0.08) with 70 reported target SNPs were searched. As the target SNPs were selected from various different populations, therefore proxy SNPS were also included from various populations.

A total of 1,612 reported and proxy SNPs were analyzed with RegulomeDB (of these 1612 variants), 412 SNPs returned a score of "No Data". Of the remaining 1200 SNPs, 187 SNPs exhibited potential regulatory functions as evidenced by a score of "less than 3" in RegulomeDB, of which 124 SNPs were category 1 variant indicating they likely effect expression of a gene while remaining 63 were category 2 variant that might regulate by effecting the binding proteins. From these 187 SNPs with suggested regulatory functions, only 13 were the reported genome-wide significant SNPs that had established association with type 1 diabetes i.e. IL27/rs4788084 (score = 1a), CENPW/rs9388489 (score = 1a), HLA/rs9272346 (score = 1b), CD69/rs4763879 (score = 1d), CLEC16A/rs12708716 (score = 1f), ERBB3/rs2292239 (score = 1f), PRKCQ/rs11258747 (score = 1f), HLA/rs9268645 (score = 1f), intergenic/rs2165738 (score = 1f), RAB5B/rs11171710 (score = 1f), CTSH/rs3825932 (score = 1f), IL10/rs3024505 (score = 2b), PTPN22/rs2476601 (score = 2b), while remaining 182 SNPs were in LD to the 70 target GWAS SNPs (Table 3).

Table 3. RegulomeDB score of genome-wide significant T1D SNPs

Sr. 1	SNP	Gene	Score
1	rs4788084	IL27	1a
2	rs9388489	CENPW	1a
3	rs9272346	HLA	1b
4	rs4763879	CD69	1d
5	rs12708716	CLEC16A	1f
6	rs2292239	ERBB3	1f
7	rs11258747	PRKCQ	1f
8	rs9268645	HLA	1f
9	rs2165738	Intergenic	1f
10	rs11171710	RAB5B	1f
11	rs3825932	CTSH	1f
12	rs3024505	IL10	2b
13	rs2476601	PTPN22	2b

Out of the 187 SNPs with potential regulatory function only 2 were identified with RegulomeDB score 1a, 11 SNPs had a score of 1b, 2 SNPs had got 1c score, 14 SNPs scored 1d, while 95 SNPs had a score of 1f, 5 SNPs scored 2a and 58 SNPs had a RegulomeDB score 2b.

The SNPs with most evident regulatory function were rs4788084 (IL27) and rs9388486 with RegulomeDB score 1a. Of which rs4788084 is a reported SNP located in the intronic region of IL27 gene, and had already established association with T1D. Four other SNPs in IL27 region had scored < 3 and all of them are eQTLs for EIF3CL and EIF3S8 and effect binding of EBF1, HNF4A, HDAC2, EP300, RXRA, SP1, PAX5 and IKZF1 protiens. While rs9388489 is also a reported SNP located in the 5′UTR region of CENPW gene and is eQTL for CREB and ATF and effect binding of various proteins i.e. POLR2A, MYBL2, CEBP, ATF2, SP4, and JUND. Out of other reported SNPs, rs9272346 located on the upstream 2 KB region of HLA-DQA1 gene, scored RegulomeDB score 1b and is a eQTL for the same gene. It also effect binding of POLR2A protein. Five other SNPs, rs272723, rs9272416, rs9272491, rs9272494 located in the HLA-DQA1 region also showed potential regulatory function by scoring 1f, 2a and 2b. Furthermore a reported SNP rs4763879 located in intronic region of CD69 gene and its proxy SNP/rs4763839 revealed its putative regulatory function by scoring 1d and 1f respectively. SNP rs4763879 is an eQTL for CLEC2B, CLEC2D and DCAL1, and effect binding of POLR2 and CREBBP proteins. While seven other reported SNPs with already establishes association to T1D (CLEC16A/rs12708716, ERBB3/rs2292239, PRKCQ/rs11258747, HLA/rs9268645, intergenic/rs2165738, RAB5B/rs11171710, CTSH/rs3825932) scored 1f and rs3024505 located in IL10 gene region had RegulomeDB score of 2b, suggesting their putative regulatory role in the transcription of the genes.

4 Discussion

Of the 128 SNPs which have a RegulomeDB score of 1, the most suggestive Regulome db score of 1a: IL27/rs4788084 and CENPW/rs9388486. Both of them are reported genome-wide significant SNPs having strong association with T1D. IL27/rs4788084 has eQTL EIF3CL and EIF3S8. EIF3CL is a component of the eukaryotic translation initiation factor 3 (eIF-3) complex, while eIF-3 complex is also required for disassembly and recycling of post-termination ribosomal complexes. The eIF-3 complex specifically targets and initiates translation of a subset of mRNAs involved in cell proliferation, including cell cycling, differentiation [7].

While *CENPW* is centromere protein W which is Component of the CENPA-NAC (nucleosome-associated) complex, a complex that plays a central role in assembly of kinetochore proteins, mitotic progression and chromosome segregation. It is one of the inner kinetochore proteins, with farther proteins binding downstream. Required for normal chromosome organization and normal progress through mitosis, hence rs388486 may regulate the expression of gene by effecting these processes [8].

Further 11 SNPs (*CTSH*/rs2870085, *CLN3*/rs149299, *SH2B1*/rs7198606, *LOC105370548*/rs194749, *CTSH*/rs11072817, ATXN2L/rs8049439, *ZPBP2*/rs11557466, *CLEC16A*/rs12917656, *CLEC16A*/rs11642631, *LOC105371082*/rs243327 and *HLA-DQA1*/rs9272346) returned RegulomeDB score of 1b, one of these SNPs rs9272346 located in *HLA-DQA1* gene region is already reported and very significantly associated GWAS SNP for T1D. In addition to that rs9272346 effect binding of POLR2A protein which is an important subunit of RNA polymerase II, it might regulate the gene expression by effecting the process of transcription. While CLN3/rs149299, SH2B1/rs7198606, CTSH/rs11072817, ZPBP2/rs11557466, CLEC16A/rs12917656 and CLEC16A/rs11642631 are relatively less studied SNPs and their potential in disease risk and pathogenesis needs to be further explored [9].

Ten SNPs in CTSH gene region on chromosome 15 came up with RegulomeDB score 1 (rs2870085/1b, rs11072817/1b, rs11855406/1d, rs11638844/1d, rs10400902/1d, rs3825932/1f, rs11856301/1f, rs11072815/1f, rs1369324/1f, rs11072818/1f). CTSH/rs2870085 has been studied as a potential expression QTL (eQTL) marker for T1D. The role of rs2870085 has also been studied in disrupting β-cell function in T1D and has shown to have structure-disrupting effects on lnc RNA [10]. One of the important binding protein for rs2870085, rs11855406, rs11856301 and rs11072815 is POLR2A which is the largest subunit of RNA polymerase II, the polymerase responsible for synthesizing messenger RNA in eukaryotes [9]. Two other important binding proteins for rs11072817, rs11855406, rs11072815, rs11072818 are CHD1 and CHD2. Chromodomain-helicase-DNA-binding protein (CHD) Proteins alter gene expression possibly by modification of chromatin structure thus altering access of the transcriptional apparatus to its chromosomal DNA template [11].

An intronic SNP rs194749 located at LOC105370548 region came up with RegulomeDB score 1b. It is a proxy SNP which is in LD with a reported SNP rs1465788. eQTL of rs194749 is not known yet but it effect the binding of proteins like MYC which is transcription factor and play role in cell cycle and apoptosis, TCF12 protein that recognize the consensus binding site (E-Box) and EBF1 protein which controls the expression of key proteins required for B cell differentiation, signal transduction and function [12].

Another reported SNP rs4763879 located in intronic region of CD69 gene, with RegulomeDB score of 1d is an eQTL for *CLEC2B*, *CLEC2D*, *DCAL1* genes and effect the binding of POLR2A and CREBBP proteins. POLR2A is subunit of RNA polymerase II, while CREBBP protein interact with transcription factors and activate transcription [9].

While two more SNPs rs12708716/*CLEC16A* and rs2292239/*ERBB3* scored 1f on RegulomeDB. Both of them are reported SNPs with already established association to T1D. Binding protein and eQTL for rs12708716/*CLEC16A* are not known yet, while it's 16 more proxy SNPs which are located in same gene region also showed strong potential regulatory function. In ERBB3 gene intronic regions rs2292239 and its two proxy SNPs (rs4759229, rs3741499) are eQTLs for *RPS26*. *RPS26* gene encode 40S ribosomal protein S26 and interact with its pre-mRNA intron I and mRNA fragment [13].

Additionally four more reported SNPs (rs11258747/*PRKCQ*, rs2165738/intergenic, rs3024505/*IL10*, rs2476601/*PTPN22*) with established association to T1D reviled their putative regulatory function in this study. Out of these five, rs11258747 is an eQTL for *GDI2,* which is involved in vesicular trafficking of molecules between cellular organelles [14]. While rs2165738 with RegulomeDB score of 1f is an eQTL for *ADCY3* gene which encode adenylyl cyclase 3 enzyme that catalyzes the formation of secondary messenger cyclic adenosine monophosphate (cAMP) [15], but the binding proteins of this SNP are not known yet. While the other two other T1D associated reported SNPs rs3024505 in the intronic region of *IL10* and rs2476601 in the intronic region of *PTPN22* scored a RegulomeDB score of 2b, which suggest their regulatory function in the expression of genes by effecting the dining of proteins required for the proper transcription and translation of genes.

All these reported SNPs for T1D and their proxy SNPs (which are in LD with the reported SNPs) with RegulomeDB score <3 suggest significant regulatory role of these SNPs, which might be playing role in the pathogenesis of type 1 diabetes. As all these SNPs are located in the intronic or intergenic region therefore this study extend the knowledge and elaborates how in addition to the exonic SNPs which directly play role in disease pathogenesis, these intronic SNPs regulate the expression of genes indirectly and play its significant role in onset of the disease.

5 Conclusion

With the help of RegulomeDB, significant regulatory function of T1D SNPs located in the intronic region of genome is revealed. This will not only help in better understanding of the disease pathogenesis but also the interaction between different biochemical pathways playing role in the onset of disease.

References

1. Hindorff, L.A., et al.: Potential etiologic and functional implications of genome-wide association loci for human diseases and traits. Proc. Natl. Acad. Sci. USA **106**(23), 9362–9367 (2009)
2. Boyle, A.P., et al.: Annotation of functional variation in personal genomes using RegulomeDB. Genome Res. **22**(9), 1790–1797 (2012)
3. Evangelou, M., et al.: A method for gene-based pathway analysis using genomewide association study summary statistics reveals nine new type 1 diabetes associations. Genet. Epidemiol. **38**(8), 661–670 (2014)
4. SNAP web portal. https://www.broadinstitute.org/mpg/snap/ldsearch.php. Accessed 14 Mar 2017
5. Johnson, A.D., et al.: SNAP: a web-based tool for identification and annotation of proxy SNPs using HapMap. Bioinformatics **24**(24), 2938–2939 (2008)
6. RegulomeDB Homepage. http://regulomedb.org/. Accessed 18 Mar 2017
7. Hao, J., et al.: Eukaryotic initiation factor 3C silencing inhibits cell proliferation and promotes apoptosis in human glioma. Oncol. Rep. **33**(6), 2954–2962 (2015)
8. Chun, Y., et al.: New centromeric component CENP-W is an RNA-associated nuclear matrix protein that interacts with nucleophosmin/B23 protein. J. Biol. Chem. **286**(49), 42758–42769 (2011)
9. Pineda, G., et al.: Proteomics studies of the interactome of RNA polymerase II C-terminal repeated domain. BMC Res Notes. **8**, 616 (2015)
10. Floyel, T., Kaur, S., Pociot, F.: Genes affecting beta-cell function in type 1 diabetes. Curr. Diabetes Rep. **15**(11), 97 (2015)
11. Siggens, L., et al.: Transcription-coupled recruitment of human CHD1 and CHD2 influences chromatin accessibility and histone H3 and H3.3 occupancy at active chromatin regions. Epigenetics Chromatin. **8**(1), 4 (2015)
12. Haberl, S., et al.: MYC rearranged B-cell neoplasms: impact of genetics on classification. Cancer Genet. **209**(10), 431–439 (2016)

13. Ivanov, A.V., Malygin, A.A., Karpova, G.G.: Human ribosomal protein S26 suppresses the splicing of its pre-mRNA. Biochim. Biophys. Acta **1727**(2), 134–140 (2005)
14. Weitzdoerfer, R., et al.: Reduction of nucleoside diphosphate kinase B, Rab GDP-dissociation inhibitor beta and histidine triad nucleotide-binding protein in fetal down syndrome brain. J. Neural Transm. Suppl. **61**, 347–359 (2001)
15. Hong, S.H., et al.: Upregulation of adenylate cyclase 3 (ADCY3) increases the tumorigenic potential of cells by activating the CREB pathway. Oncotarget **4**(10), 1791–1803 (2013)

A Perturbed Risk Model with Liquid Reserves, Credit and Debit Interests and Dividends Under Absolute Ruin

Yan Zhang[1,2(✉)], Lei Mao[2], and Bingyu Kou[2]

[1] School of Science, Nanjing University of Science and Technology, Nanjing, China
sdzyjyw@126.com
[2] College of Science, Army Engineering University of PLA, Nanjing, China

Abstract. In this paper, we consider the dividend payments in the perturbed compound Poisson risk model with liquid reserves, credit interest and debit interest. It is assumed that an insurer is allowed to borrow money at some debit interest rate when his surplus is negative and he will keep the surplus as liquid reserves if the surplus is below a certain positive level. When the surplus attains this level, the excess of the surplus above this level will earn credit interest continuously at a constant interest rate. If the surplus continues to surpass a higher level, the excess of the surplus above this higher level will be paid out as dividends to shareholders. Integro-differential equations satisfied by the Geber-Shiu function are obtained, and the closed form expressions for the Geber-Shiu function are presented. Besides, we also derive the integro-differential equations with boundary conditions satisfied by the expected discounted present value of all dividends until absolute ruin

Keywords: Credit · Debit · Liquid reserves · Dividends · Geber-Shiu function

1 Introduction

In the past decade, insurance risk models with constant interest and dividend payments have received remarkable attention in actuarial literature. For example, see Kalashnikov and Konstantinides [1], Cai and Dickson [2], Yuen et al. [3], Gao and

N. Xiong et al. (Eds.): ISCSC 2018, AISC 877, pp. 340–350, 2019.
https://doi.org/10.1007/978-3-030-02116-0_40

Liu [4], Li and Lu [5], Li and Liu [6] and Chen et al. [7]. In the classical C-L risk model, an insurer's surplus process with constant interest is described by

$$\mathrm{d}U(t) = c\mathrm{d}t + rU(t)\mathrm{d}t - \mathrm{d}S(t), \quad t \geq 0 \tag{1}$$

where $c > 0$ is the premium income rate, $r > 0$ is a constant force of credit interest, and

$$S(t) = \sum_{i=1}^{N(t)} X_i,$$

is a compound Poisson process, representing the aggregate claim amounts up to time t, where $\{N(t), t \geq 0\}$ is a Poisson claim number process with intensity $\lambda > 0$, representing the number of claims that have occurred before time t and $\{X_i, i = 1, 2, \cdots\}$ independent of $\{N(t), t \geq 0\}$ are independent and identically distributed claim-size random variables with common distribution function $P(x)$ which satisfies $P(0) = 0$ and probability density function $p(x)$. Here, $\bar{P}(x) = 1 - P(x)$ is the survival function of $P(x)$.

In the model (1), it is assumed that the insurer receives credit interest continuously at a constant force of interest r per unit of time when his surplus is positive. But, in reality, even if an insurer invests all his positive surplus into a risk-free asset (say a bank account), in certain bank accounts, the positive surplus may not receive interest described in model (1), only the excess of the surplus over a certain level can receive interest. On the other hand, an insurer may not invest all his positive surplus and he may have to keep part of his positive surplus as liquid reserves. Embrechts and Schmidli [6] proposed a risk model with a liquid reserve level, in which it is assumed that an insurer receives credit interest when his surplus is above the liquid reserve, and is allowed to borrow money when his surplus is negative. Based on Embrechts and Schmidli [8], Cai et al. [9] considered the compound Poisson risk model with liquid reserves, credit interest and dividends. In their work, assume that the surplus is kept as liquid reserves when it is below a certain level $\Delta \geq 0$, which do not earn interest. As the surplus attains the level Δ, the excess of the surplus above Δ will earn credit interest at a constant interest force r, further, if the surplus continues to surpass a higher level $b \geq \Delta$, the excess of the surplus above b will be paid out as dividends to the insurer's shareholders at a constant dividend rate α $(0 \leq \alpha \leq c + r(b - \Delta))$ and no interest will be earned on the surplus over the threshold level b, then the surplus process $\{U(t), t \geq 0\}$ with $U(0) = u$ satisfies the following stochastic differential equations:

$$\mathrm{d}U(t) = \begin{cases} (c - \alpha)\mathrm{d}t + r(b - \Delta)\mathrm{d}t - \mathrm{d}S(t), & U(t) \geq b; \\ c\mathrm{d}t + r(U(t) - \Delta)\mathrm{d}t - \mathrm{d}S(t), & \Delta \leq U(t) < b; \\ c\mathrm{d}t - \mathrm{d}S(t), & 0 \leq U(t) < \Delta; \end{cases} \tag{2}$$

It should be mentioned that the insurer is allowed to borrow an amount of money at a debit interest rate $\delta > 0$ to pay claims when his surplus turns negative, meanwhile he will repay the debts continuously from his premium income, only if the debit interest rate is reasonable, the negative surplus may turn to a positive level. But when the surplus is below some constant (it is $-c/\delta$), the insurer can't repay all his debts from his premium income, so he is no longer allowed to run his business, then absolute ruin is said to occur.

Recently, the absolute ruin problem has been receiving more and more attention in risk theory. Gerber and Yang [10] considered absolute ruin probabilities problem of a perturbed compound Poisson risk model with investment. Yuen et al. [11] studied the dividend payments in a compound Poisson risk model with debit interest. Mitric et al. [12] studied the absolute ruin problem in a Sparre Andersen risk model with constant interest. Li and Lu [5] investigated a generalized Gerber-Shiu function in a risk model with credit and debit interests, in which he defined a new stopping time including the time of absolute ruin as special case. Yu [13] who extended the risk model and results in Gao and Liu [4] studied the ruin-related quantities such as Gerber-Shiu function and the moment-generating function in a compound Poisson risk model perturbed by diffusion with investment and debit interest in the presence of a threshold dividend strategy. To the best of our knowledge, there is no work that deals with the dividend payments of a perturbed compound Poisson risk model with liquid reserves, credit and debit interests under absolute ruin. This motivates us to consider the expected discounted penalty function (called the Gerber-Shiu function) and the dividend payments in such a risk model. Inspired by Cai et al. [9] and Yu [13], in this paper we extend the model (2) and investigate the following extension model (3) which is perturbed by a Brownian motion and enriched by credit and debit interests, liquid reserves and dividends.

2 Model Formulation

Let the aggregate dividends paid in the time interval [0, t] be $D(t)$. After dividends paid, we denote the insurer's surplus at time t by $U_b(t) = U(t) - D(t)$, where $U_b(0) = u$. Then the surplus process $U_b(t)$ can be described by

$$\mathrm{d}U_b(t) = \begin{cases} c_1\mathrm{d}t + r(b-\Delta)\mathrm{d}t - \mathrm{d}S(t) + \sigma\mathrm{d}W(t), & U_b(t) \geq b; \\ c_2\mathrm{d}t + r(U_b(t)-\Delta)\mathrm{d}t - \mathrm{d}S(t) + \sigma\mathrm{d}W(t), & \Delta \leq U_b(t) < b; \\ c_2\mathrm{d}t - \mathrm{d}S(t) + \sigma\mathrm{d}W(t), & 0 \leq U_b(t) < \Delta; \\ c_2\mathrm{d}t + \delta U_b(t)\mathrm{d}t - \mathrm{d}S(t) + \sigma\mathrm{d}W(t), & -c_2/\delta \leq U_b(t) < 0; \end{cases} \tag{3}$$

Where Δ is the liquid reserve level, $\sigma > 0$ is a constant, representing the diffusion volatility parameter, and $\{W(t), t \geq 0\}$ is a standard Brownian motion with $W(0) = 0$. In this model, we assume that the surplus $U_b(t)$ will be kept as liquid reserves and don't earn interest when it is below Δ. And the insurer is allowed to pay dividends according to threshold strategy. Whenever the surplus $U_b(t)$ is below the level $b(\geq \Delta)$ no dividends are paid and the constant premium income rate is $c_2 > 0$. When $U_b(t)$ attains the level Δ, the excess of $U_b(t)$ above Δ will earn credit interest at a constant interest force r. When $U_b(t)$ continues to surpass a higher level b, the excess of $U_b(t)$ above b will be paid out as dividends to the insurer's shareholders at a constant rate $\alpha(0 \leq \alpha \leq c_2 + r(b-\Delta))$, and no interest will be earned on the surplus $U_b(t)$ over the level b. After dividend payments, the net premium rate is $c_1 = c_2 - \alpha$.

Furthermore, the insurer is allowed to borrow an amount of money at a debit interest rate $\delta > 0$ to pay claims when $U_b(t)$ is negative. Meanwhile, the insurer will repay the debts continuously from his premium income. But when $U_b(t)$ is below $-c_2/\delta$, the insurer can't repay all his debts from his premium income, so the insurer is no longer allowed to run his business.

Suppose that $X_i, N(t)$ and $W(t)$ are mutually independent. The net profit condition is given by $c_2 - r\Delta > \lambda E[X_i]$ and $c_1 + r(b - \Delta) > \lambda E[X_i]$. Denote the absolute ruin time of $U_b(t)$ by $T_b = \inf\{t \geq 0 | U_b(t) \leq -c_2/\delta\}$, and $T_b = \infty$ if $U_b(t) > -c_2/\delta$ for all $t \geq 0$. Let $\beta > 0$ be the force of interest valuation. We denote the Gerber-Shiu function in the model (3) by

$$m(u, \bar{b}) = E\{e^{-\beta T_b}\omega(U_b(T_{\bar{b}}^-), |U_b(T_b)|)I(T_b < \infty)|U_b(0) = u\}, \tag{4}$$

where $U_b(T_{\bar{b}}^-)$ is the surplus prior to absolute ruin, $|U_b(T_b)|$ is the deficit at absolute ruin. The penalty function $\omega(x, y)$ is an arbitrary nonnegative measurable function defined on $(-c_2/\delta, +\infty) \times (-c_2/\delta, +\infty)$. The vector $\bar{b} = (\Delta, b)$ is listed as an argument of the Gerber-Shiu function to emphasize that the function depends on the liquid reserves level Δ and the threshold level b. Similar to Cai et al. [9] and Yu [13], throughout this paper we assume that $m(u, \bar{b})$ are sufficiently smooth functions in u and in its domain.

3 Results

In this section, we will discuss the integro-differential equations satisfied by $m(u, \bar{b})$. Clearly, the Gerber-Shiu function $m(u, \bar{b})$ behaves differently when its initial surplus u is in different intervals of $[-c_2/\delta, 0), [0, \Delta), [\Delta, b)$ and $[b, +\infty)$. Hence, we write $m(u, \bar{b})$ as $m_3(u, \bar{b})$ for $u \geq b$; $m_2(u, \bar{b})$ for $\Delta \leq u < b$; $m_1(u, \bar{b})$ for $0 \leq u < \Delta$; $m_0(u, \bar{b})$ for $-c_2/\delta \leq u < 0$. For notational convenience, let

$$\begin{aligned} h_0(t) &= ue^{\delta t} + c_2/\delta(e^{\delta t} - 1), h_1(t) = \Delta + (u - \Delta)e^{rt} + c_2/r(e^{rt} - 1), h(t, p) \\ &= \Delta + c_2/r(e^{r(t-p)} - 1). \end{aligned}$$

Theorem 1. $m_i(u, \bar{b})(i = 0, 1, 2, 3)$ satisfies the following system of integro-differential equations

$$\begin{aligned} &\sigma^2/2m_0''(u, \bar{b}) + (u\delta + c_2)m_0'(u, \bar{b}) - (\lambda + \beta)m_0(u, \bar{b}) \\ &\qquad + \lambda[\int_0^{u + c_2/\delta} m_0(u - y, \bar{b})\mathrm{d}P(y) + \varsigma(u)] = 0, -c_2/\delta < u < 0; \end{aligned} \tag{5}$$

$$\begin{aligned} &\sigma^2/2m_1''(u, \bar{b}) + c_2 m_1'(u, \bar{b}) - (\lambda + \beta)m_1(u, \bar{b}) + \lambda[\int_0^u m_1(u - y, \bar{b})\mathrm{d}P(y) \\ &\qquad + \int_u^{u + c_2/\delta} m_0(u - y, \bar{b})\mathrm{d}P(y) + \varsigma(u)] = 0, 0 \leq u < \Delta; \end{aligned} \tag{6}$$

$$\sigma^2/2m_2''(u,\bar{b})+[r(u-\Delta)+c_2]m_2'(u,\bar{b})-(\lambda+\beta)m_2(u,\bar{b})$$
$$+\lambda[\int_0^{u-\Delta} m_2(u-y,\bar{b})\mathrm{d}P(y)+\int_{u-\Delta}^{u} m_1(u-y,\bar{b})\mathrm{d}P(y) \tag{7}$$
$$+\int_u^{u+c_2/\delta} m_0(u-y,\bar{b})\mathrm{d}P(y)+\varsigma(u)]=0, \quad \Delta\le u<b;$$

$$\sigma^2/2m_3''(u,\bar{b})+\tilde{c}m_3'(u,\bar{b})-(\lambda+\beta)m_3(u,\bar{b})+\lambda[\int_0^{u-b} m_3(u-y,\bar{b})\mathrm{d}P(y)$$
$$+\int_{u-b}^{u-\Delta} m_2(u-y,\bar{b})\mathrm{d}P(y)+\int_{u-\Delta}^{u} m_1(u-y,\bar{b})\mathrm{d}P(y) \tag{8}$$
$$+\int_u^{u+c_2/\delta} m_0(u-y,\bar{b})\mathrm{d}P(y)+\varsigma(u)]=0, \quad u\ge b.$$

Where $\tilde{c}=c_1+r(b-\Delta)$ and $\varsigma(u)=\int_{u+c_2/\delta}^{\infty}\omega(u,y-u)\mathrm{d}P(y)$.

Proof

(1) When $-c_2/\delta<u<0$, considering $U_b(t)$ in a small time interval $(0,t]$, here t is sufficiently small so that $U_b(t)$ can't reach b. By conditioning on the time and amount of the first claim and whether the claim causes absolute ruin, one gets

$$m_0(u,\bar{b})=(1-\lambda t)e^{-\beta t}Em_0(h_0(t)+\sigma W(t),\bar{b})+\lambda te^{-\beta t}E\gamma_0(h_0(t)+\sigma W(t))+o(t), \tag{9}$$

where $\gamma_0(u)=\int_0^{u+c_2/\delta} m_0(u-y,\bar{b})\mathrm{d}P(y)+\varsigma(u)$.
By Taylor's expansion, $EW(t)=0, E[W^2(t)]=t$ and replacing $e^{-\beta t}$ by $1-\beta t+o(t)$, $h_0(t)$ by $u+(u\delta+c_2)t+o(t)$, then we have

$$Em_0(h_0(t)+\sigma W(t),\bar{b})=m_0(u,\bar{b})+m_0'(u,\bar{b})(u\delta+c_2)t+\sigma^2/2tm_0''(u,\bar{b})+o(t),$$
$$E\gamma_0(h_0(t)+\sigma W(t),\bar{b})=\gamma_0(u,\bar{b})+\gamma_0'(u,\bar{b})(u\delta+c_2)t+\sigma^2/2t\gamma_0''(u,\bar{b})+o(t).$$

Substituting the above two equations into (9), and dividing both sides of (9) by t, then letting $t\to 0$, we arrive at (5).

(2) When $0\le u<\Delta$, we obtain

$$m_1(u,\bar{b})=(1-\lambda t)e^{-\beta t}Em_1(u+c_2t+\sigma W(t),\bar{b})+\lambda te^{-\beta t}E\gamma_1(u+c_2t+\sigma W(t))]+o(t). \tag{10}$$

where $\gamma_1(u)=\int_0^u m_1(u-y,\bar{b})\mathrm{d}P(y)+\int_u^{u+c_2/\delta} m_0(u-y,\bar{b})\mathrm{d}P(y)+\varsigma(u)$.

By Taylor's expansion, we get

$$Em_1(u+ct+\sigma W(t),\bar{b}) = m_1(u,\bar{b}) + ctm_1'(u,\bar{b}) + \sigma^2/2tm_1''(u,\bar{b}) + o(t),$$
$$E\gamma_1(u+ct+\sigma W(t),\bar{b}) = \gamma_1(u,\bar{b}) + ct\gamma_1'(u,\bar{b}) + \sigma^2/2t\gamma_1''(u,\bar{b}) + o(t).$$

Plugging the above two equations into (10) and then dividing both sides by t and let $t \to 0$, we obtain (6).

(3) Similarly, for $\Delta \le u < b$, in a small time interval $(0,t]$, by the law of total probability

$$m_2(u,\bar{b}) = (1-\lambda t)e^{-\beta t}Em_2(h_2(t)+\sigma W(t),\bar{b}) + \lambda te^{-\beta t}E\gamma_2(h_2(t)+\sigma W(t)) + o(t), \quad (11)$$

Where

$$\gamma_2(u) = \int_0^{u-\Delta} m_2(u-y,\bar{b})\mathrm{d}P(y) + \int_{u-\Delta}^{u} m_1(u-y,\bar{b})\mathrm{d}P(y) + \int_u^{u+c_2/\delta} m_0(u-y,\bar{b})\mathrm{d}P(y) + \varsigma(u).$$

Replacing $h_2(t)$ by $u+[r(u-\Delta)+c_2]t+o(t)$, by Taylor's Expansion, we have

$$Em_2(h_2(t)+\sigma W(t),\bar{b}) = m_2(u,\bar{b}) + m_2'(u,\bar{b})[r(u-\Delta)+c_2]t + \sigma^2/2tm_2''(u,\bar{b}) + o(t),$$
$$E\gamma_2(h_0(t)+\sigma W(t),\bar{b}) = \gamma_2(u,\bar{b}) + \gamma_2'(u,\bar{b})[r(u-\Delta)+c_2]t + \sigma^2/2t\gamma_2''(u,\bar{b}) + o(t).$$

Inserting the above two equations into (11) and dividing both sides by t, then let $t \to 0$, we obtain (7).

(4) When $u \ge b$, in a small time interval $(0,t]$, by conditioning on the time and the amount of the first claim, we obtain,

$$m_3(u,\bar{b}) = (1-\lambda t)e^{-\beta t}Em_3(u+\tilde{c}t+\sigma W(t),\bar{b}) + \lambda te^{-\beta t}E\gamma_3(u+\tilde{c}t+\sigma W(t)) + o(t), \quad (12)$$

where

$$\gamma_3(u) = \int_0^{u-b} m_3(u-y,\bar{b})\mathrm{d}P(y) + \int_{u-b}^{u-\Delta} m_2(u-y,\bar{b})\mathrm{d}P(y) + \int_{u-\Delta}^{u} m_1(u-y,\bar{b})\mathrm{d}P(y)$$
$$+ \int_u^{u+c_2/\delta} m_0(u-y,\bar{b})\mathrm{d}P(y) + \zeta(u).$$

Similar to the derivation of (6), and by (12), we can arrive at (8). □

Theorem 2. The integro-differential equations (5)–(8) can be expressed by Volterra equations

$$m_0(u,\bar{b}) = L_0(u) + \int_{-c_2/\delta}^{u} m_0(t,\bar{b})l_0(u,t)\mathrm{d}t, \quad -c_2/\delta < u < 0, \quad (13)$$

$$m_1(u,\bar{b}) = L_1(u) + \int_0^{u} m_1(t,\bar{b})l_1(u,t)\mathrm{d}t, \quad 0 \le u < \Delta, \quad (14)$$

$$m_2(u,\bar{b}) = L_2(u) + \int_{\Delta}^{u} m_2(t,\bar{b}) l_2(u,t)\mathrm{d}t, \quad \Delta \le u < b, \tag{15}$$

$$m_3(u,\bar{b}) = L_3(u) + \int_{b}^{u} m_3(t,\bar{b}) l_3(u,t)\mathrm{d}t, \quad u \ge b, \tag{16}$$

where

$$l_0(u,t) = 2/\sigma^2((\lambda+\beta+\delta)(u-t) - (\delta t + c_2) - \lambda \int_t^u P(x-t)\mathrm{d}x),$$

$$L_0(u) = m_0(-c_2/\delta,\bar{b}) + m_0'(-c_2/\delta,\bar{b})(u+c_2/\delta) - 2\lambda/\sigma^2 \int_{-c_2/\delta}^{u} \varsigma(t)(u-x)\mathrm{d}x,$$

$$l_1(u,t) = 2/\sigma^2((\lambda+\beta)(u-t) - c_2 - \lambda \int_t^u P(x-t)\mathrm{d}x),$$

$$L_1(u) = m_1(0,\bar{b}) + \left(m_1'(0,\bar{b}) + 2c_2/\sigma^2 m_1(0,\bar{b})\right)u - 2\lambda/\sigma^2 \int_0^u W_1(x)\mathrm{d}x,$$

$$W_1(x) = \int_{-c_2/\delta}^{0} m_0(v,\bar{b})[P(x-v) - P(-v)]\mathrm{d}v + \int_0^x \varsigma(v)\mathrm{d}v,$$

$$l_2(u,t) = 2/\sigma^2\left((r+\lambda+\beta)(u-t) - (r(t-\Delta)+c_2) - \lambda \int_t^u P(x-t)\mathrm{d}x\right),$$

$$L_2(u) = m_2(\Delta,\bar{b}) + \left(m_2'(\Delta,\bar{b}) + 2c_2/\sigma^2 m_2(\Delta,\bar{b})\right)(u-\Delta) - 2\lambda/\sigma^2 \int_{\Delta}^{u} W_2(x)\mathrm{d}x,$$

$$\begin{aligned} W_2(x) = &\int_0^{\Delta} m_1(v,\bar{b})[P(x-v) - P(\Delta-v)]\mathrm{d}v \\ &+ \int_{-c_2/\delta}^{0} m_0(v,\bar{b})[P(x-v) - P(\Delta-v)]\mathrm{d}v + \int_{\Delta}^{x} \varsigma(v)\mathrm{d}v, \end{aligned}$$

$$l_3(u,t) = 2/\sigma^2\left((\lambda+\beta)(u-t) - \tilde{c} - \lambda \int_t^u P(x-t)\mathrm{d}x\right),$$

$$L_3(u) = m_3(b,\bar{b}) + \left(m_3'(b,\bar{b}) + 2\tilde{c}/\sigma^2 m_3(b,\bar{b})\right)(u-b) - 2\lambda/\sigma^2 \int_b^u W_3(x)\mathrm{d}x,$$

$$\begin{aligned} W_3(x) = &\int_{\Delta}^{b} m_2(v,\bar{b})[P(x-v) - P(b-v)]\mathrm{d}v + \int_0^{\Delta} m_1(v,\bar{b})[P(x-v) - P(b-v)]\mathrm{d}v \\ &+ \int_{-c_2/\delta}^{0} m_0(v,\bar{b})[P(x-v) - P(b-v)]\mathrm{d}v + \int_b^x \varsigma(v)\mathrm{d}v. \end{aligned}$$

Proof. Replacing u by t in (5) and then integrate both sides of the equation from $-c_2/\delta$ to 0 with respect to t. Then, we get for $-c_2/\delta < u < 0$,

$$\begin{aligned}&\sigma^2/2\big(m_0'(u,\bar b)-m_0'(-c_2/\delta,\bar b)\big)+(u\delta+c_2)m_0(u,\bar b)\\&\quad=(\lambda+\beta+\delta)\int_{-c_2/\delta}^{u}m_0(t,\bar b)\mathrm{d}t-\lambda\int_{-c_2/\delta}^{u}m_0(t,\bar b)P(u-t)\mathrm{d}t-\lambda\int_{-c_2/\delta}^{u}\varsigma(t)\mathrm{d}t.\end{aligned}\tag{17}$$

Substituting x with u and integrating (17) over $(-c_2/\delta,u)$ yields

$$\begin{aligned}&\sigma^2/2\big(m_0(u,\bar b)-m_0(-c_2/\delta,\bar b)+m_0'(-c_2/\delta,\bar b)(u+c_2/\delta)\big)+\int_{-c_2/\delta}^{u}m_0(x,\bar b)(x\delta+c_2)\mathrm{d}x\\&\quad=(\lambda+\beta+\delta)\int_{-c_2/\delta}^{u}m_0(t,\bar b)(u-t)\mathrm{d}t-\lambda\int_{-c_2/\delta}^{u}\mathrm{d}x\int_{-c_2/\delta}^{x}m_0(t,\bar b)P(x-t)\mathrm{d}t-\lambda\int_{-c_2/\delta}^{u}\varsigma(t)(u-t)\mathrm{d}t.\end{aligned}\tag{18}$$

Since $\int_{-c_2/\delta}^{u}\mathrm{d}x\int_{-c_2/\delta}^{x}m_0(t,\bar b)P(x-t)\mathrm{d}t=\int_{-c_2/\delta}^{u}m_0(t,\bar b)\big(\int_t^u P(x-t)\mathrm{d}x\big)\mathrm{d}t$, inserting it into (18) yields (13). Similarly, (14) and (16) are also derived. □

Remark 1. It is well known, $L_0(u)$, $L_1(u)$, $L_2(u)$ and $L_3(u)$ are absolutely integrable, and $l_0(u,t)$, $l_1(u,t)$, $l_2(u,t)$ and $l_3(u,t)$ are all continuous. Thus $m_i(u,\bar b)(i=0,1,2,3)$ can be approximated recursively by Picard's sequence (see Mikhlin [14]), i.e.

$$\begin{aligned}&m_0(u,\bar b)=L_0(u)+\sum_{n=1}^{\infty}\int_{-c_2/\delta}^{u}l_{0n}(u,s)L_0(s)\mathrm{d}s, m_1(u,\bar b)=L_1(u)+\sum_{n=1}^{\infty}\int_0^u l_{1n}(u,s)L_1(s)\mathrm{d}s,\\&m_2(u,\bar b)=L_2(u)+\sum_{n=1}^{\infty}\int_{\Delta}^{u}l_{2n}(u,s)L_2(s)\mathrm{d}s, m_3(u,\bar b)=L_3(u)+\sum_{n=1}^{\infty}\int_b^u l_{2n}(u,s)L_3(s)\mathrm{d}s,\end{aligned}$$

where

$$\begin{aligned}l_{01}(u,s)=l_0(u,s),\ l_{0n}(u,s)&=\int_s^u l_0(u,t)l_{0,n-1}(t,s)\mathrm{d}t, n=2,3,\cdots.\\ l_{11}(u,s)=l_1(u,s), l_{1n}(u,s)&=\int_s^u l_1(u,t)l_{1,n-1}(t,s)\mathrm{d}t, n=2,3,\cdots\\ l_{21}(u,s)=l_2(u,s), l_{2n}(u,s)&=\int_s^u l_2(u,t)l_{2,n-1}(t,s)\mathrm{d}t, n=2,3,\cdots\\ l_{31}(u,s)=l_3(u,s), l_{3n}(u,s)&=\int_s^u l_3(u,t)l_{3,n-1}(t,s)\mathrm{d}t, n=2,3,\cdots.\end{aligned}$$

Therefore, the exact forms of the solutions for $m_i(u,\bar b)$ are available by Theorem 2 once $m_0(-c_2/\delta,\bar b)$, $m_1(0,\bar b)$, $m_1'(0,\bar b)$, $m_2(\Delta,\bar b)$, $m_2'(\Delta,\bar b)$, $m_3(b,\bar b)$ and $m_3'(b,\bar b)$ are found, recursively. □

Since $V(u,\bar b)$ depends on the initial surplus u, similar to $m(u,\bar b)$, we write $V(u,\bar b)$ as $V_3(u,\bar b)$ for $u\geq b$; $V_2(u,\bar b)$ for $\Delta\leq u<b$; $V_1(u,\bar b)$ for $0\leq u<\Delta$; $V_0(u,\bar b)$ for $-c_2/\delta\leq u<0$.

Theorem 3. $V_i(u,\bar b)(i=0,1,2,3)$ satisfy

$$V_0(0^-,\bar b)=V_1(0^+,\bar b);\ V_1(\Delta^-,\bar b)=V_2(\Delta^+,\bar b);\ V_2(b^-,\bar b)=V_3(b^+,\bar b).\tag{19}$$

Proof. When $-c_2/\delta < u < 0$, let t_0 be the hitting time of the level 0 if the hitting occurs before a claim arrival, in fact, t_0 is the solution to the equation of $h_0(t) = 0$. Let τ_0 be the time that the surplus reaches 0 for the first time from $u < 0$. Thus, by the strong Markov property of the surplus process, we have

$$\begin{aligned} V_0(u,\bar{b}) &= E[I(\tau_0 < T_b)D_{u,b}] + E[I(\tau_0 \geq T_b)D_{u,b}] = E[I(\tau_0 < T_b)e^{-\beta\tau_0}\int_{\tau_0}^{T_b} e^{-\beta t}\mathrm{d}D(t)] \\ &= E\left\{E[I(\tau_0 < T_b)e^{-\beta\tau_0}\int_{\tau_0}^{T_b} e^{-\beta t}\mathrm{d}D(t)\middle| U_b(\tau_0) = 0]\right\} \\ &= V_1(0,\bar{b})E[I(\tau_0 < T_b)e^{-\beta\tau_0}] \leq V_0(0,\bar{b}). \end{aligned} \tag{20}$$

On the other hand, let T_1 be the time of the first claim, then

$$\begin{aligned} V_0(u,\bar{b}) &\geq E[I(\tau_0 < T_b, \tau_0 = t_0)D_{u,b}] + E[I(\tau_0 \geq T_b)D_{u,b}] \\ &= E[I(\tau_0 < T_b)e^{-\beta\tau_0}\int_{\tau_0}^{T_b} e^{-\beta t}\mathrm{d}D(t)] = V_1(0,\bar{b})E[I(\tau_0 < T_b, \tau_0 = t_0)e^{-\beta\tau_0}] \\ &= V_1(0,\bar{b})e^{-\beta t_1}P(T_1 > t_0) = V_1(0,\bar{b})e^{-(\beta+\lambda)t_0}, \end{aligned} \tag{21}$$

Letting $u \uparrow 0$ in Eqs. (20) and (21) gives $t_0 \to 0$, $\tau_0 \to 0$, then we arrive at the first equation in (20). Similarly, the second and the third equations in (19) are derived. □

Theorem 4. $V_i(u,\bar{b})(i = 0,1,2,3)$ satisfies the following system of integro-differential equations

$$\begin{aligned} &\sigma^2/2V_0''(u,\bar{b}) + (u\delta + c_2)V_0'(u,\bar{b}) - (\lambda+\beta)V_0(u,\bar{b}) \\ &\qquad + \lambda\int_0^{u+c_2/\delta} V_0(u-y,\bar{b})\mathrm{d}P(y) = 0, \quad -c_2/\delta < u < 0; \end{aligned} \tag{22}$$

$$\begin{aligned} &\sigma^2/2V_1''(u,\bar{b}) + c_2V_1'(u,\bar{b}) - (\lambda+\beta)V_1(u,\bar{b}) + \lambda[\int_0^u V_1(u-y,\bar{b})\mathrm{d}P(y) \\ &\qquad + \int_u^{u+c_2/\delta} V_0(u-y,\bar{b})\mathrm{d}P(y)] = 0, \quad 0 \leq u < \Delta; \end{aligned} \tag{23}$$

$$\begin{aligned} &\sigma^2/2V_2''(u,\bar{b}) + [r(u-\Delta)+c_2]V_2'(u,\bar{b}) - (\lambda+\beta)V_2(u,\bar{b}) + \lambda[\int_0^{u-\Delta} V_2(u-y,\bar{b})\mathrm{d}P(y) \\ &\qquad + \int_{u-\Delta}^{u} V_1(u-y,\bar{b})\mathrm{d}P(y) + \int_u^{u+c_2/\delta} V_0(u-y,\bar{b})\mathrm{d}P(y)] = 0, \quad \Delta \leq u < b; \end{aligned} \tag{24}$$

$$\sigma^2/2V_3''(u,\bar{b})+\tilde{c}V_3'(u,\bar{b})-(\lambda+\beta)V_3(u,\bar{b})+\lambda[\int_0^{u-b}V_3(u-y,\bar{b})\mathrm{d}P(y) + \int_{u-b}^{u-\Delta}V_2(u-y,\bar{b})\mathrm{d}P(y)+\int_{u-\Delta}^{u}V_1(u-y,\bar{b})\mathrm{d}P(y) + \int_u^{u+c_2/\delta}V_0(u-y,\bar{b})\mathrm{d}P(y)]+\alpha=0, \quad u\geq b \tag{25}$$

with boundary conditions

$$V_0(-c_2/\delta,\bar{b})=0,\ \lim_{u\to\infty}V_3(u,\bar{b})=\alpha/\beta,\ V_0''(-c_2/\delta,\bar{b})=0, \tag{26}$$

$$\sigma^2/2V_0''(0^-,\bar{b})+c_2V_0'(0^-,\bar{b})=\sigma^2/2V_1''(0^+,\bar{b})+c_2V_1'(0^+,\bar{b}), \tag{27}$$

$$\sigma^2/2V_1''(\Delta^-,\bar{b})+c_2V_1'(\Delta^-,\bar{b})=\sigma^2/2V_2''(\Delta^+,\bar{b})+c_2V_2'(\Delta^+,\bar{b}), \tag{28}$$

$$\sigma^2/2V_2''(b^-,\bar{b})+[r(b-\Delta)+c_2]V_2'(b^-,\bar{b})=\sigma^2/2V_3''(b^+,\bar{b})+\tilde{c}V_3'(b^+,\bar{b})+\alpha. \tag{29}$$

Proof. The proofs of (22)–(25) are similar to the derivations of (5)–(7) and we omit them. Here we only proof the (25). When $u\geq b$, at the end of a short time interval of length t, no matter what happens, there is always a guaranteed discounted dividend payments, which is $\alpha(1-e^{\beta t})/\beta$. Thus

$$V_3(u,\bar{b})=\alpha(1-e^{\beta t})/\beta+(1-\lambda t)e^{-\beta t}V_3(u+\tilde{c}t,\bar{b})+\lambda te^{-\beta t}\gamma_4(u+\tilde{c}t)+o(t), \tag{30}$$

where

$$\gamma_4(u)=\int_0^{u-b}V_3(u-y,\bar{b})\mathrm{d}P(y)+\int_{u-b}^{u-\Delta}V_2(u-y,\bar{b})\mathrm{d}P(y)+\int_{u-\Delta}^{u}V_1(u-y,\bar{b})\mathrm{d}P(y)+\int_u^{u+c_2/\delta}V_0(u-y,\bar{b})\mathrm{d}P(y)].$$

Replacing $\alpha(1-e^{\beta t})/\beta$ by $\alpha t+o(t)$ in (30), similar to the proof (5), (22) can be got. When $u=-c_2/\delta$, the absolute ruin is immediate and there is no dividend, so the first equation in (26) holds. When $u\to\infty$, then $T_b=\infty$, so $D_{u,b}=\alpha/\beta$, the second equation in (26) holds and then letting $u\downarrow -c_2/\delta$ in (22) yields the third equation in (26). Letting $u\downarrow 0$ in (22) and $u\uparrow 0$ in (23), we get (27). Similarly, letting $u\downarrow\Delta$ in (23) and $u\uparrow\Delta$ in (24), we can obtain (25). Letting $u\downarrow b$ in (24) and $u\uparrow b$ in (24) gives (29). □

References

1. Kalashnikov, V., Konstantinides, D.: Ruin under interest force and subexponential claims: a simple treatment. Insur. Math. Econ. **27**, 145–149 (2000)
2. Cai, J., Dcikson, D.C.M.: On the expected discounted penalty function at ruin of a surplus process with interest. Insur. Math. Econ. **3**, 389–404 (2002)

3. Yuen, K.C., Wang, G.J., Li, W.K.: The Gerber-Shiu expected discounted penalty function for risk process with interest and a constant dividend barrier. Insur. Math. Econ. **40**, 104–112 (2007)
4. Gao, S., Liu, Z.M.: The perturbed compound Poisson risk model with constant interest and a threshold dividend strategy. J. Comput. Appl. Math. **233**, 2181–2188 (2010)
5. Li, S., Lu, Y.: On the generalized Gerber–Shiu function for surplus processes with interest. Insur. Math. Econ. **52**(1), 127–134 (2013)
6. Li, Y., Liu, G.: Optimal Dividend and capital injection strategies in the cramér-lundberg risk model. Math. Probl. Eng. **2015**, 16 (2015). Article ID 439537
7. Chen, S., Zeng, Y., Hao, Z.: Optimal dividend strategies with time-inconsistent preferences and transaction costs in the Cramér-Lundberg model. Insur. Math. Econ. **31**, 31–45 (2017)
8. Embrechts, P., Schmidli, H.: Ruin estimation for a general insurance risk model. Adv. Appl. Probab. **26**, 404–422 (1994)
9. Cai, J., Feng, R., Willmot, G.: Analysis of the compound Poisson surplus model with liquid reserves, interest and dividends. Astin Bull. **39**(1), 225–247 (2009)
10. Gerber, H.U., Yang, H.L.: Absolute ruin probabilities in a jump diffusion risk model with investment. North Am. Actuarial J. **11**(3), 159–169 (2007)
11. Yuen, K.C., Zhou, M., Guo, J.Y.: On a risk model with debit interest and dividend payments. Stat. Probab. Lett. **78**, 2426–2432 (2008)
12. Mitric, L.R., Badescu, A.L., Stanford, D.: On the absolute ruin problem in a Sparre Andersen risk model with constant interest. Insur. Math. Econ. **50**, 167–178 (2012)
13. Yu, W.: Some results on absolute ruin in the perturbed insurance risk model with investment and debit interests. Econ. Model. **31**, 625–634 (2013)
14. Mikhlin, S.G.: Integral Equations. Pergamon Press, London (1957)

Calculation of Target Heating by a Moving Source of Heat by the Finite Element Method Using the Splitting Scheme

Tatiana A. Akimenko(✉) and Inna V. Dunaeva(✉)

Tula State University, 92 "Lenina" prospect, Tula 300012, Russia
tantan72@mail.ru, i_w_d@mail.ru

Abstract. A numerical simulation of the process of heating a target by a moving heat source by the finite element method is presented using the scheme for splitting the three-dimensional thermal problem into two spatial directions-along the plane of the target and along its thickness. The basis of the mathematical model of thermal processes in the target the equation of non-stationary heat conduction of parabolic type is put.

Keywords: Splitting scheme · Finite element method · Discrete model Heating · Target · Target plane · Thermal problem

1 Introduction

The solution of the three-dimensional heat conduction problem in calculating the target heating process by a moving heat source that arises under the action of a laser beam is associated with an excessive expenditure of computing resources. A feature of the shape of the target is its relative thinness, which makes it possible to reduce these requirements by presenting the calculation of a three-dimensional temperature field in the form of a sequence of smaller-dimensional problems. In connection with this, let us consider the possibility of implementing the finite element method (FEM) using the scheme for splitting the three-dimensional thermal problem into two spatial directions-along the plane of the target and its thickness [1, 2].

The mathematical model of thermal processes in the target is based on the equation of non-stationary heat conductivity of parabolic type, assuming the infinity of the velocity of heat propagation in the material.

2 Mathematical Model

The mathematical description of the temperature field includes the differential heat equation, the boundary and initial conditions [3, 4]:

N. Xiong et al. (Eds.): ISCSC 2018, AISC 877, pp. 351–360, 2019.
https://doi.org/10.1007/978-3-030-02116-0_41

– differential heat equation

$$c\rho \frac{\partial T}{\partial t} = div(\lambda\, grad\, T) + q_V, \quad \forall x_m \in V, \quad t > 0;$$

– boundary conditions on the surface S of heat exchange with a gaseous medium:

$$q_{Sg} = -\sum_m \lambda_m \frac{\partial T}{\partial x_m} l_{xm}, \quad \forall x_m \in S, \quad t > 0;$$

– the conjugation conditions on the contact surfaces S_K of the layers:

$$-\sum_m \lambda_{im} \frac{\partial T}{\partial x_m} l_{xim} = \frac{1}{R_K}(T_j - T_i), \qquad \forall x_m \in S_K, \quad t > 0;$$

– initial conditions:

$$T = f(x_m), \qquad \forall x_m \in V, t = 0,$$

where T – is the temperature; t – is the time; x_m – are the spatial coordinates; λ – is the coefficient of thermal conductivity of the material; c and ρ – are the specific heat and density of the material; n – is the internal normal to the heat exchange surface S; q_{Sg} – are the surface heat fluxes from the gas flow side, including q_l – is the radiant heat flux density and $q_g = -\lambda_g \partial T_g / \partial n$ – is the density of the convective heat flux (the index “g” refers to the gas parameters on the wall); R_k – is the thermal contact resistance between the layers i and j; l_{xm} – are the direction cosines of the vector of the inner normal to the surface S with axes x_m.

Taking into account the subsequent solution of this problem by the finite element method, we represent its approximate variational formulation [3–5]. Considering in the general case a three-dimensional spatial setting in a Cartesian coordinate system, for the differential heat equation and the fixed time t, written above, the functional that needs to be minimized can be represented in the form:

$$\mathbf{\Phi}[T(x,y,z)] = \int_{xyz} \left\{ \frac{1}{2}\left[\lambda_x^* \left(\frac{\partial T}{\partial x}\right)^2 + \lambda_y^* \left(\frac{\partial T}{\partial y}\right)^2 + \lambda_z^* \left(\frac{\partial T}{\partial z}\right)^2 \right] + q_C^* T \right\} dxdydz -$$
$$- \int_S \left[q_{Sg} T + \alpha (T_g - \tfrac{1}{2} T) T \right] dS - \int_{S_k} \frac{1}{R_k} (T_j - \tfrac{1}{2} T_i)\, T dS_k,$$

where $q_c^* = c^* \rho^* (\partial T^* / \partial t)$ – is the heat flux accumulated in the material at time t due to its heat capacity, the “*” index corresponds to the fixed (“frozen”) momentary distribution of the parameter in space. For the sake of generality, the boundary conditions for convective heat transfer are presented in two versions.

The problem reduces to finding a function of the temperature field that satisfies the stationary value of the functional Φ written for the structure under study

$$\delta\Phi[T(x,y,z)] = 0.$$

Let us consider the numerical solution of the presented equations, describing the process of heat conduction, within the framework of constructing a computational experiment for the target.

3 Numerical Solution of the Heat Conduction Problem by the Finite Element Method Using the Splitting Scheme

To construct an economical multidimensional difference scheme for numerical solution of the thermal conductivity equation for MCE, we apply the componentwise splitting method with respect to coordinates by means of local splitting along two directions-thickness (one-dimensional problem) and plane (two-dimensional problem) [4].

3.1 Splitting Scheme

Let us write the differential equation of heat conduction in the form

$$c\rho\frac{\partial T}{\partial t} = \sum_{i=1}^{3} A_i T + f(x,t), \qquad x_i = \{x, y, z\},$$

where A – is a differential linear operator.

We introduce the intermediate time discretization layer and denote the solution on this layer $T_i^p (i = z, xy \text{ or } rz)$. Approximating the recorded differential equation by a purely implicit difference scheme, we obtain for the indicated directions:

$$\begin{aligned}
&\tfrac{1}{\Delta t}(T_z^p - T^k) = \Lambda_z T_z^p + \varphi_z,\\
&\tfrac{1}{\Delta t}(T_{xy}^p - T_z^p) = \Lambda_{xy} T_{xy}^p + \varphi_{xy},\\
&T^{k+1} = T_{xy}^p,
\end{aligned}$$

with boundary conditions

$$q_{Sg} + \alpha_g (T_g - T_i^p) = -\lambda \mathrm{grad} T_i^p, \qquad \frac{1}{R_K}(T_K^p - T_i^p) = -\lambda \mathrm{grad} T_i^p,$$

where Λ_i – are the difference operators in the directions z and xy, Δt – is the step of time discretization.

As the difference operators Λ_i when using the finite element method, in contrast to the finite difference method, using expressions of the type

$$\Lambda_z T_i = \frac{1}{\Delta z^2}(T_{i+1} - 2T_i + T_{i-1}),$$

$$\Lambda_{xy} T_{ij} = \frac{1}{\Delta x^2}(T_{i+1,j} - 2T_{ij} + T_{i-1,j}) + \frac{1}{\Delta y^2}(T_{i,j+1} - 2T_{ij} + T_{i,j-1})$$

We use the corresponding expression of the FEM

$$\Lambda_z T_z = [\Lambda_z]\{T\},$$
$$\Lambda_{xy} T_{xy} = [\Lambda_{xy}]\{T\},$$

where $[\Lambda_z]$, $[\Lambda_{xy}]$ are the thermal conductivity matrices of the construction in one-dimensional and two-dimensional problems; $\{T\}$ is a vector of nodal temperatures.

In accordance with this, for the written heat equation, the functionals that need to be minimized can be represented in the form:

$$\Phi_z[T_z] = \int_V \left[\frac{1}{2}\lambda^*\left(\frac{\partial T_z}{\partial z}\right)^2 + q^*_{Cz}T\right] dV -$$
$$- \int_S \left[\alpha^*_z\left(T_c - \frac{1}{2}T\right)T + q_{Sgz}T\right] dS - \int_{S_k} \left[\frac{1}{R_z}\left(T_K - \frac{1}{2}T\right)T\right] dS_k;$$

$$\Phi_{xy}[T_{xy}] = \int_V \left[\frac{1}{2}\lambda^*\left(\frac{\partial T_{xy}}{\partial x}\right)^2 + \frac{1}{2}\lambda^*\left(\frac{\partial T_{xy}}{\partial y}\right)^2 + q^*_{Cxy}T\right] dV -$$
$$- \int_S \left[\alpha^*_{xy}\left(T_c - \frac{1}{2}T\right)T + q_{Sgxy}T\right] dS - \int_{S_k} \left[\frac{1}{R_{kxy}}\left(T_K - \frac{1}{2}T\right)T\right] dS_k.$$

Using the relations introduced, the resolving FEM equations for the temperature field in both i-directions of the splitting will take the form, which coincides in form:

$$[\Lambda_i]\{T_i\} = -[C_i]\left\{\frac{\partial T_i^p}{\partial t}\right\} + \{Q_{Li}\} + \{Q_{\alpha i}\} + \{Q_{Ki}\},$$

where $\{Q\}$ are the heat flux vectors of the whole region - radiant, convective and contact; $[C]$ is the matrix of the heat capacity of the entire region.

Thus, the three-dimensional problem is reduced to a sequence of solutions at each step in time, one-dimensional (along the z axis) and two-dimensional (in the xy plane). Let's write briefly the basic formulas of discrete models for one-dimensional and two-dimensional problems, using resolving FEM relationships for the three-dimensional problem.

3.2 Numerical Solution of the One-Dimensional Problem

To calculate the temperature distribution over the thickness, we will use one-dimensional finite elements with a linear temperature distribution function, relating the specific heat of the element to the nodal points. The temperature at any point of such a finite element is determined through the nodal temperatures according to a linear law

$$T = n_1 T_1 + n_2 T_2;$$

where n_i – are the coefficients of the form of the one-dimensional finite element, determined from the dependences: $n_1 = (z_2 - z)/z_{21}$; $n_2 = (z - z_1)/z_{21}$; $z_{21} = (z_2 - z_1)$ – thickness of the element.

The heat flux absorbed during the heating of the element, due to the heat capacity of its material, is expressed through nodal heat fluxes:

$$q_c = \int_V c\frac{\partial T}{\partial t}dv = \int_V c\left[\sum_{i=1}^{2} n_i \frac{\partial T_i}{\partial t}\right]dv = \sum_{i=1}^{2}\frac{\partial T_i}{\partial t}\int_V cn_i dv = \sum_{i=1}^{2} q_{ci},$$

$$q_{ci} = \frac{\partial T_i}{\partial t}C_i, \qquad C_i = \int_V c\,n_i dv = \frac{1}{2}cS_t z_{21},$$

where $V = z_{21}\,S_t$ – the volume of the finite element, and S_t – is the cross-sectional area of the one-dimensional finite element.

Surface heat fluxes are expressed through concentrated heat fluxes at the nodes of the i-th FE: $q_{li}S_T$ – nodal heat fluxes of radiant heat transfer; $q_{\alpha i} = (T_{Ci} - T_i)\alpha S_T$ – nodal heat fluxes of convective heat transfer; $q_{Ki} = (T_{Ki} - T_i)/R_K \cdot S_T$ – nodal heat fluxes in the contact zone.

As a result of minimizing the initial functional in the class of linear temperature functions with point sources by the node quantities T_1 and T_2, we obtain the following expressions for the heat conduction matrix of a one-dimensional finite element:

$$[\lambda] = \frac{\lambda_z}{z_{21}}S_t\begin{bmatrix} 1 & -1 \\ -1 & 1 \end{bmatrix}$$

The relations obtained make it possible to form global matrices of thermal conductivity, heat capacity, heat flows of a one-dimensional heat problem, and the corresponding system of algebraic equations.

3.3 The Numerical Solution of the Two-Dimensional Problem

Introducing in each triangular element a linear temperature distribution function $T = N_1 T_1 + N_2 T_2 + N_3 T_3$, and relating the specific heat of the element to the nodal points, we obtain:

$$q_c = \int\limits_v c\frac{\partial T}{\partial t}dv = \int\limits_v c\left[\sum_{i=1}^{3} N_i\frac{\partial T_i}{\partial t}\right]dv = \sum_{i=1}^{3}\frac{\partial T_i}{\partial t}\int\limits_v cN_i dv = \sum_{i=1}^{3} q_{ci},$$
$$q_{ci} = \frac{\partial T_i}{\partial t}C_i, \quad C_i = \int\limits_v cN_i dv,$$

where q_c – is the heat flux absorbed when the element is heated due to the heat capacity of its material; N_i – coefficients of the shape of the triangular finite element, determined from the dependence

$$N_i = \left[\, x_{jk}(y - y_k) - y_{jk}(x - x_k)\right]/(2S_t)$$

S_t – area of a triangle; the indices i, j, k = 1, 2, 3 - are determined by circular substitution; $v = S_t \cdot z_{21}$ – is the volume of the triangular finite element, $S_t = (y_{21} \cdot x_{32} - y_{32} \cdot x_{21})/2$ – the area of the triangle.

Minimizing for the triangle the initial functional in the class of linear temperature functions with point sources in terms of the node quantities T_1, T_2 and T_3, we obtain the following expressions for the heat conduction matrix of a one-dimensional finite element:

$$[\lambda] = \frac{1}{4S_t}\begin{bmatrix} (\lambda_y x_{23}x_{23} + \lambda_x y_{23}y_{23}); & (\lambda_y x_{31}x_{23} + \lambda_x y_{23}y_{23}); & (\lambda_y x_{12}x_{23} + \lambda_x y_{12}y_{23}); \\ (\lambda_y x_{23}x_{31} + \lambda_x y_{23}y_{31}); & (\lambda_y x_{31}x_{31} + \lambda_x y_{23}y_{31}); & (\lambda_y x_{12}x_{31} + \lambda_x y_{12}y_{31}); \\ (\lambda_y x_{23}x_{12} + \lambda_x y_{23}y_{12}); & (\lambda_y x_{31}x_{12} + \lambda_x y_{23}y_{12}); & (\lambda_y x_{12}x_{12} + \lambda_x y_{12}y_{12}); \end{bmatrix}.$$

The presented relations make it possible to form global matrices of heat conductivity, heat capacity, heat flows of a two-dimensional thermal problem, and the corresponding system of algebraic equations.

4 Practical Accuracy of Numerical Mode

To estimate the accuracy of the relations considered and the calculation algorithm, the results of solving a number of test problems are given below. In this case, comparison of the results obtained by different methods with different degree of discretization is given. The effect of the degree of discretization of the investigated region by a finite element mesh on the accuracy of calculations is shown. The accuracy of the results obtained in solving heat conduction problems by algorithms of various dimensions is analyzed [6, 7].

In Fig. 1. the results of the numerical solution of the problem of one-dimensional heat propagation are shown (by the example of the thermal state of a rod, an infinite plate, or a cube heated on one of the surfaces) by a gas. The temperature of the hot gas T_g was set at 1000 °C, and the heat transfer coefficient was $\alpha = 4000$ W/ m^2 °C. The initial temperature T_0 and the gas temperature of the other surface were set equal to 0 °C for the same coefficient of heat transfer. Conditions of the third kind were used as boundary conditions on both boundaries. The question of the minimum degree of discretization of the structure in the direction of heat propagation necessary to provide a satisfactory error was investigated. In this case, the calculated values of the temperatures obtained by the FEM algorithms are practically the same, so the figures give unified graphs for the results of the FEM. These results are compared with the results of solving the same problem by the method of finite differences.

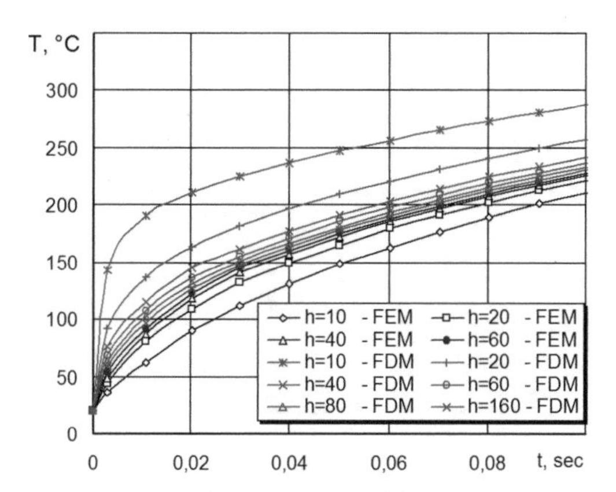

Fig. 1. Change in time of temperature of the heated surface in case of different level of sampling (h - number of steps of integration).

The analysis of the presented graphs shows that already at a discretization degree with respect to the spatial coordinate h > 20 elements, a satisfactory error in the solution of the FEM E is achieved, and at h > 40 the solution of the FEM differs slightly from the finite difference method (FDM) solution performed at a high sampling rate (h = 160).

In Figs. 2, 3 and 4 shows the results of calculations of the three-dimensional thermal state of the body by the finite element method using the splitting scheme and the finite difference method. As a test problem, we considered the problem of

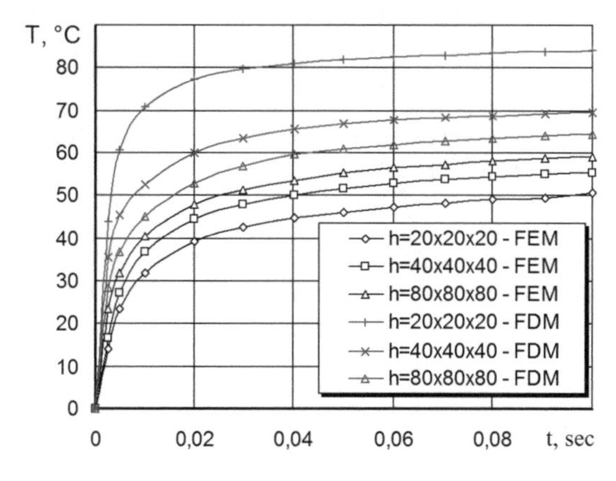

Fig. 2. Change of temperature of the heated surface in time (the three-dimensional task).

calculating a cube whose dimensions allow us to regard it as a semi-infinite space. The cube is heated in the center of one of its faces. The heating zone is a rectangular spot of a fixed size.

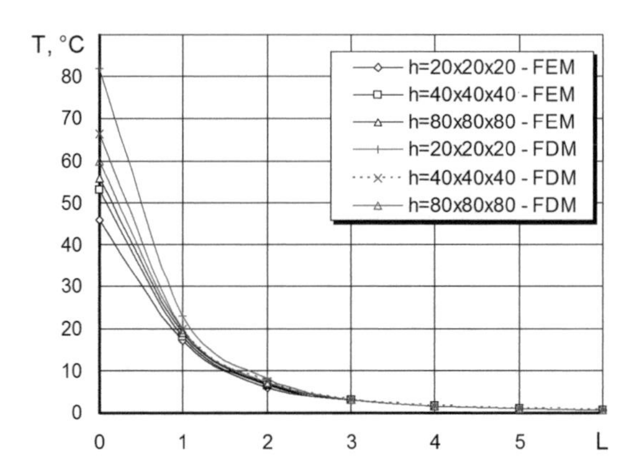

Fig. 3. Distribution of temperature on a normal to the heated surface at heating spot center in time point of t = 0,05 of second.

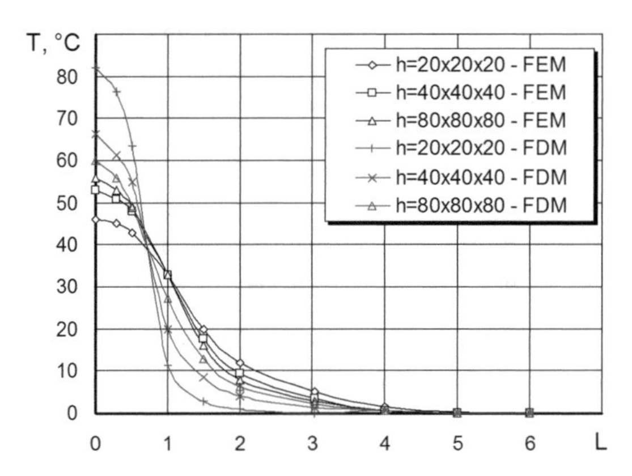

Fig. 4. Distribution of temperature on the heated surface in time point of t = 0,05 of second (the origin of coordinates corresponds to heating center).

The resulted solutions of test problems testify about an opportunity of a choice of a demanded degree of digitization for achievement of required accuracy of modeling of processes of non-stationary thermal conductivity.

5 Practical Accuracy of Numerical Mode

In the computer simulation of the process of heating a target by a moving source of heat produced by the action of a laser beam during vertical scanning, its thermal effect on the target surface was investigated. The following characteristics were specified: thermophysical properties of materials; Initial conditions: ambient temperature, power density of laser radiation, surface blowing speed. The simulation results are presented in the form of color patterns of isotherms (Fig. 5a) and in the form of isolines (Fig. 5b) for different instants of time.

The results of the calculation make it possible to determine the required sweep speed, the power of the source, and the intensity of the blow to obtain a sufficiently stable thermal picture on the target.

This article was written within the framework of project 2.3121/ГЗ "Parallel semi-Markov processes in mobile robot control systems".

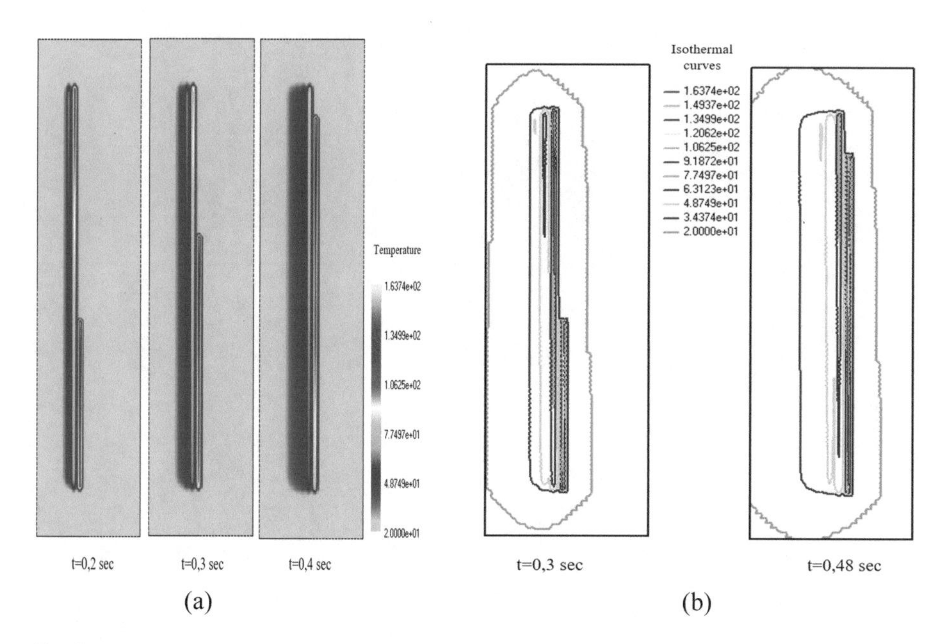

Fig. 5. Nonstationary temperature field: (a) in different time points (the third, fourth and fifth vertical steps of scanning of a ray); (b) in the form of isolines in different timepoints (the fourth and sixth vertical steps of scanning of a ray).

References

1. Shuja, S.Z., Yilbas, B.S., Shazli, S.Z.: Laser repetitive pulse heating influence of pulse duty. Heat Mass Transf. **43**, 949–955 (2007)
2. Yilbas, B.S.: Analytical solution for time unsteady laser pulse heating of semi-infinite solid. Int. J. Mech. Sci. **39**(6), 671–682 (1997)

3. Zienkiewiez, O.C.: The Finite Element Method in Engineering Science, p. 541. McGraw-Hill, London (1971)
4. Segerlind, L.: Applied Finite Element Analysis, p. 392. Wiley, New York (1976)
5. Fornberg, B.: Calculation of weights in finite difference formulas. SIAM Rev. **40**, 685–691 (1998)
6. Akimenko, T., Dunaev, V., Gorbunova, O.: Simulation of surface heating process with laser. In: NAA 2016, Sixth Conference on Numerical Analysis and Applications, Bulgaria, Lozenetz, 16–21 June, pp. 150–157 (2016)
7. Akimenko, T. Dunaev, V., Larkin, E.: Computer simulation of the surface heating process by the movable laser. In: Proceedings of 5th International Workshop on Mathematical Models and their Applications Krasnoyarsk, Russia, 7–9 November 2016, pp. 11–17 (2016)

Mathematical Model of Heating the Surface of the Laser Beam

Tatiana A. Akimenko(✉)

Tula State University, 92 "Lenina" prospect, Tula 300012, Russia
tantan72@mail.ru

Abstract. It was created a model heating of a target surface in the air by a laser beam. The system of equations describing the process of heating-cooling the target surface is solved for an axisymmetric formulation of the problem and in the case of dynamically changing the energy of the laser beam at constant parameters of air flow. The developed method allows a computer simulation of the heating surface of the target point heat source and investigate the dynamics of heating the target surface in rectilinear uniform motion of the beam at a given speed.

Keywords: Heating · Target · Target plane · Thermal problem
Finite element method · Discrete model

1 Introduction

For a thermal image of the object target surface is heated by a movable laser beam.

To determine the parameters of the laser beam (power, duty cycle, pulse parameters) to solve the problem of non-stationary thermal condition of the surface of the target under the action of a heat source. This is considered two options for setting objectives.

Total heat equation has the form [4]:

$$c\rho \frac{\partial T}{\partial t} = div(\lambda gradT), \tag{1}$$

where c - is the heat capacity; ρ - the density; T - the temperature; λ - the heat conductivity coefficient.

To determine the necessary power of the beam we consider the problem of calculating the axisymmetric temperature field in the heat for a fixed point of the laser beam operating in the pulsed-periodic mode (the first task).

To determine the duty cycle and pulse width parameters consider the problem of calculating the three-dimensional non-stationary temperature field target in rectilinear uniform motion of the beam at a given speed (the second task).

The mathematical model of thermal processes in the target is based on the equation of non-stationary heat conductivity of parabolic type, assuming the infinity of the velocity of heat propagation in the material.

N. Xiong et al. (Eds.): ISCSC 2018, AISC 877, pp. 361–370, 2019.
https://doi.org/10.1007/978-3-030-02116-0_42

2 Mathematical Model of Heating the Surface of the Fixed Laser Beam

In the case of an axisymmetric problem, the process of cooling and heating the target surface is described as follows:

$$c\rho\frac{\partial T}{\partial t} = \frac{1}{r}\frac{\partial}{\partial r}\left(r\lambda_r\frac{\partial T}{\partial r}\right) + \frac{\partial}{\partial z}\left(\lambda_z\frac{\partial T}{\partial z}\right), \qquad \forall r, z \in V, \; t > 0, \tag{2}$$

$$q_l = -\lambda_r\frac{\partial T}{\partial r}, \qquad \forall r, z \in S, \; t > 0 \tag{3}$$

$$\alpha(T_g - T) = -\lambda_r\frac{\partial T}{\partial r}, \quad \forall r, z \in S, \; t > 0, \tag{4}$$

$$T = T_0, \forall r, z \in V, t = 0, \tag{5}$$

where (2) - is the differential equation of the heat conductivity; (3) - the boundary conditions on the laser hot spot surface S; (4) - the boundary conditions of heat exchange with ambience on the surface Sa; (5) - the initial conditions; T - the temperature; t - the time; r, z - are radial and axial coordinates; λ_r - the heat conductivity coefficient in the isotropy plane in the direction of axis r; λ_z - the heat conductivity coefficient in the direction of axis z, V - the heating surface volume; α - the heat exchange coefficient; T_g - the ambient temperature; S_a - the target surface, in the $\vec{r}$ direction it's an infinite plate, in the $\vec{Z}$ direction thickness of the target is δ; T_0 - the initial temperature [6, 7].

To solve this problem, the finite element method [4, 5] is used, the approximate variational formulation of which for the differential equation of thermal conductivity written above uses the functional:

$$\Phi[T(r,z)] = \int_V \left\{\frac{1}{2}\left[\lambda_r^*\left(\frac{\partial T}{\partial r}\right)^2 + \lambda_z^*\left(\frac{\partial T}{\partial z}\right)^2\right] + q_C^* T\right\} rdrdz - \int_S \left[q_l T + \alpha\left(T_g - \frac{1}{2}T\right)\right] dS, \tag{6}$$

where $q_C^* = c^*\rho^*\frac{\partial T^*}{\partial t}$ - the heat flow, conditioned by the heat capacity of the material, index "*" corresponds to a temperature distribution in the space fixed in this moment [6, 9].

To solve this problem, it is necessary to find the functions of the temperature field satisfying the stationary value of the functional Φ, which is written for the region under study:

$$\delta\Phi\;[T(r,z,t^*)] = 0. \tag{7}$$

We use a hybrid ring element with a quadrangular cross-section as the final element. This hybrid ring element can be obtained by combining several triangles with a linear temperature distribution function [1–7].

Through the node temperature values, linearly, one can determine the temperature at any point of the final element:

$$T = N_1 T_1 + N_2 T_2 + N_3 T_3, \tag{8}$$

Where N_i - shape coefficient of triangular finite element.

The heat flux absorbed during the heating of an element due to the heat capacity of its material is expressed in terms of nodal heat fluxes:

$$\begin{gathered} q_c = \int\limits_v c\frac{\partial T}{\partial t}dv = \int\limits_v c\left[\sum_{i=1}^{3} N_i \frac{\partial T_i}{\partial t}\right]dv = \sum_{i=1}^{3}\frac{\partial T_i}{\partial t}\int\limits_v cN_i dv = \sum_{i=1}^{3} q_{ci}; \\ q_{ci} = \frac{\partial T}{\partial t}c_i;\ c_i = \int\limits_v cN_i dv, \end{gathered} \tag{9}$$

where c_i - reduced heat capacity of the node; $v = S_T \cdot 2\pi r_T$ - the volume of triangle finite element, r_T - centre of gravity of triangle.

Consider Eq. (6) (initial functional), taking into account Eq. (8) (linear temperature functions) and Eq. (9) (point sources). Since the nodal values uniquely determine the temperature at any point in the study area, this functional can be minimized by these values [5].

Introducing the conditions (8)–(9) into the initial functional and differentiating with respect to the node temperature T_i for the region of the finite element, we obtain:

$$\begin{aligned} \frac{\partial \Phi}{\partial T_1} &= \iint\limits_V \left[\lambda_z \frac{\partial T}{\partial z}\frac{\partial}{\partial T_1}\left(\frac{\partial T}{\partial z}\right) + \lambda_r \frac{\partial T}{\partial r}\frac{\partial}{\partial T_1}\left(\frac{\partial T}{\partial r}\right)\right] rdrdz + q_{c1} \\ &= [(\lambda_z r_{23} r_{23} + \lambda_r z_{23} z_{23})T_1 + (\lambda_z r_{31} r_{23} + \lambda_r z_{31} z_{23})T_2 + (\lambda_z r_{12} r_{23} + \lambda_r z_{12} z_{23})T_3]\frac{1}{4S_T} + c_1\frac{\partial T}{\partial t}; \\ \frac{\partial \Phi}{\partial T_2} &= \iint\limits_V \left[\lambda_z \frac{\partial T}{\partial z}\frac{\partial}{\partial T_2}\left(\frac{\partial T}{\partial z}\right) + \lambda_r \frac{\partial T}{\partial r}\frac{\partial}{\partial T_2}\left(\frac{\partial T}{\partial r}\right)\right] rdrdz + q_{c2} \\ &= [(\lambda_z r_{23} r_{31} + \lambda_r z_{23} z_{31})T_1 + (\lambda_z r_{31} r_{31} + \lambda_r z_{31} z_{31})T_2 + (\lambda_z r_{12} r_{31} + \lambda_r z_{12} z_{31})T_3]\frac{1}{4S_T} + c_2\frac{\partial T}{\partial t}; \\ \frac{\partial \Phi}{\partial T_3} &= \iint\limits_V \left[\lambda_z \frac{\partial T}{\partial z}\frac{\partial}{\partial T_3}\left(\frac{\partial T}{\partial z}\right) + \lambda_r \frac{\partial T}{\partial r}\frac{\partial}{\partial T_3}\left(\frac{\partial T}{\partial r}\right)\right] rdrdz + q_{c3} \\ &= [(\lambda_z r_{23} r_{12} + \lambda_r z_{23} z_{12})T_1 + (\lambda_z r_{12} r_{31} + \lambda_r z_{12} z_{31})T_2 + (\lambda_z r_{12} r_{12} + \lambda_r z_{12} z_{12})T_3]\frac{1}{4S_T} + c_3\frac{\partial T}{\partial t}. \end{aligned} \tag{10}$$

Combining the Eqs. (10) obtained, we write the result in the form:

$$\left[\frac{\partial \boldsymbol{\Phi}}{\partial T}\right] = \boldsymbol{\lambda T} + c^T \frac{\partial \boldsymbol{T}}{\partial t}, \tag{11}$$

where $\boldsymbol{\lambda}$ - is heat conduction matrix of the finite element; c^T - is heat capacity matrix of the finite element; $r_{ij} = r_i - r_j$; $\boldsymbol{T}$ - is temperature matrix of all nodal points of the body:

$$\lambda = \frac{1}{4S_T}\begin{bmatrix} (\lambda_r z_{23} z_{23} + \lambda_z r_{23} r_{23}); & (\lambda_r z_{31} z_{23} + \lambda_z r_{23} r_{23}); & (\lambda_r z_{12} z_{23} + \lambda_z r_{12} r_{23}); \\ (\lambda_r z_{23} z_{31} + \lambda_z r_{23} r_{31}); & (\lambda_r z_{31} z_{31} + \lambda_z r_{23} r_{31}); & (\lambda_r z_{12} z_{31} + \lambda_z r_{12} r_{31}); \\ (\lambda_r z_{23} z_{12} + \lambda_z r_{23} r_{12}); & (\lambda_r z_{31} z_{12} + \lambda_z r_{23} r_{12}); & (\lambda_r z_{12} z_{12} + \lambda_z r_{12} r_{12}); \end{bmatrix},$$
$$\boldsymbol{c^T} = \begin{Bmatrix} c_1 \\ c_2 \\ c_3 \end{Bmatrix}, \boldsymbol{T} = \begin{bmatrix} T_1 \\ T_2 \\ T_3 \end{bmatrix}. \tag{12}$$

Combinations of all the derivatives (10) over all finite elements into which the investigated region is discretized and equating them to zero, we obtain the final equations of the process of minimizing the initial functional with respect to the temperature at the nodal points:

$$\sum_{l=1}^{n} \sum_{i=1}^{3} \frac{\partial \boldsymbol{\Phi}}{\partial \mathrm{T}} = [0], \tag{13}$$

where n - the number of finite elements.

3 Mathematical Model of Surface Heating by a Moving Laser Beam

The heating-cooling process of the target surface is represented by the following system of equations:

$$c\rho \frac{\partial T}{\partial t} = \frac{\partial}{\partial x}\left(\lambda_{\mathrm{x}} \frac{\partial T_{\mathrm{i}}}{\partial y}\right) + \frac{\partial}{\partial z}\left(\lambda_{\mathrm{y}} \frac{\partial T}{\partial y}\right) + \frac{\partial}{\partial z}\left(\lambda_{\mathrm{z}} \frac{\partial T}{\partial z}\right), \quad \forall x, y, z \in V, \ t > 0, \tag{14}$$

$$q_l = -\lambda_z \frac{\partial T}{\partial z}, \quad \forall x, y \in S, \ t > 0, \tag{15}$$

$$\alpha(T_g - T) = -\lambda_z \frac{\partial T}{\partial z}, \quad \forall x, y \in S, \ t > 0, \tag{16}$$

$$T(x, y, z)|_{t=0} = T_0 = \boldsymbol{const}, \ \forall x, y, z \in V, \tag{17}$$

where (14) – is the differential equation of the heat conductivity; (15) – the boundary conditions on the laser hot spot surface S are of follows; (16) – the boundary conditions of heat exchange with ambience on the surface S are the next; (17) – the initial conditions for $t = 0$; T - the target surface temperature; t - the time; x, y, z - are spatial coordinates; $\lambda_x, \lambda_y, \lambda_z$ - are heat conductivity coefficients in the direction of axis x, y and z; V - target volume; $q_l = f(x,y,t,e)$ - the heat flux from the laser beam; α - the heat exchange coefficient; T_g - ambient temperature; S - target surface; T_0 - initial temperature [6, 7].

We use the finite element method for solving this problem [4, 5]. The definition of the stationary value of the following functional is found by the formula:

$$\Phi[T(r,z)] = \int_X \left\{ \frac{1}{2}\left[\lambda_x^* \left(\frac{\partial T}{\partial r}\right)^2 + \lambda_y^* \left(\frac{\partial T}{\partial y}\right)^2 + \lambda_z^* \left(\frac{\partial T}{\partial z}\right)^2 \right] + q_C^* T \right\} dxdy - \int_S \left[q_l T + \alpha \left(T_c - \frac{1}{2} T \right) T \right] dS, \quad (18)$$

where $q_C^* = c^* \rho^* \frac{\partial T^*}{\partial t}$ - the heat flux, conditioned by the heat capacity of the material, index "*" corresponds to a temperature distribution in the space fixed at the moment [7].

The definition of change of the temperature distribution over the volume of the target as a function of time:

$$\delta \Phi[T(x, y, z, t^*)] = 0. \quad (19)$$

A cubic hybrid element is used as a final element. It is formed by the union of several tetrahedra. The temperature distribution function is a linear function, and the heat capacity of the element is concentrated at the nodes [4, 5, 7] (Fig. 1):

Linear values of nodal temperatures determine the temperature at any point of the finite element:

$$T(x, y, z) = (N_1 T_1 + N_2 T_2 + N_3 T_3 + N_4 T_4), \quad (20)$$

where N_l - shape coefficient of tetrahedral finite element; T_l - the temperature of the nodal points of a finite element; l = 1, 2, 3, 4.

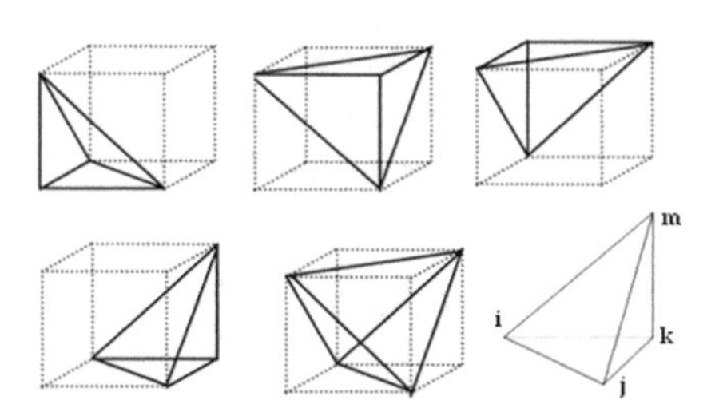

Fig. 1. Scheme of the finite element.

The linear function of approximating the temperature (20) in this element is written in the form:

$$T(x,y,z) = [1,\ x,\ y,\ z]\begin{bmatrix}\alpha_1\\ \alpha_2\\ \alpha_3\\ \alpha_4\end{bmatrix}; \tag{21}$$

where α_1, α_2, α_3, α_4 - are the coefficients of the approximating function.

The vector of the nodal temperature parameters in the tetrahedron will have the form:

$$\boldsymbol{T} = \begin{bmatrix}T_1\\ T_2\\ T_3\\ T_4\end{bmatrix} = \begin{bmatrix}1 & x_i & y_i & z_i\\ 1 & x_j & y_j & z_j\\ 1 & x_k & y_k & z_k\\ 1 & x_m & y_m & z_m\end{bmatrix}\begin{bmatrix}\alpha_1\\ \alpha_2\\ \alpha_3\\ \alpha_4\end{bmatrix} = \boldsymbol{C\alpha}.$$

Inversion of matrix $\boldsymbol{C}$ permits to determine the coefficients of the tetrahedron N_i:

$$N_i = \frac{1}{6V_T}\sum_{i=1}^{4}(a_i + b_i x + c_i y + d_i z);\ \frac{1}{6V_T}\begin{bmatrix}a_1 & a_2 & a_3 & a_4\\ b_1 & b_2 & b_3 & b_4\\ c_1 & c_2 & c_3 & c_4\\ d_1 & d_2 & d_3 & d_4\end{bmatrix} = \boldsymbol{C}^{-1};\ V_T = \frac{1}{6}\det\boldsymbol{C},$$

where V_T - volume of the tetrahedron; x_l, y_l, z_l - coordinates of the nodal points; indexes a_l, b_l, c_l, d_l - the coefficients of the approximating function of temperature in the amount of finite element; $l \in \{i, j, k, m\}$; i, j, k, m =1, 2, 3, 4 are obtained by cyclic permutation in the sequence [7].

The heat flux absorbed during the heating of an element due to the heat capacity of its material is expressed in terms of nodal heat fluxes:

$$\begin{aligned} \mathrm{q_c} &= \int_V \mathrm{c}\frac{\partial \mathrm{T}}{\partial \mathrm{t}}\mathrm{dv} = \int_V \mathrm{c}\left[\sum_{\mathrm{l}=1}^{4}\mathrm{N_l}\frac{\partial \mathrm{T}_l}{\partial \mathrm{t}}\right]\mathrm{dv} = \sum_{\mathrm{l}=1}^{4}\frac{\partial \mathrm{T_l}}{\partial \mathrm{t}}\int_v \mathrm{cN_l dv} \\ &= \sum_{\mathrm{l}=1}^{4}\mathrm{q_{cl}},\quad \mathrm{q_{cl}} = \frac{\partial \mathrm{T_l}}{\partial \mathrm{t}}\mathrm{C_l},\quad \mathrm{C_l} = \int_V \mathrm{cN_l dv}, \end{aligned} \tag{22}$$

where c - the specific heat capacity of the target material; v - the volume of finite element; C_l - reduced heat capacity of the node.

Summation over the nodal points of the boundary side of the finite element is carried out. In this case, in the form of concentrated nodal values, the surface heat flux on the boundary surfaces is represented:

$$q_l = \int_S q_l\, dS_l + \int_s \alpha(T_g - T)dS_g = \sum_l q_l + \sum_l q_{gl} - \sum_l q_{\alpha l}. \tag{23}$$

Since the nodal values uniquely determine the temperature at any point in the study region, it is possible to minimize the initial functional in the class of linear temperature functions (22) [4].

We introduce the conditions (20)–(23) into the initial functional (18) and differentiate with respect to the node temperatures of the tetrahedron T_i *(i = 1,2,3,4)* for the region of the finite element:

$$\frac{\partial \boldsymbol{\Phi}}{\partial T_l} = \iint\limits_{v} \left[\lambda_x \frac{\partial}{\partial x}\frac{\partial}{\partial T_l}\left(\frac{\partial T}{\partial x}\right) + \lambda_y \frac{\partial}{\partial y}\frac{\partial}{\partial T_l}\left(\frac{\partial T}{\partial y}\right) + \lambda_z \frac{\partial}{\partial z}\frac{\partial}{\partial T_l}\left(\frac{\partial T}{\partial z}\right) \right] dxdydz + q_{cl}. \quad (24)$$

Combining the equations obtained, we write the result in the form:

$$\left[\frac{\partial \boldsymbol{\Phi}}{\partial T_l}\right] = \boldsymbol{\lambda T} + {}^4\boldsymbol{C}\frac{\partial \boldsymbol{T}}{\partial t}, \quad (25)$$

where $\boldsymbol{\lambda}$ - is the heat conduction matrix of the finite elements; ${}^4\boldsymbol{C}$ - is the heat capacity diagonal matrix of size 4×4; $\boldsymbol{T}$ - vector of temperature in all nodal points [4, 7].

The resulting equation for the process of minimizing the initial functional with respect to temperature at the node points is obtained by combining all the derivatives (25) over all finite elements into which the investigated region is discretized and equating them to zero.

When the corresponding members of the thermal conductivity matrices, heat capacity and heat fluxes of the finite elements are added, thermal conductivity matrices, heat capacity matrices, and thermal node flow vectors of the entire structure can be obtained. Then for the case under consideration the resolving equation of the finite element method takes the form:

$$\boldsymbol{\Lambda T} = -{}^n\boldsymbol{C}\left[\frac{\partial T}{\partial t}\right] + \boldsymbol{Q}_\alpha + \boldsymbol{Q}_L, \quad (26)$$

where $\boldsymbol{\Lambda}$, ${}^n\mathbf{C}$ - is the global heat conductivity matrix and heat capacity matrix correspondingly of size $n \times n$; $\boldsymbol{Q}_L$, $\boldsymbol{Q}_\alpha$ - are the global vectors of the nodal heat flow of radiant and convective heat exchange correspondingly;

$$\boldsymbol{\Lambda} = \begin{bmatrix} \Lambda_{11} & \Lambda_{12} & \dots & \Lambda_{1n} \\ \Lambda_{21} & \Lambda_{22} & \dots & \Lambda_{2n} \\ \dots & \dots & \dots & \dots \\ \Lambda_{n1} & \Lambda_{n1} & \dots & \Lambda_{n1} \end{bmatrix}; {}^n\boldsymbol{C} = \begin{bmatrix} C_{11} & 0 & \dots & 0 \\ 0 & C_{22} & \dots & 0 \\ \dots & \dots & \dots & \dots \\ 0 & 0 & \dots & C_{nn} \end{bmatrix}; \boldsymbol{Q}_L = \begin{bmatrix} Q_{L1} \\ Q_{L2} \\ \dots \\ Q_{Ln} \end{bmatrix}; \boldsymbol{Q}_\alpha = \begin{bmatrix} Q_{\alpha 1} \\ Q_{\alpha 2} \\ \dots \\ Q_{\alpha n} \end{bmatrix}. \quad (27)$$

4 Numerical Calculations

An implicit difference scheme of the solution for numerical calculations of Eq. (26) is used. In order to exclude fluctuations in the numerical values of the temperature for pronounced heating transitions and to obtain a design scheme for the transition temperature field suitable for constructing the method of numerical investigation of the thermal state of the target under investigation, we use a finite element with a concentrated heat capacity (Fig. 1).

Using the finite-difference expression of the time derivative, we get

$$\Lambda \boldsymbol{T}^{p+1} = -\boldsymbol{C}\left(\boldsymbol{T}^{p+1} - \boldsymbol{T}^{p}\right)\frac{1}{\Delta t} + \boldsymbol{A}\left(\boldsymbol{T}_{g}^{p+1} - \boldsymbol{T}^{p+1}\right) + \boldsymbol{Q}_{\alpha}^{p+1} + \boldsymbol{Q}_{L}^{p+1}, \tag{28}$$

where p - is the index corresponding to the time; $\boldsymbol{T}^{p+1}$, $\boldsymbol{T}^{p}$ - are vectors of nodal temperatures in current and subsequent time; $\boldsymbol{T}_{g}^{p+1}$ - ambient temperature.

To combine all the terms containing unknown quantities on the left-hand side of the equation, we select in the matrix $\boldsymbol{Q}_{\alpha}$ the unknown temperatures of the nodal points of the construction:

$$\boldsymbol{Q}_{\alpha} = \boldsymbol{A}\left(\boldsymbol{T}_{g} - \boldsymbol{T}\right), \tag{29}$$

where $\boldsymbol{A}$ - is the diagonal matrix of convective heat exchange on surface of target.

Shifting to the left side of the Eq. (31) the terms containing the unknown $\boldsymbol{T}^{p+1}$, one can get:

$$\left(\Lambda + \frac{1}{\Delta t}\boldsymbol{C} + \boldsymbol{A}\right)\boldsymbol{T}^{p+1} = \frac{1}{\Delta t}\boldsymbol{C}\boldsymbol{T}^{p} + \boldsymbol{A}\left(\boldsymbol{T}_{g}^{p+1} - \boldsymbol{T}^{p+1}\right) + \boldsymbol{Q}_{L}^{p+1}, \tag{30}$$

The vector $\boldsymbol{T}$ of size n at the initial time is determined from the initial conditions. The solution of resulting linear system of algebraic equations is carried out by the method of adjoint gradient [4, 5, 7].

5 Example Calculation

To solve the problem of the non-stationary thermal state of the target surface under the action of a heat source, computer simulations were carried out. For the numerical solution of the problem, the following characteristics of the target are necessary: target material (steel); initial conditions (ambient temperature, laser radiation power density, airflow velocity) [6, 7].

The results are presented as graphs of the heating temperature of the target surface as a function of time of laser emission pulses exposure (Fig. 2).

Having carried out a number of investigations: different operating conditions of a pulsed-periodic laser, different in the time of heating and cooling of the target surface, were chosen, it is possible to determine the optimum operating mode of such a laser.

The results are shown on Fig. 3 as isotherms. Computer experiment was held for target of 10 mm thick, the material has the conductivity 0.5 W/mK, grid of size 300 × 100 × 20 vas used [7].

With the help of a theoretical and computational complex of computer modeling and visualization of heat exchange processes is carried numerical simulation of the process [6, 7].

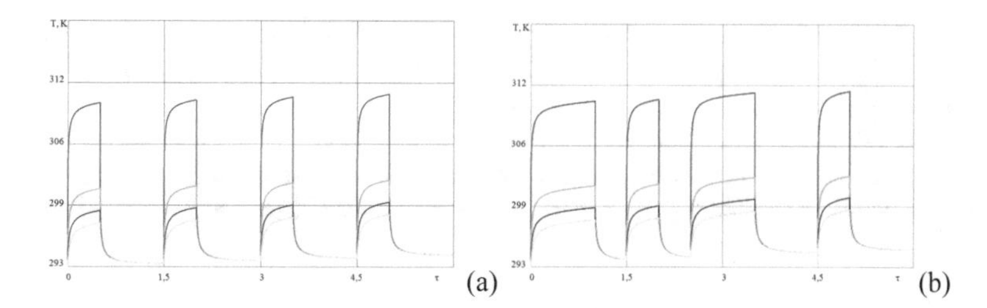

Fig. 2. The results of the computer simulation of the target surface heating by the laser beam: (a) the pulse duration is two times less than the spacing interval $t_1 = 0{,}5t_2$; (b) the pulse duration is selected in random order

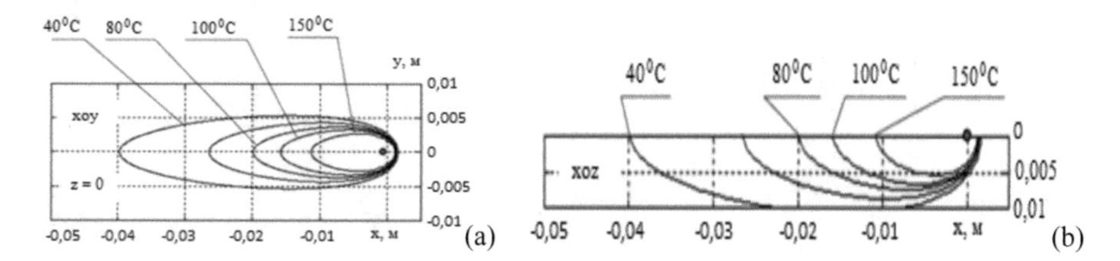

Fig. 3. Isotherms in the plane *x-y* and *x-z* of the plate: (a) target surface; (b) target cross section

The developed method of numerical solution of the task of heating the target surface by a laser beam moving in accordance with a given law and constant parameters of the air flow makes it possible to carry out a computer simulation of the heating-cooling of the target surface by a laser and to investigate the dynamics of heating the surface of a target irradiated by a mobile pulsed-periodic laser.

This article was written within the framework of project 2.3121/Г3 “Parallel semi-Markov processes in mobile robot control systems”.

References

1. V International Workshop on Mathematical Models and their Applications 2016 IOP Publishing IOP Conf. Series: Materials Science and Engineering, vol. 173, 012002 (2017) https://doi.org/10.1088/1757-899x/173/1/012002
2. Numerical Analysis and Its Applications. Springer Nature (2017)
3. Chen, Z., He, C., Zheng, Y., et al.: A novel thermodynamic model and temperature control method of laser soldering systems. Mathematical Problems in Engineering, 2015 (2015)
4. Zienkiewiez, O.C.: The Finite Element Method in Engineering Science, 541 p. McGraw-Hill, London (1971)
5. Segerlind, L.: Applied Finite Element Analysis, 392 p. Wiley, New York/London/Sydney/Toronto (1976)

6. Akimenko, T., Dunaev V., Gorbunova, O.: Simulation of surface heating process with laser. In: NAA 2016 (Sixth Conference on Numerical Analysis and applications), Bulgaria, Lozenetz, 16–21 June, pp. 150–157 (2016)
7. Akimenko, T., Dunaev, V., Larkin, E.: Computer simulation of the surface heating process by the movable laser. In: Proceedings of 5th International Workshop on Mathematical Models and their Applications Krasnoyarsk, Russia, 7–9 November 2016, pp. 11–17 (2016)

General Fractal Dimension Calculation of Surface Profile Curve of Shearing Marks Based on Correlation Integral

Bingcheng Wang[1(✉)] and Chang Jing[2]

[1] Center Laboratory for Forensic Science, Shenzhen University, Shenzhen, China
wbc8631@163.com
[2] Department of Technology, Guangdong Police College, Guangzhou, China
wbc8636@163.com

Abstract. The morphology of shear traces is a form of rough surface. Its complexity determines that it needs to be described in a multi-scale way. Therefore, the multi scale fractal method is introduced in the study, and the calculation principle of the generalized fractal dimension is discussed. For reconstruction system phase space of a series of observation data, the correlation integral is obtained, and a q order correlation integral is defined. In this way, the method of calculating the generalized fractal dimension based on the correlation integral is obtained. The calculation method is compared with other methods to calculate the generalized fractal dimension. Because the two quantities of the delay factor and the embedding dimension are reasonably selected, the calculated values are better reflecting the dynamic characteristics of the system. The content of the study is to deal with the contour curve of the shear mark as a one-dimensional signal, and the fractal dimension of the contour curve of the shear mark surface is calculated by using the generalized fractal dimension based on the correlation integral. The sum of squares of generalized fractal dimension is used as characteristic parameter. The purpose of the study is to explore a new method to provide the basis for the feature extraction and recognition of contour curves.

Keywords: Shearing marks · Multi-scale fractal
Generalized fractal dimension · Correlation integral

1 Introduction

Fractal dimension is an important parameter to describe the microscopic characteristics of the surface. After applying fractal dimension to study surface topography and obtaining more research results, further research found that some rough surfaces formed by machining such as grinding and wear have multi-scale and random multi-scale fractal characteristics, that is, with different fractal parameter values in different scale ranges. Therefore, the single fractal dimension [1] can only represent the whole object as a whole, but this representation is too simplistic and can not be more comprehensive and detailed. Especially for some statistical self similar fractals and non-uniform fractal

N. Xiong et al. (Eds.): ISCSC 2018, AISC 877, pp. 371–377, 2019.
https://doi.org/10.1007/978-3-030-02116-0_43

problems, the characterization of single fractal dimension is not enough to describe its properties. Therefore, it involves the description of multi-scale and multi parameters of fractal set, and makes multi scale fractal spectrum line to get more information from it. Multi-scale fractal concept is established in such a demand.

The shear trace morphology is a form of rough surface morphology, the mechanism of formation is similar to that of machined or worn surfaces. In the process of introducing fractal theory and using fractal dimension as the parameter of the surface topography of shear marks, it is also found that its complexity needs to be described in multiple scales. Therefore, the multi scale fractal method is introduced in the study, and the calculation principle of the generalized fractal dimension is analyzed. For reconstruction system phase space of a series of observation data, the correlation integral is obtained, and a q order correlation integral is defined. In this way, the method of calculating the generalized fractal dimension based on the correlation integral is obtained. When $q = 2$, the calculation formula at this time is the correlation dimension calculation method. The calculation method of generalized fractal dimension based on associated integral is compared with other methods of calculating generalized fractal dimension, such as number box method, fixed radius method, fixed mass method, etc., because the two factors of delay factor and embedding dimension are reasonably selected, the calculated dimensional values can better reflect the dynamic characteristics of the system.

The surface contour curve of shear marks is treated as a one-dimensional signal [2], and the fractal dimension of the contour curve of the shear marks is calculated by using the generalized fractal dimension based on the associated integral. The square sum of generalized fractal dimension is used as the characteristic parameter to realize the feature extraction and recognition of contour curves.

2 Making Sample and Observation Data Collection

On the test bench, Scissors, wire pliers and wire broken clipper were used as the shear tool. The lead wire with a diameter of 6 mm was selected as the cutting object to make three kinds of shear trace samples. The test sample is inspected under a stereo microscope, the samples data is collected which can reflect the machining and using feature of cutting edge of the shearing tool.

A large three-dimensional surface topography instrument is used as a tool for data acquisition of the surface topography of the forceps-scissors trace. The magnification of the object lens is 10 times, and the sampling vertical resolution is 1.0 μm. The data of the pliers - scissors samples were collected to get a trace surface represented by three dimensional data. By using the marking function of the instrument, the contour curve perpendicular to the trace surface is drawn, and the discrete data of the contour curve is obtained. The three contours formed by scissors, wire pliers and wire broken clipper are shown in Fig. 1(a), (b), and (c), respectively.

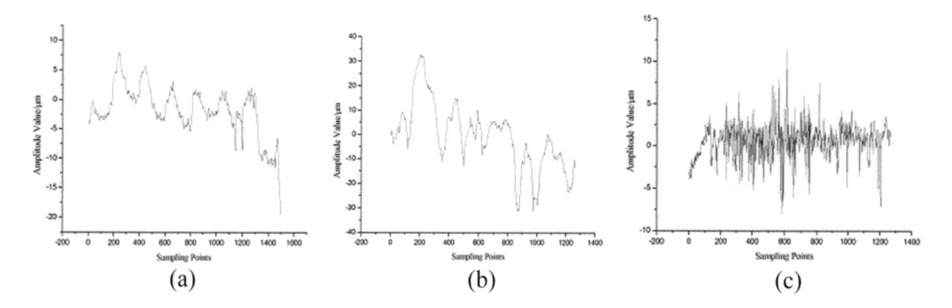

Fig. 1. Three kinds of profile curves

3 Selections of Two Parameters

First, the problem of determining the delay factor is analyzed. There are many ways to determine the delay factor. The method is different and its characteristics are different. The mutual information method [3] is characterized by clear geometric meaning and small amount of calculation. The mutual information method is used to calculate the delay factor τ.

Let $\{x_k\}$ be a set of observations data, and then the mutual information relationship between the $\{x_k\}$ observation variable and the $\{x_{k+\tau}\}$ observation variable is defined as:

$$I(\tau) = \sum_{k=1}^{N} \beta(x_k, x_{k+\tau}) \ln[\beta(x_k, x_{k+\tau})/\beta(x_k)\beta(x_{k+\tau})] \tag{1}$$

$\beta(x_k)$, $\beta(x_{k+\tau})$ and $\beta(x_k, x_{k+\tau})$ in the formula are probability functions. The value of the known variable $\{x_k\}$, the probability function value and the value of the mutual information function $I(\tau)$ can be calculated. The curve of function $I(\tau)$ is plotted, and the value of τ corresponding to the first minimum point on the curve is the delay factor. The delay factor τ of the state space reconstruction of the three contour curves is calculated by using Eq. (1), as shown in Fig. 2.

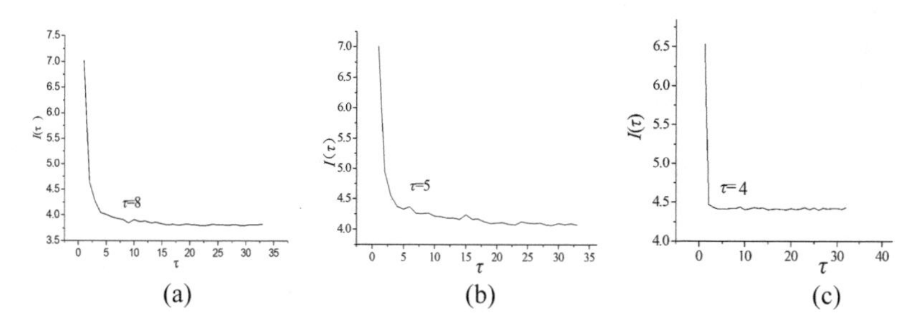

Fig. 2. Delay factor τ of different profile curves

The delay factors τ of the three contour curve state space reconstructions are 8, 5 and 4, respectively, and they are all different.

When the delay parameter is selected, it is now time to analyze the embedded dimension d. There are many ways to determine the embedding dimension. The method is different and its characteristics are different. The characteristics of pseudo nearest neighbor method [4] are clear geometric meaning and small amount of computation. The pseudo-nearest neighbor method is used to determine the embedding dimension, the expression is:

$$E_1(d) = T(d+1)/T(d) \tag{2}$$

Where, $T(d) = \frac{1}{N-d\tau}\sum_{i=1}^{N-d\tau} \lambda(i.d)$, $\lambda(i,d) = \frac{\left\|K_i(d+1)-K_{n(i,d)}(d+1)\right\|}{\left\|K_i(d)-K_{n(i,d)}(d)\right\|}$.

Obviously, $E_1(d)$ is the binary function of variables τ and d. After knowing the variable τ, $E_1(d)$ is a unary function of variable d, that is, function $E_1(d)$ changes only with variable d. Draw the curve of the function $E_1(d)$. When the $E_1(d)$ value does not increase with d, the corresponding d is the minimum embedding dimension.

According to the function expression, the $E_1(d)$ curves of the three observation data are respectively shown in Fig. 3. Using the eye observation function $E_1(d)$ curve, the minimum embedding dimension of the three contour curves (three observation data) are obtained, and the values are 7, 6, and 7, respectively.

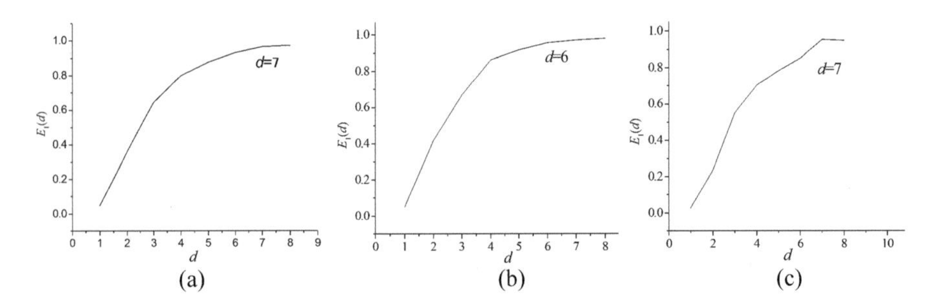

Fig. 3. The minimum embedding dimension of three profile curves

4 Calculation of Generalized Fractal Dimension Based on Correlation Integral

For a nonlinear system, only a set of observations $\{x_k\}$ is often measured. Such a set of observations does not reflect the phase space dimension of the system, and its dynamic form cannot be truly reflected. The dimension of the attractor is generally lower than the dimension of the system phase space. As long as the embedding dimension is high enough, the dynamic form of the original system can be restored in the sense of topology equivalence. In order to show the dimension of his phase space in seemingly one-dimensional observation data, phase space reconstruction technology has been developed.

A set of observation data is $\{x_n\}$ $n = 1, 2, \ldots .N$, and the phase space of the observed data is reconstructed. The reconstruction result is recorded as $X_n(m, \tau) = (x_n, x_{n+1}, \cdots, x_{n+(m-1)\tau})$. Among them, $\tau = K\Delta t$ is the time factor, Δt is the sampling interval of the data, k is any integer, and Δt is the dimension of the reconstructed phase space, $N_m = N - (m-1)\tau$.

Select a reference point X_i randomly from N_m points and calculate the distance from the remaining $N_m - 1$ points to X_i:

$$r_{ij} = d(X_i, X_j) = [\sum_{l=0}^{m-1} (X_{i+l\tau} - X_{j+l\tau})^2]^{\frac{1}{2}} \tag{3}$$

Repeat this process for all $\{x_n\}$, defining q-order correlation integral.

$$C_q(r) = \{\frac{1}{N_m} \sum_{j=1}^{Nm} [\frac{1}{N_m} \sum_{i=1}^{Nm} H(r - r_{ij})]^{q-1}\}^{\frac{1}{q-1}} \tag{4}$$

The r is m dimensional hyper sphere radius and H is Heaviside function.

$$H(r - r_{ij}) = \begin{cases} 1 & (r - r_{ij}) \geq 0 \\ 0 & (r - r_{ij}) < 0 \end{cases} \tag{5}$$

Generalized dimension is obtained by q-order correlation integral:

$$D_q = \lim_{r \to 0} \frac{\ln C_q(r)}{\ln r} \tag{6}$$

It can be seen from Eq. (6) that when q = 2, the calculation formula is the correlation dimension calculation method. Therefore, it can be considered as the extension of the correlation dimension D_2 method based on the correlation integral. Compared with other methods of calculating generalized fractal dimensions [5–9] (such as number box method, fixed radius method, fixed mass method, etc.), the calculated dimensional values are more powerful to reflect the dynamic characteristics of the system. According to formula (6), the generalized fractal dimension of the three contour curves is calculated, as shown in Fig. 4.

As can be seen from Fig. 4, the distribution of generalized dimension and maximum and the minimum fractal dimension are not same. Sensitivity of generalized dimension to profile curve is explained. Therefore, the features of contour curves can be recognized by using generalized dimension.

For some statistical self similar fractals and non-uniform fractal problems, the characterization of single fractal dimension is not enough to describe its properties. Therefore, it involves the description of multi-scale and multi parameters of fractal set, and makes multi scale fractal spectrum line to get more information from it. Therefore, all dimension values in the generalized dimension spectrum are used as characteristic quantities to identify different contour curves, that is: D_1, D_2, D_3 … D_m. In order to

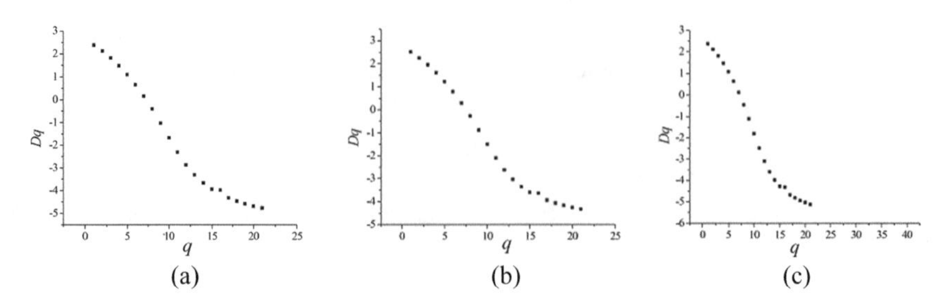

Fig. 4. The generalized dimension of three profile curves

facilitate the application, the sum of squares of each dimension value is defined as the identification characteristic quantity, which is used to distinguish the different contour curves. It is

$$\lambda = \sum D_i^2 \tag{7}$$

By using the formula (7), the characteristic quantity λ is calculated in Table 1.

Table 1. Characteristic quantity of different profile curves

Type	Scissors	Wire cutters	Wire broken clipper
λ	28.3568	25.0689	31.7627

As can be seen from Table 1, the characteristic quantity λ of the shear mark formed by wire broken pliers is large, and the characteristic quantity λ of the shear mark formed by wire cutters is smaller. The three characteristic quantities are obviously different, and the differentiation degree is high, which is convenient for the analysis of trace characteristics.

5 Conclusions

The morphology of shear traces is a form of rough surface. Its complexity determines that it needs to be described in a multi-scale way. Therefore, the introduction of multi scale fractal method is more than necessary. After analyzing the algorithm for calculating the generalized fractal dimension, the generalized fractal dimension based on the associated integral is used to calculate the fractal dimension of the surface contour curve of the shear mark. Compared with other methods, this method first rationally selects two quantities of the delay factor and the embedding dimension. At the same time, it pays attention to the sensitivity of the correlation integral to the calculation of the fractal dimension. The calculated value can reflect the dynamic characteristics of the system. The multi-scale fractal spectrum of different contour curves is calculated by this algorithm, and more information is obtained from it. The squared sum of all dimension values in the fractal spectrum is defined as the fractal dimension eigenvalue, which is used to distinguish different contour curves. That is, multi-scale and multi-parameters are used to describe the surface features of the traces, and the feature quantity is relatively reliable and accurate. A method for quantitative analysis of shear trace surface is provided.

Acknowledgments. This work is supported by the Nature Science Fund of China (NSFC), No.: 61571307.

References

1. Lv, J., Lu, J., Chen, S.: Analysis and Application of Chaotic Time Series. Wuhan University Press, China (2002)
2. Wang, A., Yang, C.: The calculation methods for the fractal characteristics of surface topography. China Mech. Eng. **13**(8), 714–718 (2002). (in Chinese)
3. Fraser, A.M.: Information and entropy in strange attractors. IEEE Trans. Inf. Theory **35**, 245–262 (1989)
4. Cao, L.Y.: Practical method for determining embedding dimension of a scalar time series. Physica D **110**(5), 43–50 (1977)
5. Su, F., Xie, W.X., Dong, J.: Using multi-fractal to analyze time series. J. Data Acquis. Process. **12**(3), 178–181 (1997). (in Chinese)
6. Wang, B.C., Wang, Z.Y., Jing, C.: Study on examination method of striation marks based on fractal theory. Appl. Mech. Mater. **740**, 553–556 (2015)
7. Yuxiu, X., Guo, W., Zhang, J.: Application of multi-fractal theory to coupling fault diagnosis and classification. Mech. Sci. Technol. **25**(3), 363–366 (2006). (in Chinese)
8. Wang, B.C., Jing, C.: Study on surface characteristics of shearing marks based on correlation dimension. In: 10th International Symposium on Computational Intelligence and Design (ISCID), pp. 12–15 (2017)
9. Wang, B.C., Jing, C.: Quantitative inspection of shear mark based on Lyapunov dimension. In: 2017 Chinese Intelligent Systems Conference. Lecture Notes in Electrical Engineering, vol. 460, pp. 267–275 (2018)

A Comparative Study of the Tone Systems of Maguan Dialect, Dai Language and Thai Language

Yan Xu[1,2,4](✉) and Jin Yang[3,5]

[1] Comparative Linguistics EISU, College of Humanities and Communications, Shanghai Normal University, Shanghai, China
494012753@qq.com

[2] The School of International Relations, Yunnan Open University, Kunming, China

[3] Department of Foreign Languages, University of Shanghai for Science and Technology, Shanghai, China

[4] Princeton University, Princeton, NJ 08544, USA

[5] Springer Heidelberg, Tiergartenstr. 17, 69121 Heidelberg, Germany

Abstract. This paper, based on the Four tones and eight tones principle of original Tai language raised by Li Fanggui, conducted an comparative study of Maguan dialect and Thai language with phonetic experiments, which has not been done before on the tones of these two languages.

Keywords: Tone · Maguan dialect · Thai language

1 The Tone System of Kam-Tai Language

There are four tone types in Chinese, ping, shang, qu and ru, each divided again into yin and yang due to whether the initial is voiceless or voiced, So forming the Four tones and eight tones system (Table 1).

Li Fang-kuei claimed that there were four types of tones in the Proto-Tai language: A, B, C and D. The tonal systems of Dai and Thai languages conform to Li's above rules. To facilitate a historical comparative study of Sino-Tibetan language, experts on Kam-Tai language use the four tones eight tones system to describe Kam-Tai language. Since the ancient initial of Zhuang-Dong language has a voiceless-voiced contrastive system and according to the ten-tone classification of Tai language listed by Li Fang-kuei, Tone 1, 3, 5, 7 refer to ancient voiceless initial syllables while Tone 2, 4, 6, 8 refer to ancient voiced initial syllables.

Li's theory is of theoretical and practical significance to languages of Zhuang-Dong language family. Since Zhuang-Dong languages have contrastive voiceless and voiced ancient initials, Tone 1, 3, 5, 7 are used to label ancient voiceless tone types while Tone 2, 4, 6, 8 are used to label ancient voiced tone types. Most languages of Zhuang-Dong language family have over four tones, lax or tense according to pronunciation duration. In terms of Dai and Thai languages, Tone 1, 2, 3, 4, 5, 6 are lax tones while Tone 7, 8

N. Xiong et al. (Eds.): ISCSC 2018, AISC 877, pp. 378–387, 2019.
https://doi.org/10.1007/978-3-030-02116-0_44

Table 1. The four tones and eight tones system

Middle tone	Middle initial	Tone name	Tone type
Level tone	voiceless	*yinping*	1
	voiced	*yangping*	2
Falling-rising tone	voiceless	*yinshang*	3
	voiced	*yangshang*	4
Falling tone	voiceless	*yinqu*	5
	voiced	*yangqu*	6
Entering tone	voiceless	*yinru*	7
	voiced	*yangru*	8

are tense tones or entering tones. Lax tones contain vowel coda or zero coda. Tense tones contain /-p/, /-t/ or /-k/ coda, some a few contain /-ʔ/ coda. The odd Tone 1, 3, 5, 7 form contrast with the even Tone 2, 4, 6, 8, as ancient voiceless initial versus voiced ones.

1. Tones of maguan dialect. According to Li's theory, there are six tone values and ten tone types. Tone 1, 2, 3, 4, 5, 6 are lax tones while Tone 7, 8, 9, 10 are tense tones. The tones of yinshang and yangqu, due to whole voiced initial become voiceless, Tone 6 and Tone 3 merge and are both labelled as Tone 3 with a tone value of 42. Tone 1, due to the initial become voiceless or sub-voiceless, divides into two tones, Tone 1′ valued 22 including initials being unaspirated stops while Tone 1 valued 24 including initials being aspirated stops. The tone value of tense tones, Tone 7, 8, 9, 10 equal respectively with those of lax tones, Tone 3, 4, 5, 1′, namely 42, 35, 21 and 22 (Table 2).

Table 2. Tone system of Maguan dialect

Tone type	Tone name	Coda	Tone value	Example
1	*yinping*	Lax Tone Coda	24	Head: ho^{1}/wolf: sɯk^{1}/back: laŋ^{1}
1'			22	Mud:din$^{1'}$/steel:kaŋ$^{1'}$/month:dɯn$^{1'}$
2	*yangping*		44	Hand: mɯ^{2}/human-being, each (quantifier): kun^{2}/smoke: jɛn^{2}
3	*yinshang*		42	Pit: kɔŋ^{3}/craftsman: tsaŋ^{3}/ground: hai^{3}
4	*yangshang*		35	Water: nam^{4}/little, few: nɔi^{4}/tree, wood, bamboo: mai^{4}
5	*yinqu*		21	Bean; tʰo^{5}/age, year: hai^{5}/chicken: kai^{5}
6	*yangqu*		42	Merge into Tone 3
7	*yinru*	Tense Coda	42	Cabbage:pʰak^{7}/iron:liak^{7}/six: hok^{7}
8	*yangru*		35	Bird: nɔk^{8}/meet: tʰɯp^{8}/ant: mɯt^{8}
9	*yinru*		21	Bone: dok^{9}/shell: kɛp^{9}/blind: bɔt^{9}
10	*yangru*		22	Otter:mak^{10}/leech:tak^{10}/blood:lɯt^{10}

Different tones differ a lot in the tone figure. We hence define those that locate in the upper part of the tone domain as high tones while those in the lower part of the tone domain as low tones. Mainly based on Zhu Xiaonong's normalization principle, we get the tone system of Maguan dialect as follows: all recordings here are from field work. It is conducted in a quiet room by using laptop, Praat software and Mbox external audio card and collar microphone. The tones are drawn by following the undermentioned procedures:

First, note in the table the average pitch of all recorded token samples of each tone type. The unit is Hz. Praat software is used to extract pitch value. The main procedure is as follows: to set up a "temp" folder in hard disk C; put in "extract pitch" procedure and all sound files to be analyzed in "temp" folder; read the sound files with Praat; click "periodicity – to pitch"; choose to set pitch range; highlight pitch appeared; then click "down to PitchTier"; the "PitchTier" file is then write to text file to "temp" folder; then "highlight sound", click "annotate" to "to textgrid", choose "Mary" for upper column "all tier name" and delete the lower two, rename the lower column to "mary" before labelling. Then draw "File" and "write to textgrid" to "temp" folder, highlight file sound and file textgrid. Then click "extract pitch file" in the emerged dialog box and start running the procedure, key in 11 to substitute 10 in the dialog box. Finally in "temp" folder there are four files with same name but different format. Open the pitch data file in excel and it's done.

Second, turn the above data to logarithm and normalize by Lz-score. Calculate the speaker's average normalized μ and standard deviationσ. While calculating average value and standard deviation, not all values in the table are to be used. Those tones that are in the vicinity of 0% and the falling tone near 100% are deleted and those deleted data are not to be calculated in Excel. Therefore we can obtain different tone maps. For example, according to Step 1, we can obtain the following tone data of Dai language and use these data to draw the tone map of Dai language (x axis stands for time with second as unit while y axis stands for pitch with Hz as unit).

According to the pitch and location in the tone map obtained, we hereby conclude it be a 42 tone and analyze, according to the Four Tones Eight Tones system claimed by Mr. Li Fangkuei, this tone type to be Tone 3.

However, the analysis here on judging tones does not include the following situation where physical factors matter. One is the high-level tone has a rising initial. Second is the low-level tone has a falling ending. It is due to the fact that voice initial stance is a 3 tone in a four-tone labelling system. Therefore, the high-level tone has a voice initial of 3 tone and 355 tone. The low-level tone also has a voice initial of 3 tone and 322 tone. On these two occasions, 3 tone initial is a redundant element that doesn't matter in identifying tones and is therefore not necessary to be labelled in a four tones labelling system. Third, the long level tone usually takes a natural attenuation which is due to two reasons, resuming to voice initial 3 tone and natural pitch attenuation caused by nonlinguistic physical factors such as fatigue, idling and effort saving preference. Fourth, consonant initial will influence tone initial, voiceless consonant correspond to a higher initial while voiced consonant correspond to a lower initial.

Maguan dialect Tone 3 (Fig. 1) features that the initial point is high in the tone domain and near Tone 2, the terminal point is in the low of the tone domain and near Tone 1. The tonal contour is protruding and is a high-falling tone. Tone 3, Tone 6 and

Tone 7 have the same tonal value. Tone 7 is a tense tone while Tone 3 and Tone 6 are lax tones.

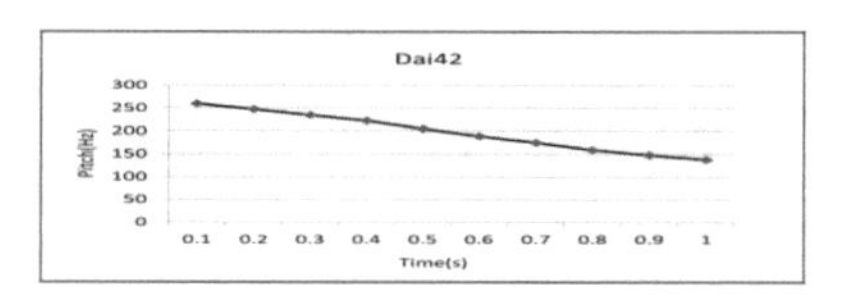

Fig. 1. Dai language (42 tone)

Tone 1 of Maguan dialect (Fig. 2) is a low-rising tone. The initial point is in the low of the tone domain and is near Tone 1. The terminal point is in the high of tone domain and is near Tone 2. The contour curve lies in the middle of the tone domain and takes a concave shape.

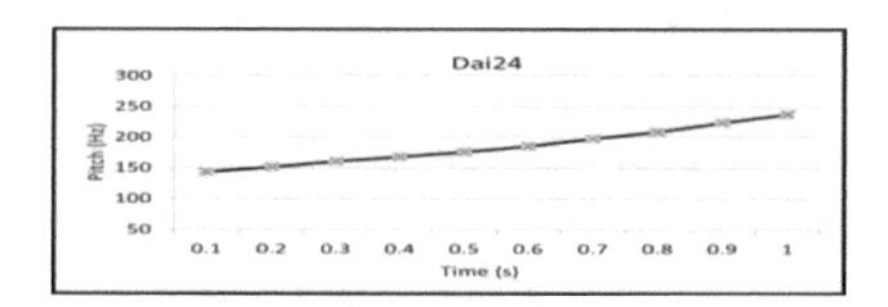

Fig. 2. Dai language (24 tone)

Tone 4 of Maguan dialect (Fig. 3) is a high-rising tone and the tonal contour is in the high of tone domain. The tone curve is protruding.

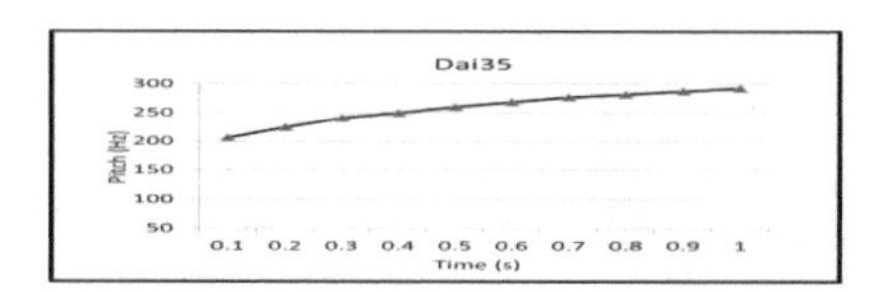

Fig. 3. Dai language (35 tone)

The initial point is in the medium-high of the tone domain, the middle fairly stable and the ending approaching horizontal and is near the top of the tone domain. Tone 8 and Tone 4 have equal tone value. Tone 8 is a tense tone while Tone 4 is a lax tone.

Tone 1 of Maguan Dialect (Fig. 4) is a low-level tone. It is at low of the tone domain, is a little higher than Tone 5 and is near the bottom of the tone domain. It's a semitone away from Tone 5. The initial point is near Tone 5. It's falling more moderate than Tone 5 and takes a tonal contour of a little concaving. The initial point and the terminal one are close. The whole tonal contour is quite stable and is therefore an ideal level tone. The tone value of Tone 10 and Tone 1 is equal. Tone 10 is a tense tone while Tone 1 is a lax tone. They differ only in terms of duration.

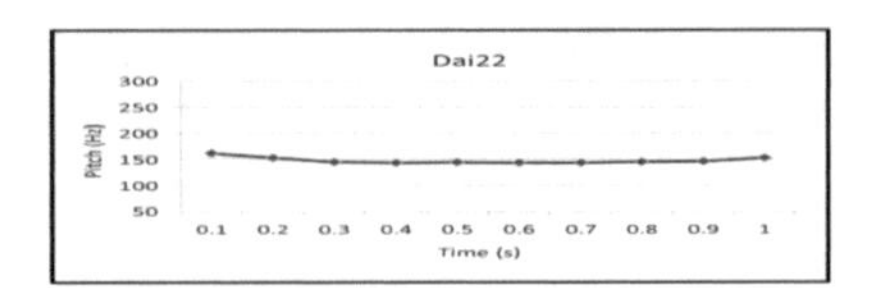

Fig. 4. Dai language (22 tone)

Tone 2 of Maguan Dialect (Fig. 5) is a high-level lax tone. The tonal contour is in the high of the tonal domain. Its actual pitch is four semitone higher than that of Tone 1. The initial and the terminal is quite level and don't differentiate much. The whole tonal contour is quite stable and is an ideal level tone.

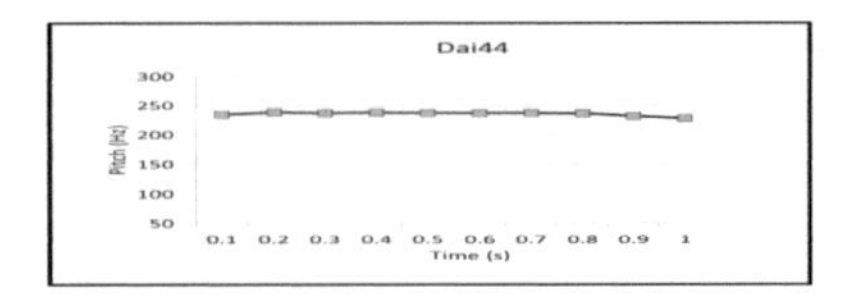

Fig. 5. Dai language (44 tone)

Tone 5 of Maguan Dialect (Fig. 6) is a low-falling tone. The whole tonal contour is at low of the tonal domain. The initial part is quite high then fall. The middle part is quite stable. The terminal part declines obviously to near the bottom of the tonal domain. The tone value of Tone 9 and Tone 5 is equal. Tone 9 is a tense tone while Tone 5 is a lax tone.

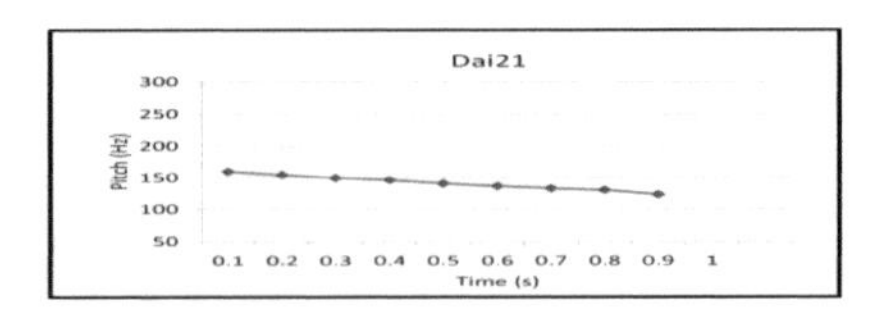

Fig. 6. Dai language (21 tone)

We can come into the following figure according to the above tone figures (Fig. 7).

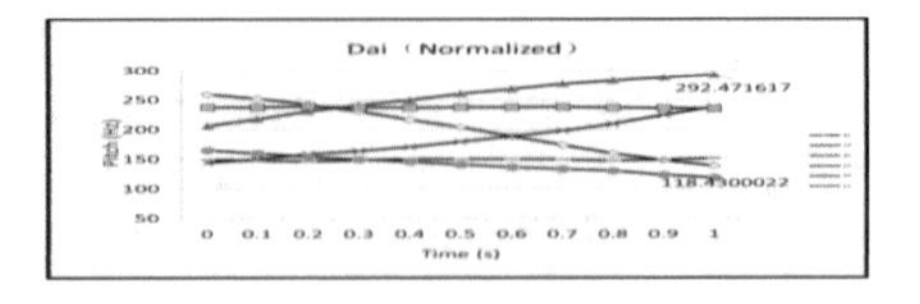

Fig. 7. Dai language (Normalized)

As is shown in the above figure, Maguan dialect spans from 292 Hz to 118 Hz. In this figure, the tone value of Maguan dialect range from 21 to 35. Tone categories differentiate well in Fig. 1. We hence define those that are in the upper part of the tone range and that has high pitch as high tones, while those in the lower part of the tone range and has low pitch as low tones. Therefore, there are two level tones, two falling tones and two rising tones in the six total of Maguan dialect. Since Tone 1corresponds to tone value 24 and 22 and to make it easier to differentiate, we record Tone 1 of tone value 24 as Tone 1 while that of tone value 22 as Tone 1′. Tone 1′ only appears at a few words with initials and Chinese borrowings and is only limited to words with falling-rising tone and a few yinping words.

For some words of Tone 3, initials correspond to low initial consonants in Thai language, therefore are considered low initial consonants, too, for example, place "hai^{3}" brother, sister "pi^{3}" location, place "ti^{3}". These initials are voiced and are supposed to be labelled Tone 6. However in Zhuang-Dai language family, the yin and yang tone of falling-rising tone and falling tone have merged due to that the whole voiced initial has become voiceless. Yangqu merged into yinshang and Tone 6 merged into Tone 3. Therefore voiced initial they are, they are labelled Tone 3 the odd tone.

2. Tones of Thai language. Devoicing is a common rule of the languages of Sino-Tibetan language family. First, initial consonants and tones correlate closely. The odd tones, i.e., Tone 1, 3, 5, 7 and 9 correlate with ancient voiceless initials, and the even tones, i.e., Tone 2, 4, 6, 8 and 10 correlate with ancient voiced initials, hence Tone yin and yang of ancient Chinese tones. yin tones correlate with ancient voiceless initials while yang tones correlate with ancient voiced initials. This is in accordance with the initial consonants system of Sino-Tibetan language family. Second, the high and middle consonants of the Thai language belong to ancient voiceless initials while the low consonant belong to ancient voiced initials. The middle consonant is derived from the high consonant. According to the above characteristics of Thai language and in accordance with the ten tone labelling system, Thai can be classified into the following ten tones. thus, middle Thai language has six lax tones and four tense tones. Yangqu merge into yinshang due to that whole voiced initials become voiceless. Tense Tone 7, 9 boasts a same tone value of 32. The tone value of tense Tone 7, 8, 9, 10 is 32, 45, 32 and 53 and corresponds respectively with those of lax Tone 5, 4, 5, 3 (Table 3)

Based on Zhu Xiaonong's normalization principle and use Bangkok pronunciation as standard Thai pronunciation, we get the tone system of Thai language as follows:

Tone 4 of Thai language (Fig. 8) is a high-rising tone whose tone contour is in the high of the tone domain. The initial point is a little higher than the middle of the domain. The middle part of the contour is rather stable. The ending part is rising to near the top of the domain. The tone value of Tone 8 and Tone 4 are equal. Tone 8 is a tense tone while Tone 4 is a lax tone.

As is shown in the above figure, we can clearly see the tonal type of Tone 4. Besides, the tonal pattern is described and reflected in the above part, here and hereafter.

The observation we have conducted and the results obtained are therefore satisfactory.

Table 3. Tone system of Thai language

Tone type	Tone name	Coda	Tone value	Example
1	*yinping*	Lax Tone Coda	323	Pig: mu:^{1}/dog: ma^{1}/bear: mi:^{1}
2	*yangping*		33	Fire: fai^{2}/low: loŋ^{2}/hand: mɯ:^{2}/eye: ta$^{1-2}$
3	*yinshang*		53	Face:na:^{3}/high:kʰɯn^{3}/banana:klu:i^{3}
4	*yangshang*		45	Water: nam^{4}/bad: ra:i^{4}/sky: fa:^{4}
5	*yinqu*		32	Each: ta:ŋ^{5}/well: bɔ^{5}/empty: plau^{5}
6	*yangqu*		53	Thing:rɯ:aŋ$^{6-3}$/granny:ja:$^{6-3}$/sit: naŋ$^{6-3}$
7	*yinru*	Tense Coda	32	Spirit: ʦit^{7}/duck: pet^{7}/mushroom: het^{7}
8	*yangru*		45	Think: nɯk^{8}/usually: mak^{8}/accept: rap^{8}
9	*yinru*		32	Mouth: pa:k^{9}/check: sɯ:p^{9}/peel: pɔ: k^{9}
10	*yangru*		53	Plough: kʰra:t^{10}/knife: mi:t^{10}/quick: ri:p^{10}

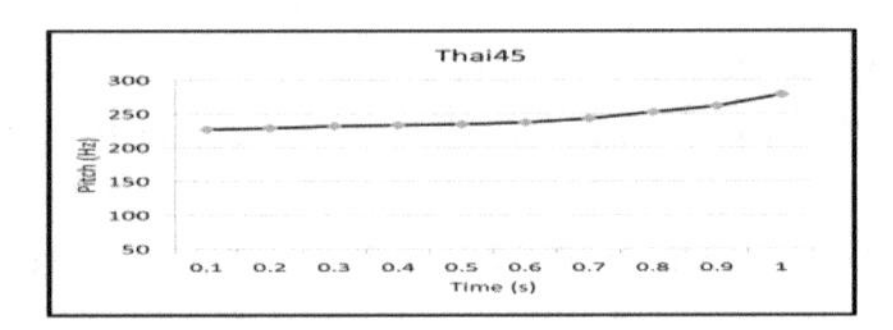

Fig. 8. Thai language (45 tone)

Tone 2 of Thai language (Fig. 9) is a mid-level tone. The whole tone lies at the middle of the domain, with near initial and terminal points.

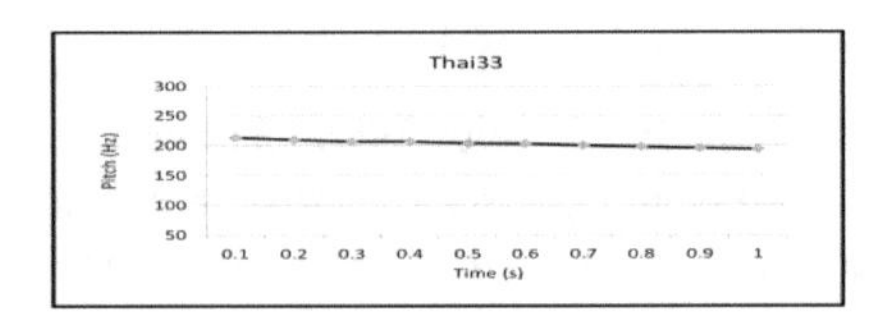

Fig. 9. Thai language (33 tone)

The whole tone contour is quite stable and is therefore an ideally balanced level tone.

Tone 1 of Thai language (Fig. 10) is a contour tone. The whole tone lies in the high of the domain. The initial point is near the top of the domain and the contour is a concaving one.

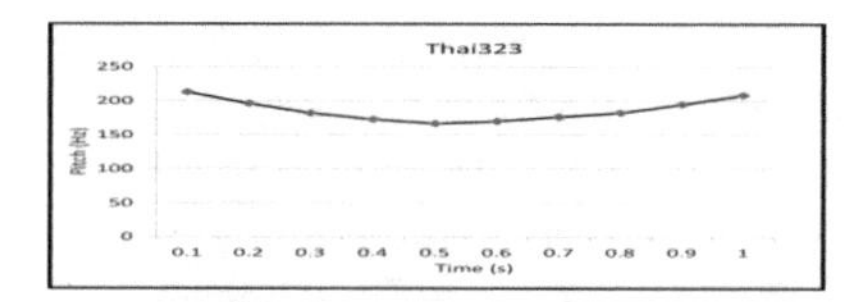

Fig. 10. Thai language (323 tone)

Tone 5 of Thai (Fig. 11) is a mid-falling tone. The whole tone contour lies in the low of the domain. The initial point is a little high, then decline. The middle part is quite stable. The ending part fall obviously and is near the bottom of the domain. The tone value is Tone 7, Tone 9 and Tone 5. Tone 7 is a tense tone with short vowel. Tone 9 is a tense tone with long vowel. Tone 5 is a lax tone.

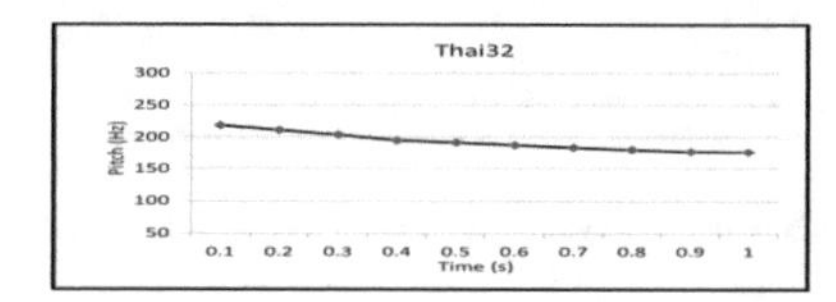

Fig. 11. Thai language (32 tone)

Tone 3 of Thai language (Fig. 12) is a high-falling tone. The initial point is in the high of the domain and the terminal is in the middle of the domain. The tonal contour is protruding and is therefore a high-falling tone.

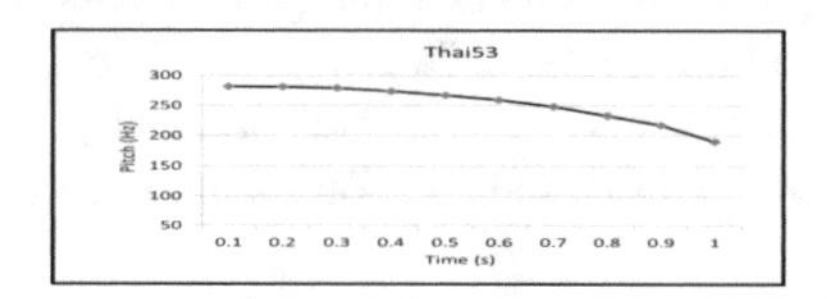

Fig. 12. Thai language (53 tone)

The tone value of Tone 3, Tone 6 and Tone 10 are equal. Tone 10 is a tense tone with long vowels. Tone 3 and Tone 6 are lax tones. Tone 3 is spelled with high consonant and open syllables. Tone 6 is to spell low consonant open syllables. According to the above tone figures, we can therefore conclude as follows (Fig. 13).

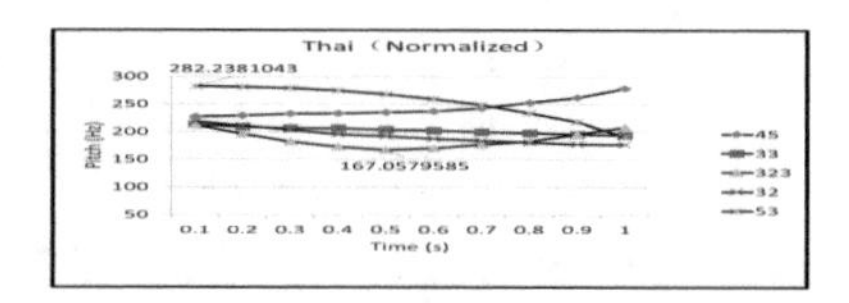

Fig. 13. Thai language (Normalized)

As is shown in the figure, Thai spans from 282 Hz to 167 Hz. The tone value of Thai language range from 32 to 45, a level tone of 33, a low falling tone of 32, a high falling tone of 53, a high rising tone of 45 and a low contour tone of 323. These tone value are calculated from Thai speakers in the author's experiment. The calculated tone value will not influence tone types in Kam-Tai language.

3. Tones of Maguan Dialect vs. Thai A comparison of the tones of Maguan dialect and Thai is given as follows: Fig. 14 shows a higher Thai tone over that of Dai. The pitch of the further centers around 167–282 Hz while that of the latter around 118–292 Hz. Devoicing is a common rule of the languages of Sino-Tibetan language family. Initial consonants and tones of Dai and Thai correlate closely.

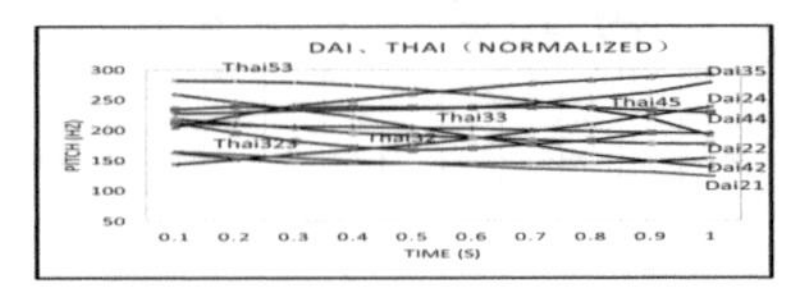

Fig. 14. Dai vs. Thai language (Normalized)

In summary, Tone 1 words whose initials are unaspirated stops, stops or fricatives in Dai language have been merged into Tone 2 in Thai language. Tone 6 merge into Tone 3 and Tone 9 merge into Tone 7 in Thai. In comparison, these merged tone types are considered to be identical.

While labelling, numbers to the right side of the decimal point belong to ancient tone type. Sometimes the nasal ending ʔ after short vowels drop.

For mid-Thai language, although Tone 6 merge into Tone 3, Tone 9 versus Tone 7, Tone 10 versus Tone 8 still differentiate. For Maguan dialect, Tone 6 merge into Tone 3 but Tone 10 doesn't merge into Tone 8 and not many Tone 9 into Tone 7, for example, heavy nak^7 is still Tone 9 nak^9 in Thai language (Table 4).

Table 4. Tone system of Dai and Thai languages

Symbol	Tone type	Tone name	Coda	Tone value	
				Maguan	Thai
A	1	*yinping*	Lax Tone Coda	24	323
	1′			22	–
	2	*yangping*		44	33
B	5	*yinqu*		21	32
	6	*yangqu*		42	53
C	3	*yinshang*		42	53
	4	*yangshang*		35	45
D	7	*yinru*	Tense Coda	42	32
	8	*yangru*		35	45
	9	*yinru*		21	32
	10	*yangru*		22	53

2 Conclusion

Yangqu of Maguan dialect and Thai language merge into *yinshang*, but they differ in "entering tones assigned to the other three tones". In Maguan dialect and Thai language, *yinru* (Tone 9) and *yinqu* (Tone 5) has equal tone value; *yangru* (Tone 8) and *yangshang* (Tone 4) has equal tone value. In Maguan dialect, *yangru* (Tone 10) and *yinping* (Tone 1') has equal tone value. In Thai language, *yangru* (Tone 10) and *yinshang* (Tone 3) has equal tone value. As language develop, when entering tone disappear, *yinru* and *yangru* will most likely merge into "the other three tones" that is of the same tone value of them. Then, as coda disappear, words of the same distinctive featured tone will have to differentiate and realize a precise and contact language system, besides phonetic system. It is highly probably to yield vowel duration contrast and realize a new cycle of "concise – complex – concise" in phonetic features. It is worth and needs further studies and exploration.

References

1. Li, F.-K.: The tone system of original Tai language. Li Zhaoxiang translator. Luo Meizhen revisor. National language information collection (1986)
2. Li, F.-K.: An Collection of Kam-Tai Language Studies. Tsinghwa University Press, June 2005
3. Ni, D.: An Introduction of Kam-Tai Languages. National Press, January 2010
4. Wang, J., Chen, X.: A Study of the Tones of Dai Language. Chinese Section, National Institute of Yunnan Province, October 1983
5. Luo, M.: A Comparison of Dai and Thai Words. National language, February 1988
6. Zhu, X.: Phonetics. Commercial Press (2010)
7. Yang, G.: A Study of the Phonetic System of Dai and Thai Languages in 13th Century. National Press (2007)
8. Yang, G.: Early initial system and some phonetic changes of Dai and Thai languages. Yunnan Natl. Coll. J. (1991)
9. Yang, J.: Lectures on Yinyun Studies. Fudan University Press, March 2012
10. Xu, Y.: A Study of Maguan Dialect in Dai Language. Yunnan National Press, April 2017
11. Zeng, X.: On tonal correspondence of ancient Chinese borrowings in Zhuang, Dai, Dong and Sui languages. National language (1) (2003)
12. Cai, R.: The Vowel and Tone Framework of Dai Language. Sichuan University Press, April 2007

Some Results in Hyper BE-algebras

Xiao Yun Cheng[1], Jun Tao Wang[2(✉)], and Wei Wang[3]

[1] School of Science, Xi'an Aeronautical University, Xi'an 710077, China
[2] School of Science, Xi'an Shiyou University, Xi'an 710065, China
[3] Department of Basic Courses, Shaanxi Railway Institute, Wei Nan 714000, China
chengxiaoyun2004@163.com, 15829065086@163.com, wangwei135420@163.com

Abstract. In this paper, the generated hyper filters in hyper BE-algebras are represented. Also, the concept of multipliers over hyper BE-algebras is introduced and the existence of multipliers is discussed. In particular, it is proven that the set of all right multipliers is commutative on a self-distributive commutative hyper BE-algebra.

Keywords: Hyper BE-algebra · Commutative hyper BE-algebra (generated) Hyper filter · (left) Right multiplier

1 Introduction

The hyper structure theory (or super structure theory) was put forward by Marty [1] at the Eighth Congress of Mathematicians in Scandinavia. Since then many researchers have introduced hyper algebras such as hyper BCK-algebras [2], hyper BCI-algebras [3], hyper K-algebras [4], hyper MV-algebras [5] and hyper equality algebras [6], etc. Radfar introduced hyper BE-algebras in [7] and then studied commutative hyper BE-algebras in [8]. Cheng [9] investigated the filter theory in hyper BE-algebras. Now hyperstructures have been extensively applied such as geometry, hypergraph, lattice, probabilities, cryptography, fuzzy sets, combinatorics, automata and artificial intelligence and probability, etc (see [10],[11]).

The concept of multipliers on commutative semigroups was introduced by Larsen [12] in 1971. Now multipliers have been applied to many algebraic systems in order to investigate the properties and characterize the structures of them. For example, Cornish [13] defined multipliers for distribute lattices. Khorami and Saeid [14] introduced multipliers on BL-algebras and study some relationships between multipliers and other special operators such as closure operators and derivations. In 2013, Rao [15] gave the concept of multipliers on hyper semilattices where the author defined f-invariant subsets and f-fixed sets and investigated some properties of them. Moreover the author proved that the set of all f-invariant subsets of a hyper semilattice constitutes a hyper semilattice.

The present paper will further give some results about hyper BE-algebras including generated hyper filters and multipliers on hyper BE-algebras.

N. Xiong et al. (Eds.): ISCSC 2018, AISC 877, pp. 388–395, 2019.
https://doi.org/10.1007/978-3-030-02116-0_45

2 Preliminaries

In this section, we recollect some definitions and results which will be used in the following.

Definition 1. [7] *Let $H \neq \emptyset$ with a hyperoperation $\diamondsuit$ and a constant* 1. *Then $(H, \diamondsuit, 1)$ is called a hyper BE-algebra if it satisfies the following conditions: for all $x, y \in H$,*

(HBE1) $x \ll 1, x \ll x$;
(HBE2) $x\diamondsuit(y\diamondsuit z) = y\diamondsuit(x\diamondsuit z)$;
(HBE3) $x \in 1\diamondsuit x$;
(HBE4) $1 \ll x \Rightarrow x = 1$,

where $x \ll y$ iff $1 \in x\diamondsuit y$. Moreover for any $\emptyset \neq A, B \subseteq H$, $A \ll B$ means: there are $a \in A, b \in B$ such that $a \ll b$, and $A \leq B$ means: for any $a \in A$, there is $b \in B$ such that $a \ll b$. Also, $A\diamondsuit B = \bigcup_{a\in A, b\in B} a\diamondsuit b$.

Example 1. [7] Let $H = \{a, b, 1\}$. Define hyper operations $\diamondsuit_1, \diamondsuit_2$ on H as follows

$\diamondsuit_1$	a	b	1
a	$\{a, b, 1\}$	$\{b\}$	$\{1\}$
b	$\{a, b\}$	$\{b, 1\}$	$\{1\}$
1	$\{a\}$	$\{b\}$	$\{1\}$

$\diamondsuit_2$	a	b	1
a	$\{a, b, 1\}$	$\{b\}$	$\{1\}$
b	$\{a, b, 1\}$	$\{a, b, 1\}$	$\{b, 1\}$
1	$\{a, b\}$	$\{b\}$	$\{1\}$

Then $(H, \diamondsuit_1, 1)$ and $(H, \diamondsuit_2, 1)$ are hyper BE-algebras.

Proposition 1. [7,9] *Let $(H, \diamondsuit, 1)$ be a hyper BE-algebra. Then for any $x, y, z \in H, \emptyset \neq A, B, C \subseteq H$:*

(1) $A\diamondsuit(B\diamondsuit C) = B\diamondsuit(A\diamondsuit C)$;
(2) $1 \ll A \Rightarrow 1 \in A$;
(3) $A \ll B \Leftrightarrow 1 \in A\diamondsuit B$;
(4) $A \subseteq 1\diamondsuit A$;
(5) $A \subseteq B \Rightarrow A \ll B, A \subseteq B \Rightarrow A \leq B$;
(6) $A \ll B, 1 \in A \Rightarrow 1 \in B$;
(7) $x \ll y\diamondsuit x, A \leq B\diamondsuit A$;
(8) $x \ll (x\diamondsuit y)\diamondsuit y$;
(9) $y \in 1\diamondsuit x \Rightarrow y \ll x$.

Definition 2. [7,9] *For any $x \in H$, we call a hyper BE-algebra $(H, \diamondsuit, 1)$*

- *R-hyper BE-algebra, if $1\diamondsuit x = \{x\}$;*
- *C-hyper BE-algebra, if $x\diamondsuit 1 = \{1\}$;*
- *D-hyper BE-algebra, if $x\diamondsuit x = \{1\}$;*
- *RC-hyper BE-algebra, if H is a R-hyper and C-hyper BE-algebra.*

Given a hyper BE-algebra $(H, \diamondsuit, 1)$, we call $\emptyset \neq A \subseteq H$ a hyper BE-subalgebra if $x\diamondsuit y \subseteq A$, for any $x, y \in A$.

Definition 3. [7] *Let* $(H, \diamondsuit, 1)$ *be a hyper BE-algebra and* $\emptyset \neq F \subseteq H, 1 \in F$. *For any* $x, y \in H$, *F is said to be a*

- *weak hyper filter, if* $x \diamondsuit y \subseteq F, x \in F \Rightarrow y \in F$;
- *hyper filter, if* $x \diamondsuit y \bigcap F \neq \emptyset, x \in F \Rightarrow y \in F$, *for all* $x, y \in H$.

In a hyper BE-algebra, it is well-known that every hyper filter is a weak hyper filter and moreover every hyper filter satisfies
(F) $x \ll y, x \in F \Rightarrow y \in F$.

Definition 4. [8] *A hyper BE-algebra H is called commutative provided that* $(x \diamondsuit y) \diamondsuit y = (y \diamondsuit x) \diamondsuit x$ *for any* $x, y \in H$.

3 The Generated Hyper Filters

In the following sequel, we will denote a hyper BE-algebra $(H, \diamondsuit, 1)$ by H, unless otherwise stated.

Definition 5. *H is said to be good if for any* $x, y, z \in H, x \ll y \Rightarrow z \diamondsuit x \leq z \diamondsuit y$.

Proposition 2. *If* $H = \{1, x, y\}$ *with* $x \ll y$ *is a hyper BE-algebra, then H is good.*

Proof. If $x \ll y$, then $1 \in x \diamondsuit y$ and so $x \diamondsuit x \leq x \diamondsuit y$. Since $1 \in y \diamondsuit y$, we have $y \diamondsuit x \leq y \diamondsuit y$. Hence it follows from $y \in 1 \diamondsuit y$ and $1 \notin 1 \diamondsuit x$ that $1 \diamondsuit x \leq 1 \diamondsuit y$. If $x \ll 1$, then by $1 \in a \diamondsuit 1$, $a \diamondsuit x \leq a \diamondsuit 1$ for any $a \in H$.

In Example 1 $(H, \diamondsuit_2, 1)$ is a good hyper BE-algebra. The following example indicates that there exists hyper BE-algebras no good.

Example 2. Let $H = \{a, b, c, 1\}$. Define hyperoperation $\diamondsuit$ on H as follows:

$\diamondsuit$	a	b	c	1
a	$\{1\}$	$\{a\}$	$\{b, c\}$	$\{1\}$
b	$\{1\}$	$\{1\}$	$\{1\}$	$\{1\}$
c	$\{1\}$	$\{a\}$	$\{b, c, 1\}$	$\{1\}$
1	$\{a\}$	$\{b\}$	$\{c\}$	$\{1\}$

Then $(H, \diamondsuit, 1)$ is a hyper BE-algebra [7]. Clearly, $b \ll c$, but $a \ll b, a \ll c$ is not true. Hence $a \diamondsuit b \leq a \diamondsuit c$ is not true. It implies that H is not good.

Let S be a nonempty subset and F be a hyper filter of H. Denote by $[S)$ the least hyper filter of H containing S which is called as the hyper filter generated by S. In particular, if $S = \{a\}$, we write $[\{a\}) = [a)$. In addition, we use $[F \cup \{x\})$ to denote the hyper filter generated by F and x, where $x \in H - F$. The following are some results about the generated hyper filters.

Theorem 1. *Given a good hyper BE-algebra H and* $\emptyset \neq S \subseteq H$, $[S) = \{x \in H : 1 \in a_1 \diamondsuit (a_2 \diamondsuit (\cdots (a_n \diamondsuit x) \cdots)), \exists a_1 \cdots, a_n \in S, n, m \geq 1\}$.

Proof. Denote the right side of the above equation by A. According to Theorem 3.9 of [9], $[S) \supseteq A$. Now we prove $[S) \subseteq A$, that is, A is a hyper filter including S. Firstly, since $1 \in a\diamondsuit 1$ and $1 \in x\diamondsuit x$ for any $a, x \in S$, we have $1 \in A$ and $S \subseteq A$.

Let $x \in A$ and $x\diamondsuit y \cap A \neq \emptyset$. Then there exist $a_1, a_2, \cdots, a_n, b_1, b_2, \cdots, b_m, \in S$ such that $1 \in a_1\diamondsuit(a_2\diamondsuit(\cdots(a_n\diamondsuit x)\cdots))$ and $1 \in b_1\diamondsuit(b_2\diamondsuit(\cdots(b_m\diamondsuit(x\diamondsuit y)\cdots))$. Hence by (HBE2) and (1) of Proposition 1, $b_1\diamondsuit(b_2\diamondsuit(\cdots(b_m\diamondsuit(x\diamondsuit y)\cdots)) = x\diamondsuit(b_1\diamondsuit(b_2\diamondsuit(\cdots(b_m\diamondsuit y)\cdots))$. This shows $1 \in x\diamondsuit(b_1\diamondsuit(b_2\diamondsuit(\cdots(b_m\diamondsuit y)\cdots))$ and so $x \ll b_1\diamondsuit(b_2\diamondsuit(\cdots(b_m\diamondsuit y)\cdots)$. Since H is a good, we obtain $a_1\diamondsuit(a_2\diamondsuit(\cdots(a_n\diamondsuit x)\cdots)) \leq a_1\diamondsuit(a_2\diamondsuit(\cdots(a_n\diamondsuit(b_1\diamondsuit(\cdots(b_m\diamondsuit y)\cdots))$.
Thus $1 \in a_1\diamondsuit(a_2\diamondsuit(\cdots(a_n\diamondsuit(b_1\diamondsuit(\cdots(b_m\diamondsuit y)\cdots))$, which implies $y \in A$ and therefore A is a hyper filter of H.

Combining the above arguments, $[S) = A$.

Corollary 1. *Given a good hyper BE-algebra H and $a \in H$, $[a) = \{x \in H : 1 \in a^n\diamondsuit x, n \geq 1\}$.*

Theorem 2. *Given a hyper filter F in a good hyper BE-algebra H and $a \in H - F$, $[F \cup \{a\}) = \{x \in H : a^n\diamondsuit x \cap D \neq \emptyset, n \geq 1\}$.*

Proof. Denote the right side of the above equation by A. Firstly, $[F \cup \{a\}) \supseteq A$. In fact, let $x \in A$. Then $a^n\diamondsuit x \cap F \neq \emptyset$ and further $a^n\diamondsuit x \cap [F \cup \{a\}) \neq \emptyset$. Since $a \in [F \cup \{a\})$ and $[F \cup \{a\})$ is a hyper filter, then $x \in [F \cup \{a\})$. Next we show $[F \cup \{a\})_s \subseteq A$. Since $a\diamondsuit 1 \cap F \neq \emptyset$ and $a\diamondsuit a \cap F \neq \emptyset$, we get $1 \in A$ and $a \in A$. Now let $x \in F$. Then $a\diamondsuit x \cap F \neq \emptyset$ from $x \ll a\diamondsuit x$. Using H good, $a\diamondsuit x \leq a\diamondsuit(a\diamondsuit x) = a^2\diamondsuit x$. Hence $a^2\diamondsuit x \cap F \neq \emptyset$. Repeating the above argument we obtain $a^n\diamondsuit x \cap F \neq \emptyset$. Consequently $x \in A$ and so $F \subseteq A$. Let $x \in [F \cup \{a\})$. Then by Theorem 1 there are $a_1, a_2, \cdots, a_n \in D \cup \{a\}$ such that $1 \in a_1\diamondsuit(a_2\diamondsuit(\cdots(a_n)\diamondsuit x)\cdots))$.

Case 1. If there exists i such that $a_i = a$, then $1 \in a_1\diamondsuit(a_2\diamondsuit(\cdots(a^r\diamondsuit x)\cdots))$. Since $a_j \in F(j \neq i)$ and F is a hyper filter, then $a^r\diamondsuit x \cap F \neq \emptyset$, namely, $x \in A$.

Case 2. If for all i such that $a_i \neq a$, then $a_i \in F$. Since F is a hyper filter of H, it follows $x \in F$ and hence $x \in A$.

Combing the above argument, $[F \cup \{a\}) = A$.

4 Left (Right) Multipliers

Definition 6. *A mapping $f_l : H \to H$ ($f_r : H \to H$) on H is called as a left (right) multiplier if $f_l(x\diamondsuit y) \subseteq f_l(x)\diamondsuit y$ ($f(x\diamondsuit y) \subseteq x\diamondsuit f(y)$) for any $x, y \in H$. In particular, f_l (f_r) is called regular if $f_l(1) = 1$ ($f_r(1) = 1$).*

Notice that $f_l(A) = \{f_l(a) : a \in A\}$ ($f_r(A) = \{f_r(a) : a \in A\}$) for any $A \subseteq H$ and denote by $M_r(H)$ the collection of all right-multipliers on H.

Example 3.

(1) Define a mapping $1_H : H \to H$ by $1_H(x) = 1$ for all $x \in H$. Then 1_H is a right-multiplier on H.

(2) Define a mapping $1_H : H \to H$ by $Id_H(x) = x$ for all $x \in H$. Then 1_H is a left-multiplier (right-multiplier) on H, which is called the trivial left-multiplier (right-multiplier) on H.

Example 4. Consider the hyper BE-algebra $(H; \diamondsuit_2, 1)$ in Example 1. Define a mapping f on H by $f(1) = 1, f(a) = f(b) = b$. One can check that f is a regular left-multiplier on H.

Example 5. Let $H = \{a, b, 1\}$. Define hyperoperation $\diamondsuit$ on H as follows

$\diamondsuit$	a	b	1
a	$\{1\}$	$\{b\}$	$\{1\}$
b	$\{a, 1\}$	$\{1\}$	$\{1\}$
1	$\{1\}$	$\{b\}$	$\{1\}$

Then $(H, \diamondsuit, 1)$ is a hyper BE-algebra [7]. Define a mapping f on H by $f(b) = f(1) = 1, f(a) = a$. One can calculate that f is a regular right-multiplier on H.

The following propositions can be obtained immediately.

Proposition 3. *Suppose that f is a left-multiplier on H. Then for any $x, y \in H$:*

(1) $x \ll y \Rightarrow f(1) \in f(x)\diamondsuit y$. If f is regular, then $f(x) \ll y$;
(2) $f(1) \in f(1)\diamondsuit 1$. If H is a C-hyper BE-algebra, then f is regular;
(3) $f(1) \in f(x)\diamondsuit x$. If f is regular, then $f(x) \ll x$, that is, f is contractive.

Proposition 4. *Suppose that f is a right-multiplier on H. Then for any $x, y \in H$:*

(1) $x \ll y \Rightarrow f(1) \in x\diamondsuit f(y)$. If f is regular, then $x \ll f(y)$;
(2) $f(1) \in f(1)\diamondsuit f(1)$. If H is a D-hyper BE-algebra, then f is regular;
(3) $f(1) \in x\diamondsuit f(x)$. In particular, if f is regular, then $x \ll f(x)$, that is, f is expansive.

Proposition 5. *Suppose that $(H, \diamondsuit, 1)$ is a hyper BE-algebra and $f_1, f_2 \in M_l(H)\,(M_r(H))$. Define a mapping $f_1\diamondsuit f_2 : H \to H$ by for any $x \in H, (f_1\diamondsuit f_2)(x) = f_1(f_2(x))$. Then $f_1\diamondsuit f_2$ is a left-multiplier (right-multiplier) on H.*

Theorem 3. *(1) There is no non-regular left-multiplier in every C-hyper BE-algebra.*
(2) There is no non-trivial regular left-Multiplier in every R-hyper BE-algebra.

Proof. (1) Assume that H is a C-hyper BE-algebra and f is a self-mapping on H satisfying $f(1) \neq 1$. Then $x\diamondsuit 1 = \{1\}$ for any $x \in H$. Since $f(x\diamondsuit 1) = f(1) \neq 1$ and $f(x)\diamondsuit 1 = \{1\}$, we have that $f(x\diamondsuit 1)$ is not included in $f(x)\diamondsuit 1$.
(2) Assume that H is a R-hyper BE-algebra and f is a regular left-multiplier on H. Then $1\diamondsuit x = \{x\}$ for any $x \in H$. Since $\{f(x)\} = f(1\diamondsuit x) \subseteq f(1)\diamondsuit x = 1\diamondsuit x = \{x\}$, we have $f(x) = x$ and so $f = Id_H$.

Example 6. Consider the hyper BE-algebra H in Example 5. It is clear that it is a RC-hyper BE-algebra. Define a mapping f on H by $f(1) = 1, f(a) = f(b) = b$. Then f is not a regular left-multiplier, since $f(a\diamondsuit b) = \{b\}$ and $\{1\} = f(a)\diamondsuit b$.

Assume that f is a self-mapping on a hyper BE-algebra H. Define the fixed set of f by $Fix_f(H) = \{x \in H : f(x) = x\}$. By the proof of Theorem 3, we have the following theorem.

Theorem 4. *Assume that f is a left-multiplier on a R-hyper BE-algebra H. The following are equivalent:*

(1) f is regular;
(2) $H = Fix_f(H)$;
(3) $f = Id_H$.

Definition 7. *H is called self-distribute if for all $x, y, z \in H, (x\diamondsuit y)\diamondsuit(x\diamondsuit z) \subseteq x\diamondsuit(y\diamondsuit z)$.*

Example 7. Let $H = \{a, 1\}$. Define hyperoperation $\diamondsuit$ on H as follows:

$\diamondsuit$	a	1
a	$\{1\}$	$\{a, 1\}$
1	$\{a\}$	$\{1\}$

Then $(H, \diamondsuit, 1)$ is a self-distribute hyper BE-algebra.

For $f_1, f_2 \in M_r(H)$, define hyperoperation $\triangle$ on $M_r(H)$ by $(f_1 \triangle f_2)(x) = f_1(x)\diamondsuit f_2(x)$ for all $x \in H$.

Lemma 1. *For any $A \subseteq H, f_1, f_2 \in M_r(H), (f_1 \triangle f_2)(A) \subseteq f_1(A)\diamondsuit f_2(A)$.*

Proof. Since $(f_1 \triangle f_2)(A) = \{(f_1 \triangle f_2)(a) : a \in A\} = \{(f_1(a)\diamondsuit f_2(a) : a \in A\}$ and $f_1(A)\diamondsuit f_2(A) = \{(f_1(a)\diamondsuit f_2(b) : a, b \in A\}$, we can obtain $(f_1 \triangle f_2)(A) \subseteq f_1(A)\diamondsuit f_2(A)$.

Theorem 5. *Suppose that $(H, \diamondsuit, 1)$ is a self-distribute commutative hyper BE-algebra. Then $((M_r(H), \triangle, 1_H)$ is a commutative hyper BE-algebra, where for $f_1, f_2 \in M_r(H)$, $f_1 \ll f_2$ mean that $f_1(x) \ll f_2(x)$ for all $x \in H$.*

Proof. Firstly, $f_1 \triangle f_2 \subseteq M_r(H)$. Let $f_1, f_2 \in M_r(H)$ and $x, y \in H$. By Lemma 1 and H self-distribute, $(f_1 \triangle f_2)(x\diamondsuit y) \subseteq f_1(x\diamondsuit y)\diamondsuit f_2(x\diamondsuit y) \subseteq (x\diamondsuit f_1(y))\diamondsuit(x\diamondsuit f_2(y))$
$\subseteq x\diamondsuit(f_1(y)\diamondsuit f_2(y)) = x\diamondsuit((f_1 \triangle f_2)(y))$. Hence $f_1 \triangle f_2 \subseteq M_r(H)$.

In the following, we shall prove that $(M_r(H), \triangle, 1_H)$ is a hyper BE-algebra.

(HBE1) For any $f \in M_r(H)$ and $x \in H$, $f(x) \ll 1 = 1_H(x)$, it implies that $f \ll 1_H$, namely, $f \ll f$.

(HBE2) For any $f_1, f_2, f_3 \in M_r(H)$ and $x \in H$, since $(f_1 \triangle (f_2 \triangle f_3))(x) = f_1(x)\diamondsuit(f_2 \triangle f_3)(x) = f_1(x)\diamondsuit(f_2(x)\diamondsuit f_3(x))$ and $(f_2 \triangle (f_1 \triangle f_3))(x) = f_2(x)\diamondsuit(f_1 \triangle f_3)(x) = f_2(x)\diamondsuit(f_1(x) \triangle f_3)(x)$, then $(f_1 \triangle (f_2 \triangle f_3))(x) = (f_2 \triangle (f_1 \triangle f_3))(x)$. This implies that $f_1 \triangle (f_2 \triangle f_3) = f_2 \triangle (f_1 \triangle f_3)$.

(HBE3) For any $f \in M_r(H)$ and $x \in H$, $f(x) \in 1\diamondsuit f(x) = 1_H(x)\diamondsuit f(x) = (1_H \triangle f)(x)$.

(HBE4) For any $f \in M_r(H)$ and $x \in H$, if $1_H \ll f$, then $1_H(x) \ll f(x)$ and thus $1 \ll f(x)$. Thus $f(x) = 1$ for any $x \in H$ and so $f = 1_H$.

Finally, we shall show that $(M_r(H), \triangle, 1_H)$ is commutative. Indeed, if $f_1, f_2, f_3 \in M(H)$ and $x \in H$, then

$((f_1 \triangle f_2) \triangle f_2)(x) = (f_1 \triangle f_2)(x)\diamondsuit f_2(x) = (f_1(x)\diamondsuit f_2(x))\diamondsuit f_3(x)$
$= (f_2(x)\diamondsuit f_1(x))\diamondsuit f_1(x) = (f_2 \triangle f_1)(x)\diamondsuit f_1(x) = ((f_2 \triangle f_1) \triangle f_1)(x)$.

Therefore $(f_1 \triangle f_2) \triangle f_2 = (f_2 \triangle f_1) \triangle f_1$.

If f is a self-mapping on H, define the kernel of f by $Kerf = \{x \in H : f(x) = 1\}$.

Theorem 6. *Suppose that f is a regular right-multiplier on a C-hyper BE-algebra H. Then $Kerf$ is a hyper BE-subalgebra of H.*

Proof. Clearly, $1 \in Kerf$ and hence $Kerf \neq \emptyset$. If $x, y \in Kerf$, then $f(x) = f(y) = 1$. Thus $f(x\diamondsuit y) \subseteq x\diamondsuit f(y) = x\diamondsuit 1 = \{1\}$, which indicate $x\diamondsuit y \subseteq Kerf$.

Theorem 7. *Suppose that f is a regular right-multiplier on H. Then $Kerf$ is a (weak) hyper filter of H.*

Proof. It suffices to prove that $Kerf$ is a hyper filter of H. It is clear that $1 \in Kerf$. Let $x \in Kerf, x\diamondsuit y \cap Kerf \neq \emptyset$ for any $x, y \in H$. Then $f(x) = 1$ and there is $x \in x\diamondsuit y$ such that $a \in Kerf$. Therefore $1 = f(a) \in f(x\diamondsuit y) \subseteq f(x)\diamondsuit y = 1\diamondsuit y$, which shows $1 \ll y$. It follows that $y = 1$ and further $y \in Kerf$.

5 Conclusions

In this paper, we give some results in hyper BE-algebras. We investigate the generated representations of hyper filters in hyper BE-algebras. Also, by virtue of the fixed set we deliver a characterization of the regular left-multiplier on a R-hyper BE-algebra. In addition, we show that the set of all right multipliers $(M_r(H), \triangle, 1_H)$ forms a commutative hyper BE-algebra whence H is a self-distribute commutative hyper BE-algebra.

Acknowledgments. This research is supported by a grant of Shaanxi Railway Institute Research Fund Project (Ky2017-093).

References

1. Marty, F.: Surune generalization de la notion de group. In: The 8th Congress Math. Scandinaves, Stockholm, pp. 45–49 (1934)
2. Jun, Y.B., Zahedi, M.M., Xin, X.L., Borzooei, R.A.: On hyper BCK-algebras. Ital. J. Pure Appl. Math.-N. **8**, 127–136 (2000)
3. Xin, X.L.: Hyper BCI-algebras. Discussiones Mathematicae Gen. Algebr. Appl. **26**, 5–19 (2006)

4. Borzooei, R.A., Hasankhani, A., Zahedi, M.M., Jun, Y.B.: On hyper K-algebras. Math. Japonicae **52**, 113–121 (2000)
5. Ghorbani, S., Hasankhani, A., Eslami, E.: Hyper MV-algebras. Set-Valued Math. Appl. **1**, 205–222 (2008)
6. Cheng, X.Y., Xin, X.L., Jun, Y.B.: Hyper equality algebras. In: Quantitative Logic and Soft Computing 2016, pp. 415–428. Springer, Cham (2017)
7. Radfar, A., Rezaei, A., Saeid, A.B.: Hyper BE-algebras. Novi. Sad. J Math. **44**(2), 137–147 (2014)
8. Rezaei, A., Radfar A., Saeid, A.B.: On commutative hyper BE-algebras. Facta Universitatis (NIŠ). **30**(4), 389–399 (2015)
9. Cheng, X.Y., Xin, X.L.: Filter theory on hyper BE-algebras. Ital. J. Pure Appl. Math.-N. **35**, 509–526 (2015)
10. Corsini, P.: Prolegomena of hypergroup theory. Aviani (1994)
11. Corsini P., Leoreanu V.: Applications of hyperstructure theory. Springer, US (2013)
12. Larsen, R.: An Introduction to the Theory of Multipliers. Springer, Heidelberg (1971)
13. Cornish, W.H.: The multiplier extension of a distributive lattice. J. Algebr. **32**(2), 339–355 (1974)
14. Khorami, R.T., Saeid, A.B.: Multiplier in BL-algebras. Iran. J. Sci. Technol. (Sci.) **8**(2), 95–103 (2014)
15. Rao, M.S.: Multiplier of hypersemilattices. Int. J. Math. Soft Comput. **3**(1), 29–35 (2013)

Distance and Similarity Measures for Generalized Hesitant Fuzzy Soft Sets

Chen Bin(✉)

School of Mathematical Sciences, University of Jinan, Jinan 250022, China
jnchenbin@126.com

Abstract. Distance and similarity measures are very important topics in fuzzy set theory. In this paper, we propose a variety of distance measures for generalized hesitant fuzzy sets, based on which the corresponding similarity measures can be obtained. We investigate the connections of the mentioned distance measures and further develop a number of generalized hesitant ordered weighted distance measures and generalized hesitant ordered weighted similarity measures.

Keywords: Generalized hesitant fuzzy set · Distance · Similarity

1 Introduction

Distance and similarity measures have attached a lot of attention in the last decades due to the fact that they can be applied to many areas such as pattern recognition, clustering analysis, approximate reasoning, image processing, medical diagnosis and decision making.

Recently, Xu and Xia [1] originally developed a series of distance measures for hesitant fuzzy sets. Although the distance and similarity measures are significant, there are no distance and similarity measures between GHFSs. This paper is organized as follows. In Sect. 2, we list the notions of GHFS. In Sects. 3, 4 and 5, we present some new geometric distance measures between GHFSs base on geometric distance model and set-theory approach.

2 Preliminaries

Definition 1. ([2]). *Let X be a set, an IFS A on X is given by two functions $\mu : X \to [0,1]$ and $\nu : X \to [0,1]$, which satisfies the condition $0 \leq \mu(x)+\nu(x) \leq 1$, for all $x \in X$.*

Definition 2. ([3],[4]). *Let X be a set, then a T-HFS A on X in terms of a function h is $A = \{\langle x, h_A(x)\rangle \mid x \in X\}$.*

Given a hesitant fuzzy set h, we define below its lower and upper bound,

- lower bound: $h^-(x) = minh(x)$ and
- upper bound: $h^+(x) = maxh(x)$.

Gang Qian et al. [5] defined the generalized hesitant fuzzy set.

N. Xiong et al. (Eds.): ISCSC 2018, AISC 877, pp. 396–403, 2019.
https://doi.org/10.1007/978-3-030-02116-0_46

Definition 3. ([5]). *Let X be a fixed set, then a generalized hesitant fuzzy set (GHF set) G on X is described as:*

$G = \{\frac{x}{h(x),g(x)}|x \in X\}$,

$0 \leq \mu_i(x), \nu_i(x) \leq 1, 0 \leq \mu_i(x) + \nu_i(x) \leq 1,\ 1 \leq i \leq N_x = |h(x)| = |g(x)|$,

where $\mu_i(x) \in h(x)$, $\nu_i(x) \in g(x)$, for all $x \in X$. And $|h(x)|$ denote the cardinality of the set $h(x)$.

Definition 4. *Let M, N be two GHF set on X. Then, M is a generalized hesitant fuzzy subset of N, if for each $x \in X$, $1 \leq \sigma(j) \leq l_x$, we have $\mu_M^{\sigma(j)}(x) \leq \mu_n^{\sigma(j)}(x)$ and $\nu_M^{\sigma(j)}(x) \geq \nu_n^{\sigma(j)}(x)$. And denote by $M \sqsubseteq N$.*

Example 1. Let $X = \{x_1, x_2, x_3\}$ be the given set, and

$M = \{\frac{x_1}{(0.4),(0.5)}, \frac{x_2}{(0.4,0.5),(0.5,0.3)}, \frac{x_3}{(0.2,0.3,0.5,0.6),(0.7,0.6,0.4,0.3)}\}$,

$N = \{\frac{x_1}{(0.5,0.7),(0.4,0.2)}, \frac{x_2}{(0.6),(0.3)}, \frac{x_3}{(0.5,0.6),(0.3,0.2)}\}$.

be two GHFS sets on X. Then,

$M = \{\frac{x_1}{(0.4,0.4),(0.5,0.5)}, \frac{x_2}{(0.4,0.5),(0.5,0.3)}, \frac{x_3}{(0.2,0.3,0.5,0.6),(0.7,0.6,0.4,0.3)}\}$,

$N = \{\frac{x_1}{(0.5,0.7),(0.4,0.2)}, \frac{x_2}{(0.6,0.6),(0.3,0.3)}, \frac{x_3}{(0.5,0.6,0.6,0.6),(0.3,0.2,0.2,0.2)}\}$.

We can find that $\mu_M^{\sigma(j)}(x_i) \leq \mu_N^{\sigma(j)}(x_i)$ and $\nu_M^{\sigma(j)}(x_i) \geq \nu_N^{\sigma(j)}(x_i)$, for each $x_i \in X$ and each $1 \leq \sigma(j) \leq l_{x_i}$. Then, M is a generalized hesitant fuzzy subset of N and denote by $M \sqsubseteq N$.

3 Distance Between Two Generalized Hesitant Fuzzy Sets

Definition 5. *Let M, N be two GHFS sets on X and $d_{max} = max\{d(A, B), A, B \in GHFS\}$. Then the distance measure between M, N is defined as $d(M, N)$:*

(D1) $0 \leq d(M, N) \leq d_{max}$

(D2) $d(M, N) = 0 \Leftrightarrow M = N$

(D3) $d(M, N) = d(N, M)$

(D4) Let P be a GHFS set on X. If $M \sqsubseteq N \sqsubseteq P$. Then, $d(M, N) \leq d(M, P)$ and $d(N, P) \leq d(M, P)$.

If (D1)′ replaces (D1), the $d(M, N)$ is called a normalized distance measure, where (D1)′ $0 \leq d(M, N) \leq 1$.

Definition 6. *GH normalized Hamming distance:*

$d_1(M, N) = \frac{1}{2n}\sum_{i=1}^{n}[\frac{1}{l_{x_i}}\sum_{j=1}^{l_{x_i}}|\mu_M^{\sigma(j)}(x_i) - \mu_N^{\sigma(j)}(x_i)| + \frac{1}{l_{x_i}}\sum_{j=1}^{l_{x_i}}|\nu_M^{\sigma(j)}(x_i) - \nu_N^{\sigma(j)}(x_i)|$

Definition 7. *GH normalized Euclidean distance:*

$d_2(M, N) = [\frac{1}{2n}\sum_{i=1}^{n}[\frac{1}{l_{x_i}}\sum_{j=1}^{l_{x_i}}|\mu_M^{\sigma(j)}(x_i) - \mu_N^{\sigma(j)}(x_i)|^2 + \frac{1}{l_{x_i}}\sum_{j=1}^{l_{x_i}}|\nu_M^{\sigma(j)}(x_i) - \nu_N^{\sigma(j)}(x_i)|^2]^{\frac{1}{2}}$

where $\mu_M^{\sigma(j)}(x_i)$, $\nu_M^{\sigma(j)}(x_i)$, $\mu_N^{\sigma(j)}(x_i)$ and $\nu_N^{\sigma(j)}(x_i)$ are the jth values in $\mu_M(x_i)$, $\nu_M(x_i)$, $\mu_N(x_i)$ and $\nu_N(x_i)$, respectively, which will be used thereafter.

Definition 8. *generalized GH normalized distance:*
$d_3(M,N) = [\frac{1}{2n}\sum_{i=1}^{n}[\frac{1}{l_{x_i}}\sum_{j=1}^{l_{x_i}}|\mu_M^{\sigma(j)}(x_i)-\mu_N^{\sigma(j)}(x_i)|^\alpha+\frac{1}{l_{x_i}}\sum_{j=1}^{l_{x_i}}|\nu_M^{\sigma(j)}(x_i)-\nu_N^{\sigma(j)}(x_i)|^\alpha]^{\frac{1}{\alpha}}$, *where* $\alpha > 0$.

Theorem 1. *$d_3(M,N)$ is a normalized distance measure between M and N.*

Proof. We prove (D4).
Let $M \sqsubseteq N \sqsubseteq P$, for each $x_i \in X$ and each $1 \leq \sigma(j) \leq l_{x_i}$
$\mu_M^{\sigma(j)}(x_i) \leq \mu_N^{\sigma(j)}(x_i) \leq \mu_P^{\sigma(j)}(x_i)$ and $\nu_M^{\sigma(j)}(x_i) \geq \nu_N^{\sigma(j)}(x_i) \geq \nu_P^{\sigma(j)}(x_i)$.
It follows that: $|\mu_M^{\sigma(j)}(x_i) - \mu_N^{\sigma(j)}(x_i)| \leq |\mu_M^{\sigma(j)}(x_i) - \mu_P^{\sigma(j)}(x_i)|$ and
$|\nu_M^{\sigma(j)}(x_i) - \nu_N^{\sigma(j)}(x_i)| \leq |\nu_M^{\sigma(j)}(x_i) - \nu_P^{\sigma(j)}(x_i)|$.
Then, $\frac{1}{l_{x_i}}\sum_{j=1}^{l_{x_i}}(|\mu_M^{\sigma(j)}(x_i) - \mu_N^{\sigma(j)}(x_i)|^\alpha + |\nu_M^{\sigma(j)}(x_i) - \nu_N^{\sigma(j)}(x_i)|^\alpha)$
$\leq \frac{1}{l_{x_i}}\sum_{j=1}^{l_{x_i}}(|\mu_M^{\sigma(j)}(x_i) - \mu_P^{\sigma(j)}(x_i)|^\alpha + |\nu_M^{\sigma(j)}(x_i) - \nu_P^{\sigma(j)}(x_i)|^\alpha)$
And we have $d_3(M,N) \leq d_3(M,P)$. By the same analysis, we have $d_3(N,P) \leq d_3(M,P)$. Thus the property (D4) is obtained.

Corollary 1. *$d_1(M,N)$ and $d_2(M,N)$ are also normalized distance measures between M and N.*

Definition 9. *generalized GH normalized Hausdorff distance:*
$d_4(M,N) = [\frac{1}{n}\sum_{i=1}^{n} max_j\{|\mu_M^{\sigma(j)}(x_i) - \mu_N^{\sigma(j)}(x_i)|^\alpha, |\nu_M^{\sigma(j)}(x_i) - \nu_N^{\sigma(j)}(x_i)|^\alpha\}]^{\frac{1}{\alpha}}$, *where* $\alpha > 0$.

(1) If $\alpha = 1$, then Def 3.4 becomes:
$d_5(M,N) = \frac{1}{n}\sum_{i=1}^{n} max_j\{|\mu_M^{\sigma(j)}(x_i) - \mu_N^{\sigma(j)}(x_i)|, |\nu_M^{\sigma(j)}(x_i) - \nu_N^{\sigma(j)}(x_i)|\}$
(2) If $\alpha = 2$, then Def 3.4 becomes:
$d_6(M,N) = [\frac{1}{n}\sum_{i=1}^{n} max_j\{|\mu_M^{\sigma(j)}(x_i) - \mu_N^{\sigma(j)}(x_i)|^2, |\nu_M^{\sigma(j)}(x_i) - \nu_N^{\sigma(j)}(x_i)|^2\}]^{\frac{1}{2}}$

Theorem 2. *$d_4(M,N)$ is a normalized distance measure between M and N.*

Proof. We prove (D4).
Let $M \sqsubseteq N \sqsubseteq P$, for each $x_i \in X$ and each $1 \leq \sigma(j) \leq l_{x_i}$
$\mu_M^{\sigma(j)}(x_i) \leq \mu_N^{\sigma(j)}(x_i) \leq \mu_P^{\sigma(j)}(x_i)$ and $\nu_M^{\sigma(j)}(x_i) \geq \nu_N^{\sigma(j)}(x_i) \geq \nu_P^{\sigma(j)}(x_i)$.
It follows that: $|\mu_M^{\sigma(j)}(x_i) - \mu_N^{\sigma(j)}(x_i)| \leq |\mu_M^{\sigma(j)}(x_i) - \mu_P^{\sigma(j)}(x_i)|$ and
$|\nu_M^{\sigma(j)}(x_i) - \nu_N^{\sigma(j)}(x_i)| \leq |\nu_M^{\sigma(j)}(x_i) - \nu_P^{\sigma(j)}(x_i)|$.
Then, $|\mu_M^{\sigma(j)}(x_i) - \mu_N^{\sigma(j)}(x_i)|^\alpha \leq |\mu_M^{\sigma(j)}(x_i) - \mu_P^{\sigma(j)}(x_i)|^\alpha$ and
$|\nu_M^{\sigma(j)}(x_i) - \nu_N^{\sigma(j)}(x_i)|^\alpha \leq |\nu_M^{\sigma(j)}(x_i) - \nu_P^{\sigma(j)}(x_i)|^\alpha$.
And $max_j\{|\mu_M^{\sigma(j)}(x_i) - \mu_N^{\sigma(j)}(x_i)|^\alpha, |\nu_M^{\sigma(j)}(x_i) - \nu_N^{\sigma(j)}(x_i)|^\alpha\}$
$\leq max_j\{|\mu_M^{\sigma(j)}(x_i) - \mu_P^{\sigma(j)}(x_i)|^\alpha, |\nu_M^{\sigma(j)}(x_i) - \nu_P^{\sigma(j)}(x_i)|^\alpha\}$
Then, $[\frac{1}{n}\sum_{i=1}^{n} max_j\{|\mu_M^{\sigma(j)}(x_i) - \mu_N^{\sigma(j)}(x_i)|^\alpha, |\nu_M^{\sigma(j)}(x_i) - \nu_N^{\sigma(j)}(x_i)|^\alpha\}]^{\frac{1}{\alpha}}$
$\leq [\frac{1}{n}\sum_{i=1}^{n} max_j\{|\mu_M^{\sigma(j)}(x_i) - \mu_P^{\sigma(j)}(x_i)|^\alpha, |\nu_M^{\sigma(j)}(x_i) - \nu_P^{\sigma(j)}(x_i)|^\alpha\}]^{\frac{1}{\alpha}}$
And we have $d_4(M,N) \leq d_4(M,P)$. By the same analysis, we have $d_4(N,P) \leq d_4(M,P)$. Thus the property (D4) is obtained.

Corollary 2. *$d_5(M,N)$ and $d_6(M,N)$ are also normalized distance measures between M and N.*

Definition 10. *hybrid GH normalized Hamming distance:*

$$d_7(M,N) = \frac{1}{4n}\sum_{i=1}^{n}[\frac{1}{l_{x_i}}\sum_{j=1}^{l_{x_i}}|\mu_M^{\sigma(j)}(x_i) - \mu_N^{\sigma(j)}(x_i)| + \frac{1}{l_{x_i}}\sum_{j=1}^{l_{x_i}}|\nu_M^{\sigma(j)}(x_i) - \nu_N^{\sigma(j)}(x_i)|] + \frac{1}{2n}\sum_{i=1}^{n} max_j\{|\mu_M^{\sigma(j)}(x_i) - \mu_N^{\sigma(j)}(x_i)|, |\nu_M^{\sigma(j)}(x_i) - \nu_N^{\sigma(j)}(x_i)|\}$$

Definition 11. *hybrid GH normalized Euclidean distance:*

$$d_8(M,N) = [\frac{1}{4n}\sum_{i=1}^{n}[\frac{1}{l_{x_i}}\sum_{j=1}^{l_{x_i}}|\mu_M^{\sigma(j)}(x_i) - \mu_N^{\sigma(j)}(x_i)|^2 + \frac{1}{l_{x_i}}\sum_{j=1}^{l_{x_i}}|\nu_M^{\sigma(j)}(x_i) - \nu_N^{\sigma(j)}(x_i)|^2] + \frac{1}{2n}\sum_{i=1}^{n} max_j\{|\mu_M^{\sigma(j)}(x_i) - \mu_N^{\sigma(j)}(x_i)|^2, |\nu_M^{\sigma(j)}(x_i) - \nu_N^{\sigma(j)}(x_i)|^2\}]^{\frac{1}{2}}$$

Definition 12. *generalized hybrid GH normalized distance:*

$$d_9(M,N) = [\frac{1}{4n}\sum_{i=1}^{n}[\frac{1}{l_{x_i}}\sum_{j=1}^{l_{x_i}}|\mu_M^{\sigma(j)}(x_i) - \mu_N^{\sigma(j)}(x_i)|^\alpha + \frac{1}{l_{x_i}}\sum_{j=1}^{l_{x_i}}|\nu_M^{\sigma(j)}(x_i) - \nu_N^{\sigma(j)}(x_i)|^\alpha] + \frac{1}{2n}\sum_{i=1}^{n} max_j\{|\mu_M^{\sigma(j)}(x_i) - \mu_N^{\sigma(j)}(x_i)|^\alpha, |\nu_M^{\sigma(j)}(x_i) - \nu_N^{\sigma(j)}(x_i)|^\alpha\}]^{\frac{1}{\alpha}},$$ *where* $\alpha > 0$.

Theorem 3. *$d_9(M,N)$ is a normalized distance measure between M and N.*

Corollary 3. *$d_7(M,N)$ and $d_8(M,N)$ are also normalized distance measures between M and N.*

Definition 13. *generalized GH weighted distance:*

$$d_{10}(M,N) = [\frac{1}{2}\sum_{i=1}^{n}\omega_i(\frac{1}{l_{x_i}}\sum_{j=1}^{l_{x_i}}|\mu_M^{\sigma(j)}(x_i) - \mu_N^{\sigma(j)}(x_i)|^\alpha + \frac{1}{l_{x_i}}\sum_{j=1}^{l_{x_i}}|\nu_M^{\sigma(j)}(x_i) - \nu_N^{\sigma(j)}(x_i)|^\alpha)]^{\frac{1}{\alpha}},$$ *where* $\alpha > 0$.

and a generalized GH weighted Hausdorff distance:

Definition 14. *generalized GH weighted Hausdorff distance:*

$$d_{11}(M,N) = [\sum_{i=1}^{n}\omega_i max_j\{|\mu_M^{\sigma(j)}(x_i) - \mu_N^{\sigma(j)}(x_i)|^\alpha, |\nu_M^{\sigma(j)}(x_i) - \nu_N^{\sigma(j)}(x_i)|^\alpha\}]^{\frac{1}{\alpha}},$$ *where* $\alpha > 0$.

If $\alpha = 1$, then:

$$d_{12}(M,N) = \frac{1}{2}\sum_{i=1}^{n}\omega_i(\frac{1}{l_{x_i}}\sum_{j=1}^{l_{x_i}}|\mu_M^{\sigma(j)}(x_i) - \mu_N^{\sigma(j)}(x_i)| + \frac{1}{l_{x_i}}\sum_{j=1}^{l_{x_i}}|\nu_M^{\sigma(j)}(x_i) - \nu_N^{\sigma(j)}(x_i)|)$$

and:

$$d_{13}(M,N) = \sum_{i=1}^{n}\omega_i max_j\{|\mu_M^{\sigma(j)}(x_i) - \mu_N^{\sigma(j)}(x_i)|, |\nu_M^{\sigma(j)}(x_i) - \nu_N^{\sigma(j)}(x_i)|\}$$

If $\alpha = 2$, then:

$$d_{14}(M,N) = [\frac{1}{2}\sum_{i=1}^{n}\omega_i(\frac{1}{l_{x_i}}\sum_{j=1}^{l_{x_i}}|\mu_M^{\sigma(j)}(x_i) - \mu_N^{\sigma(j)}(x_i)|^2 + \frac{1}{l_{x_i}}\sum_{j=1}^{l_{x_i}}|\nu_M^{\sigma(j)}(x_i) - \nu_N^{\sigma(j)}(x_i)|^2)]^{\frac{1}{2}}$$

and:

$$d_{15}(M,N) = [\sum_{i=1}^{n}\omega_i max_j\{|\mu_M^{\sigma(j)}(x_i) - \mu_N^{\sigma(j)}(x_i)|^2, |\nu_M^{\sigma(j)}(x_i) - \nu_N^{\sigma(j)}(x_i)|^2\}]^{\frac{1}{2}}$$

Theorem 4. *$d_{10}(M,N)$ is a normalized distance measure between M and N.*

Corollary 4. *$d_{12}(M,N)$ and $d_{14}(M,N)$ are also normalized distance measures between M and N.*

Theorem 5. *$d_{11}(M,N)$ is a normalized distance measure between M and N.*

Proof. Similar to the proof of Theorem 3.2.

Corollary 5. *$d_{13}(M,N)$ and $d_{15}(M,N)$ are also normalized distance measures between M and N.*

Definition 15. *generalized hybrid GH weighted distance:*

$d_{16}(M,N)=[\frac{1}{2}\sum_{i=1}^{n}\omega_i(\frac{1}{l_{x_i}}\sum_{j=1}^{l_{x_i}}|\mu_M^{\sigma(j)}(x_i)-\mu_N^{\sigma(j)}(x_i)|^{\alpha}+\frac{1}{l_{x_i}}\sum_{j=1}^{l_{x_i}}|\nu_M^{\sigma(j)}(x_i)-\nu_N^{\sigma(j)}(x_i)|^{\alpha})+\sum_{i=1}^{n}\omega_i max_j\{|\mu_M^{\sigma(j)}(x_i)-\mu_N^{\sigma(j)}(x_i)|^{\alpha},|\nu_M^{\sigma(j)}(x_i)-\nu_N^{\sigma(j)}(x_i)|^{\alpha}\}]^{\frac{1}{\alpha}}$, *where $\alpha>0$.*

When $\alpha=1,2$, we get:

$d_{17}(M,N)=\frac{1}{2}\sum_{i=1}^{n}\omega_i(\frac{1}{l_{x_i}}\sum_{j=1}^{l_{x_i}}|\mu_M^{\sigma(j)}(x_i)-\mu_N^{\sigma(j)}(x_i)|+\frac{1}{l_{x_i}}\sum_{j=1}^{l_{x_i}}|\nu_M^{\sigma(j)}(x_i)-\nu_N^{\sigma(j)}(x_i)|)+\sum_{i=1}^{n}\omega_i max_j\{|\mu_M^{\sigma(j)}(x_i)-\mu_N^{\sigma(j)}(x_i)|,|\nu_M^{\sigma(j)}(x_i)-\nu_N^{\sigma(j)}(x_i)|\}$

$d_{18}(M,N)=[\frac{1}{2}\sum_{i=1}^{n}\omega_i(\frac{1}{l_{x_i}}\sum_{j=1}^{l_{x_i}}|\mu_M^{\sigma(j)}(x_i)-\mu_N^{\sigma(j)}(x_i)|^2+\frac{1}{l_{x_i}}\sum_{j=1}^{l_{x_i}}|\nu_M^{\sigma(j)}(x_i)-\nu_N^{\sigma(j)}(x_i)|^2)+\sum_{i=1}^{n}\omega_i max_j\{|\mu_M^{\sigma(j)}(x_i)-\mu_N^{\sigma(j)}(x_i)|^2,|\nu_M^{\sigma(j)}(x_i)-\nu_N^{\sigma(j)}(x_i)|^2\}]^{\frac{1}{2}}$.

Theorem 6. *$d_{16}(M,N)$ is a distance measure between M and N.*

Corollary 6. *$d_{17}(M,N)$ and $d_{18}(M,N)$ are also distance measures between M and N.*

Definition 16. *continuous GH weighted Hamming distance:*

$d_{19}(M,N)=\frac{1}{2}\int_a^b\omega(x)(\frac{1}{l_x}\sum_{j=1}^{l_x}|\mu_M^{\sigma(j)}(x)-\mu_N^{\sigma(j)}(x)|+\frac{1}{l_x}\sum_{j=1}^{l_x}|\nu_M^{\sigma(j)}(x)-\nu_N^{\sigma(j)}(x)|)dx$

Definition 17. *continuous GH weighted Euclidean distance:*

$d_{20}(M,N)=[\frac{1}{2}\int_a^b\omega(x)(\frac{1}{l_x}\sum_{j=1}^{l_x}|\mu_M^{\sigma(j)}(x)-\mu_N^{\sigma(j)}(x)|^2+\frac{1}{l_x}\sum_{j=1}^{l_x}|\nu_M^{\sigma(j)}(x)-\nu_N^{\sigma(j)}(x)|^2)dx]^{\frac{1}{2}}$

Definition 18. *generalized continuous GH weighted distance:*

$d_{21}(M,N)=[\frac{1}{2}\int_a^b\omega(x)(\frac{1}{l_x}\sum_{j=1}^{l_x}|\mu_M^{\sigma(j)}(x)-\mu_N^{\sigma(j)}(x)|^{\alpha}+\frac{1}{l_x}\sum_{j=1}^{l_x}|\nu_M^{\sigma(j)}(x)-\nu_N^{\sigma(j)}(x)|^{\alpha})dx]^{\frac{1}{\alpha}}$, *where $\alpha>0$.*

If $\omega(x)=\frac{1}{b-a}$, for any $x\in[a,b]$, we get:

Definition 19. *continuous GH normalized Hamming distance:*

$d_{22}(M,N)=\frac{1}{2(b-a)}\int_a^b(\frac{1}{l_x}\sum_{j=1}^{l_x}|\mu_M^{\sigma(j)}(x)-\mu_N^{\sigma(j)}(x)|+\frac{1}{l_x}\sum_{j=1}^{l_x}|\nu_M^{\sigma(j)}(x)-\nu_N^{\sigma(j)}(x)|)dx$

Definition 20. *continuous GH normalized Euclidean distance*
$d_{23}(M,N) = [\frac{1}{2(b-a)}\int_a^b(\frac{1}{l_x}\sum_{j=1}^{l_x}|\mu_M^{\sigma(j)}(x) - \mu_N^{\sigma(j)}(x)|^2 + \frac{1}{l_x}\sum_{j=1}^{l_x}|\nu_M^{\sigma(j)}(x) - \nu_N^{\sigma(j)}(x)|^2)dx]^{\frac{1}{2}}$

Definition 21. *generalized continuous GH normalized distance:*
$d_{24}(M,N) = [\frac{1}{2(b-a)}\int_a^b(\frac{1}{l_x}\sum_{j=1}^{l_x}|\mu_M^{\sigma(j)}(x) - \mu_N^{\sigma(j)}(x)|^\alpha + \frac{1}{l_x}\sum_{j=1}^{l_x}|\nu_M^{\sigma(j)}(x) - \nu_N^{\sigma(j)}(x)|^\alpha)dx]^{\frac{1}{\alpha}}$, *where* $\alpha > 0$.

Definition 22. *generalized continuous GH weighted Hausdorff distance:*
$d_{25}(M,N)=[\int_a^b \omega(x)max_j\{|\mu_M^{\sigma(j)}(x)-\mu_N^{\sigma(j)}(x)|^\alpha, |\nu_M^{\sigma(j)}(x)-\nu_N^{\sigma(j)}(x)|^\alpha\}dx]^{\frac{1}{\alpha}}$, *where* $\alpha > 0$.
$d_{26}(M,N) = \int_a^b \omega(x)max_j\{|\mu_M^{\sigma(j)}(x) - \mu_N^{\sigma(j)}(x)|, |\nu_M^{\sigma(j)}(x) - \nu_N^{\sigma(j)}(x)|\}dx$
$d_{27}(M,N) = [\int_a^b \omega(x)max_j\{|\mu_M^{\sigma(j)}(x) - \mu_N^{\sigma(j)}(x)|^2, |\nu_M^{\sigma(j)}(x) - \nu_N^{\sigma(j)}(x)|^2\}dx]^{\frac{1}{2}}$

If $\omega(x) = \frac{1}{b-a}$, for any $x \in [a,b]$, we get:
$d_{28}(M,N) = [\frac{1}{b-a}\int_a^b max_j\{|\mu_M^{\sigma(j)}(x)-\mu_N^{\sigma(j)}(x)|^\alpha, |\nu_M^{\sigma(j)}(x)-\nu_N^{\sigma(j)}(x)|^\alpha\}dx]^{\frac{1}{\alpha}}$, where $\alpha > 0$.
$d_{29}(M,N) = \frac{1}{b-a}\int_a^b max_j\{|\mu_M^{\sigma(j)}(x) - \mu_N^{\sigma(j)}(x)|, |\nu_M^{\sigma(j)}(x) - \nu_N^{\sigma(j)}(x)|\}dx$
$d_{30}(M,N) = [\frac{1}{b-a}\int_a^b max_j\{|\mu_M^{\sigma(j)}(x) - \mu_N^{\sigma(j)}(x)|^2, |\nu_M^{\sigma(j)}(x) - \nu_N^{\sigma(j)}(x)|^2\}dx]^{\frac{1}{2}}$

Definition 23. *generalized hybrid continuous GH weighted distance:*
$d_{31}(M,N) = [\int_a^b \omega(x)[\frac{1}{4l_x}\sum_{j=1}^{l_x}|\mu_M^{\sigma(j)}(x) - \mu_N^{\sigma(j)}(x)|^\alpha + \frac{1}{4l_x}\sum_{j=1}^{l_x}|\nu_M^{\sigma(j)}(x) - \nu_N^{\sigma(j)}(x)|^\alpha$
$\frac{1}{2}max_j\{|\mu_M^{\sigma(j)}(x) - \mu_N^{\sigma(j)}(x)|^\alpha, |\nu_M^{\sigma(j)}(x) - \nu_N^{\sigma(j)}(x)|^\alpha\}]dx]^{\frac{1}{\alpha}}$, *where* $\alpha > 0$.

If $\omega(x) = \frac{1}{b-a}$, for any $x \in [a,b]$, we get:

Definition 24. *generalized hybrid continuous GH normalized distance*
$d_{32}(M,N) = [\frac{1}{b-a}\int_a^b[\frac{1}{4l_x}\sum_{j=1}^{l_x}|\mu_M^{\sigma(j)}(x) - \mu_N^{\sigma(j)}(x)|^\alpha + \frac{1}{4l_x}\sum_{j=1}^{l_x}|\nu_M^{\sigma(j)}(x) - \nu_N^{\sigma(j)}(x)|^\alpha$
$+\frac{1}{2}max_j\{|\mu_M^{\sigma(j)}(x) - \mu_N^{\sigma(j)}(x)|^\alpha, |\nu_M^{\sigma(j)}(x) - \nu_N^{\sigma(j)}(x)|^\alpha\}]dx]^{\frac{1}{\alpha}}$, *where* $\alpha > 0$.

Let $\alpha = 1, 2$, we get:
$d_{33}(M,N) = \int_a^b \omega(x)[\frac{1}{4l_x}\sum_{j=1}^{l_x}|\mu_M^{\sigma(j)}(x) - \mu_N^{\sigma(j)}(x)| + \frac{1}{4l_x}\sum_{j=1}^{l_x}|\nu_M^{\sigma(j)}(x) - \nu_N^{\sigma(j)}(x)|$
$+\frac{1}{2}max_j\{|\mu_M^{\sigma(j)}(x) - \mu_N^{\sigma(j)}(x)|, |\nu_M^{\sigma(j)}(x) - \nu_N^{\sigma(j)}(x)|\}]dx$
and
$d_{34}(M,N) = [\int_a^b \omega(x)[\frac{1}{4l_x}\sum_{j=1}^{l_x}|\mu_M^{\sigma(j)}(x) - \mu_N^{\sigma(j)}(x)|^2 + \frac{1}{4l_x}\sum_{j=1}^{l_x}|\nu_M^{\sigma(j)}(x) - \nu_N^{\sigma(j)}(x)|^2$
$+\frac{1}{2}max_j\{|\mu_M^{\sigma(j)}(x) - \mu_N^{\sigma(j)}(x)|^2, |\nu_M^{\sigma(j)}(x) - \nu_N^{\sigma(j)}(x)|^2\}]dx]^{\frac{1}{2}}$
respectively.

Let $\omega(x) = \frac{1}{b-a}$, for any $x \in [a, b]$, we get:
$d_{35}(M,N) = \frac{1}{b-a}\int_a^b [\frac{1}{4l_x}\sum_{j=1}^{l_x} |\mu_M^{\sigma(j)}(x) - \mu_N^{\sigma(j)}(x)| + \frac{1}{4l_x}\sum_{j=1}^{l_x} |\nu_M^{\sigma(j)}(x) - \nu_N^{\sigma(j)}(x)|$
$+\frac{1}{2}max_j\{|\mu_M^{\sigma(j)}(x) - \mu_N^{\sigma(j)}(x)|, |\nu_M^{\sigma(j)}(x) - \nu_N^{\sigma(j)}(x)|\}]dx$:
$d_{36}(M,N) = [\frac{1}{b-a}\int_a^b [\frac{1}{4l_x}\sum_{j=1}^{l_x} |\mu_M^{\sigma(j)}(x) - \mu_N^{\sigma(j)}(x)|^2 + \frac{1}{4l_x}\sum_{j=1}^{l_x} |\nu_M^{\sigma(j)}(x) - \nu_N^{\sigma(j)}(x)|^2$
$+\frac{1}{2}max_j\{|\mu_M^{\sigma(j)}(x) - \mu_N^{\sigma(j)}(x)|^2, |\nu_M^{\sigma(j)}(x) - \nu_N^{\sigma(j)}(x)|^2\}]dx]^{\frac{1}{2}}$
respectively.

4 Similarity Between Two Generalized Hesitant Fuzzy Sets

It is known that the similarity measure and distance measure are dual concept. Hence we may use distance measures to define similarity measures.

Definition 25. *Let M, N be two GHFS sets on X and $d_{max} = max\{d(A,B), A, B \in GHFS\}$. Then the similarity measure between M, N is defined as $s(M,N)$, which satisfies the following properties:*
(D1) $0 \leq s(M,N) \leq 1$
(D2) $s(M,N) = 1 \Leftrightarrow M = N$
(D3) $s(M,N) = s(N,M)$
(D4) Let P be A GHFS set on X. If $M \sqsubseteq N \sqsubseteq P$. Then, $s(M,P) \leq s(M,N)$ and $s(M,P) \leq s(N,P)$.

5 Similarity Measures Based on the Set-Theoretic Approach

The set-theoretic approach is used usually to similarity measures for fuzzy sets [6] and intuitionistic fuzzy sets [7]. Thus we also define a similarity measure between two generalized hesitant fuzzy sets M and N from the point of set-theoretic views as follows:

$$s(M,N) = \frac{1}{n}\sum_{i=1}^{n} \frac{\sum_{j=1}^{l_{x_i}} (min_j\{\mu_M^{\sigma(j)}(x_i), \mu_N^{\sigma(j)}(x_i)\} + min_j\{\nu_M^{\sigma(j)}(x_i), \nu_N^{\sigma(j)}(x_i)\})}{\sum_{j=1}^{l_{x_i}} (max_j\{\mu_M^{\sigma(j)}(x_i), \mu_N^{\sigma(j)}(x_i)\} + max_j\{\nu_M^{\sigma(j)}(x_i), \nu_N^{\sigma(j)}(x_i)\})}$$

Theorem 7. *$s(M,N)$ is a similarity measure of generalized hesitant fuzzy sets M and N.*

Proof. $s(M,N)$ satisfies the properties (S1)-(S3). We prove (S4).
Let $M \sqsubseteq N \sqsubseteq P$, for each $x_i \in X$ and each $1 \leq \sigma(j) \leq l_{x_i}$
$\mu_M^{\sigma(j)}(x_i) \leq \mu_N^{\sigma(j)}(x_i) \leq \mu_P^{\sigma(j)}(x_i)$ and $\nu_M^{\sigma(j)}(x_i) \geq \nu_N^{\sigma(j)}(x_i) \geq \nu_P^{\sigma(j)}(x_i)$.
Then we have
$\frac{\sum_{j=1}^{l_{x_i}} (min_j\{\mu_M^{\sigma(j)}(x_i), \mu_P^{\sigma(j)}(x_i)\} + min_j\{\nu_M^{\sigma(j)}(x_i), \nu_P^{\sigma(j)}(x_i)\})}{\sum_{j=1}^{l_{x_i}} (max_j\{\mu_M^{\sigma(j)}(x_i), \mu_P^{\sigma(j)}(x_i)\} + max_j\{\nu_M^{\sigma(j)}(x_i), \nu_P^{\sigma(j)}(x_i)\})}$

$$= \frac{\sum_{j=1}^{l_{x_i}}(\mu_M^{\sigma(j)}(x_i)+\nu_P^{\sigma(j)}(x_i))}{\sum_{j=1}^{l_{x_i}}(\mu_P^{\sigma(j)}(x_i)+\nu_M^{\sigma(j)}(x_i))} \le \frac{\sum_{j=1}^{l_{x_i}}(\mu_M^{\sigma(j)}(x_i)+\nu_N^{\sigma(j)}(x_i))}{\sum_{j=1}^{l_{x_i}}(\mu_N^{\sigma(j)}(x_i)+\nu_M^{\sigma(j)}(x_i))}$$

$$= \frac{\sum_{j=1}^{l_{x_i}}(min_j\{\mu_M^{\sigma(j)}(x_i),\mu_N^{\sigma(j)}(x_i)\}+min_j\{\nu_M^{\sigma(j)}(x_i),\nu_N^{\sigma(j)}(x_i)\})}{\sum_{j=1}^{l_{x_i}}(max_j\{\mu_M^{\sigma(j)}(x_i),\mu_N^{\sigma(j)}(x_i)\}+max_j\{\nu_M^{\sigma(j)}(x_i),\nu_N^{\sigma(j)}(x_i)\})}$$

Thus, $s(M,P) \le s(M,N)$. Similarly, we have
$s(M,P) \le s(N,P)$.

Let see each element $x_i \in X$, then we obtain

$$s(M,N) = \sum_{i=1}^{n} \omega_i \frac{\sum_{j=1}^{l_{x_i}}(min_j\{\mu_M^{\sigma(j)}(x_i),\mu_N^{\sigma(j)}(x_i)\}+min_j\{\nu_M^{\sigma(j)}(x_i),\nu_N^{\sigma(j)}(x_i)\})}{\sum_{j=1}^{l_{x_i}}(max_j\{\mu_M^{\sigma(j)}(x_i),\mu_N^{\sigma(j)}(x_i)\}+max_j\{\nu_M^{\sigma(j)}(x_i),\nu_N^{\sigma(j)}(x_i)\})}$$

where $\omega_i \in [0,1]$ and $\Sigma_{i=1}^{n}\omega_i$ =1. Specially, if
$\omega_i = \frac{1}{n}, (i = 1, 2, , n)$, then the Eq. (32) are reduced to the Eq. (31).

Let the weight of element $x \in X = [a,b]$ be $\omega(x)$, where $\omega(x) \in$ [0,1] and $\int_a^b \omega(x)dx$ =1, then we define the continuous similarity measures corresponding to Eq. (32) as follow:

$$s(M,N) = \int_a^b \omega_i \frac{\sum_{j=1}^{l_{x_i}}(min_j\{\mu_M^{\sigma(j)}(x_i),\mu_N^{\sigma(j)}(x_i)\}+min_j\{\nu_M^{\sigma(j)}(x_i),\nu_N^{\sigma(j)}(x_i)\})}{\sum_{j=1}^{l_{x_i}}(max_j\{\mu_M^{\sigma(j)}(x_i),\mu_N^{\sigma(j)}(x_i)\}+max_j\{\nu_M^{\sigma(j)}(x_i),\nu_N^{\sigma(j)}(x_i)\})} dx$$

Especially, if $\omega(x) = \frac{1}{b-a}$, for any $x \in [a,b]$, then Eq. (33) become

$$s(M,N) = \frac{1}{b-a} \int_a^b \frac{\sum_{j=1}^{l_{x_i}}(min_j\{\mu_M^{\sigma(j)}(x_i),\mu_N^{\sigma(j)}(x_i)\}+min_j\{\nu_M^{\sigma(j)}(x_i),\nu_N^{\sigma(j)}(x_i)\})}{\sum_{j=1}^{l_{x_i}}(max_j\{\mu_M^{\sigma(j)}(x_i),\mu_N^{\sigma(j)}(x_i)\}+max_j\{\nu_M^{\sigma(j)}(x_i),\nu_N^{\sigma(j)}(x_i)\})} dx$$

It is obvious that $si(A,B)(i = 11, 12, , 14)$ also satisfies the properties (S1)–(S4).

Acknowledgment. This paper is supported by the Project of Shandong Province Higher Educational Science and Technology Program 2016(J16LI08).

References

1. Xu, Z.S., Xia, M.M.: Distance and similarity measures for hesitant fuzzy sets. Inf. Sci. **181**(11), 2128–2138 (2011)
2. Atanassov, K.T.: Intuitionistic fuzzy sets. Fuzzy Sets Syst. **20**(1), 87–96 (1986)
3. Torra, V.: Hesitant fuzzy sets. Int. J. Intell. Syst. **25**(6), 529–539 (2010)
4. Torra, V., Narukawa, Y.: On hesitant fuzzy sets and decision. In: 2009 IEEE International Conference on Fuzzy Systems, pp. 1378-1382 (2009)
5. Qian, G., Wanga, H., Feng, X.: Generalized hesitant fuzzy sets and their application in decision support system. Knowl.-Based Syst. **37**, 357–365 (2013)
6. Pal, S.K., King, R.A.: Image enhancement using smoothing with fuzzy sets. IEEE Trans. Syst. Man Cybern. **11**, 495–501 (1981)
7. Hung, W., Yang, M.: On similarity measures between intuitionistic fuzzy sets. Int. J. Intell. Syst. **23**, 364–383 (2008)

The Improvement of Business Processing Ability of Bank Business Hall by the Queuing Theory

Jianhua Dai(✉) and Xinyang Wang

Communication University of China, Beijing 100024, China
daijianhua66@163.com

Abstract. Bank business hall queuing problem is both a common phenomenon and a tricky problem. The key issue in solving the queuing of the banking business hall is how to adjust the relationship between service demand and service supply. This paper introduces the prevalence of queuing problems and the impact of queuing on customers, and then find out the root of the problem, using the relevant knowledge of queuing theory for case analysis. From the adjustment of the supply and demand of services to start by changing the customer's psychological perception to solve the problem of bank queuing, and ultimately improve customer satisfaction. The purpose of this paper is to solve the problem of bank queuing through the adjustment of service supply and demand and the change of customer's psychological perception, and ultimately improving customer satisfaction.

Keywords: Queuing theory · Business processing capacity · Business hall Customer diversion

1 Introduction

In the current fast-paced modern life, banks as the service industry to provide people with convenient and efficient service is the most basic requirements. The contradiction between the expansion of bank customers and the business processing capacity become more and more obvious. Waiting behavior has a strong impact on the customer experience. Waiting for service is an important factor in the perception of customer service quality. In general, the longer the waiting time, the lower the customer's evaluation of the quality of service. So shortening the queuing time is an urgent problem needed to be solved for the bank.

2 Queuing Management Theory

The idea of solving customer queuing is usually to increase the number of service facilities, but the greater the number of increases, the greater the human and material resources, and even the waste. If the service facilities are too few, the customer will wait a long time. In fact, the banks should not only guarantee a certain quality of

N. Xiong et al. (Eds.): ISCSC 2018, AISC 877, pp. 404–412, 2019.
https://doi.org/10.1007/978-3-030-02116-0_47

service indicators, but also make the cost of service facilities economically reasonable. How to properly solve the contradiction between customer queuing time and the cost of service facilities is the problem solving by queuing theory.

The general queuing system consists of three basic parts. First, the input process means that the customer arrives at the queuing system [1]. Whether the customer is limited or infinite, the time at which the customer arrives in succession is deterministic or random, whether the customer arrives is independent or relevant, the input process may be smooth or not stable. Second, the queuing rules can be divided into: first come first served, after the first service, random service and priority service. Third, the service organization [2]. It includes the time probability distribution required for each customer service, the number of service stations, and the arrangement of the service desk (series, parallel, etc.). The queuing system is shown as follow:

The queuing theory has several performance indicators: the average queue length in the system L_q, the average waiting time of the customer in the system W_q, the average waiting time of the customer in the system W_s, the average number of customers in the system L_s. The average number of commonly used indicators: average arrival rate λ, average service rate μ, the number of parallel service stations in the system S, service desk strength, that is, the average service time ρ per unit time interval. Steady-state probability P_0 and busy probability of the system P.

3 The Application of Queuing Theory in Banks

In general, the level of bank service will reduce the customer's waiting costs, but it will increase the cost of the bank. Our goal is to minimize the sum of the costs of the two, so as to achieve the optimal level of service [3]. In general, the cost of banking services can be calculated or estimated. The waiting cost of customers can also be calculated. For example, three types of service systems can help us to Fig. 1 out and solve the problem. The level of service is generally expressed by the average service rate. Queuing Theory of Economic Essence: service cost is equal to the cost of waiting, that is, the minimum total cost of services, as shown in Fig. 2.

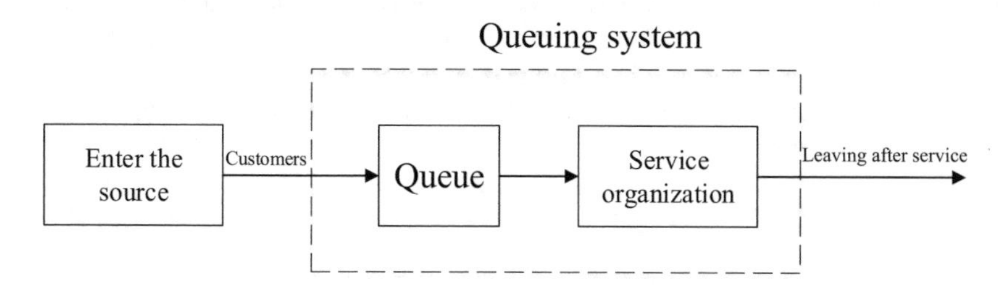

Fig. 1. Queuing system.

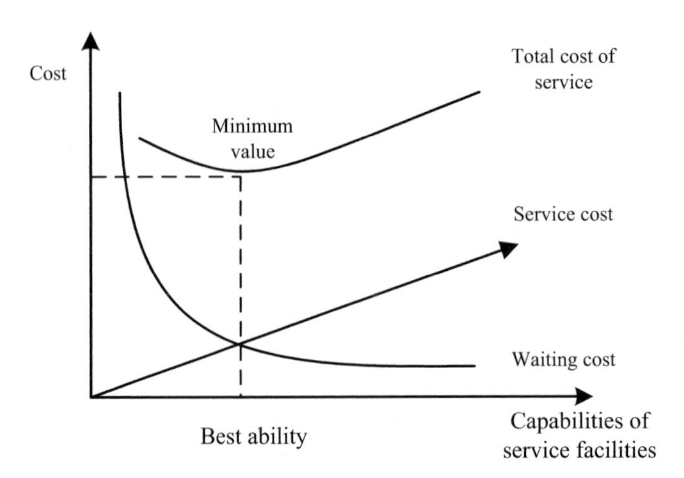

Fig. 2. Cost expenses function curve.

The real meaning of Fig. 2 is that when the cost of service is equal to the cost of waiting, the total cost of the service is minimal, and the best service capability is achieved. According to the queuing theory, for the queuing problem, we have some suggestions for determining the waiting time that the customer can accept and distracting his attention within the waiting time [4]. Promptly informing the customer the real situation. Do not let the customer see the staff rest. The relevant classification of customers. Staff training to deal with queuing phenomenon. Encourage customers to arrive at non-queuing peak, and it can effectively avoid too much waiting time. Develop a feasibility plan to eliminate queuing.

4 M/M/C/∞/∞ Queuing System

First introduce the M/M/C queuing system. The M/M/C queuing system exists in a service: the customer arrives the Poisson distribution with the parameter λ; the customer's service time follows the exponential distribution of the parameter μ; there are C service stations [5]. The customer receives the service in the order of arrival. When the customers arrives, customers will immediately get service if there is a free service station, and the customers will be arranged in a queue waiting for service if all the service stations are busy.

M/M/C/∞/∞ queuing system is the customer source and system capacity are infinite system M/M/C/∞/∞ system, generally referred to as standard M/M/C/∞/∞ systems [6].

5 Analysis of Bank Case

A professional branch of CCB of the existing site can be set up to 6 single temporary counter [7]. The branch can provide up to 6 staff members. The daily expected business volume is 6 million CNY. According to estimates, the amount of work per person per day is 200 million. CCB provides a single deposit and withdrawal business processing time limit of less than 3 min. The customer arrived in the specific selection of the customer to reach a more concentrated representative of the time period for 15 days of survey statistics, the frequency shown in Table 1. Each additional one single temporary cabinet work room will need an additional investment of 100,000 CNY. According to the characteristics of the work of the savings, combined with the customer waiting for service expectations, queuing system indicators of the standard reference value can be the following values (Fig. 3):

$$P_0 = 0.4,\ L_s = 2,\ W_s = 3$$

In general, the customer will form the Poisson flow in the process of arrival, and the negative exponential probability distribution can better describe the probability distribution of service time in the queuing system. CCB's expected daily business volume of 6 million CNY per person per day to complete the workload of 200 million, so the number of service units range [3, 6]. Therefore, the bank's queuing model belongs to the M/M/C/∞/∞ model. Solution: first, to determine the average number of customers arriving per unit time. Second, to determine the average service rate. Third, calculate

Table 1. Customer arrival frequency table.

Number of arrived customers	21	24	33	38	42	45	48	49	52	53	61
Days	1	1	2	1	2	2	1	1	2	1	1

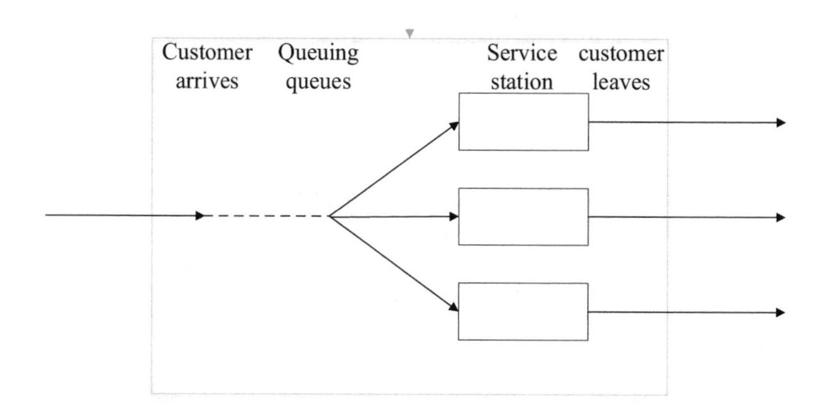

Fig. 3. M/M/C queuing system.

the C points to determine the best number of service units. Fourth, synthesize the amount of investment.

Calculate the average number of customers arriving per unit of time λ:

$$\lambda = \frac{1}{N}\sum nf_n \tag{1}$$

According to the data in Table 1, it can be obtained $\lambda = 0.71$. Calculate the average service rate μ: the maximum time limit for service is 3 min, so it can be assumed that the system processes 0.3 users on average, i.e. the average service rate $\mu = 0.3$. Calculate the values of P_O, L_q, W_q, W_S, L_S when C is 3, 4, 5, 6, and compare with the standard reference:

The calculation of each quantity index is as follows:

$$\mathrm{P}_0 = \frac{1}{\sum_{n=0}^{s-1} \frac{(\lambda/\mu)^n}{n!} + \frac{(\lambda/\mu)^c}{c!}\left(\frac{c\mu}{c\mu-\lambda}\right)} \tag{2}$$

$$L_q = \frac{(\lambda/\mu)^c \lambda\mu}{(c-1)!(c\mu-\lambda)^2} P \tag{3}$$

$$L_s = L_q + \frac{\lambda}{\mu};\ W_q = \frac{L_q}{\lambda};\ W_s = W_q + \frac{1}{\mu} \tag{4}$$

P_O, L_q, W_q, W_S, L_S values shown in Table 2 when C is 3, 4, 5, 6.

Table 2. The value of the number of indicators when the C take different values.

C	P_O	L_q	W_q	W_S	L_S
3	0.056	2.291	3.080	3.226	4.338
4	0.083	0.409	1.001	0.577	1.410
5	0.089	0.100	0.574	0.142	0.808
6	0.090	0.025	0.420	0.036	0.591

Finally, determine the best number of service stations. Compared with the standard reference value $P_O = 0.4$, $L_S = 2$, $W_S = 3$, it can be found that when $C \geq 4$, the requirements of P_O, L_S and WS are met. From an intuitive point of view, each additional service desk needs to spend more than 100,000 CNY. And the value of the indicators is no significant improvement compared with the C take 4 when C take 5, 6. So the service desk should be set up 4 units.

6 Customer Diversion Strategy

The root cause of the queuing phenomenon is that the service demand does not match the service supply, that is, the service demand is greater than the service supply [8]. So queuing can be done by adjusting service requirements and adjusting service offerings. At the same time, the management of customer queuing because customers do not want to wait, its purpose is to increase customer awareness of the quality of service, thereby enhancing customer satisfaction. And the customer's perception will be different because of the specific circumstances of waiting, so the third way to solve customer queuing problem is to adjust the psychological perception of customer.

6.1 Adjust Service Requirements

Adjust customer demand is through a variety of channels to customers divert to other channels, or encourage customers to do banking business at leisure time, to achieve the purpose of reducing the peak demand for services.

Introducing Customers to Other Business Offices. Drawing an online business hall "electronic map", and the business hall entity queuing called system connection. It can be real-time display of the business outlets of the queuing information. Customers will be able to go through the network to check the business hall of the queue waiting to avoid the flow of people peak. This initiative will effectively achieve the customer's diversion, to avoid too many customers to the same business hall congestion.

Distributing Customers to Other Channels. Try to increase the publicity of electronic channels, guide the customer effectively, including the electronic channel of their own publicity, the use of media and so on. Business hall can also introduce and promote relevant information through the online business hall, SMS business hall, self-service business hall and other channels.

Strengthening Assignments of the Network Lobby Manager for Customer Guide. For larger business outlets, the branch needs to be equipped with a full-time lobby manager. Lobby manager can be the first time to identify customers, and actively take the initiative to do the work area guide work. Knowing the needs of customers in advance and then effectively shunting will help to minimize the waiting time for customers to do business.

6.2 Adjusting Service Supply

By tapping the potential of the business hall, the rational allocation of resources and improving the efficiency of resource use can increase the service supply of the business hall to meet the service needs. According to the specific circumstances of the business hall, the strategy is as follows:

Dynamic Shift Management. Dynamic scheduling refers to historical data based on business volume. Using the forecast method to calculate the business volume of the next stage business hall. And then according to the changes in traffic, calculate the number of staff required for each period, business hall instructors, the number of staff, other staff to do business hall shift and desk arrangements, the number of tellers per day and the required number of tellers [9]. Dynamic scheduling is to achieve the business hall business volume, traffic and the number of seats, service personnel and other operational indicators to achieve the best allocation of reasonable conditions.

There are many ways to schedule a job. The commonly used method is to use the collection of historical business acceptance data to calculate. The brief description is as follows:

First, clear the number of business processes for each time period. Second, count the number of each hour of each business for a unit. According to the provisions of each business processing time, calculate the time for each business for all the time required. Use this time divided by 60 min, that is, the number of seats required to be scheduled for this period.

The formula for calculating the number of seats is as follows:

The number of seats to be scheduled for this period = ($\sum$ The transactions in a period of time) $\times$ the length of time required to process the business/60 min.

Implementation of "Flexible Working Hours" and "Flexible Window Measures". For seasonal, phased, regional characteristics, customer queuing problems can be solved through the implementation of "flexible working hours", "flexible window settings" and other measures.

Simplify Business Processes. Now people go to the bank to handle the business tends to be diversified and complicated. And the bank's services can't keep up with the time, so the emergence of the bank long queues phenomenon. To solve the problem of bank queuing is necessary to do customer segmentation. Bank of China has been piloted in Shanghai [10]. The public areas of the bank counters are classified. Paid counters are all used to do the payment business, the other bank agents counter are used for specialized agents business.

Increase Electronic Self-service Equipment and Improve Its Use and Security. The long-term effective way solving the contradiction between the supply and demand of financial services is to replace the manual services by increasing the number of electronic self-service devices (mainly telephone banking, online banking and self-service banks), with lower cost of modern means to solve the problem. Banks not only need to increase the number of electronic self-service equipment, but also to improve the electronic self-service equipment business capacity. The first is to improve the technical content, so that the general counter business can be quickly completed with self-service equipment or online banking. Second, it is necessary to solve the electronic banking security issues, so that customers believe that the bank's electronic products, diverting customers and reducing the pressure on the counter.

Strengthen Staff Training and Improve Business Skills. With the increasingly fierce competition in the industry, new products and new business constantly updated. If the teller business training is not enough, there will be unskilled operator, affecting the business processing speed. Therefore, commercial banks should pay attention to the training of tellers. When the new products and new business launch, the bank should promptly train the staff. Teller skills training should also be strengthened in the regular time for the teller technical level assessment rating. The combination of rating assessment and assessment will ensure the efficiency of the teller business. Eventually, the customer waiting time is shortened by improving the efficiency of the teller business.

7 Conclusion

Customer evaluation of the service depends on the customer's expectations and service perception, rather than the actual enjoyment of the service. So customers perceive that waiting is usually more important than the actual waiting time. In the case that the customer must wait in line, the waiting time in the customer's sense can be reduced in an appropriate manner so that the customer does not feel that the wait is intolerable and the customer is "willing" and comfortable to wait. We can use a series of means to reduce the customer's waiting time.

Establishing a fair queuing rules. Now a lot of banks will generally give priority to VIP customers to do business, reflecting the high value of the outstanding customer service. But if the customer see the later do their business earlier, the waiting customer will increase anxiety, and even feel anger because of unfair treatment. So bank should try to avoid ordinary customers when the use of the highest priority law on the VIP customer priority service. Fairness also comes from the attitude of the teller. And when the customers feel that the salesperson does not know that they are waiting, or they see the tellers are not at work, they will feel unfairly treated. So the tellers need to strengthen coordination and management that they know customers are waiting.

Acknowledgment. The helpful comments of the editor and two anonymous referees are gratefully acknowledged. The work has been supported by National Social Science Fund (17BC059) and the research tasks of Communication University of China (3132014XNG1458, 3132018XNG1863, CUC18QB12, CUC16A17, CUC17A16, CUC18QB12). Moreover, this work has been partly performed in the project "A Study on Interconnection Charges during China's Integration of Telecommunications, Broadcasting and Internet Networks" supported by National Natural Science Fund (71203202).

References

1. Wüchner, P., Sztrik, J.: Finite-source M/M/S retrial queue with search for balking and impatient customers from the orbit. Comput. Netw. **53**(8), 1264–1273 (2009)
2. Fiems, D., Maertens, T.: Queueing systems with different types of server interruptions. Eur. J. Oper. Res. **188**(3), 838–845 (2008)
3. Van Do, T.: M/M/1 retrial queue with working vacations. Acta Inform. **47**(1), 67–75 (2010)
4. Hühn, O., Markl, C., Bichler, M.: On the predictive performance of queueing network models for large-scale distributed transaction processing systems. Inf. Technol. Manag. **10**(2), 135–149 (2009)
5. Gershkov, A., Schweinzer, P.: When queueing is better than push and shove. Int. J. Game Theory **39**(3), 409–430 (2010)
6. Grassmann, W.K., Tavakoli, J.: Transient solutions for multi-server queues with finite buffers. Queueing Syst. **62**(1), 35–49 (2009)

7. Amador, J., Artalejo, J.R.: Transient analysis of the successful and blocked events in retrial queues. Telecommun. Syst. **41**(4), 255–265 (2009)
8. Creemers, S., Lambrecht, M.: Queueing models for appointment-driven systems. Ann. Oper. Res. **178**(1), 155–172 (2010)
9. Qiu, W., Dai, D., Zuo, H.: Business Hall Full-Service Operation and Management, pp. 169–180. People Post Press (2010)
10. Tang, B.: Analysis and solution of bank queuing problem. Spec. Econ. Zone (3) (2008)

Celastrol Binds to HSP90 Trigger Functional Protein Interaction Network Against Pancreatic Cancer

Guang Zheng[1(✉)], Chengqiang Li[1], Xiaojuan He[2], Jihua Wang[2], and Hongtao Guo[3(✉)]

[1] School of Information Science and Engineering, Lanzhou University, Lanzhou 730000, People's Republic of China
forzhengguang@163.com
[2] Institute of Basic Research in Clinical Medicine, China Academy of Chinese Medical Sciences, Beijing 100700, China
[3] The first affiliated hospital of Henan University of TCM, Zhengzhou 450000, People's Republic of China
guoht2009@126.com

Abstract. Celastrol is a bioactive compound that can be extracted from Chinese herbal medicine named *Tripterygium wilfordii* Hook f. It had been reported that celastrol exhibited potential anti-tumor effects against pancreatic cancer. However, with limited evidences, it is hard to understand its therapeutic mechanism. In this study, focused on the very binding target protein HSP90, functional regulation network of celastrol against pancreatic cancer were established by integrating functional protein interaction data. In this functional regulation network, celastrol first binds to HSP90, then AKT1/2/3, IL6, IKBKB, TSC2, MTOR, NOTCH1 and CDH1 may further regulated against pancreatic cancer.

Keywords: Celastrol · Pancreatic cancer · HSP90 · AKT · Regulation network

1 Introduction

Extracted from Chinese herbal medicine named *Tripterygium wilfordii* Hook f., celastrol is a natural bioactive compound with potential therapeutic effects on pancreatic cancer [1, 2]. Primary experiments reported that celastrol exhibited anti-tumor effects via binding to target protein HSP90 [3].

However, with limited and scattered reports, it is difficult to build the functional protein interacting network regulated by celastrol against pancreatic cancer. As the functional protein interacting network may shed light on the understanding in therapeutic mechanism of celastrol, it is important to construct this functional protein network.

As being reported, HSP90 was identified as the binding target of celastrol [4]. Then, things left for the network construction involves integrating functional protein interactions. The integrated protein regulation relationships including four major types e.g., *activation*, *up-regulation*, *inhibition*, and *down-regulation* which can be categorized into 2: *activation* and *inhibition*.

N. Xiong et al. (Eds.): ISCSC 2018, AISC 877, pp. 413–419, 2019.
https://doi.org/10.1007/978-3-030-02116-0_48

Begin with HSP90, the creation of functional protein interaction network targeting pancreatic cancer and tissue pancreas was launched. Although HSP90 can be bound by celastrol. However, functional protein interaction data contains no knowledge on HSP90. Further literature reviewing demonstrated that in mammalian cells, there are several genes encoding homologues of HSP90 [5]. Besides, they share large portion sequence identity (>85%) [5]. To be specified, there are five functional human genes encoding Hsp90 protein isoforms: HSP90AA1, HSP90AA2, HSP90AB1, HSP90B1, and TRAP1 [6].

Started with these five genes, functional protein interactions towards pancreas and pancreatic cancer were extracted from protein interaction knowledge base. However, there is no direct interactions from HSP90s to pancreas and pancreatic cancer. Based on the experimental result, there must be some functional protein interactions to deliver its anti-cancer effects which might have been found, reported and collected indirectly. With this hypothesis hold, additional bridging protein interactions were taken into consideration. These bridging protein interactions connect HSP90s from the left and deliver interactions towards pancreas and pancreatic cancer to the right.

As a result, this network is large. There are 86 edges started from HSP90s to pancreatic cancer associated genes, and 5 edges from HSP90s to proteins expressed in pancreas. It was hard to figure out its meaning which contains 33 nodes and 91 edges. In order to simplify the functional network, nodes with higher connection degree (incoming degree $\geq$ 2 and outgoing degree $\geq$ 2) were included. The resulting network is simplified with relatively high connection degrees (13 nodes, 30 edges) and demonstrated in Fig. 1.

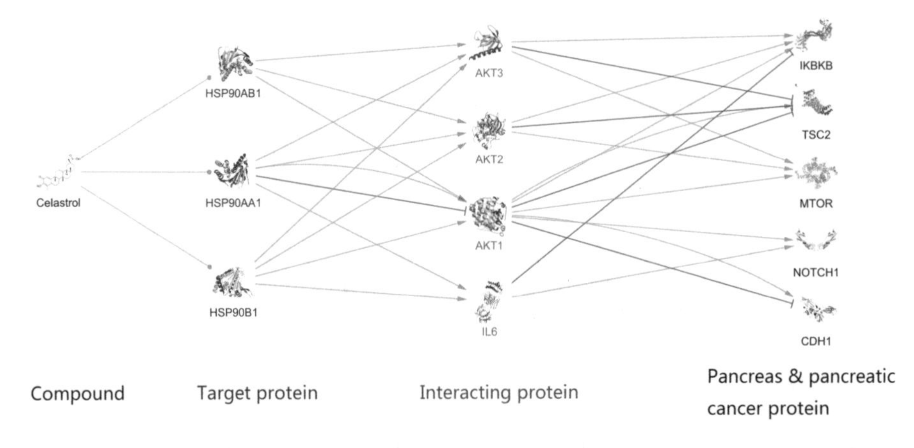

Fig. 1. The regulation network of celastrol targeting pancreatic cancer. On querying in PubMed with keyword "celastrol", "activation" and "inhibition", scattered experimental results reported revealed that proteins of MAPK1/3, PIK3R1, CHUK, TLR4, AKT1, MYC, PTGS2, CASP7/8, HSF1, MTOR and IKBKB (middle) can be either activated or inhibited by celastrol (left). Interestingly, two pancreas and pancreatic cancer proteins MTOR and IKBKB (middle-right) can be inhibited both directly by celastrol or intermediated by PIK3R1, CHUK, TLR4, AKT1 and CASP8. Other four pancreas and pancreatic cancer proteins TSC2, CDH1, BAG3 and PARP1 (right) can also be regulated via proteins reported in PubMed with experimental validations.

With the established protein interaction network with functional relationship restricted within activation and inhibition, the general meaning of it was done with pathway enrichment analysis in KEGG (http://www.genome.jp/kegg/). KEGG returned supportive results of pathway in pancreatic cancer (Fig. 2) which means the resulting network may lead the attack against pancreatic cancer triggered by celastrol binds to HSP90s.

This study provides an approach of establishing functional protein interaction network with only one limited binding information of compound that can be extracted from Chinese herbal medicine against a disease.

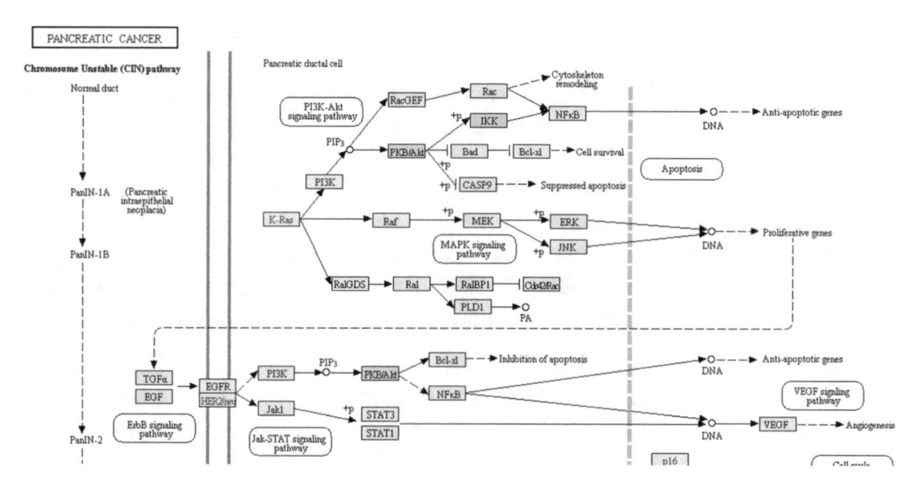

Fig. 2. David annotation of proteins in the functional PPI network of celastrol against pancreatic cancer. In this figure, KEGG pathway chart of pancreatic cancer was enriched with P-value = 1.6E−7. In this char, 6 gene/proteins matched and highlighted with red stars. These proteins participated in the process of anti-apoptotic and proliferations which are vital for the development of pancreatic cancer.

2 Data and Method

2.1 Data

This part contains data collected for the construction of functional PPI network of celastrol binds to HSP90s targeting tissue pancreas and pancreatic cancer.

Tissue Protein and Disease Associated Gene. Tissue proteins expressed in pancreas were extracted from UniProt as it provides the scientific community with a high-quality, comprehensive, and freely accessible knowledgebase of protein sequence together with functional information [7]. Focused on pancreas, UniProt returned 1,557 proteins expressed in it.

As for the genes associated with pancreatic cancer, they were fetched from OMIM as it is a database contains human genetic disorders associated with disease. OMIM particularly focuses on the gene-phenotype relationship [8]. For pancreatic cancer, OMIM returned 250 associated genes.

Functional Protein Interaction Knowledge. The functional protein interaction knowledge was collected from STRING (http://string-db.org/) as it is a leading protein-protein interaction database [9]. In order to collect confidential functional protein interaction knowledge. The searching conditions were strictly restricted with `confidence_score` greater than `700`, `a_is_acting` is true, `mode` is not `NULL`, and `action` in `{activation, inhibition}`.

2.2 Method

Simplification of Functional Protein Interaction Network. When queried, STRING returned 86 functional protein interactions started from HSP90s to pancreatic cancer associated genes, and 5 functional protein interactions from HSP90s to proteins expressed in pancreas. Thus, the functional protein interaction network contains 33 nodes and 91 edges which is large.

In order to simplify the network, the cropping was done via graph node degree. Bridging functional proteins were left only if they have total degree ≥ 4 and incoming degree ≥ 2. Pancreas and pancreatic cancer proteins were restricted with degree ≥ 2.

The Demonstration of Functional Protein Interaction Network. The simplified functional protein interaction network was visualized in Cytoscape (version 3.5.1, http://cytoscape.org/) as it is an important software in visualizing complex bio-medical networks [10].

Pathway Enrichment Analysis. The biological process annotation of functional protein interaction network was carried out with pathway enrichment analysis in KEGG as it is a knowledgebase for understanding macro-level bio-medical functions, systematically [11].

3 Result

In this study, the functional protein interaction network of celastrol binds to HSP90s against pancreatic cancer was constructed with known target binding protein, UniProt tissue protein information, and OMIM disease genes.

3.1 Simplified Densely Connected Functional Protein Interaction Network of HSP90s Bound by Celastrol Against Pancreatic Cancer Was Established

After proteins with less functional interactions were moved away, in Fig. 1, only 13 proteins with 30 functional interactions were left in the functional protein interaction network regulated after HSP90s bound by celastrol against pancreatic cancer.

The bridging proteins are AKT1/2/3 and IL6. Further regulated proteins including IKBKB, TSC2, MTOR, NOTCH1 and CDH1 which are either pancreas residence proteins or pancreatic cancer associated genes. Taken together, they form the simplified functional protein interaction network.

3.2 The Functional Protein Interaction Network Is Composed with Proteins Enriched in Pancreatic Cancer Pathway

With proteins in network uploaded onto KEGG for pathway enrichment analysis, KEGG returned supportive result that pancreatic cancer pathway is enriched with AKT1/2/3 and IKBKB which are highlighted with red background in Fig. 2.

4 Discussion

4.1 AKT1/2/3 and IL6 Are the Bridging Proteins to Deliver Celastrol's Anti-pancreatic Cancer Effect

AKT, with isoforms of AKT1/2/3, is involved in the PI3K/AKT/mTOR pathway and other signaling pathways [12, 13]. When being regulated, signaling pathway AKT/PI3K plays a candidate potential therapeutic role against pancreatic cancer [14]. When activated and cooperate with KRas, AKT1 can to accelerate pancreatic tumor onset and progression in vivo [12].

IL6 plays crucial role in many bio-medical activities e.g., immune responds. It was reported as a key role in the development and progression of pancreatic cancer [15].

4.2 IKBKB, TSC2, MTOR, NOTCH1 and CDH1 Are Targets Regulated by Bridging Proteins Against Pancreatic Cancer

IKBKB is the activator of NF-κB pathway [16]. As NF-κB plays an essential role in both acute and chronic inflammation, [17] then, inhibit IKBKB may induce anti-pancreatic cancer effects.

When mutated, TSC2 gene may induce susceptibility to pancreatic cancer which can be taken as therapeutic targets on PI3K-mTOR pathway [18].

MTOR, a pancreas-residence protein, plays a central role in connecting intracellular/extracellular growth signals and metabolism which is associated with both innate and adaptive immune responses [19].

When associated with and enhanced the signaling of NOTCH1, cancer cell can survival and process invasion [20]. It was also argued that when being down-regulated by inhibitors or natural compounds, suppressing the NOTCH signaling pathway be a new approach for the treatment against pancreatic cancer [21].

As for CDH1, abnormal expression of cytoplasmic CDH1 and HDAC3 in nuclear can lead to low quality prognosis in pancreatic cancer patients [22]. It was also argued the associations between the risk of pancreatic cancer and CDH1 polymorphisms in Chinese Han population [23].

Taken together, both 4 bridging functional proteins and 5 targeted proteins in tissue pancreas and pancreatic cancer, when connected with functional regulations against pancreatic cancer, they form a dynamic regulation network which may shed light on the mechanism of pancreatic cancer.

5 Conclusion

In this study, a simplified functional protein interaction network of celastrol against pancreatic cancer was carried out. In this network, focused on 3 isoforms of HSP90 bound by celastrol, inter-linked with 4 bridging functional proteins, 5 targeted proteins of tissue pancreas and pancreatic cancer were filtered out with threshold of interaction degrees.

The correctness of this functional protein interaction network can be validated by both biomedical literatures and bioinformatic tools. Yet this network contains new knowledge for both the pathology of pancreatic cancer and the therapeutic effects of celastrol against it.

This study proposes a new way to construct the functional protein interaction network with only one binding information against a specified disease.

Acknowledgement. The authors should thank for the financial support from projects: 158360004 and 142300410386.

References

1. Yadav, V.R., Sung, B., Prasad, S., et al.: Celastrol suppresses invasion of colon and pancreatic cancer cells through the downregulation of expression of CXCR4 chemokine receptor. J. Mol. Med. **88**(12), 1243–1253 (2010)
2. Sung, B., Park, B., Yadav, V.R., Aggarwal, B.B.: Celastrol, a triterpene, enhances TRAIL-induced apoptosis through the down-regulation of cell survival proteins and up-regulation of death receptors. J. Biol. Chem. **285**(15), 11498–11507 (2010)
3. Chen, X., Wang, L., Feng, M.: Effect of Tripterine on mRNA expression of c-myc and platelet derived growth factor of vascular smooth muscle cell in rats. Zhongguo Zhong Xi Yi Jie He Za Zhi **18**(3), 156–158 (1998)
4. Zanphorlin, L.M., Alves, F.R., Ramos, C.H.: The effect of celastrol, a triterpene with antitumorigenic activity, on conformational and functional aspects of the human 90kDa heat shock protein Hsp90α, a chaperone implicated in the stabilization of the tumor phenotype. Biochim. Biophys. Acta **1840**(10), 3145–3152 (2014)
5. Chen, B., Zhong, D., Monteiro, A.: Comparative genomics and evolution of the HSP90 family of genes across all kingdoms of organisms. BMC Genom. **7**(1), 156 (2006)
6. Chen, B., Piel, W.H., Gui, L., Bruford, E., Monteiro, A.: The HSP90 family of genes in the human genome: insights into their divergence and evolution. Genomics **86**(6), 627–637 (2005)
7. Pundir, S., Martin, M.J., O'Donovan, C.: UniProt protein knowledgebase. In: Methods in Molecular Biology, pp. 41–55. Humana Press, New York (2017)

8. Amberger, J.S., Bocchini, C.A., Schiettecatte, F., Scott, A.F., Hamosh, A.: OMIM.org: Online Mendelian Inheritance in Man (OMIM®), an online catalog of human genes and genetic disorders. Nucleic Acids Res. **43**(Database issue), D789–D798 (2015)
9. Szklarczyk, D., Franceschini, A., Wyder, S., et al.: STRING v10: protein-protein interaction networks, integrated over the tree of life. Nucleic Acids Res. **43**(Database issue), D447–D452 (2015)
10. Shannon, P., Markiel, A., Ozier, O., et al.: Cytoscape: a software environment for integrated models of biomolecular interaction networks. Genome Res. **13**(11), 2498–2504 (2003)
11. Kanehisa, M., Furumichi, M., Tanabe, M., et al.: KEGG: new perspectives on genomes, pathways, diseases and drugs. Nucleic Acids Res. **45**(D1), D353–D361 (2017)
12. Albury, T.M., Pandey, V., Gitto, S.B., et al.: Constitutively active Akt1 cooperates with KRas(G12D) to accelerate in vivo pancreatic tumor onset and progression. Neoplasia **17**(2), 175–182 (2015)
13. Marchbank, T., Mahmood, A., Playford, R.J.: Pancreatic secretory trypsin inhibitor causes autocrine-mediated migration and invasion in bladder cancer and phosphorylates the EGF receptor, Akt2 and Akt3, and ERK1 and ERK2. Am. J. Physiol. Renal Physiol. **305**(3), F382–F389 (2013)
14. Ebrahimi, S., Hosseini, M., Shahidsales, S., et al.: Targeting the Akt/PI3K signaling pathway as a potential therapeutic strategy for the treatment of pancreatic cancer. Curr. Med. Chem. **24**(13), 1321–1331 (2017)
15. Holmer, R., Goumas, F.A., Waetzig, G.H., et al.: Interleukin-6: a villain in the drama of pancreatic cancer development and progression. Hepatobiliary Pancreat. Dis. Int. **13**(4), 371–380 (2014)
16. Schmid, J.A., Birbach, A.: IkappaB kinase beta (IKKbeta/IKK2/IKBKB)–a key molecule in signaling to the transcription factor NF-kappaB. Cytokine Growth Factor Rev. **19**(2), 157–165 (2008)
17. Catrysse, L., van Loo, G.: Inflammation and the metabolic syndrome: the tissue-specific functions of NF-κB. Trends Cell Biol. **27**(6), 417–429 (2017)
18. Navarrete, A., Armitage, E.G., Musteanu, M., et al.: Metabolomic evaluation of Mitomycin C and rapamycin in a personalized treatment of pancreatic cancer. Pharmacol. Res. Perspect. **2** (6) (2014)
19. Jones, R.G., Pearce, E.J.: MenTORing immunity: mTOR signaling in the development and function of tissue-resident immune cells. Immunity **46**(5), 730–742 (2017)
20. Liu, H., Zhou, P., Lan, H., et al.: Comparative analysis of Notch1 and Notch2 binding sites in the genome of BxPC3 pancreatic cancer cells. J. Cancer **8**(1), 65–73 (2017)
21. Gao, J., Long, B., Wang, Z.: Role of Notch signaling pathway in pancreatic cancer. Am. J. Cancer Res. **7**(2), 173–186 (2017)
22. Jiao, F., Hu, H., Han, T., Zhuo, M., et al.: Aberrant expression of nuclear HDAC3 and cytoplasmic CDH1 predict a poor prognosis for patients with pancreatic cancer. Oncotarget **7** (13), 16505–16516 (2016)
23. Zhao, L., Wang, Y.X., Xi, M., et al.: Association between E-cadherin (CDH1) polymorphisms and pancreatic cancer risk in Han Chinese population. Int. J. Clin. Exp. Pathol. **8**(5), 5753–5760 (2015)

Identification of Key Genes and Pathways Involved in Compensatory Pancreatic Beta Cell Hyperplasia During Insulin Resistance

Amnah Siddiqa[1], Jamil Ahmad[1(✉)], Rehan Zafar Paracha[1], Zurah Bibi[1], and Amjad Ali[2]

[1] Research Center for Modeling and Simulation, National University of Sciences and Technology, H-12, Islamabad, Pakistan
dr.ahmad.jamil@gmail.com

[2] Atta-ur-Rehman School for Applied Biosciences, National University of Sciences and Technology, H-12, Islamabad, Pakistan

Abstract. Restoration of pancreatic beta cell mass is anticipated as beneficial treatment for both types of Diabetes Mellitus (DM) (I & II). Previously, compensatory increase of pancreatic beta cell mass has been observed in various insulin resistance conditions. Investigation of regulatory mechanism behind the pancreatic beta cell hyperplasia through the liver derived growth factors have profound implications in treatment of DM. Therefore, this study is conducted with a specific aim to identify the key regulatory elements and pathways involved in the compensatory increase of pancreatic beta cells during insulin resistance in liver. For this purpose, three transcriptomics datasets of different insulin resistant mouse models (S961 induced, high fat diet induced and db/db) are analyzed in this study. Common differentially expressed genes (DEGs) and significantly altered pathways in these datasets are obtained. These genes and pathways represent the key elements of the gene regulatory circuit involved in compensatory pancreatic beta cell hyperplasia during insulin resistance. These findings have increased our current knowledge regarding the genes and gene circuits involved collectively in different insulin resistance conditions.

Keywords: Insulin resistance · Pancreatic beta cell replication
Pathway analysis

1 Introduction

Diabetes Mellitus (DM) has become a major disease burden in both developing and developed countries and its prevalence is predicted to reach 366 million (approximately) by 2030, worldwide [1]. In this regard, targeting the proliferation of insulin producing pancreatic beta cells is a promising anticipated therapy

N. Xiong et al. (Eds.): ISCSC 2018, AISC 877, pp. 420–427, 2019.
https://doi.org/10.1007/978-3-030-02116-0_49

for both types of DM [2]. The pancreatic beta cells have the ability to replicate rapidly under certain physiological (such as gestation) and pathological (such as obesity and diabetes) insulin resistance challenges. This compensatory increase of pancreatic beta cells observed in obesity, diabetes and gestational periods has been witnessed in both the mouse models of obesity (leptin gene mutant also known as ob/ob), diabetes (leptin receptor gene mutant also known as db/db), High Fat Diet (HFD) induced obesity and gestational diabetes as well as in equivalent human scenarios [3–7]. Several studies demonstrated the existence of liver derived systemic growth factors involved in this proliferative process. Identification of common genes and biological processes of different insulin resistance states in liver could aid in identification and targeting of the biological molecules (genes/proteins) which might play a role in gene regulatory circuits involved in the interorgan (liver-pancreas) crosstalk for therapeutic purposes [8–10]. Therefore, the aim of the present study is to identify common set of genes expressed in different insulin resistance mouse models to increase our understanding towards the biological processes involved in pancreatic beta cell hyperplasia in these states.

2 Methodology

The complete methodology workflow adopted in this study is illustrated in Fig. 1.

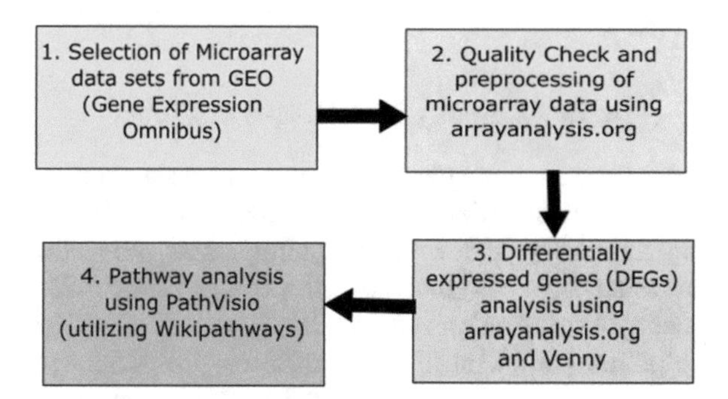

Fig. 1. Methodology: GEO (Gene Expression Omnibus) was searched for appropriate transcriptomics data sets. Each data set was individually checked for quality control and preprocessing using on-line microarray data analysis server "arrayanalysis.org". DEGs (Differentially expressed genes) were obtained by performing statistical analysis using arrayanalysis.org, from each data set. Next, the pathway analysis was performed in order to identify the significantly overrepresented pathways from each dataset using PathVisio (pathway visualization and analysis software).

2.1 Transcriptomics Data Selection

In this study, we searched the public genomics data repository Gene Expression Omnibus (GEO) to find the transcriptomics datasets which fulfill the following criteria: (i) they must be the insulin resistance mouse models of obesity (ob/ob), diabetes (db/db), insulin receptor inhibitor drugs (S961 etc) treatment, HFD treatment or gestational diabetes (ii) the studied tissue should be liver (iii) the biological replicates must be greater than or equal to three (iv) studies must provide raw data files belonging to any microarray platform (affymetrix, agilent, illumina etc). Three different datasets were selected based on the inclusion criteria described above. An overview of features of these datasets is provided in Table 1. The selected datasets include an insulin resistance mouse model developed through S961 (insulin receptor inhibitor drug) treatment (GEO ID: GSE45694), HFD treated (GEO ID: GSE57425) mice and diabetic (db/db) mice (GEO ID: GSE59930) compared with their control counterparts. S961 treated mouse model is a recently designed and robust insulin resistance model whose details can be viewed in [10].

Table 1. Selected transcriptomics datasets

GEO ID	Array platform	Number of samples (abbreviations: C=Controls, HFD=High fat diet treated, db/db=Diabetic)
GSE45694	Illumina MouseRef-8 v2.0	8 (C:4; S961 treated:4)
GSE57425	Affymetrix Mouse Genome 430 2.0	6 (C:3; HFD:3)
GSE59930	Affymetrix Mouse Genome 430 2.0	9 (C:5; db/db:4)

2.2 Data QC (Quality Check) and Preprocessing

Data QC and preprocessing of microarray data is necessary for removing the outliers and technical noise (details can be followed in [11]). The raw data for each selected dataset was downloaded which was proceeded by QC analysis and preprocessing. Two affymetrix datasets i.e. GSE57425 and GSE59930 were analyzed using ArrayAnalysis.org which is an online microarray QC and pre-processing pipeline [11]. The illumina dataset (GSE45694) was read using the R/Bioconductor package "beadarray" in the R environment.

2.3 DEGs (Differentially Expressed Genes) Analysis

The gene expression difference between control and treatment samples in transcriptomics datasets is used to identify the DEGs (Differentially Expressed Genes). In the current study, the criteria for considering the genes to be differentially expressed in each data set was as follows: (i) absolute log2 fold change (FC) greater than 0.58 (equivalent to FC equals to 1.5) and (ii) p-value less than 0.05. The number of upregulated and downregulated DEGs common among all datasets was determined using the logical intersection.

2.4 GO (Gene Ontology) Enrichment Analysis

GO-Elite is an application for performing GO (Gene Ontology) enrichment analysis [12]. It can determine the minimal set of non-redundant biologically distinct ontology terms describing any given list of genes. Three types of GO annotation terms including cell compartment (CC), molecular function (MF), and biological process (BP) overrepresented for each gene are identified in this analysis. In the current study, GO-Elite (v 1.2.5) was used to compare the set of DEGs found common in all the three data sets against all investigated genes in these datasets. The threshold settings of GO-Elite used include Z-score greater than 1.96, p-value less than 0.05 and minimum number of changed genes equal to 3.

2.5 Pathway Analysis

Pathway analysis of the expression data allows to identify the overrepresented signalling pathways. In this study, PathVisio (v3.2.4) was used to perform pathway analysis for the DEGs of each data set and is freely available at [1]. The Mus musculus pathway collection containing 174 pathways from Wikipathways was used for the analysis. The criteria for overrepresentaion of pathways included: (i) Z-score greater than 1.96, (ii) p-value less than 0.05 and (iii) minimum number of changed genes is 3.

3 Results and Discussion

3.1 QC (Quality Check) and Preprocessing

Several QC tests were performed on the raw and normalized data to asses the data quality. These checks revealed acceptable quality of all the samples in these datasets. The boxplots of the log intensities after normalization for each dataset is illustrated in Fig. 2(a, b and c). These plots demonstrate an overview of the data quality indicating the successful normalization leading to comparable distributions between all arrays in each dataset.

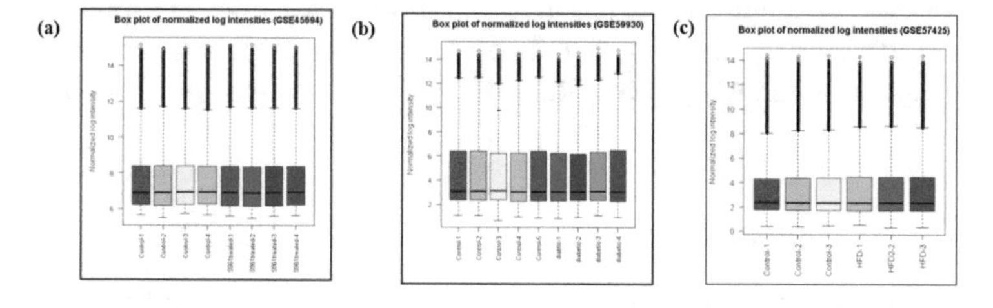

Fig. 2. QC Plots: (a), (b) and (c) depicts the boxplots of log intensities after normalization for datasets GSE45694, GSE59930 and GSE57425, respectively.

[1] http://www.pathvisio.org.

3.2 DEGs Analysis

The DEGs identified for the three datasets analyzed in this study are summarized in the Table 2. For each datasets GSE45694, GSE57425 and GSE59930 a total of 224, 1364 and 1252 DEGs were obtained after data preprocessing, respectively. A set of eleven upregulated and eleven downregulated genes was obtained after comparison of lists of upregulated and downregulated genes from the three datasets, respectively (Table 3).

Table 2. Differentially expressed genes identified in the analyzed datasets

GEO ID of data set	Total no of DEGs	No of upregulated DEGs	No of downregulated DEGs
GSE45694	224	130	94
GSE57425	1364	922	442
GSE59930	1252	602	650

Table 3. Common set of up (↑) and down (↓) regulated DEGs

Genes name	Expression
Pla2g15, Gdf15, Treh, Mvd, Gpc1, Pdk4, Tmie ,Apoa4, Dusp6, G0s2, Angptl8	↑
Mcm10, Nudt7, Serpina4-ps1, Foxq1, Atxn1, Lifr, Egfr, Kegl, Wdr45b, Hsd3b5, Cbfb	↓

Among the upregulated DEGs, eight genes (DUSP6, ANGPTL8, Pla2g15, Treh, Mvd, PDK4, GDF15, Apoa4 and G0s2) encodes secretory proteins and two genes (GPC1 and Tmie) encode different membrane/transmembrane proteins. All of these genes have function in lipid/glucose metabolism and/or cellular proliferation. The downregulated DEGs are involved in glucose/lipid metabolic processes (such as Nudt7, kegl, Hsd3b5, foxq1, and serpina4-ps1), growth/proliferation processes (such as mcm10, cbfb, lifr, and egfr) and others (such as atxn1 and wdr45b).

3.3 GO Enrichment Analysis

The GO analysis performed on DEGs (n=22:Table 2) revealed the associated key GO terms (see Table 4). The results reveal the enrichment of BP (Biological Processes) terms related to metabolism such as carbohydrate metabolic process (GO:0005975), steroid metabolic process (GO:0008202), cellular lipid metabolic process (GO:0044255), regulation of lipid metabolic process (GO:0019216) and alcohol metabolic process (GO:0006066) as well as related to growth

Table 4. GO Enrichment Analysis of DEGs

Ontology-ID	Ontology name	Related genes
Overrepresented GO biological processes		
GO:0048522	Positive regulation of cellular process	Apoa4,Atxn1,Cbfb,Dusp6, Egfr,G0s2,Gpc1,Lifr,Mvd
GO:0044707	Single-multicellular organism process	Apoa4,Atxn1,Cbfb,Dusp6, Egfr,Foxq1,Tmie
GO:0009892	Negative regulation of metabolic process	Apoa4,Atxn1,Dusp6,Egfr, Foxq1,Gm6484
GO:0006796	Phosphate-containing compound metabolic process	Apoa4,Dusp6,Egfr,Nudt7, Pdk4,Pla2g15
GO:0005975	Carbohydrate metabolic process	Egfr,Gm6484,Hsd3b5,Pdk4, Treh
GO:0044255	Cellular lipid metabolic process	Apoa4,Egfr,Gm6484,Mvd, Pla2g15
GO:0030154	Cell differentiation	Cbfb,Dusp6,Gm6484,Gpc1, Nudt7
GO:0006790	Sulfur compound metabolic process	Egfr,Gpc1,Nudt7,Pdk4
GO:0048878	Chemical homeostasis	Apoa4,Egfr,Gm6484,Pdk4
GO:0042981	regulation of apoptotic process	Dusp6,Egfr,G0s2,Pdk4
GO:0009653	Anatomical structure morphogenesis	Egfr,Foxq1,Tmie,Treh
GO:0080184	Response to phenylpropanoid	Apoa4,Egfr,Hsd3b5
GO:0072593	Reactive oxygen species metabolic process	Apoa4,Egfr,Pdk4
GO:0008202	Steroid metabolic process	Apoa4,Hsd3b5,Mvd
GO:0019216	Regulation of lipid metabolic process	Apoa4,Gm6484,Pdk4
GO:0006066	Alcohol metabolic process	Apoa4,Mvd,Pla2g15
GO:0007167	Enzyme linked receptor protein signaling pathway	Egfr,Lifr,Pdk4
GO:0010565	Regulation of cellular ketone metabolic process	Apoa4,Egfr,Pdk4
Overrepresented GO cellular components		
GO:0005739	Mitochondrion	G0s2,Hsd3b5,Keg1,Pdk4,Pla2g15
GO:0005576	Extracellular region	Apoa4,Gm6484,Lifr,Pla2g15
Overrepresented GO molecular function		
GO:0005102	Receptor binding	Egfr,Gdf15,Gm6484,Lifr,Nudt7
GO:0042802	Identical protein binding	Apoa4,Atxn1,Egfr,Mvd
GO:0019838	Growth factor binding	Egfr,Gpc1,Lifr
GO:0016788	Hydrolase activity, acting on ester bonds	Dusp6,Nudt7,Pla2g15

such as cell differentiation (GO:0030154), anatomical structure morphogenesis (GO:0009653), and regulation of apoptotic process (GO:0042981). Furthermore, the genes (Cbfb, Dusp6,Gm6484,Gpc1,Nudt7) involved in cell differentiation represent invaluable targets for further investigation.

3.4 Pathway Analysis

Pathway analysis of the DEGs obtained from each datasets revealed several significantly changed pathways. These pathways are identified with occurrence of maximum number of DEGs from each study based on statistical significance and thus represent most important gene regulatory/signaling events with respect to the treatment condition under observation. Twelve pathways were found significantly altered in GSE45694 Furthermore, twenty two and twenty four pathways were found significantly altered in GSE59930 and GSE57425 respectively. The intersection of the overrepresented pathways obtained from all the datasets resulted in three pathways including "Complement Activation Classical Pathway","Statin Pathway" and "PPAR signaling pathway". The "Complement Activation Classical Pathway" is a signalling cascade which is a part of innate immune system aiding antibodies to clear pathogens from an organism."Statins pathway" is a signalling cascade representing the role of statins (cholesterol lowering drugs) in cholesterol biosynthesis and plasma lipoprotein remodeling mechanism. The "PPAR signaling pathway" represents the signaling of PPARs (Peroxisome proliferator-activated receptors) which belong to nuclear hormone receptors class activated by fatty acids and their derivatives. These receptors encode multiple genes especially involved in lipid metabolic processes.

These three pathways represent core elements of the regulatory circuit involved in compensatory pancreatic beta cell proliferation. Several studies have demonstrated the link between regulatory elements of immune system and lipids metabolic pathways in insulin resistance [13–16]. In this study, the results of pathway analysis provides insights on the possible role of these links in pancreatic beta cell hyperplasia as well.

4 Conclusion

This study was designed in order to investigate the key genes and molecular signalling pathways significantly altered in different models of insulin resistance. Twenty two key genes were identified as DEGs from all of these datasets. Furthermore, three significant pathways (Complement Activation Classical Pathway, Statin Pathway and PPAR signaling pathway) were identified as a part of regulatory circuit which might be involved in compensatory increase of pancreatic beta cells in these insulin resistant states. These genes and pathways represents key targets for further exploration to aid pancreatic beta cell restoration in DM. Overall results of this study reveals possible crosstalk mechanism between lipid metabolic pathways in liver and cellular proliferative pathways in pancreas being involved in compensatory pancreatic beta cell hyperplasia during insulin resistance.

References

1. Rathmann, W., Giani, G.: Global prevalence of diabetes: estimates for the year 2000 and projections for 2030. Diabetes Care **27**(10), 2568–2569 (2004)
2. Meier, J.J.: Beta cell mass in diabetes: a realistic therapeutic target? Diabetologia **51**(5), 703–713 (2008)
3. Tomita, T., Doull, V., Pollock, H.G., Krizsan, D.: Pancreatic islets of obese hyperglycemic mice (ob/ob). Pancreas **7**(3), 367–375 (1992)
4. Gapp, D.A., Leiter, E.H., Coleman, D.L., Schwizer, R.W.: Temporal changes in pancreatic islet composition in C57BL/6J-db/db (diabetes) mice. Diabetologia **25**(5), 439–443 (1983)
5. Karnik, S.K., Chen, H., McLean, G.W., Heit, J.J., Gu, X., Zhang, A.Y., Kim, S.K.: Menin controls growth of pancreatic ß-cells in pregnant mice and promotes gestational diabetes mellitus. Science **318**(5851), 806–809 (2007)
6. Parsons, J.A., Brelje, T.C., Sorenson, R.L.: Adaptation of islets of Langerhans to pregnancy: increased islet cell proliferation and insulin secretion correlates with the onset of placental lactogen secretion. Endocrinology **130**(3), 1459–1466 (1992)
7. Saisho, Y., Butler, A.E., Manesso, E., Elashoff, D., Rizza, R.A., Butler, P.C.: ß-Cell mass and turnover in humans. Diabetes care **36**(1), 111–117 (2013)
8. Michael, M.D., Kulkarni, R.N., Postic, C., Previs, S.F., Shulman, G.I., Magnuson, M.A., Kahn, C.R.: Loss of insulin signaling in hepatocytes leads to severe insulin resistance and progressive hepatic dysfunction. Mol. cell **6**(1), 87–97 (2000)
9. El Ouaamari, A., Kawamori, D., Dirice, E., Liew, C.W., Shadrach, J.L., Hu, J., Katsuta, H., Hollister-Lock, J., Qian, W.J., Wagers, A.J., Kulkarni, R.N.: Liver-derived systemic factors drive ß cell hyperplasia in insulin-resistant states. Cell Rep. **3**(2), 401–410 (2013)
10. Yi, P., Park, J.S., Melton, D.A.: Betatrophin: a hormone that controls pancreatic ß cell proliferation. Cell **153**(4), 747–758 (2013)
11. Eijssen, L.M., Jaillard, M., Adriaens, M.E., Gaj, S., de Groot, P.J., Müller, M., Evelo, C.T.: User-friendly solutions for microarray quality control and pre-processing on ArrayAnalysis. org. Nucleic Acids Res. **41**(W1), W71–W76 (2013)
12. Zambon, A.C., Gaj, S., Ho, I., Hanspers, K., Vranizan, K., Evelo, C.T., Conklin, B.R., Pico, A.R., Salomonis, N.: GO-Elite: a flexible solution for pathway and ontology over-representation. Bioinformatics **28**(16), 2209–2210 (2012)
13. Glass, C.K., Olefsky, J.M.: Inflammation and lipid signaling in the etiology of insulin resistance. Cell Metab. **15**(5), 635–645 (2012)
14. Shi, H., Kokoeva, M.V., Inouye, K., Tzameli, I., Yin, H., Flier, J.S.: TLR4 links innate immunity and fatty acid-induced insulin resistance. J. Clin. Investig. **116**(11), 3015–3025 (2006)
15. Shoelson, S.E., Lee, J., Goldfine, A.B.: Inflammation and insulin resistance. J. Clin. Investig. **116**(7), 1793–1801 (2006)
16. Boden, G., She, P., Mozzoli, M., Cheung, P., Gumireddy, K., Reddy, P., Xiang, X., Luo, Z., Ruderman, N.: Free fatty acids produce insulin resistance and activate the proinflammatory nuclear factor-B pathway in rat liver. Diabetes **54**(12), 3458–3465 (2005)

Music and Color Synaesthesia with Visual Saliency Information

Hongbao Han[1] and Xiaohui Wang[2](✉)

[1] Suzhou Nuclear Power Research Institute, Suzhou, China
hanhongbao@cgnpc.com.cn
[2] School of Mechanical Engineering, Institute of Artificial Intelligence, University of Science and Technology Beijing, Beijing 100083, China
wangxh14@ustb.edu.cn

Abstract. Music and color synesthesia is a stimulus in the acoustic modality will consistently and automatically trigger concurrent percepts in the visual modality. In this paper, we study the music and color synesthesia with visual saliency information in the emotional level by the eye tracking technique. We organize an experiment, letting participants look at four different color combinations on the screen when listen to the music with eight emotions. And the eye tracker is adopted to catch the eye movement of participants during the experiments. From the metrics of the total fixation time and the scan paths, we obtain the relations between music and colors in the emotional level by this objective way. Besides, a questionnaire about the music and color synaesthesia is provided for participants to fill in. So that we can compare the objective and subjective cognition styles.

Keywords: Music · Synaesthesia · Visual saliency

1 Introduction

Synesthesia, a condition in which a stimulus in one sensory modality consistently and automatically triggers concurrent percepts in another modality, provides a window into the neural correlates of cross-modal associations [9]. Music and color synesthesia is the neural correlates between the acoustic modality and the visual modality, which is a stimulus in the acoustic modality will consistently and automatically trigger concurrent percepts in the visual modality. For example, when listening to the music "The Blue Danube", the blue color and the sad emotion will be linked together.

Music and color synesthesia has always been a hot spot for research. Some researchers used the psychophysical experiments, while some objective methods are adopted. For example, some studies uses functional magnetic resonance imaging (FMRI) to mine the brain functions related with music-color synaesthesia. In this paper, we use another new objective method, the eye tracking techniques.

N. Xiong et al. (Eds.): ISCSC 2018, AISC 877, pp. 428–435, 2019.
https://doi.org/10.1007/978-3-030-02116-0_50

Eye tracking has been widely used in many research areas and achieved very good results, such as user behavior analysis, point of regard measure between schizophrenic patients and normals. Eye tracking can catch the natural physiological phenomenon. Based on the eye movement data, we can obtain the visual saliency information by the metrics of the total fixation time and the scan paths. Based on this visual saliency information, we can obtain the relations between music and colors in the emotional level.

We use the eye tracker to automatically detect the colors with most attentions and least attentions when listening to the music. This method is more objective and accurate. The pipeline of the research on music and color synaesthesia with visual saliency information is shown in Fig. 1. We choose the familiar songs which is related to the common eight emotions, including happy, sad, warm, sweet, comfortable, quiet, mania and passionate. For each music, four color combinations are shown on the screen. Eye trackers catch the eye movement of each participant during the entire experiments. By the eye-movement metrics total fixation time and scan paths, we can obtain the relationship between the music and colors in the emotional level. After the objective eye movement experiments, a questionnaire about the music and color synaesthesia is provided for participants to fill in. So that we can compare the results between the eye tracking data and the questionnaire, to further compare the two cognition styles.

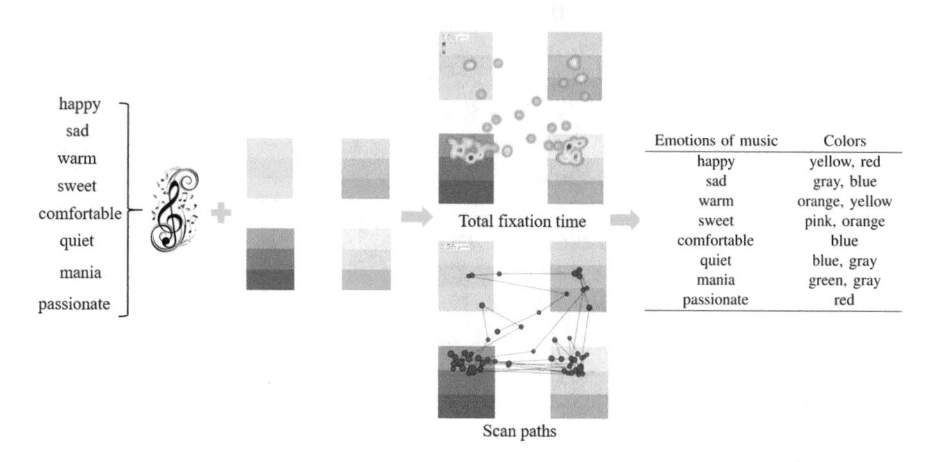

Fig. 1. The pipeline of the research on music and color synaesthesia with visual saliency information.

2 Related Work

2.1 Study of Music-Color Synaesthesia

Many researchers studied diverse cross-modal experiences of color while listening to the music. Some researchers used the psychophysical experiments to study the relations between colors and music. For example, the participants are asked to choose the five colors that were most consistent with the music, which is classical orchestral music by Bach, Mozart, and Brahms [3]. Ward studied the relations between single colors and the single notes from different instruments [11]. Some researchers used functional magnetic resonance imaging (FMRI) to mine the brain functions related with music-color synaesthesia. For example, E. M. Hubbard identified the brain regions mainly related to colored music during non-linguistic sound perception (chords and pure tones) in synaesthetes and non-synaesthetes [7]. Neufeld used FMRI to find the functional connectivity on auditory-visual synesthetes [8]. Zamm used FMRI to find the structural connectivity in color-music synesthesia [12]. In this paper, we used eye tracking techniques to study the relations between music and colors in the emotional level. The gaze data reflects the natural physiological phenomenon, which is more objective.

2.2 Eye Tracking

In early years, eye tracking are studied to find that schizophrenic patients' smooth pursuit eye-tracking patterns are different strikingly from the generally smooth eye-tracking seen in normals [10]. In recent researches, eye trackers are used to measure point of regard [4]. Eye trackers are also used to analyze user behavior, for example, in WWW search [5].

Eye tracking is adopted to catch the eye movement. There are some eye-movement metrics, for example, first fixation latency, first fixation duration, total fixation time, scan paths, etc. [2]. The first gaze represents the most attractive part of the image. The total fixation time indicts the most important part of the image. To visualize the fixations, heap map is adopted to summarize large quantities of data in an intuitive way. A heap map uses different colors to show the number of fixations participants made in certain areas of the image or for how long they fixated within that area. Red usually indicates the highest number of fixations or the longest time, and green the least, with varying levels in between [1]. The scan paths reflect the direction, the order and the length of the fixations. To visualize scan paths, gaze plot is used to display a static view of the gaze data for each image of the stimuli. In the gaze plot, each fixation is illustrated with a dot where the radius represents the length of the fixation.

In our experiments, the participants are asked to watch various color combinations while listening to the music, then select the color combination which is most related with the music by the questionnaire. The eye tracker catches the gaze data of each participant while listening to the music. The gaze data is objective, and the questionnaire is subjective. We will compare the music and color synaesthesia results between the gaze data and the questionnaire.

2.3 Music-Color Synaesthesia Applications

Some applications were developed based on the researches of color-music synesthesia. Zhang explored the converter relationship between digital audio and digital image based on synesthesia, and applied it to design a identify the red-green traffic light prototype system for color-blind people [13]. Kang implement an interactive interface that converts a users drawing into music [6]. In this paper, we also develop an interactive application based on our conclusion.

3 Data Preparation

The music contains eight representative songs with eight emotions, that is happy, sad, warm, sweet, comfortable, quiet, mania and passionate. There are several positive emotions in the eight emotions, because people have stronger abilities to distinguish between positive emotions. For each music, four color combinations are shown on the screen for participants, which is demonstrated in Fig. 2. Colors with high saturations are easy to give the participants an uncomfortable feelings, while colors with low saturations are easy to confused with other colors, even if it looks more comfortable. So we adopt color combinations, which contains three colors with the same hue and three different saturations.

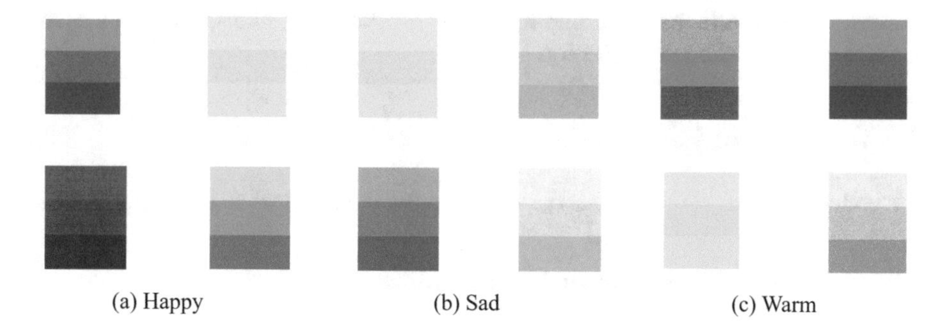

Fig. 2. Color combinations for eight emotions in the experiments.

We prepare a questionnaire to the cognition, which contains eight questions. One example is as follows:

Image when you listen to the happy music, which color can represent the happy emotion.

A. Blue B. Yellow C. Red D. Gray

4 Experiments

We invited 30 participants (15 females and 15 males) with ages ranging from 20 to 35, each of which had normal vision and hearing. Each participant was given the same eight music to listen to and eight questions to answer. The emotions of the eight music are happy, sad, warm, sweet, comfortable, quiet, mania and passionate. The orders of the eight music in the test are random.

First, before listening to the music, the experiment description is shown on the screen. The experiment description is "When you are listening to the music, four color combinations are displayed on the screen. You should choose a color combination which is most consistent with the music, that is to say, when you are listening to this music, looking at this color combination is the most comfortable."

Then, the participant listens to the eight music in a random order. The corresponding color combinations are displayed on the screen. The color combinations for the eight music in the test are shown in Fig. 2. The eye tracker catch the eye movement during the test. In our experiments, we use the Tobii eye tracker. From the eye tracking data, we can get the natural physiological phenomenon rather than the high level of human cognition.

Finally, after each music is over, a corresponding question in the questionnaire is shown on the screen. For example, when the sad music is over, the second question in the questionnaire is shown. We want to obtain the human cognition data by the question, so there is the emotion of the music in the question to guide the participant to think the relationship between the emotion of the music and the colors from their long-established cognitive system. So that we can compare the results between the eye tracking data and the questionnaire, to further compare the two cognition styles.

5 Results and Analysis

We collected the eye movement data of 30 participants during the experiments.

5.1 Relations Between Music and Colors

We find the relations between music and colors in the emotional level by the eye movement data. Taking the sad music as an example, the average fixations are shown in Fig. 3(a). The larger the fixation point, the longer the gaze is. Besides, red indicates the highest number of fixations, and green the least, with varying levels in between. So from the result shown in Fig. 3(a), we can see that the fixation points are mainly concentrated in the blue and gray colors. This means that participants tend to choose blue and gray colors when they listen to the sad music. Besides there are several red areas on the blue and gray colors, while there are only a few sporadic green points on the yellow and green colors. And the area of these sporadic green points are similar with that of white parts on the screen. This shows that the time which participants stay in yellow and green colors is very short. That is to say, participants do not hesitate to make the choice and this relation is very obvious.

The above conclusions can also be drawn from the metric of scan paths. The scan paths of a typical participant are shown in Fig. 3(b). There are many small circles in the purple, and there is a number in each small circle. The numbers indicate the orders of the eye movement. From Fig. 3(b), we can see that the first gaze is in the middle of the screen, and the orders of the eye movement are

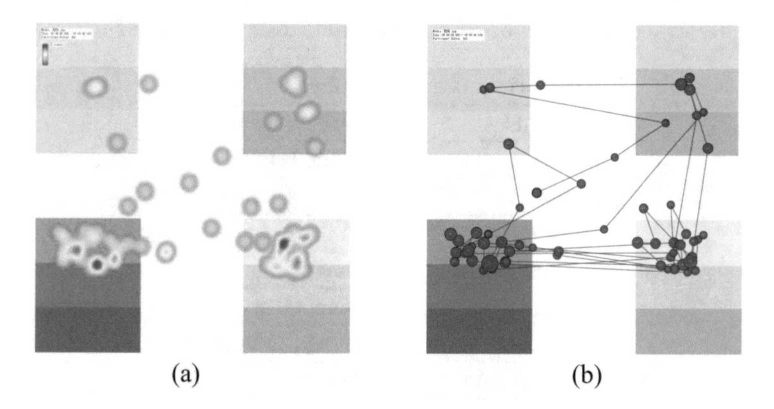

Fig. 3. (a) The total fixation time of the sad music. (b) The scan paths of the sad music.

green color, yellow color, blue color and gray color alternately. This is consistent with the common sense that people are first easily attracted by bright colors. But then people will be tangled in two similar emotional colors. If there is no obvious difference in the metrics of eye movement, it indicates that both colors are related to the music in the emotional level.

From the total fixation time and scan paths for the eight subtests, we summarize the relations of music with eight emotions and colors, which is shown in Table 1. Even if happy, warm, sweet, comfortable, quiet mania and passionate are all positive emotions, colors are different with the different degrees in positive scale. The passionate emotion is relatively intense, so the red color is the most relevant. The sweet emotion is relatively soft, so the pink color and the orange color are more relevant.

From the relations in Table 1, we can see that one color does not just correspond to one emotion, and there may be exactly the opposite of the emotions. For example, the blue color is related with the sad emotion and the quiet emotion at the same time. This is also consistent with our common sense that when listening to the music The Blue Danube, the blue color and the sad emotion will be linked together, while we feel quiet when looking at the sky.

We collected 30 questionnaires and counted the results. Then we compare the results between the questionnaires and the eye tracking data. We find that the two results are the same in Table 1. This means that the music and color synaesthesia in the two cognition styles are consistent.

5.2 Cognitive Differences Between Females and Males on Music-Color Synaesthesia

There are some cognitive differences between females and males to identify the emotions of colors and music. Figure 4 demonstrated the total fixation time of the happy music with the female and male participants. The results shows that the female participants are more tangled in the emotional level, while the male is more decisive.

Table 1. The relations between music and colors in the emotional level

Emotions of music	Colors
Happy	Yellow, Red
Sad	Gray, Blue
Warm	Orange, Yellow
Sweet	Pink, Orange
Comfortable	Blue
Quiet	Blue, Gray
Mania	Green, Gray
Passionate	Red

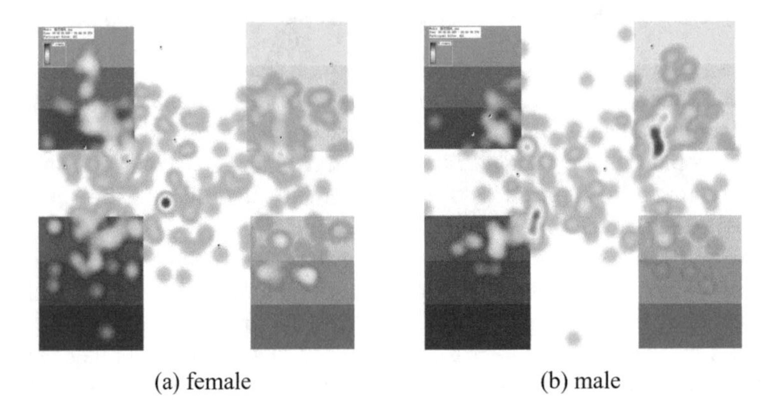

Fig. 4. Cognitive differences between females and males on music-color synaesthesia. (a) The total fixation time of the happy music with the female participants. (b) The total fixation time of the happy music with the male participants.

6 Conclusions

In this paper, we study the music and color synaesthesia with visual saliency information. We organize an experiment, letting participants look at four different color combinations on the screen when listen to the music with eight emotions. The eye tracker is adopted to catch the eye movement of participants during the experiments. From the metrics of the total fixation time and the scan paths, we obtain the relations between music and colors in the emotional level. Besides, we find that one color can correspond to more than one emotion, The results shows that there are cognitive differences between females and males on music and color synaesthesia. In the future work, we will continue to study the music and color synaesthesia by both the qualitative and quantitative methods and develop interesting applications based on our findings.

Acknowledgment. This work was supported by the Natural Science Foundation of China (61602033), USTB- NTUT Joint Research Program (TW2018004).

References

1. Bojko, A.A.: Informative or Misleading? Heatmaps Deconstructed. Springer, Heidelberg (2009)
2. Duque, A., Vzquez, C.: Double attention bias for positive and negative emotional faces in clinical depression: evidence from an eye-tracking study. J. Behav. Ther. Exp. Psychiatry **46**, 107–114 (2015)
3. Palmera, S.E., Schlossa, K.B., Xua, Z., PradoLenbLilia, L.: MusicCcolor associations are mediated by emotion. PNAS **110**(22), 8836–8841 (2006)
4. Goldberg, H.J., Wichansky, A.M.: Eye tracking in usability evaluation: a practitioners guide. In: The Minds Eye: Cognitive and Applied Aspects of Eye Movement Research, Amsterdam, pp. 493–516 (2003)
5. Granka, L.A., Joachims, T., Gay, G.: Eye-tracking analysis of user behavior in www search. In: Proceedings of the 27th Annual International ACM SIGIR Conference on Research and Development in Information Retrieval, SIGIR, New York, NY, USA, pp. 478–479 (2004)
6. Kang, L., Gu, T., Gay, G.: Harmonic paper: interactive music interface for drawing. In: CHI Extended Abstracts on Human Factors in Computing Systems, New York, NY, USA, pp. 763–768 (2013)
7. Hubbard, E.M., Ramachandran, V.: Neurocognitive mechanisms of synesthesia. Neuron **48**(3), 509–520 (2005)
8. Neufeld, J., Sinke, C., Zedlera, M., Dillo, W., Emrich, H., Bleich, S., Szycik, G.: Disinhibited feedback as a cause of synesthesia: evidence from a functional connectivity study on auditory-visual synesthetes. Neuropsychologia **50**(7), 1471–1477 (2012)
9. Neufelda, J., Sinkea, C., Dillo, W., Emricha, H., Szycika, G., Dimab, D., Bleicha, S., Zedler, M.: The neural correlates of coloured music: a functional mri investigation of auditory-visual synaesthesia. Neuropsychologia **50**(1), 85–89 (2012)
10. Holzman, P.S., Proctor, L.R., Hughes, D.W.: Eye-tracking patterns in schizophrenia. Science **181**(4095), 179–181 (1973)
11. Ward, J., Huckstep, B., Tsakanikos, E.: Sound-colour synaesthesia: to what extent does it use cross-modal mechanisms common to us all. Cortex **42**(2), 264–280 (2006)
12. Zamm, A., Schlauga, G., Eagleman, D.M., Louia, P.: Pathways to seeing music: enhanced structural connectivity in colored-music synesthesia. NeuroImage **74**(1), 359–366 (2013)
13. Zhang, Y., Liu, X.: The research on digital audio-visual synesthesia. In: Proceedings of the 10th International Conference on Computer Science and Education, ICCSE, pp. 186–189. Cambridge University, UK (2015)

The Comparative Analysis on Users' Information Demand in Online Medical Platforms

Junhao Shen, Jie Wang(✉), Yan Peng, and Jinyi Xue

Capital Normal University, Beijing 100048, China
Wangjie_cnu@126.com

Abstract. This paper focused on the research of the users' consultation in online medical platforms. The patients' questions regarding to hypertension in three different most frequently used Chinese medical platforms were crawled and have been analyzed comparatively. Web spider program written by Jsoup was used for crawling data from 39.net, xywy.com and haodf.com and 12500 data was collected from these websites. We used Chinese text-processing tool ROST to analyze the collected text data. The analyzing results indicate the hypertensive patients demanding different information from these three platforms. On 39.net, the patients' questions are mostly related to daily management of disease. On xywy.com, the patients would more likely to get advice and primary diagnosis. On haodf.com, most patients have had offline diagnosis previously and seek for further diagnosis; the frequent usage of professional terminology will contribute doctors with further precise online diagnosis. The findings of this study improve the service of online medical platforms and the understanding of users' information demand precisely.

Keywords: Online medical platforms · Hypertension · Text-processing
Word frequency analysis

1 Introduction

The rapid development of the Internet has contributed to make online medical treatment one of the most convenient ways to obtain medical information and treatment. In the 39th China Statistical Report on Internet Development [1], the scale of Chinese online medical users is 195 Million, which represents 26.6% of total Internet users. The top three most frequently used features are medical information query 10.8%, online booking register 10.4%, online consulting & diagnosis 6.4%. In the field of online consulting & diagnosis, there are already scaled online platforms. With the increasing capital, the tendency that the transition of these online platforms towards the offline diagnosis and treatment is increasingly obvious, which indicates that certain shortcomings still needed to reinforce by offline methods.

N. Xiong et al. (Eds.): ISCSC 2018, AISC 877, pp. 436–443, 2019.
https://doi.org/10.1007/978-3-030-02116-0_51

2 The Review of Related Research

Previous research about User behavior on online platforms tended to study the mature and developed online medical forums. The research on the early stage mostly used questionnaires to learn about why people use online health care communities, what their demand on online health care communities and what benefits they can get from online health care communities [2]. For example, some Korean researchers did a survey with 1000 residents in Korea, and found that the respondents' functional health literacy had positive effects on the scope of health information sources and health information self-efficacy but not health information-seeking intention. Respondents' social capital had positive effects on the scope of health information sources, health information efficacy, and health information-seeking intention [3]. Some researchers aimed to the certain content of a specific topic, and investigated terms of consumers diabetes based on a log from the Yahoo! Answers social question and answers (Q&A) forum, they used the log analysis method, data coding method, and visualization multiple-dimensional scaling analysis method for analysis. The authors found that there were 12 categories in the diabetes related schema, which emerged from the data coding analysis [4]. There was another study focused on the critical factors affecting Online Health Communities members loyalty from a perspective of IS success model and social support, the study found that the quality of the system and information have significant impact on member satisfaction, appraisal and emotional support on a sense of belonging [5]. Some other researchers used the method of the simulation experiment and questionnaires, selecting college student as the experimental object, studied the influence mechanism of health literacy and search experience on user's behavior [6]. In recent years, Web text mining has been used widely, some researchers apply text-mining techniques to text processing in studies of online communities. A study crawled hypertensive questions information from Baidu Zhidao (Q&A forum), analyzed users health behavior in online social Q&A communities [7]. However, studies analyzing and comparing the users demand for information on online medical platform are scarce in existing material.

3 Research Method Description

3.1 Selection of Data Source

39.net is the oldest and biggest health service platform with the most users and medical content. Some parts of 39.net such as Health News, Medicine Information Query, online medical counseling, are Industry leading. Its counseling platform has mass base, covers various questions of the patients and a vast user base. Xywy.com, established in 2001, had the purpose of constructing a service platform, on which public health and professional medical practitioners could be combined. The Q&A section of this website, as the online counseling platform, covers information for many kinds of disease.

Haodf.com, founded in 2006, is a leading online medical service platform. Patients who have already been diagnosed and are still looking for further diagnosis mostly use its counseling service. It also has the service of simple preliminary diagnosis for those patients who have not yet been diagnosed. Its online diagnosis is relatively comprehensive and information like previous diseases or complications are included. Moreover, users' personal information is partly hidden, in order to protect the privacy.

In this paper, we used Java Web-crawler depends on Jsoup to grab data from the three representative Chinese online medical platforms introduced above. These medical platforms have different UIs and the types of users provided information are different, however, they all contain the basic information of the patients and the description of the diseases. Therefore, different strategies were applied to gain access to the date from different websites. However, they all contain the basic information of the patients and the description of the diseases. Therefore, different strategies were applied to gain access to the date from different websites. In total, 9547 data items from 39.net, 2036 from xywy.com and 917 from haodf.com were collected in each section regarding to hypertension.

4 Result and Discussion

4.1 High-Frequency Words Counting Result

At the beginning of the words frequency counting process, the authors found various high-frequency words which have the same meaning or represent a certain type of object, such as "pregnancy" and "with child", which actually have similar meanings. There are many names of medicine words which contain the suffix "-dipine". So, the authors merged and clustered these words and found more precise results. of the diseases. In addition, different strategies were applied to gain access to the date from different websites. In total, 9547 data items from 39.net, 2036 from xywy.com and 917 from haodf.com were collected in each section regarding to hypertension.

4.2 Text-Processing

In this study, we use python and import jieba (Chinese text segmentation module) to deal with segment text document files. In the text-segmentation process, the user lexicon is mainly based on THUOCLs (THU Open Chinese Lexicon) medical lexicon combined with clinic hypertension medicine terms which were obtained in primary text segmentation and NLM (National Library of Medicine)'s MeSH (Medical Subject Headings) to ensure appropriate word particles. The authors also eliminated some noisy data and merged some oral Expressions with similar meaning.

The study used ROST CM (ROST Content Mining), a content mining tool, to segment the collected material and count the frequency of the words. The study also use a coding schema developed from a previous study, which analyzed content of health question [8] for reference. The authors adjust some part of the coding schema and make it suitable for the study of hypertension-related questions. The following three tables are the results of the analysis of three websites (Tables 1, 2 and 3).

Table 1. The top 24 high-frequency words list from 39.net

Key words	Frequency	Key words	Frequency
Hypertension	18301	Dizzy	818
Blood pressure	11232	Mother	774
Dipine	2318	Cholesterol	760
Medical history	1935	Elderly	760
High blood pressure	1925	Sleep	724
Heart	1384	Diastolic pressure	695
Vessel	1332	Effect	614
Check	1305	Artery	606
Hypotensor	1122	Systolic pressure	570
Secondary hypertension	912	Captopril	560
Hypotension	907	Age	550
Cardiogram	861	Low blood pressure	479

Table 2. The top 24 high-frequency words list from xywy.com

Key words	Frequency	Key words	Frequency
Hypertension	1749	Effect	115
Blood pressure	986	Operation	104
Check	407	Reduce blood pressure	99
High blood pressure	380	Father	95
Dizzy	296	Comfortable	92
Hypotensor	270	Body check	91
Headache	222	Body	91
Dipine	167	Pregnancy	89
Heart	153	Heart rate	85
Mother	133	Eyes	83
Drugs	116	Hypertension caused	77
Heartbeat	115	Hospitalized	76

4.3 The Analysis of High-Frequency Words Counting Result

It is obvious that the words like "elderly/ seniors", "mother", "father" appear frequently and repetitively in the statistic. The elderly belong to a high-risky group of getting hypertensions or hypertension-related complications. Moreover, considering the fact that most of the elderly in China don't have the access to the Internet, or don't know how to utilize these platforms, their offspring instead is mostly likely the one who helps them ask questions online. Further studies focusing on the real needs of this specific group need to be conducted.

On 39.net, "secondary hypertension" was a remarkable high-frequency word, which indicates the big number of questions from the patients. Theoretically, secondary hypertension has clear causes. It can either be cured or certainly relieved. But

Table 3. The top 24 high-frequency words list from haodf.com

Key words	Frequency	Key words	Frequency
Hypertension	713	Stenosis	64
Blood pressure	609	Father	61
Check	232	Drugs	61
Dizzy	182	Effect	60
High blood pressure	141	Chest distress	60
Operation	133	Stent	59
Diabetes	119	Taking medicine	54
Hospitalized	106	Pain	51
Hypotensor	106	Body check	51
Heart	101	Headache	49
Dipine	81	Stable	48
Leave hospital	74	Apoplexy	48

unfortunately, the true causes of secondary hypertension are usually neglected, thus to delay the further diagnosis. Therefore, the online diagnosing and treating patients with secondary hypertension can be a topic for further studies.

On 39.net and xywy.com, questions focus on medication and their effects, while many questions on haodf.com relate to the treatment after an operation, like conservative treatment. After comparing with most frequent words from all 3 online platforms, following features can also be discovered: The users on haodf.com generally use more professional terminologies to describe their conditions, which contributes to the effectiveness and efficiency of the further online diagnoses. Most of these patients have been diagnosed by doctors offline, and a considerable amount of them have had a heart stents operation before or have received doctors' recommendation for a heart stent operation in the future. On 39.net and xywy.com, users tend to use more subjective descriptions like vague spoken language for their symptoms, which causes negative effect on online diagnosis. A considerable number of users on 39.net and xywy.com raised questions about pregnancy, which means many active users would choose online methods to seek help from doctors before the offline diagnoses. Many questions are also related to the effect of hypertension drugs on fetus or whether hypertensives could be pregnant. Therefore, pregnancy-related words were identified and merged, in order to get a more precious result. At the beginning of the high-frequency words counting process, many names of the drug with the suffix of "-dipine" could be identified. Therefore, a group of words belonging to this category was also identified and merged.

4.4 The Social and Semantic Network Analysis - Demographic Information

As mentioned above, the offspring of the elderly often ask questions on behalf of their parents on online medical platforms. In this study, the authors used the social and semantic network analysis feature in ROST CM to illustrate the matrixes and selected the word items which included keywords like "elderly", "papa", "mom", "father", "mother" and so on as research objects. The "elderly/ old people" related questions social and semantic network matrix. In addition, the result is presented as below (Figs. 1, 2 and 3):

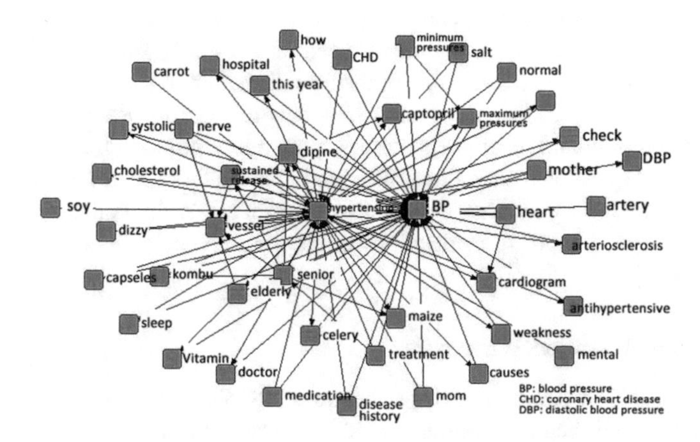

Fig. 1. 39.net "The elderly" related matrix

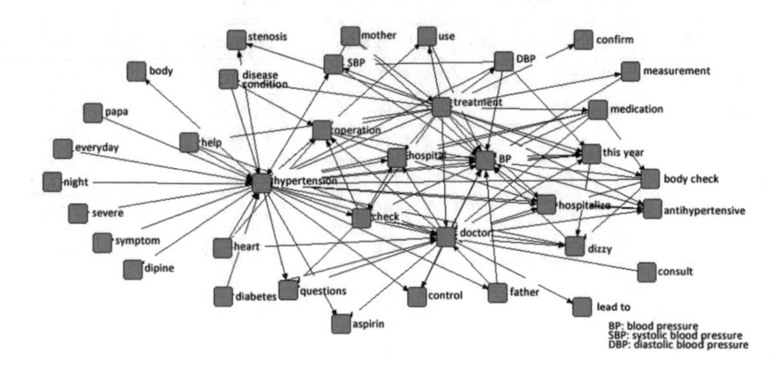

Fig. 2. xywy.com "The elderly" related matrix

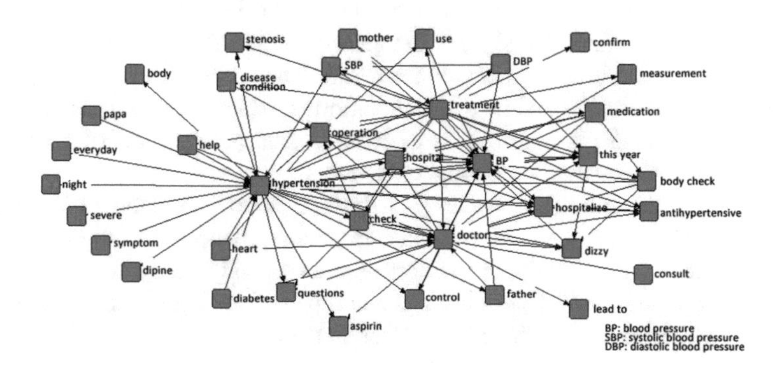

Fig. 3. haodf.com "The elderly" related matrix

According to the matrixes listed above, most of the questions on 39.net are related to daily diet, for example, words like “carrot”, “celery”, “soy” and so on appear frequently. The elderly patients tend to care more about what kind of effect their daily diet will have on hypertension. The elderly patients’ questions on xywy.com are more related to the medical treatment and changes of their symptoms. Despite the fact that many patients on haodf.com have already been diagnosed by doctors offline, most of them still cast doubt and seek for further confirmation on online platforms. For example, in previous offline diagnosis, if a patient had an artery stenosis related symptom, a doctor would recommend a surgery. In this condition, those patients tended to raise further questions on online platforms. The authors also found “dipine” in results of 39.net and xywy.com, which also contributes to the conclusion that many old hypertensive people’s medication counseling is related to medicines of “dipine-category”.

4.5 Emotional Tendency Analysis

The authors used the emotion analysis feature of ROST CM, and analyzed the emotional tendency and factors in online medical counseling (Table 4).

Table 4. Emotional tendency analysis

Medical platform	Positive emotion	Neutral emotion	Negative emotion	Total items
39.net	42.15%	13.64%	44.21%	9547
Xywy.com	50.73%	17.46%	31.81%	4999
Haodf.com	53.26%	15.89%	30.85%	1838
39.net	42.15%	13.64%	44.21%	9547
Xywy.com	50.73%	17.46%	31.81%	4999
Haodf.com	53.26%	15.89%	30.85%	1838

5 Conclusion

In recent years, the rapid development of the Internet brings online medical service into a new era, many online medical platforms have also developed their own patients counseling sections, which try to meet the growing demand of remote diagnosis who didn’t obtain a satisfying diagnosis or treatment. The study aimed to comparatively analyze three representative online medical platforms, which contain massive user data. The authors identified the main topics of hypertensive patients’ questions and comparatively analyzed patients’ emotional tendency.

According to the emotional tendency analysis, a polarized emotional tendency can be identified. Therefore, the doctors should pay more attention to the patients’ emotional expressions. In this study, the authors used a relatively simple method to classify users’ emotional tendency. Considering the complexity of emotions, a more precise method still needs to be introduced in further studies.

References

1. The 39th Statistical Report of China Internet Development. http://www.cnnic.net.cn/hlwfzyj/hlwxzbg/hlwtjbg/201701/t20170122_66437.htm. Accessed 8 June 2017
2. Colineau, N., Paris, C.: Talking about your health to strangers: understanding the use of online social networks by patients. New Rev. Hypermedia Multimed. **16**(1–2), 141–160 (2010)
3. Kim, Y.C., Lim, J.Y., Park, K.: Effects of health literacy and social capital on health information behavior. J. Health Commun. **20**(9), 1084–1094 (2015)
4. Zhang, J., Zhao, Y.: A user term visualization analysis based on a social question and answer log. Inf. Process. Manag. **49**(5), 1019–1048 (2013)
5. Zhang, X., Chen, X., Xia, H.S., Wang, L.: A study on factors affecting online health communities members loyalty from a perspective of is success and social support. Inf. Sci. **34**(3), 133–138 (2016)
6. Zhang, M., Nie, R., Luo, M.F.: Analysis on the effect of health literacy on users' online health information seeking behavior. Libr. Inf. Serv. **60**(7), 103–109 (2016)
7. Deng, S.L., Liu, J.: Understanding health information behavior in Q&A community based on text mining: taking baidu knows as an example. J. Inf. Resour. Manag. **6**(3), 25–33 (2016)
8. Oh, S., Zhang, Y., Park, M.S.: Health information needs on diseases: a coding schema development for analyzing health questions in social Q&A. Proc. Am. Soc. Inf. Sci. Technol. **49**(1), 1–4 (2012)

The Countermeasures Study of the Disposal of the Crude Oil Pipeline Leaks

Lian rui Jiang, Jing Liu(✉), Chen Liu, and Yang Wei

Department of Fire Commanding, The Chinese People's Armed Police Force Academy, Langfang, China
xiadengyou@126.com

Abstract. This paper takes the crude oil pipeline transmission process as the breakthrough point, in which makes an analysis of the crude oil pipeline leakage cases for the past decade with the method of random sampling, finding out the causes of oil pipeline leakage and leakage patterns. On the basis of the crude oil pipeline transmission process, it makes the risk analysis of the crude oil pipeline and the oil station, then using the accident tree safety risk assessment method to make the risk identification on pipeline accidents. It gives study on the technology and scheme of oil pipeline leakage disposal from the accident emergency response, process disposal measures, and on-site plugging technology.

Keywords: Firefighting · Crude oil pipeline · Leakage
Disposal countermeasure

1 Introduction

With the increasing demand for crude oil in China, the construction and development of crude oil pipeline is rapid. Only in Beijing, there are 74 oil and gas pipelines monitored and managed by oil and gas dispatching center, the total length of 53282 km, of which 9872 km crude oil pipeline, annual oil capacity of more than 100 million tons. Pipeline oil is the most economical way, but the possibility of an accident is increasing with time [1]. Figure 1 reflects the relationship between the probability of pipe accident and pipeline use time [2].

Once crude oil pipes had a sudden leakage, they are prone to fire, explosion, poisoning and casualties, not only will cause significant economic losses, but also endanger public safety.

N. Xiong et al. (Eds.): ISCSC 2018, AISC 877, pp. 444–452, 2019.
https://doi.org/10.1007/978-3-030-02116-0_52

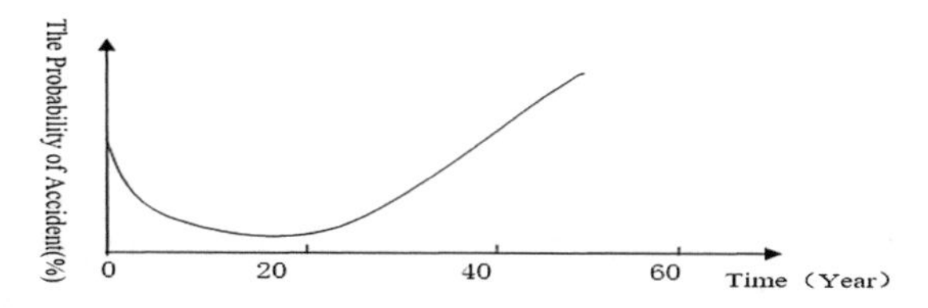

Fig. 1. The relationship between the probability of pipe accident and pipeline use time

2 Risk Identification of Crude Oil Pipeline Transmission Process

2.1 Current Status of Crude Oil Transportation Pipeline

It is necessary to grasp the danger of the delivery of crude oil pipelines and to protect, guard the pipeline and dispose of the accident.

Pipeline running time shown in Fig. 2, 44% of the pipeline running time of more than 30 years, 70% of the crude oil pipeline has more than 20 years of age. With the increase in pipe age, the risk of leakage is also increasing.

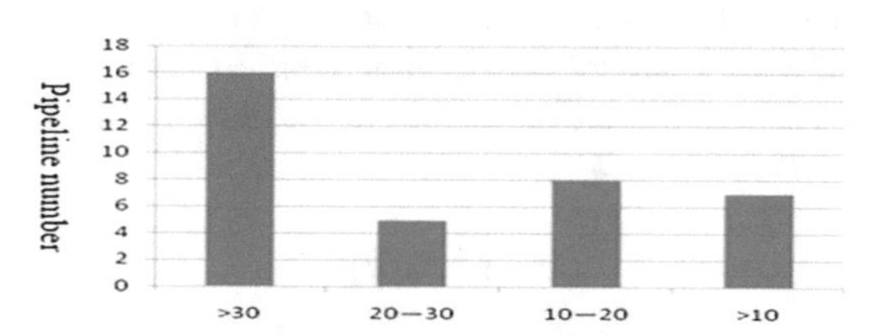

Fig. 2. The running time of crude oil pipeline

2.2 Case Analysis of Crude Oil Pipeline Leakage

Crude oil pipeline leakage cases are more, this article selected nearly a decade of typical crude oil pipeline leakage accident through the Internet with a random sampling method. From the cases, about 17% of the accident occurred in the oil transmission are pipeline station, 81% of the cases occurred in the oil transmission pipeline. So it should focus on long-distance pipeline. The causes that lead to the accident as shown in Fig. 3:

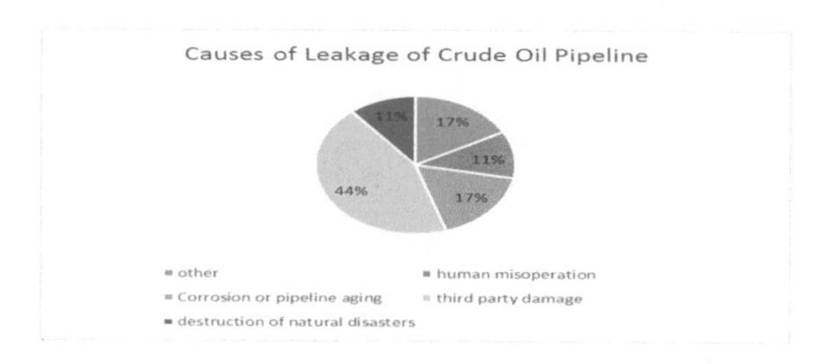

Fig. 3. The cause of the accident

According to statistics, the crude oil storage tanks are quite concentrated, the process pipeline is more intensive, complex, equipment pressure is relatively high, and electrostatic sparks and construction of more fire, in the event of crude oil leakages, can easily lead to fire and explosion [3]. The high risk of the station equipment and areas are oil tanks, oil pump shed, pig and valve area, variable distribution, fuel pump, auxiliary production area [4]. The danger of the tank area includes oil leakage, explosion and fire, and the evacuation causes the tank to crack and deform. Key equipment risk factors: Organic pumps, heating equipment, electrical equipment.

2.3 Risk Identification of Crude Oil Pipeline Transmission

This paper, by use of accident tree safety risk assessment method [5], make the crude oil pipeline leakage as the top of the accident tree incident, combined with the case and the results of the investigation of crude oil pipeline leakage in recent years, to identify the cause of the basic events, and made the accident Sexual analysis. The basic event is shown in Table 1.

Through the Boolean algebraic simplification of the structure of the accident tree of the crude oil pipeline, the minimum path set and the structural importance degree are obtained, and the cause, mode and influence degree of the accident are determined.

From the structure of the accident tree on the analysis of crude oil pipeline leakage, get the influence of each basic event on the leakage of crude oil pipeline, which is convenient for the prevention and disposal of the accident. The coefficient of structural importance of the accident tree is:

$$I = \sum_{X_i \in P_j} (1/2)^{N-1} \tag{1}$$

In the above formula:

I_i—X_i are approximate Judgment of Structural Importance Coefficient;
N—X_i are the number of basic events in the smallest path set P_j

According to the smallest set of the accident tree, X1X2X3X4X5X6 and X7X8X9X10X11X12X13X14 are concentrated in 27 size set domain, X15X16X17X18, X19X20X21X22X23X24, X25X26X27X28X29X30X31X32 are concentrated in 9 size set domain, X33X34X35X36/X37X38 are concentrated in 9 size set domain, According to the structural importance of the need to calculate the structural importance of I1, I2, I3, I4, I5, I6.

$$I_1 = I_2 = 9/2^{19} + 9/2^{21} + 9/2^{23}$$
$$I_3 = 9/2^{19}$$
$$I_4 = 9/2^{21}$$
$$I_5 = 9/2^{23}$$
$$I_6 = 3/2^{19} + 3/2^{21} + 3/^{23}$$

Table 1. Basic event

Seq-uence	Basic event	Seq-uence	Basic event	Seq-uence	Basic event
A_1	Third party damage	X_5	Pipeline signs are not obvious	X_{22}	Sulfide is in soil
A_2	Natural disasters damage	X_6	The law is not perfect	X_{23}	Soil water content is high
A_3	Corrosion	X_7	Earthquake	X_{24}	Soil acidity is strong
A_4	The pipeline itself is faulty	X_8	Floods	X_{25}	Anti-corrosion layer falls off
B_1	The pipe is perforated for Stealing oil	X_9	Debris flow	X_{26}	Anti-corrosion layer damaged
B_2	Construction or man-made damage	X_{10}	Collapse	X_{27}	Anti-corrosion layer is aging
B_3	Internal corrosion	X_{11}	Crude oil and impurities have corrosion on the pipeline	X_{28}	Anti-corrosion layer is too thin
B_4	External corrosion	X_{12}	Corrosion inhibitor don't work	X_{29}	There are inside water in Anti-corrosion layer
B_5	Pipeline installation problems	X_{13}	The inner coating is thinner	X_{30}	Anti-corrosion layer is too brittle
B_6	Equipment failured	X_{14}	The effect of cleaning pipeline is poor	X_{31}	The sticking force of Anti-corrosion layer drops
C_1	Cathodic protection failured	X_{15}	Cathode material failured	X_{32}	Anti-corrosion layer was destroyed by external forces
C_2	Soil natural factors	X_{16}	Cathode protection potential is high	X_{33}	Sealing problems
C_3	Anti-corrosion layer failured	X_{17}	Cathodic protection distance is not enough	X_{34}	Laying does not meet the standard
X_1	Check the work along the pipeline is not strict	X_{18}	There is stray current	X_{35}	Compressive strength of pipeline is insufficient
X_2	Alarm system failured	X_{19}	Soil salinity is too high	X_{36}	Detection control failured

(continued)

Table 1. (*continued*)

Seq-uence	Basic event	Seq-uence	Basic event	Seq-uence	Basic event
X_3	Legal awareness is weak	X_{20}	Soil oxidation and reduction potential is high	X_{37}	Loss of piping due to fatigue work
X_4	Social relations handled not properly	X_{21}	Soil bacterial content is high	X_{38}	Problems of run accessory equipment

The structural importance of the crude oil pipeline leakage is calculated:

$$I_1 = I_2 > I_3 > I_6 > I_4 > I_5 \tag{2}$$

According to the above order of structural importance, we can conclude that the reasons for the leakage of crude oil pipeline are: third party damage, destruction of natural disasters, corrosion, pipeline itself failure.

3 Study on Countermeasures for Disposal of Crude Oil Pipeline Leakage

3.1 Emergency Response Procedure for Oil Pipeline Leakage Accident

According to the form of crude oil leakage accidents and the form of leakage, emergency forces to the scene, should be judged before the accident analysis of the accident tree analysis of what kind of accident, according to pipeline transport oil characteristics, combined with technical measures to implement on-site safety operations.

Emergency forces should be deployed in the upper wind direction, higher terrain, leakage back, roadside or open location. Command and emergency medical points should be set up eye-catching signs. The establishment of the alert area should be based on the concentration of toxic gases and explosive gas concentration range of warning, according to toxic gas detector and combustible gas detector. Rescue site according to the need to wear a filter or isolated respirator.

3.2 Crude Oil Pipeline Leakage Treatment Process Measures

The Identifying Cases of Leakage. Exclude the interface, the pig plug, pump abnormalities, control valve abnormalities, instrument failure and other factors. When three of the following seven conditions occur simultaneously, it can be handled by leakage accident. (reference drop pressure 0.1 Mpa while Operating process, fluctuating flow 30 m^3/h, stop when the reference drop pressure 0.05 Mpa while stop transmission, fluctuating flow 30 m^3/h):

(a) the pressure drop abnormally;
(b) the upstream pressure of the detection point is abnormally decreased;
(c) the downstream pressure of the detection point is abnormally decreased;
(d) the upstream of the detection point increases;
(e) the downstream of the flow rate detection point decreases;
(f) the rotating speed of speed pump abnormal increase;
(g) the trunk line of control valve opening exception increases.

Some pipelines are equipped with a leak detection system at the regional dispatch center. In the process of pipeline leakage identification, it can also be consulted the regional dispatch center to assist in identifying pipeline leaks.

Leakage Treatment

(a) make the central control room immediately execute the "central ESD command" to the upstream pumping station which near the leak pipeline, and emergency stop all line.
(b) close the leak point of the upstream and downstream trunk valve; if there are the upper and lower valve room between the sub-transmission, the sub-transmission should be increased, as far as possible to reduce the leakage point near the pipe section pressure.
(c) timely calculation of leaks and full line batch interface location.

Note: The positioning formula reference, the location of the leak point x: $x = (L - v\Delta t)/2$

Where x is the distance from the upstream station where the leak is detected;

L is the distance between the upstream and downstream pressure detection points;

V is the propagation velocity of the pressure wave of the crude oil pipeline (the empirical value is generally 1.1 km/s); Δt is the time difference between the upstream and the downstream station of leakage point receives the pressure change.

Report to Duty dispatcher the accident situation, And inform the relevant units upstream and downstream.

3.3 With Temperature and Pressure Plugging Technology

Commonly used with temperature and pressure plugging technology shown in Table 2 [6–8].

3.4 On-Site Disposal of Different Types of Leakage

The site disposal point where the pipe is perforated for Stealing oil, The on-site disposal flow chart is shown in Fig. 4.

Table 2. Comparison of technical parameters with temperate pressure plugging

Plugging technology	Applicable pressure (MPa)	Applicable temperature (°C)	Whether preprocessing is required	Whether need be ignited
Welding Technology	Below 0.1–2.0	The temperature range should be suitable for welding person approaching	Yes	Yes
Injection Plugging	0–35	Generally for-198–1000	No	Yes
Bonding technology	Below 1.6	Determined by the performance of the binder, generally for-65–150	Yes	No

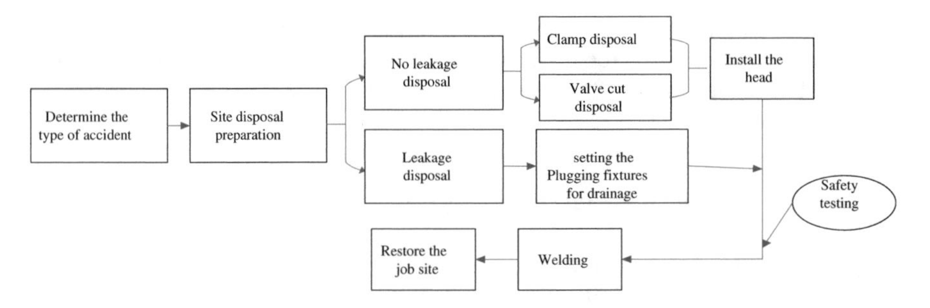

Fig. 4. Site disposal flow chart

3.5 Different Regions Crude Oil Pipeline Leakage Site Disposal

Crude oil pipeline leakage site disposal where pipeline across the river at, as shown in Fig. 5.

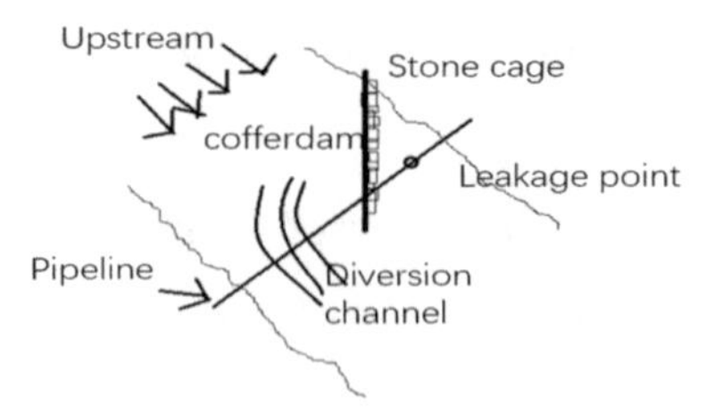

Fig. 5. Site disposal diagram

After the leak is completely exposed, Using a fixture to block. If the pipeline is broken or does not meet the plugging conditions, should be lap in the field bypass line, and change the pipeline, on-site disposal process shown in Fig. 6.

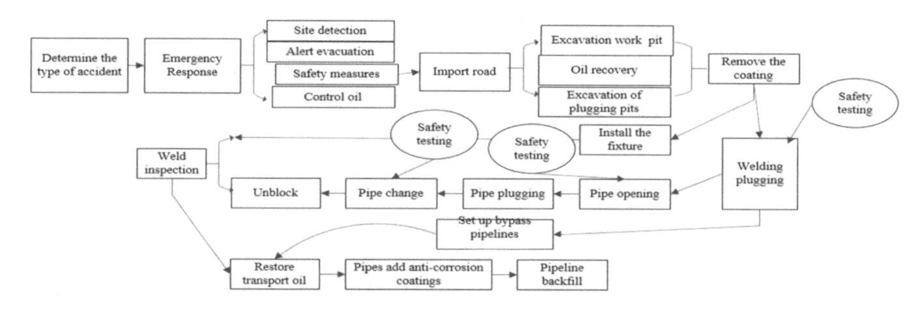

Fig. 6. On-site disposal process

4 Conclusion

In this paper, the crude oil pipeline leakage accident as the research object, around the accident emergency rescue measures carried out a more systematic study, and achieved the following conclusions:

A. Through field research and access to a large number of literature, this paper systematic combs the crude oil pipeline transportation process, including the regulation of crude oil pipeline and emergency treatment process.
B. Through analysis the case of China's crude oil pipeline leakage in recent years. The reasons for the leakage of crude oil pipeline are: third party damage, destruction of natural disasters, corrosion, pipeline itself failure.
C. Combined with the crude oil transportation process, put forward different leakage disposal measures for different levels of pipeline. And put forward disposal plan for the different leakage patterns and the situation of regional terrain.

References

1. Feng, P.X.: Numerical Simulation of Three-dim tensional Soil Temperature Field Due to Buried Thermal Pipeline Leakage. Daqing Petroleum Institute, Daqing (2006)
2. Ming-jun, C.Y.-T.D., Bao-dong, C., Qing-Jie, L.: Numerical Simulation of Perforation Leakage Diffusion Concentration of Burid Gas Pipeline. Academy of Oil and Natural Gas Engineering, Liaoling Petrochemical University, Fushun 113001 China (2010)
3. State Bureau of Technical Supervision: Classification of locations for electrical installations at petroleum facilities (1995)
4. State Bureau of Technical Supervision: Code for design of electrical installations for explosion and fire hazard environments (1992)
5. Chen, X.: System Security Engineering. China Building Materials Industry Press, Beijing (2006)

6. Chen J., Liu H., et al.: Development and application od leakage plugging device for transportation pipeline in oil field CPM **37**(2), 50–52 (2009)
7. Niu, Y., Yin, Y., Yu, Z.: A new method of plate in-line plugging for long-distance oil product pipeline. OGST **28**(11), 71–72 (2009)
8. Wang L.: Key Emergency Technology Research on the Leakage of Long Distance Pipeline, Beijing University of Technology (2012)

Attention Effects on the Processing of Task-Relevant and Task-Irrelevant Conflict

Pingyan Zhou[1], Kai Wang[2], Cai Zhang[1], and Ping Ren[1](✉)

[1] National Innovation Center of Assessment of Basic Education Quality, Beijing Normal University, Beijing 100875, China
xietong2017@163.com

[2] Institute of Psychology, Chinese Academy of Sciences, Beijing 100101, China

Abstract. In the current study, a flanker task was used to explore whether the resolution of cognitive conflict is gated by top-down attention using event-related potentials. Task-relevant and task-irrelevant processing were evoked using flanker stimuli that were either congruent or incongruent with the target. The results found that both N200 and P300 amplitude differences for color were enhanced only during task-relevant conflict, as was gender-conflict processing, indicating that cognitive conflict processing is under the control of top-down attention. Attention to the relevant conflict stimulus increases the processing of that particular attribute and unwanted information will be filtered out.

Keywords: Cognitive conflict · Top-down attention · N200 · P300

1 Introduction

One can read the newspaper in a restaurant and do not disturbed by the conversation of others nearby. This kind of ability is needed in human daily life, which enables the individual to select some useful characteristics, and to rule out the conflict or distraction stimulus, resulting in the completion of adapted behavior. The emergence of cognitive conflict can motivate the individual's top-down attention, prompting the mind to keep focused on the general awareness and perception. This ability is essential for human survival and adaptation.

Cognitive conflict has been studied extensively using classic stimulus-response compatibility (SRC) tasks like the Stroop and Flanker paradigms. Performance is faster and more accurate for compatible tasks versus incompatible tasks. Stroop N450 and flanker N200 are associated with detecting and resolving response conflict [1]. Knowledge-driven, top-down attention could enhance the neuronal activity of relevant sensory input. The discrimination between signal and distractors can be facilitated and the subject toward particular locations will be biased the competition among the multiple distractors [2]. Mangun et al. suggested that individuals can be aware of the attended stimulus and filter out unwanted information via focal (conscious) processing with limited attentional resources [3].

N. Xiong et al. (Eds.): ISCSC 2018, AISC 877, pp. 453–459, 2019.
https://doi.org/10.1007/978-3-030-02116-0_53

Desimone et al. proposed the theory of biased competition to explain such phenomena. The theory suggested that simultaneously presented stimuli compete with each other for limited attentional resources. Top-down attention is one of the ways to win the competition. The processing of a stimulus can be facilitated when it is attended so that limited processing resources are preferentially allocated to the stimulus [4]. This ability is critical for human adaptation, especially in the environment with conflict information. However, the cognitive research and neural mechanisms of the attentional modulation of cognitive conflict is still relatively under systematic. Until now, there is a few event-related potential (ERP) studies that have examined whether cognitive conflict processing is modulated by top-down attention only in visual S1–S2 matching task, which required participants to determine whether a second stimulus (S2) was the same as the first stimulus (S1) [5]. Whether there are different interactions between top-down attention and other kinds of cognitive conflict is not clear. Since conflicts due to the paradigm diversity may have different neural mechanisms. For example, the N270 component was associated with detecting conflicting information in visual S1–S2 matching task. The N270 component was most robust at the prefrontal cortex [5], while anterior cingulate cortex (ACC) was played a critical role in the generation of flanker N200 [6]. Thus, the relationship between top-down attention and cognitive conflict should be further investigated.

The current study adopted a flanker task to investigate the time course of attentional modulation of conflict processing. In the current study, neutral faces were artificially painted red and blue. Therefore the flanker faces could be congruent or incongruent from the target face along either color dimension or gender dimension or both. Two dimensions were combined to generate two levels of processing, task-relevant and task-irrelevant. Thus the top-down attentional modulation could be examined. For example, when the task was to identify the color of the target, color processing and gender processing became task-relevant and task-irrelevant, respectively. We could then examine whether color or gender processing was under the control of top-down attention feedback. If the hypothesis is correct, by varying color or gender in the task dimension, the SRC effect would be larger for the relevant conflict compared to the irrelevant conflict.

The flanker N200, a larger negativity (incongruent minus congruent) around 200 ms after stimulus onset, generated in the anterior cingulate cortex, has been shown to reflect conflict monitoring in a flanker task [7]. The amplitude of the N200, influenced by attention and task difficulty, is also sensitive to the degree of conflict level. The central-parietal P300 occurred approximately 600 ms after stimulus presentation, is a more positive-going wave in incongruent trials compared to congruent trials [8]. The amplitude of the P300 was more positive in harder task, which is also modulated by top-down attention feedback [9]. If cognitive conflict processing is modulated by attention, we predicted that the amplitudes of N200 and P300 would be enhanced in the incongruent condition compared to the congruent condition during task-relevant conflict, and that this effect would attenuate or be eliminated during task-irrelevant conflict for both color- and gender-conflict processing.

2 Material and Method

2.1 Participants

Twenty healthy college students (aged 22.55 ± 1.76; 11 women) completed the experiment. All participants were native speakers of Chinese, right-handed and reported normal or corrected-to-normal vision. A signed informed consent was obtained from each participant before the experiment.

2.2 Stimuli and Task

The stimuli were 12 faces (6 female) and the flanker task was adopted from our previous work [10]. During the flanker task, the central target face was flanked by two other distractor faces. Participants' task was to identify either the color or the gender of the central face while ignoring the two surrounding distractor faces at a viewing distance of 60 cm. All stimuli were displayed on a dark background. Following a central fixation of 100–400 ms, the two flanker stimuli appeared 100 ms prior to the target. The target and flankers were then displayed simultaneously for 600 ms. Each trial ended with a blank screen for 1400–1700 ms. Participants were instructed to respond as quickly and accurately as possible.

2.3 ERP Procedure and Data Analysis

Electroencephalogram (EEG) was recorded continuously by a set of 64 sintered Ag/AgCl electrodes according to the 10–20 system and referenced to the left mastoid. Both EEG and EOG were sampled at 500 Hz, with 0.05–100 Hz band pass using a Neuroscan NuAmps digital amplifiers system. Electrode impedance was kept below 5 KΩ throughout the experiment. The EEG data was segmented into 800 ms epochs with a pre-target baseline of −300 to −100 ms. Trials contaminated with artifacts exceeding ±100 μV were excluded from averaging.

A 5-way repeated-measures ANOVA was conducted on reaction times (RTs) and accuracy. For the N200, two time windows (230–300 ms, 300–370 ms) were selected after stimulus onset. A 5-way ANOVA was conducted to examine the effects of time windows (230–300 ms, 300–370 ms), electrodes (Fz, FCz, and Cz), task (attend color, attend gender), gender congruency (congruent, incongruent), and color congruency (congruent, incongruent) in relation to N200 amplitude. For the P300, a time window from 350 ms to 600 ms post-stimulus onset was chosen. A 4-way ANOVA was conducted including electrode (CPz, Pz, and POz), task (attend color, attend gender), gender congruency (congruent, incongruent), and color congruency (congruent, incongruent). The significance level was set at $p < .05$, and post-hoc comparisons were made using Bonferroni corrections.

3 Results

3.1 Behavioral Effects

The results revealed significant interaction between task and color congruency for RTs [F (1, 19) = 77.03, $p < .01$, $\eta p^2 = .80$] and between task and gender congruency [F (1, 19) = 22.23, $p < .01$, $\eta p^2 = .54$]. Post-hoc comparisons indicated that color SRC effects were significant when color was either task-relevant condition or task-irrelevant condition. However, the former effect was greater than the latter. Gender SRC effect was also greater for task-relevant conflict compared to task-irrelevant conflict.

3.2 N200

For the mean amplitude of N200, four-way interactions of time window, electrode, task and gender congruency [F (2, 38) = 4.00, $p < .05$, $\eta p^2 = .17$] and of time window, electrode, task and color congruency [F (2, 38) = 7.97, $p < .01$, $\eta p^2 = .30$] were significant. Post-hoc comparisons revealed that the N200 amplitude was more negative for the color-incongruent compared to the color-congruent condition only in task-relevant conflict during 230–300 ms time window. Gender N200 effect appeared at all three frontal-central electrodes (Fz, FCz, and Cz). Gender N200 effect was significant when participants attended to gender at frontal-central electrodes FCz and Cz, and disappeared when attention was diverted away from gender only in 300–370 ms time window (see Fig. 1A).

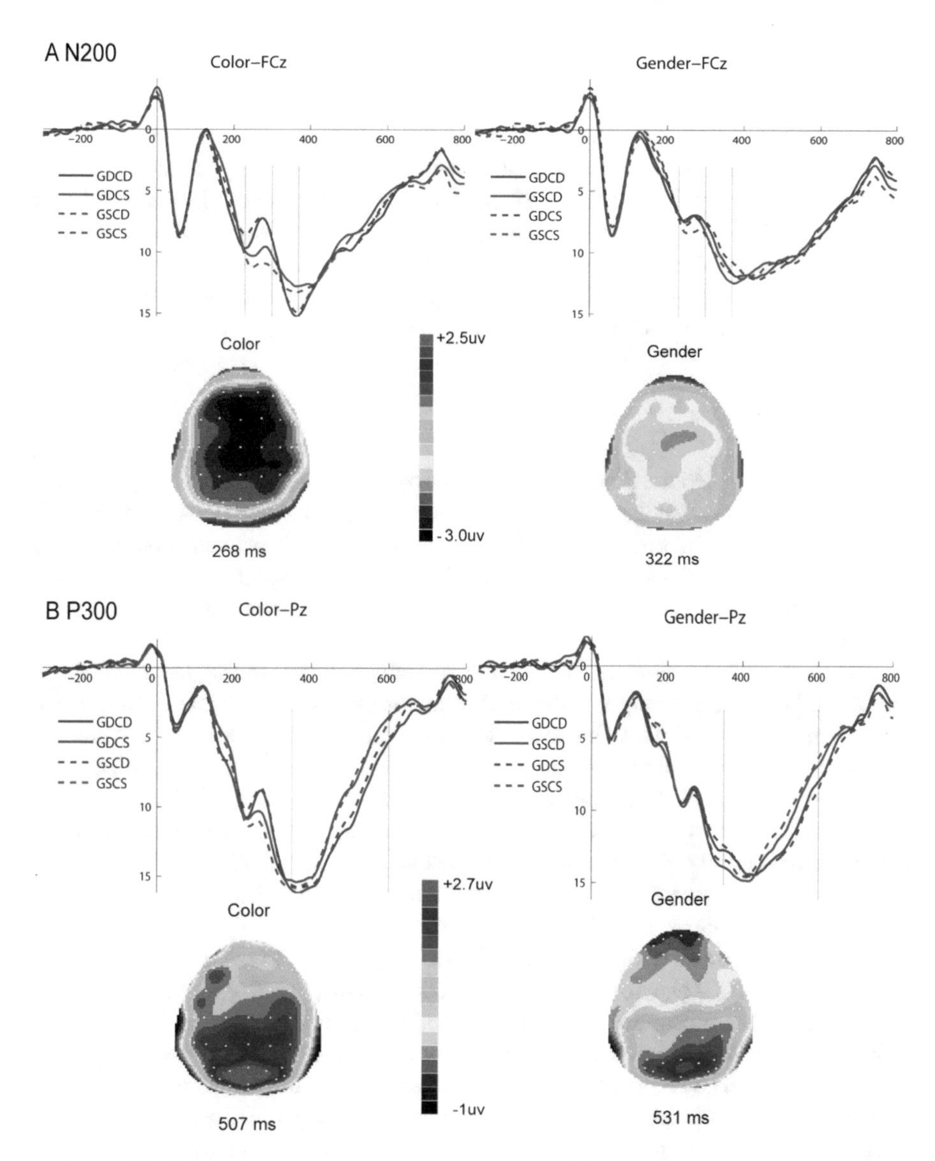

Fig. 1. N200 waveforms at electrode FCz (A) and P300 waveforms at electrode Pz (B).

(A) N200 waves in response to gender different/color different (GDCD), gender same/color different (GSCD), gender different/color same (GDCS), gender same/color same (GSCS) trials. Topographic maps of scalp voltage at 268 ms and 322 ms in two task-relevant conditions, obtained by subtracting two congruent conditions from two incongruent conditions at peak difference post-stimulus onset. (B) P300 waves in response to GDCD, GSCD, GDCS, and GSCS trials. Topographic maps of scalp voltage at 507 ms and 531 ms in the two task-relevant conditions.

3.3 P300

The P300 amplitude for color-incongruent trials was more positive than for color-congruent trials when color was attended in all three electrodes (CPz, Pz, POz,), but not when the color was unattended. The results indicated by a two-way interaction of task, and color congruency [F (1, 19) = 8.22, $p < .05$, $\eta p^2 = .30$] and of electrode and color congruency [F (2, 38) = 2.19, $p > .05$, $\eta p^2 = .10$]. A three-way interaction of electrode, task, and gender congruency was significant [F (2, 38) = 4.42, $p < .05$, $\eta p^2 = .16$]. Post-hoc comparisons revealed that the gender P300 effect appeared only when gender was attended in electrodes CPz and Pz (see Fig. 1B).

4 Discussion

This study investigated the time course of attentional modulation of conflict processing. Conflict SRC effects was greater in task-relevant versus task-irrelevant conflict in the behavioral data. N200 effects were found only during task-relevant conflict at the central-frontal electrodes (Fz, FCz, Cz). P300 effects also appeared only during task-relevant conflict. These results show that cognitive conflict processing is gated by top-down attention.

The results of current study found that N200 effects appear during the task-relevant conflict and eliminate during the task-irrelevant conflict, indicating that cognitive conflict processing is modulated by top-down attention. This is in line with Wang and colleagues' previous findings of the attentional modulation of cognitive processing, which suggested that N270 component was elicited when S2 was in conflict with S1, and that the amplitude of the N270 was more negative for task-relevant conflict compared to task-irrelevant conflict. Forster et al. posited that task-relevant conflict sourced response selection, and that perceptual conflict was the source of task-irrelevant conflict. The conflict level was higher for task-relevant conflict compared to task-irrelevant conflict [11]. Greater target-distractor incompatibility would give rise to higher levels of conflict, as indexed by both behavior and the N200 component [11]. Cognitive conflict processing was under the control of top-down attention in the current study was complied with the biased competition theory proposed by Desimone et al. The processing of the target stimulus can be facilitated when it is attended and the distractors were filtered out so that limited processing resources are preferentially allocated to the stimulus [4], resulting in the extending to cognitive conflict processing in this study.

In the current study, the N200 effect disappeared in the task-irrelevant conflict for reduced flanker interference when attention was diverted away. However, in Wang and colleagues' studies, N270 effects appeared in both task-relevant and task-irrelevant conflicts, which is attenuated but not disappeared in the latter condition [5]. One possible explanation for the discrepancy between these findings may be the task difficulty. In the current study, identifying the target color or gender is more difficult than the task in studies of Wang et al., in which participants were required to determine whether the two successively presented stimuli were identical [12].

Many existing findings have proved that incongruent trials are more difficult than congruent trials, as reflected by longer RTs and lower accuracy rates. More effort is required to resolve the difficult conflict which can be reflected by a more positive-going P300 wave in the central-parietal cortex [8]. In the current study, this phenomenon was confirmed again both in the behavioral performance and in the neural activity. The P300 amplitude of the color incongruent condition was more positive than for the color congruency condition only when color was attended, indicating that the task difficulty in the color incongruent condition was higher. While the enhanced gender P300 effect appeared only during the task-relevant conflict, indicating that target identification was disturbed by distractors and gender conflict resolution also required more attentional resources. The amplitude of P300 for both color and gender task were significantly enhanced only in task-relevant conflict, indicating that cognitive conflict was preferentially processed for attended condition compared to unattended condition.

In conclusion, the current data demonstrate that cognitive conflict processing was modulated by top-down attention, reflected by N200 and P300 effects only appeared for the task-relevant conflict. This cognitive control (top-down) of conflict can help people to ensure the rapid resolution of conflict to achieve coherent goal-directed behavior.

Acknowledgments. This work was supported by the Young Teachers fund projects in 2015 Grant funded by Beijing Normal University of China (310422102). Experimental data of this study may be requested via email.

References

1. Augustinova, M., et al.: Behavioral and electrophysiological investigation of semantic and response conflict in the Stroop task. Psychon. Bull. Rev. **22**(2), 543–549 (2015)
2. Baluch, F., Itti, L.: Mechanisms of top-down attention. Trends Neurosci. **34**(4), 210–224 (2011)
3. Mangun, G.R.: Neural mechanisms of visual selective attention. Psychophysiology **32**(1), 4–18 (1995)
4. Desimone, R., Duncan, J.: Neural mechanisms of selective visual attention. Annu. Rev. Neurosci. **18**(1), 193–222 (1995)
5. Wang, Y., et al.: The N270 component of the event-related potential reflects supramodal conflict processing in humans. Neurosci. Lett. **332**(1), 25–28 (2002)
6. Veen, V., Carter, C.S.: The anterior cingulate as a conflict monitor: fMRI and ERP studies. Physiol. Behav. **77**(4), 477–482 (2002)
7. Folstein, J.R., Van Petten, C.: Influence of cognitive control and mismatch on the N2 component of the ERP: a review. Psychophysiology **45**(1), 152–170 (2008)
8. Coderre, E., Conklin, K., van Heuven, W.J.B.: Electrophysiological measures of conflict detection and resolution in the Stroop task. Brain Res. **1413**, 51–59 (2011)
9. Hajcak, G., MacNamara, A., Olvet, D.M.: Event-related potentials, emotion, and emotion regulation: an integrative review. Dev. Neuropsychol. **35**(2), 129–155 (2010)
10. Zhou, P.Y., Liu, X.: Attentional modulation of emotional conflict processing with Flanker tasks. PLoS ONE **8**(3), 1–7 (2013)
11. Forster, S.E., et al.: Parametric manipulation of the conflict signal and control-state adaptation. J. Cogn. Neurosci. **23**(4), 923–935 (2011)
12. Mao, W., Wang, Y.: The active inhibition for the processing of visual irrelevant conflict information. Int. J. Psychophysiol. **67**(1), 47–53 (2008)

Heun-Thought Aided ZD4NgS Method Solving Ordinary Differential Equation (ODE) Including Nonlinear ODE System

Yunong Zhang[1,2,3,4](✉), Huanchang Huang[1,2,3,4], Jian Li[1,2,3,4], Min Yang[1,2,3,4], and Sheng Wu[1,2,3,4]

[1] School of Information Science and Technology, Sun Yat-sen University, Guangzhou 510006, People's Republic of China
zhynong@mail.sysu.edu.cn, ynzhang@ieee.org, jallonzyn@sina.com
[2] Key Laboratory of Autonomous Systems and Networked Control, Ministry of Education, Guangzhou 510640, People's Republic of China
[3] SYSU-CMU Shunde International Joint Research Institute, Shunde, Foshan 528300, People's Republic of China
[4] Key Laboratory of Machine Intelligence and Advanced Computing (Sun Yat-sen University), Ministry of Education, Guangzhou, People's Republic of China

Abstract. In this paper, a Zhang discretization (ZD) formula with high accuracy, termed ZD 4-node g-square (ZD4NgS) formula, is presented. Resorted to this formula, a numerical method for solving the initial value problem (IVP) of ordinary differential equation (ODE) is firstly presented. Such a method is thus considered as ZD4NgS method. Then, to improve accuracy, a Heun-thought aided modification of ZD4NgS method is proposed and investigated. For comparison, the conventional Euler method is also presented. In addition, numerical experiments are carried out, and numeric results show that the modified ZD4NgS method has a best final global error (FGE) among the three methods, substantiating that the accuracy of ZD4NgS method can indeed be improved after modification.

Keywords: Zhang discretization · Numerical method
Initial value problem · Heun thought · Final global error

1 Introduction

A differential equation is an equation that contains one or more derivatives of an unknown function. Specifically, A differential equation is an ordinary differential equation (ODE) if it involves an unknown function of only one variable [1].

Nowadays, many practical problems arising in science and engineering fields [1–3], such as artificial intelligence, robotics and even system biology, can be reconstructed into certain systems of ODE. For example, the famous recurrent neural network, Hopfield neural network, and many other artificial intelligent

N. Xiong et al. (Eds.): ISCSC 2018, AISC 877, pp. 460–471, 2019.
https://doi.org/10.1007/978-3-030-02116-0_54

tools can be described in the form of ODE. Thus, in research domain of natural computation, the significance and necessity of dealing with ODE are non negligible. However, it is worth pointing out that, in practice, only in some simple situations can the exact theoretical solutions of ODE be derived. Thus, the need arises for approximating the solutions of ODE. In this case, numerical methods are resorted to obtain useful approximations satisfying a specified tolerance.

With the development of computational methods, numerical methods for solving ODE have attracted intensive attention from many mathematicians and scientists, and have been studied from diverse perspectives for a very long time. Traditional numerical methods for ODE include Euler method [4–6], Adams method [7–9] and Runge-Kutta method [10–12]. Note that the shortcomings of the above methods should be considered; for example, the accuracy of Euler method fails to satisfy vast majority of applications despite of its simplicity; the Runge-Kutta method with order four needs a large amount of computation, which is time consuming in practical applications.

In this paper, based on the delicate work of Zhang et al. [13,14], a numerical method for ODE, which resorts to ZD4NgS formula, with Heun-thought aided, is presented, compared and investigated. To substantiate the efficacy of the presented discrete-time solution of ODE, the initial value problem (IVP) of ODE is solved as an illustration. Before ending this introductory section, we recap the main contributions of this paper as follows.

- With the thought of Heun method, an improved numerical solution for solving ODE, which resorts to ZD4NgS formula is presented and investigated in the paper.
- The efficacy and accuracy of the new method in application of discrete-time solution for ODE are substantiated by experimental results, which also means that the presented method can be developed into algorithms for discretizing continuous-time systems.

For better legibility, we briefly organize this paper as follows. Section 2 introduces the background of ODE including nonlinear ODE systems. In Sect. 3, classic Euler method and the modified Euler method (i.e., Heun method) are presented. In Sect. 4, the numerical method using ZD4NgS formula and its modification improved with Heun thought are shown. In Sect. 5, numerical experimental results are displayed and discussed. In the end, Sect. 6 presents the final conclusion of this paper.

2 ODE Including Nonlinear ODE System

In order to study a physical or real-life problem, we usually reformulate the problem into mathematical terms. Many problems are concerned with relationships among quantities and their changing rates represented by derivatives. Thus, mathematical models, which involve equations referring to an unknown function and its derivatives, are obtained. Such equations are differential equations. Note that a differential equation is an ODE if it contains the derivative(s) with respect

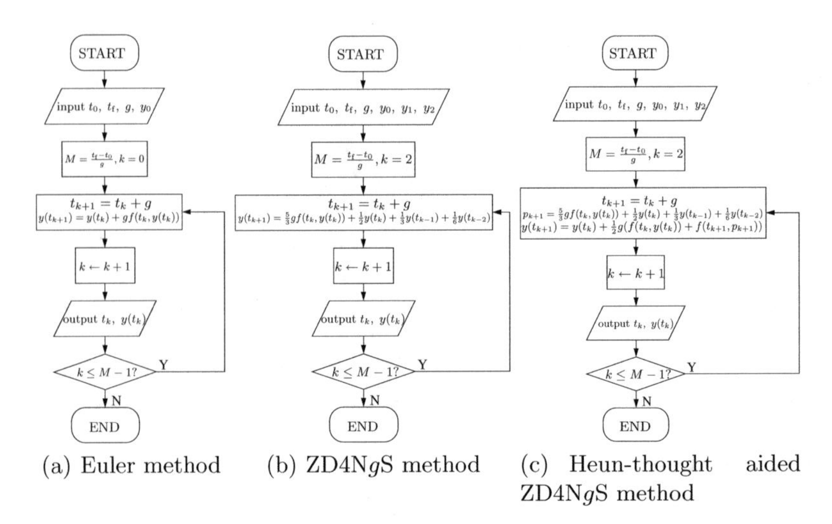

(a) Euler method (b) ZD4NgS method (c) Heun-thought aided ZD4NgS method

Fig. 1. Flowchart descriptions of Euler method, ZD4NgS method, and Heun-thought aided ZD4NgS method solving IVP of ODE.

to only one independent variable (usually, time t). A linear first-order ODE can be written as

$$\dot{y} = f(t, y) = p(t) + q(t)y. \tag{1}$$

A first-order ODE can not be written as the form of Eq. (1) is a nonlinear one.

As IVP of ODE plays a significant role in science and engineering research [1,15], this paper focuses mainly on numerical (specifically, discrete-time) solutions of IVP of ODE. The IVP of ODE can be described as

$$\begin{cases} \dot{y}(t) = f(t, y), \text{ with } t \in [t_0, t_f], \\ y(t_0) = y_0, \end{cases} \tag{2}$$

and a theoretical solution (or say, analytical solution) to IVP (2) over the interval $[t_0, t_f]$ is a differentiable function $y = y^*(t)$, with

$$\begin{cases} \dot{y}^*(t) = f(t, y^*), \text{ with } t \in [t_0, t_f], \\ y^*(t_0) = y_0. \end{cases}$$

Notice that the solution curve starts from the initial node (t_0, y_0).

To solve an ODE numerically, the existence and uniqueness of its solution should be guaranteed. The following theorem establishes this fundamental condition for ODE.

Theorem 1. *[1] Given a closed region* $\mathrm{R} = \{(t, y) : t_0 \le t \le t_f, c \le y \le d\}$, *let* $f(t, y)$ *be defined and continuous on* R. *The function* f *is considered to satisfy a Lipschitz condition with respect to* y:

$$|f(t, y_1) - f(t, y_2)| \le L|y_1 - y_2|, \text{ for any } t,\ y_1,\ y_2 \in \mathrm{R},$$

where constant $L < \infty$ is called a Lipschitz constant for function f. Then IVP (2) has a unique solution on a certain interval containing point t_0. □

Lemma 1. *Any $f(t, y)$ that is differentiable with respect to y and such that $|f_y| \leq L < \infty$ on R, satisfies the Lipschitz condition.*

Remark 1. IVP of ODE can be applied to solving control problems, intelligent control problems and a Hopfield network, which usually satisfy a Lipschitz condition when encountered in practical applications.

3 Euler Method and Heun Method

In science and engineering fields, when IVP (2) is encountered and can not be solved analytically, it is necessary to resort to numerical methods to obtain useful approximations to a solution. As known, Euler method and its modifications are the most frequently utilized simple ones.

3.1 Euler Method

Euler forward formula, named after Leonhard Euler, is one of the simplest numerical differentiation formulas and has been widely applied since proposed in 1768. Euler method, which takes advantage of Euler forward formula, is a classical numerical solution to IVP in ODE.

Let g denote the sampling gap. Based on Euler forward formula

$$\dot{y}(t_k) = \frac{y(t_{k+1}) - y(t_k)}{g} + O(g),$$

the approximation to the theoretical solution of IVP (2) can be numerically obtained:

$$y(t_{k+1}) = y(t_k) + gf(t_k, y(t_k)) + gO(g), \tag{3}$$

that is

$$y(t_{k+1}) = y(t_k) + gf(t_k, y(t_k)) + O(g^2), \tag{4}$$

with $O(g^2)$ representing that the local error in each step of Euler method is $O(g^2)$. Suppose that $y^*(t)$ is the unique solution to IVP (2) and $\{t_k, y(t_k)\}_{k=0}^{M}$ is the sequence of approximations generated by Euler method, then the final global error (FGE) $E(y(t_\text{f}), g) = |y^*(t_\text{f}) - y(t_\text{f})|$ is $O(g)$. Note that Euler method is the simplest numerical method for solving IVP of ODE. For reading convenience as well as for paper completeness, detailed steps of Euler method are displayed in a flowchart as Fig. 1(a).

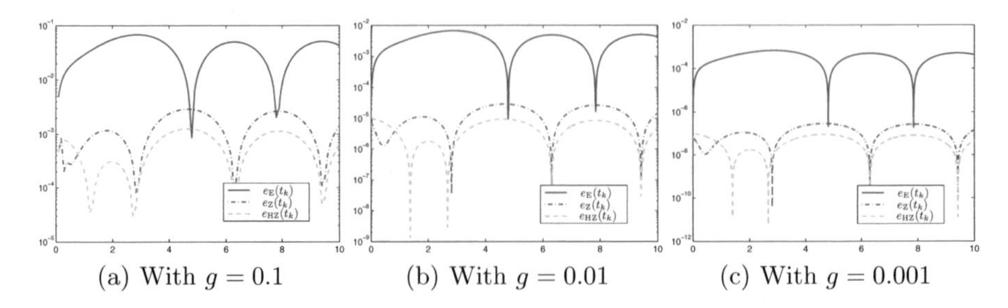

(a) With $g = 0.1$ (b) With $g = 0.01$ (c) With $g = 0.001$

Fig. 2. Absolute solution errors synthesized by Euler method, ZD4NgS method, and Heun-thought aided ZD4NgS method with different g values for solving IVP of ODE in Example 1.

Table 1. FGE of Euler method, ZD4NgS method, and Heun-thought aided ZD4NgS method with different g values solving IVP in Example 1.

Sampling gap	FGE of different numerical method		
	$E_{\mathrm{E}}(y(t_{\mathrm{f}}), g)$	$E_{\mathrm{Z}}(y(t_{\mathrm{f}}), g)$	$E_{\mathrm{HZ}}(y(t_{\mathrm{f}}), g)$
$g = 0.1$	$4.28388443 \times 10^{-2}$	$1.41433222 \times 10^{-3}$	$6.08421475 \times 10^{-4}$
$g = 0.01$	$4.20827672 \times 10^{-3}$	$1.44894879 \times 10^{-5}$	$4.70019290 \times 10^{-6}$
$g = 0.001$	$4.20093764 \times 10^{-4}$	$1.45269137 \times 10^{-7}$	$4.55742046 \times 10^{-8}$

3.2 Heun Method

Heun method, as a modification of Euler method, introduces a new thought for constructing an algorithm to solve IVP (2) numerically. In each iteration step of Heun method, Euler method is used as a prediction, and then the trapezoidal rule [1] is utilized to make a correction to obtain the approximation value, that is

$$\begin{aligned} p_{k+1} &= y(t_k) + g f(t_k, y(t_k)), \\ y(t_{k+1}) &= y(t_k) + \frac{1}{2} g (f(t_k, y(t_k)) + f(t_{k+1}, p_{k+1})). \end{aligned} \tag{5}$$

4 ZD4NgS Method and Heun-Thought Aided Modification

As a unification, outlook and extension of this research, a new class of finite-difference methods and formulas (especially, with one step ahead, or say, with one node ahead) is proposed and termed Zhang discretization (ZD), which can be used for the stable, convergent and accurate discretization of ZNN (i.e., Zhang neural networks) and other continuous-time systems. More specifically, Euler forward formula is the first one and also the simplest one of ZD that has been proposed. In this section, another formula within ZD framework, labeled as ZD4NgS formula [14], is presented. Besides, based on this formula, a numerical method and its Heun-thought aided modification are presented.

4.1 ZD4NgS Method

Let $y \in \mathrm{C}^n[t_a, t_b]$ denote the set of function y such that y and its first n derivatives are continuous on $[t_a, t_b]$. By assuming $y \in \mathrm{C}^3[t_0, t_\mathrm{f}]$, as well as t_{k-2}, t_{k-1}, t_k and $t_{k+1} \in [t_0, t_\mathrm{f}]$, ZD4N$g$S formula can be written as

$$\dot{y}(t_k) = \frac{6y(t_{k+1}) - 3y(t_k) - 2y(t_{k-1}) - y(t_{k-2})}{10g} + O(g^2). \tag{6}$$

To develop a numerical (specifically, discrete-time) solution for IVP (2), formula (6) can be rewritten into

$$y(t_{k+1}) = \frac{5}{3}gf(t_k, y(t_k)) + \frac{1}{2}y(t_k) + \frac{1}{3}y(t_{k-1}) + \frac{1}{6}y(t_{k-2}) + O(g^3). \tag{7}$$

The method which we use to solve well-posed IVP of ODE with the above numerical differentiation formula is thus named as ZD4NgS method, of which the complete steps are illustrated through flowchart in Fig. 1(b). From (7), it can be observed that the local error in each step of ZD4NgS method is $O(g^3)$ and the FGE $E(y(t_\mathrm{f}), g) = |y^*(t_\mathrm{f}) - y(t_\mathrm{f})|$ is theoretically $O(g^2)$, which can be roughly explained by $O(g^2) = O(g^3) \cdot M = O(g^3) \cdot O(1/g) = O(g^2)$ with $M = (t_\mathrm{f} - t_0)/g$ denoting the total step number, as the error would be accumulated after M steps have been made.

4.2 Heun-Thought Aided ZD4NgS Method

With the goal of improving the performance of ZD4NgS method, Heun thought of prediction and correction is employed for the above ZD4NgS method in this subsection.

Firstly, ZD4NgS method is used as a prediction:

$$p_{k+1} = \frac{5}{3}gf(t_k, y(t_k)) + \frac{1}{2}y(t_k) + \frac{1}{3}y(t_{k-1}) + \frac{1}{6}y(t_{k-2}) + O(g^3). \tag{8}$$

Secondly, trapezoidal rule is exploited for a correction step:

$$y(t_{k+1}) = y(t_k) + \frac{1}{2}g(f(t_k, y(t_k)) + f(t_{k+1}, p_{k+1})). \tag{9}$$

For better readability and convenience, the above modified method is labeled as Heun-thought aided ZD4NgS method with its flowchart description depicted in Fig. 1(c).

Remark 2. As both ZD4NgS method and its Heun-thought aided modification make use of four node when solving IVP of ODE in every iteration, three initial states (i.e., y_0, y_1 and y_2) are required to start the computing program on MATLAB platform. In consideration of this situation, the Euler method is used to initiate the iterative computation, which means that y_1 and y_2 for ZD4NgS method are obtained by Euler method (with a sufficiently small sampling gap) based on the given initial condition $y(t_0) = y_0$.

5 Numerical Experiments

In this section, in order to exemplify that the accuracy of ZD4NgS method for ODE solving can be improved with Heun-thought modification, numerical experiments are conducted on different IVP. Note that according to Theorem 1 in Sect. 2, each IVP example in this section has a unique solution. For comparison and illustration, experimental results of three numerical methods (i.e., Euler method, ZD4NgS method, and Heun-thought aided ZD4NgS method) are displayed.

Absolute solution errors, defined as $e_{\mathrm{E}}(t_k) = |y^*(t_k) - y_{\mathrm{E}}(t_k)|$, $e_{\mathrm{Z}}(t_k) = |y^*(t_k) - y_{\mathrm{Z}}(t_k)|$, and $e_{\mathrm{HZ}}(t_k) = |y^*(t_k) - y_{\mathrm{HZ}}(t_k)|$ for the three methods respectively, are shown in the log scale in Figs. 2, 3, 4, 5, 6 and 7. In addition, FGE of each method solving IVP in following examples, defined as $E_{\mathrm{E}}(y(t_{\mathrm{f}}), g) = |y^*(t_{\mathrm{f}}) - y_{\mathrm{E}}(t_{\mathrm{f}})|$, $E_{\mathrm{Z}}(y(t_{\mathrm{f}}), g) = |y^*(t_{\mathrm{f}}) - y_{\mathrm{Z}}(t_{\mathrm{f}})|$ and $E_{\mathrm{HZ}}(y(t_{\mathrm{f}}), g) = |y^*(t_{\mathrm{f}}) - y_{\mathrm{HZ}}(t_{\mathrm{f}})|$ respectively, are listed in Tables 1, 2 and 3 with different sampling gap values.

5.1 Example 1

In this example, an IVP of linear ODE is considered:

$$\begin{cases} \dot{y}(t) = f(t, y) = -y - 2\sin t, \\ y(t_0) = 3, \end{cases} \tag{10}$$

with $t \in [0, 10]$. Note that $|f_y| = |-1| \leq 1$. In accordance with Lemma 1, IVP (10) has a unique solution, which can be further derived to be $y = 2\exp(-t) + \cos t - \sin t$.

The experimental results of the above IVP are illustrated in Fig. 2 and Table 1. The corresponding absolute solution errors are illustrated in Fig. 2. It can be observed that solution errors generated by ZD4NgS method and Heun-thought aided ZD4NgS method follow the error pattern of $O(g^2)$ in the contrast to that generated by Euler method following the pattern of $O(g)$.

In this example, the maximal solution errors of Euler method are of order 10^{-2}, 10^{-3} and 10^{-4}, with g being 0.1, 0.01 and 0.001, respectively. On the other hand, the maximal solution errors obtained by ZD4NgS method and Heun-thought aided ZD4NgS method are both of order 10^{-3}, 10^{-5} and 10^{-7}. It is worth pointing out that the solution error of ZD4NgS method can be reduced after modified with Heun thought. For further investigation, FGE of the three methods are shown in Table 1 with sampling gap values being 0.1, 0.01 and 0.001. The fact that Heun-thought aided ZD4NgS method outperforms naive ZD4NgS method can be observed.

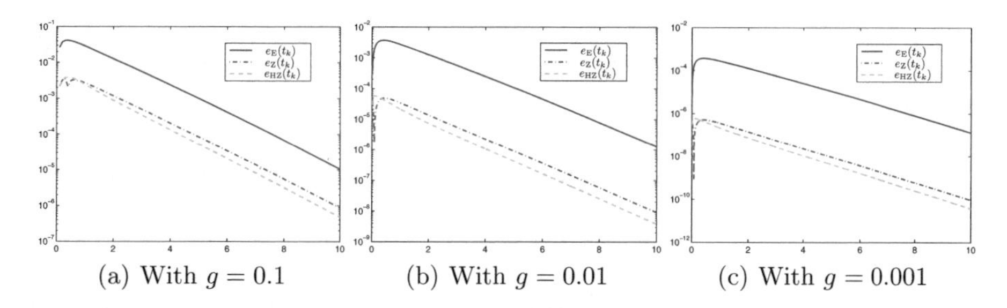

(a) With $g = 0.1$ (b) With $g = 0.01$ (c) With $g = 0.001$

Fig. 3. Absolute solution errors synthesized by Euler method, ZD4NgS method, and Heun-thought aided ZD4NgS method with different g values for solving IVP of ODE in Example 2.

Table 2. FGE of Euler method, ZD4NgS method, and Heun-thought aided ZD4NgS method with different g values solving IVP in Example 2.

Sampling gap	FGE of different numerical method		
	$E_E(y(t_f), g)$	$E_Z(y(t_f), g)$	$E_{HZ}(y(t_f), g)$
$g = 0.1$	$1.04312397 \times 10^{-5}$	$8.72532672 \times 10^{-7}$	$4.79524331 \times 10^{-7}$
$g = 0.01$	$1.26533613 \times 10^{-6}$	$9.11242775 \times 10^{-9}$	$3.81066601 \times 10^{-9}$
$g = 0.001$	$1.28966820 \times 10^{-7}$	$9.16845425 \times 10^{-11}$	$3.71804318 \times 10^{-11}$

5.2 Example 2

In this example, an IVP of nonlinear ODE is considered:

$$\begin{cases} \dot{y}(t) = f(t, y) = -y - y^2, \\ y(t_0) = 1, \end{cases} \tag{11}$$

with $t \in [0, 10]$. Note that $|f_y| = |-1-2y| \leq 3$. According to Lemma 1, IVP (11) has a unique solution, which can be further derived to be $y = 1/(2\exp(t) - 1)$.

To compare the performance of Euler method, ZD4NgS method and Heun-thought aided ZD4NgS method in nonlinear ODE solving, corresponding numeric results are shown in Fig. 3 and Table 2.

Similarly, absolute solution errors of three numerical methods with different sampling gap values are illustrated in Fig. 3. One can observe that the error pattern of Euler method is $O(g)$ while the ones of ZD4NgS method and Heun-thought aided ZD4NgS method are $O(g^2)$. Moreover, FGE of these three methods in solving IVP (11) are shown in Table 2, which can further confirm the following two facts.

- FGE of Euler method, (i.e., $E_{\mathrm{E}}(y(t_{\mathrm{f}}), g) = |y^*(t_{\mathrm{f}}) - y_{\mathrm{E}}(t_{\mathrm{f}})|$) has an pattern of $O(g)$ while FGE of the other two methods have patterns of $O(g^2)$.
- FGE of Heun-thought aided ZD4NgS method is the best among those of the three methods, which means that, after applying Heun thought in modifying ZD4NgS method, the performance in ODE solving is indeed improved.

5.3 Example 3

In this example, we conduct numerical experiments on the following IVP of system of ODE:

$$\begin{cases} \dot{y}_1(t) = f_1(t, y) = y_2^2 - y_3^2, y_1(t_0) = 0, \\ \dot{y}_2(t) = f_2(t, y) = -0.5y_3, y_2(t_0) = 1, \\ \dot{y}_3(t) = f_3(t, y) = 0.5y_2, y_3(t_0) = 0, \\ \dot{y}_4(t) = f_4(t, y) = -y_1, y_4(t_0) = 1, \end{cases} \tag{12}$$

in which $t \in [0, 20]$. Following Lemma 1, IVP (12) definitely has a unique solution, which can be obtained as

$$\begin{cases} y_1(t) = \sin t, \\ y_2(t) = \cos(0.5t), \\ y_3(t) = \sin(0.5t), \\ y_4(t) = \cos t, \end{cases}$$

with $t \in [0, 20]$.

The absolute solution errors and FGE generated by the three methods with different g values are illustrated in Figs. 4 through 7 and Table 3, respectively. Specifically, solution errors of $y_1(t_k)$ with g = 0.1, 0.01 and 0.001 depicted in Fig. 4 substantiate the efficacy of both ZD4NgS method and its modification in ODE solving. Errors of $y_2(t_k)$, $y_3(t_k)$ and $y_4(t_k)$ are depicted in Figs. 5, 6 and 7. In addition, FGE shown in Table 3 further confirm the fact that Heun-thought aided ZD4NgS method is more accurate than ZD4NgS method.

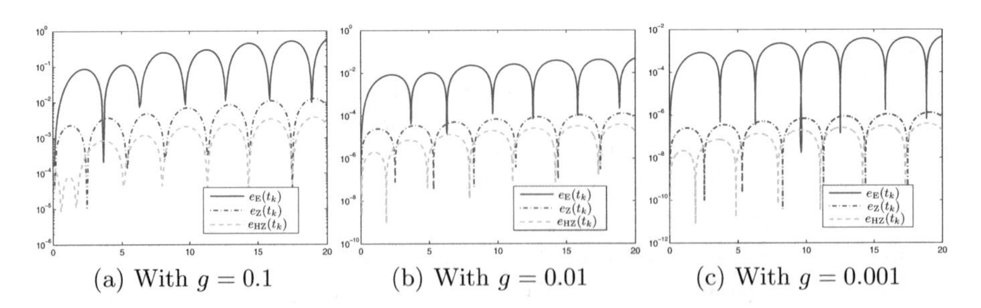

(a) With $g = 0.1$ (b) With $g = 0.01$ (c) With $g = 0.001$

Fig. 4. Absolute solution errors of $y_1(t_k)$ synthesized by Euler method, ZD4NgS method, and Heun-thought aided ZD4NgS method with different g values for solving IVP of ODE in Example 3.

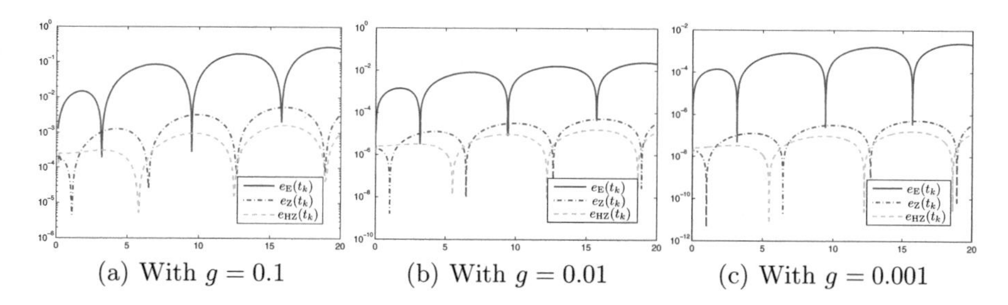

(a) With $g = 0.1$ (b) With $g = 0.01$ (c) With $g = 0.001$

Fig. 5. Absolute solution errors of $y_2(t_k)$ synthesized by Euler method, ZD4NgS method, and Heun-thought aided ZD4NgS method with different g values for solving IVP of ODE in Example 3.

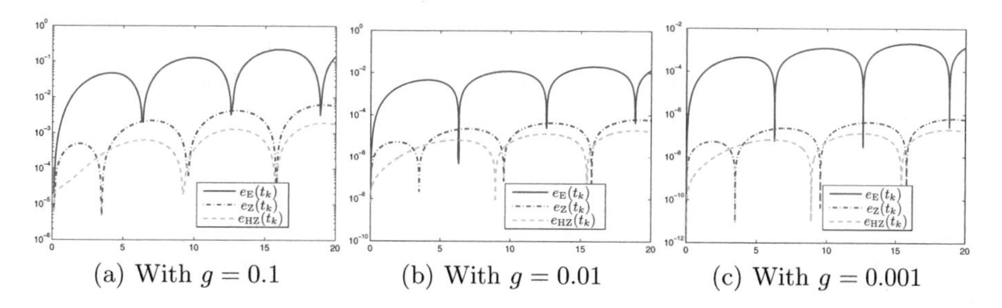

(a) With $g = 0.1$ (b) With $g = 0.01$ (c) With $g = 0.001$

Fig. 6. Absolute solution errors of $y_3(t_k)$ synthesized by Euler method, ZD4NgS method, and Heun-thought aided ZD4NgS method with different g values for solving IVP of ODE in Example 3.

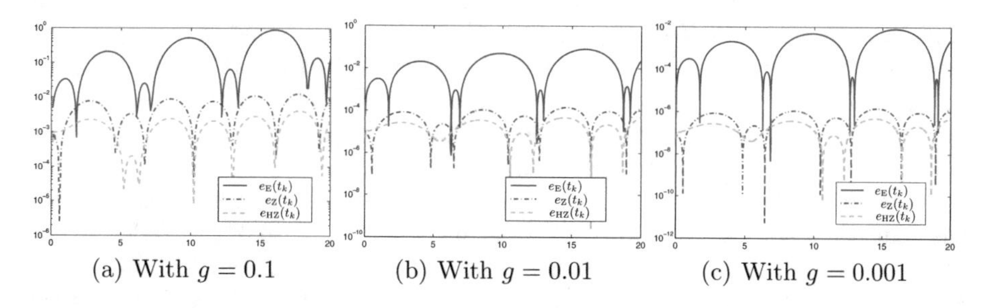

(a) With $g = 0.1$ (b) With $g = 0.01$ (c) With $g = 0.001$

Fig. 7. Absolute solution errors of $y_4(t_k)$ synthesized by Euler method, ZD4NgS method, and Heun-thought aided ZD4NgS method with different g values for solving IVP of ODE in Example 3.

In summary, the above numeric results of three different ODE substantiate that the presented ZD4NgS method and its modification, which is named as Heun-thought aided ZD4NgS method, are effective for solving IVP of ODE. Besides, the accuracy of ZD4NgS method is improved after Heun thought is adopted.

Table 3. FGE of Euler method, ZD4NgS method, and Heun-thought aided ZD4NgS method with different g values solving IVP in Example 3.

Sampling gap	Solution	FGE of different numerical method		
		$E_{\mathrm{E}}(y(t_{\mathrm{f}}), g)$	$E_{\mathrm{Z}}(y(t_{\mathrm{f}}), g)$	$E_{\mathrm{HZ}}(y(t_{\mathrm{f}}), g)$
$g = 0.1$	$y_1(t_k)$	$5.86540050 \times 10^{-1}$	$7.40485792 \times 10^{-3}$	$2.29333249 \times 10^{-3}$
	$y_2(t_k)$	$2.43754828 \times 10^{-1}$	$3.41655541 \times 10^{-3}$	$1.01573152 \times 10^{-3}$
	$y_3(t_k)$	$1.45311853 \times 10^{-1}$	$5.69097857 \times 10^{-3}$	$1.77091338 \times 10^{-3}$
	$y_4(t_k)$	$1.35320287 \times 10^{-1}$	$9.62849219 \times 10^{-3}$	$2.03151393 \times 10^{-3}$
$g = 0.01$	$y_1(t_k)$	$4.81566728 \times 10^{-2}$	$7.59328709 \times 10^{-5}$	$1.86481473 \times 10^{-5}$
	$y_2(t_k)$	$2.12874071 \times 10^{-2}$	$3.46200100 \times 10^{-5}$	$1.31147250 \times 10^{-5}$
	$y_3(t_k)$	$1.37000921 \times 10^{-2}$	$5.69537597 \times 10^{-5}$	$1.63019819 \times 10^{-5}$
	$y_4(t_k)$	$2.41081433 \times 10^{-2}$	$1.35631427 \times 10^{-4}$	$4.92654855 \times 10^{-5}$
$g = 0.001$	$y_1(t_k)$	$4.72288336 \times 10^{-3}$	$7.60777458 \times 10^{-7}$	$1.81951568 \times 10^{-7}$
	$y_2(t_k)$	$2.10075733 \times 10^{-3}$	$3.46871422 \times 10^{-7}$	$1.33999011 \times 10^{-7}$
	$y_3(t_k)$	$1.36105311 \times 10^{-3}$	$5.69576941 \times 10^{-7}$	$1.61389547 \times 10^{-7}$
	$y_4(t_k)$	$2.49407999 \times 10^{-3}$	$1.39672458 \times 10^{-6}$	$5.21575506 \times 10^{-7}$

6 Conclusion

In this paper, in order to improve the performance of the presented ZD4NgS method in solving IVP of ODE numerically, a modification aided with Heun thought has been proposed and labeled as Heun-thought aided ZD4NgS method. The modified ZD4NgS method utilizes ZD4NgS formula and trapezoidal rule in each step for prediction and correction, respectively. In addition, numerical experiments on different ODE have been carried out based on MATLAB platform. The experimental results have substantiated that numerical solution of IVP obtained by the modified method is indeed improved compared with the one generated by naive ZD4NgS method.

Besides, at the end of the paper, it is worth noting the difference between the numerical solution and the discrete-time solution of ODE or ODE system. The former is generated possibly using future time instants and function values at future time instants, e.g., $f(t_{k+1}, p_{k+1})$ in Eqs. (5), (9) and Fig. 1(c). On the other hand, the latter (i.e., the discrete-time solution) is generated only using current or previous time instants and function values, e.g., $f(t_k, y(t_k))$ or $y(t_{k-1})$ in Fig. 1(a) and (b). The discrete-time solution is more useful for real-time system control and simulation.

Acknowledgment. This work is supported by the National Natural Science Foundation of China (with number 61473323). Besides, Yunong would like to thank the coauthors of this paper by confirming their jointly first authorship.

References

1. Mathews, J.H., Fink, K.D.: Numerical Methods Using MATLAB. Pearson Prentice Hall, New Jersy (2004)
2. Spencer, S.L., Berryman, M.J., Garcia, J.A., Abbott, D.: An ordinary differential equation model for the multistep transformation to cancer. J. Theor. Biol. **231**(4), 515–524 (2004)
3. Kim, J.: Validation and selection of ODE models for gene regulatory networks. Chemometr. Intell. Lab. **157**, 104–110 (2016)
4. Blagouchine, I.V.: Expansions of generalized Euler's constants into the series of polynomials in π^{-2} and into the formal enveloping series with rational coefficients only. J. Number Theory **158**, 365–369 (2016)
5. Zhu, A., Xu, Y., Downar, T.: Stability analysis of the Backward Euler time discretization for the pin-resolved transport transient reactor calculation. Ann. Nucl. Energy **87**, 252–266 (2016)
6. Qu, X.: On delay-dependent exponential stability of the split-step backward Euler method for stochastic delay differential equations. In: Proceedings of the 6th International Conference on Intelligent Control and Information Processing, pp. 123–127. IEEE, Wuhan (2015)
7. Iavemaro, F., Mazzia, F.: Solving ordinary differential equations by generalized Adams methods: properties and implementation techniques. Appl. Numer. Math. **28**, 107–126 (1998)
8. Blasik, M.: A variant of Adams-Bashforth-Moulton method to solve fractional ordinary differential equation. In: Proceedings of the 20th International Conference on Methods and Models in Automation and Robotics, pp. 1175–1178. IEEE, Poland (2015)
9. Akinfenwa, O.A., Yao, N., Jator, S.N.: A self starting block Adams methods for solving stiff ordinary differential equation. In: Proceedings of the 14th IEEE International Conference on Computational Science and Engineering, pp. 127–136. IEEE, Dalian (2011)
10. Zhang, G.: Stability of Runge-Kutta methods for linear impulsive delay differential equations with piecewise constant arguments. J. Comput. Appl. Math. **297**, 41–50 (2016)
11. Wang, X., Weile, D.S.: Implicit Runge-Kutta methods for the discretization of time domain integral equations. IEEE T. Antenn. Propag. **59**(12), 4651–4663 (2011)
12. Tang, W., Sun, Y.: Construction of Runge-Kutta type methods for solving ordinary differential equations. Appl. Math. Comput. **234**, 179–191 (2014)
13. Zhang, Y., Jin, L., Guo, D., Yin, Y., Chou, Y.: Taylor-type 1-step-ahead numerical differentiation rule for first-order derivative approximation and ZNN discretization. J. Comput. Appl. Math. **273**, 29–40 (2014)
14. Zhang, Y., Huang, H., Shi, Y., Li, J., Wen, J.: Numerical experiments and verifications of ZFD formula 4NgSFD for first-order derivative discretization and approximation. In: Proceedings of the 15th International Conference on Machine Learning and Cybernetics, pp. 87–92. IEEE, Jeju Island (2016)
15. Hughes, T.J.R., Evans, J.A., Reali, A.: Finite element and NURBS approximations of eigenvalue, boundary-value, and initial-value problems. Comput. Method. Appl. M. **272**, 290–320 (2014)

Formative Assessment System of College English Based on the Big Data

Jiang-Hui Liu(✉), Wen-Xin Liang, and Xiao-Dan Li

Guangdong University of Foreign Studies,
Guangzhou 510006, People's Republic of China
247031690@qq.com

Abstract. It is an era of rapid development of network information technology. The informatization becomes one of the most prominent features nowadays. And the field of education has also been considerably influenced. People pay more and more attention to the innovation and reform of the modern teaching mode, connecting the new technology with education to promote the teaching quality. Based on the development of Big Data and more attention on the assessment of whole process of learning, this article will research about formative assessment system in support of Big Data and put forward a learning platform in order to improve the English learning efficiency and effectiveness of the college students. Furthermore, some college students will be chosen to see their learning records through the platform for finding their study situation. All in all, formative assessment and Big Data are advantageous to College English.

Keywords: Formative assessment · Big Data · College English

1 Introduction

The national Ministry of education in 2004 promulgated the teaching requirements of College English curriculum, advocating more scientific, more comprehensive, more reasonable assessment of the learning process with learning records, classroom activities records, interviews, discussions and other formative assessments [1, 2]. At the same time, because of the progress of network information technology, the social gradually entered the era of Big Data. All walks of life come to realize the value of Big Data and make use of them. As a typical social activity, education is also gradually penetrated by it [3]. Therefore, more and more researchers begin to set store by and explore the system of formative assessment with Big Data in foreign language teaching and learning.

2 Characteristics of the Era of Big Data

The researches of Big Data become deeper; the range of its use becomes wider. Certainly, academic circles also care about Big Data that uses in the field of education [4]. During the process of teaching and learning, data are generated, namely all kinds of learning behavior data which are traces students left in online learning. These can be

N. Xiong et al. (Eds.): ISCSC 2018, AISC 877, pp. 472–480, 2019.
https://doi.org/10.1007/978-3-030-02116-0_55

used to analyze the learning situation each of the students to support students' autonomic learning and teachers' individualized teaching, and even to be conducive to promoting educational reform and development. In Nov. 2015, the National Center for Educational Technology and Intel Corporation jointly held a conference about education informatization, whose theme is that Big Data boosts educational reform, discussing what opportunities and challenges the cloud computing, Big Data and other innovative technology bring to the educational informationization reform. It also started a research project of Big Data in education.

3 Formative Assessment System

In the process of education, it is necessary to estimate the value of educational activities in order to understand the situation of education clearly, feed back in time and adjust the teaching schedule, which will add the education value. This process of judgment called educational assessment [5], including two models, summative assessment and formative assessment.

3.1 Formative Assessment

Formative assessment was proposed by the philosopher G. F. Scriven in his book, The Methodology of Evaluation. Formative assessment provides feedback on ongoing educational activities by diagnosing problems in educational programs, plans, educational processes or activities, thereby improving the quality of education. After that, this assessment method was introduced into the classroom teaching by American educator Bloom [6]. Simply speaking, formative assessment is a student-centered dynamic assessment model, which not only pays attention to the result of the study examination, but also concerns and more inclines to the whole process of students learning so as to obtain the feedback information of all aspects [7]. It means that teacher will no longer know the learning situation of students only through test scores, but from the whole learning process in detail, finding specific and pivotal learning problems, grubbing the advantages and strengths of students and specially changing the teaching methods and help students to optimize learning method [8]. What's more, it contributes to the cultivation of learners about self-discovery, active rectification and autonomic learning ability, and transforms learners from asking me to learn into I want to learn [9].

3.2 Summative Assessment

Summative assessment is an assessment model that detects the ultimate education effect at the end of the activity. It mainly quantifies the students' academic performance and complex educational phenomenon through paper test to screen and select outstanding students and appropriate education methods [10]. In other words, only the final result will be noticed, but the existing abilities or latent talents that students have in the process of development are neglected [11].

3.3 Current Situation of Two Assessment Models

In our country, it is relatively late to apply the formative assessment to the study of College English. In 2014, the Ministry of Education promulgated College English Curriculum Requirements (Trial). It clearly put forward to strengthen formative assessment on College English. Obviously, there was a rapid increase of relevant researches after that [12]. These studies mainly focused on the assessment of students, including writing ability assessment, oral English learning assessment, autonomic learning ability assessment and other various assessment methods.

According to the retrieved literatures of 10 core foreign language journals in the past thirty years, the study of summative assessment which is dominated by large-scale test, has been developing steadily, as is shown in Table 1.

Table 1. Distribution of articles about educational assessment in each period.

	1980–1984	1985–1989	1990–1994	1995–1999	2000–2004	2005–2009	2010–2014
Test assessment	1.7%	5.9%	7.6%	11.5%	16.3%	13.1%	18.7%
Non-test assessment	0	0	0	0	0.5%	5.7%	19.0%

In twenty-first Century, the teaching assessment researches show a trend of diversification, the non-test assessment come to be concerned with each passing day, especially the formative assessment research, but it is still in a weak.

4 The Combination of Formative Assessment and Big Data in College English Education

4.1 Problems of Formative Assessment System

The formative assessment has the following forms. The first is the online quiz that are arranged according to the corresponding teaching content, with timely online automated assessment. Second is the paper exercise whose purpose is to enhance and strengthen the knowledge. Third, after each unit of study, there is a unit test. Online automatic assessment can reflect the learning situation of the unit in time. Furthermore, students are asked to participate in paper mid-term examination regularly. In addition to these, there are some other ways, like class presentation, group discussion, project research and so on. Besides, students' attendance and performance are also included. Mostly, the formative assessment of student accounts for about forty percent of his final score.

Based on current research, summative assessment has been still the main assessment methods in China. And being in the immature stage, the formative assessment has a lot of problems. There are two more prominent issues.

For one thing, the credibility of results in formative assessment system should be improved. The above assessment forms and indexes are random. Some of them are difficult to be judged, like attendance, performance and learning attitude. Because the system is imperfect and the form is monotonic, the formative assessment actually becomes a way to send students grades.

For another thing, it is difficult to establish the formative assessment system. The amount of learning information is large and mussy. Lots of time and efforts should be taken to deal with it. What's worse, it is hard to determine index for all kinds of capabilities. It is a big obstacle. Additionally, the disproportionality between teachers and students makes more serious to achieve tracking assessment to students.

4.2 College English Formative Assessment System

The formative assessment system mainly lacks technical support. It is the Big Data technology that can effectively realize it. Big Data can excavate, transform and filter huge messy learning information, and then automatically extract and analyze the valuable data, and finally feedback to users. And limited human cannot do that. Formative assessment with Big Data has huge impacts on both the autonomic learning and teaching. Some imagine about formative assessment system will be put forward under Big Data and applied to college English learning, as is shown in Fig. 1.

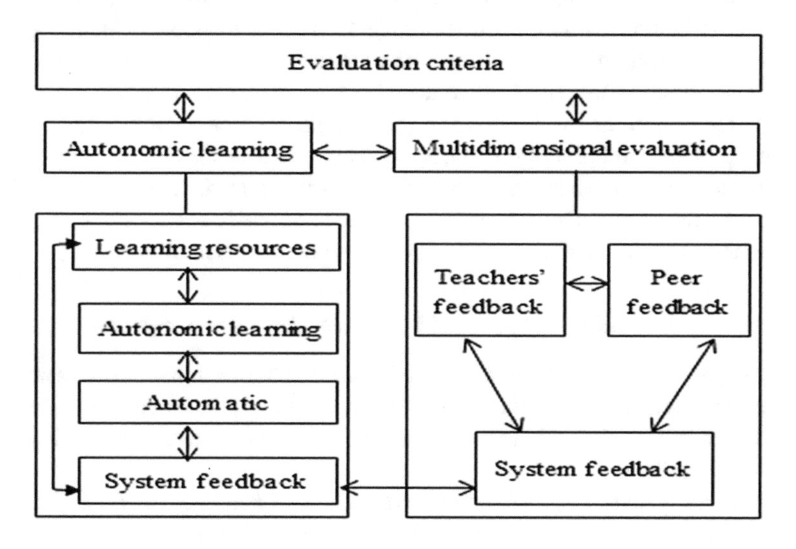

Fig. 1. Learning platform model based on Big Data and formative evaluation system.

First of all, different assessment criteria should be enacted in different period of study, according to the analysis results of Big Data, namely learning feedback, in order to make students obtain knowledge step by step. In the whole learning mode, the first is to make early demand analysis. And then according to the requirement, specific objectives and standards and a detailed assessment scheme should be determined.

Compared with the traditional writing process, the new model makes more clear goals. For example, students can use assessment criteria to arrange autonomic learning and teachers can more accurately organize the teaching tasks.

Furthermore, Big Data technology can provide the reasonable teaching contents and corresponding resources. On the one hand, the search engine, online dictionary, online translation software, English learning website, and digital video courses, as well as scoring system and online writing corpus resources, can be integrated into an information bank and connected to the English learning platform. According to the teaching plan, all students at different levels and different learning processes can rely on the learning platform to get corresponding support. On the other hand, through a lot of users' assessment about teaching resources, useful and high-quality teaching resources can be found out at a faster speed.

Students can use Big Data to analyze themselves automatically. The quantization and visualization of informal learning performance in the platform can let them know themselves about knowledge scotoma and modify it. For example, in the writing task, no matter how many times students repeatedly modify their articles, feedbacks are so timely that motivate them to correct mistake again and again. Thus a virtuous cycle of practicing, modifying and improving is formed, which can more effectively promote study by assessment. In contrast, teachers, due to their constraints about time and energy, are slightly inferior in reference to feedback and timeliness.

Last but not list, the analysis of the students' learning process through Big Data makes a contribution to guiding teaching. It is helpful to make scientific teaching decisions, appropriate academic designs and meticulous curriculum management. On the one hand, teachers use Big Data to grub learning data, conclude students' learning rules, and identify common problems in their English learning for a unified explanation. On the other hand, through the analysis of the data, teachers also can understand concrete condition of each student, so that they will give students individual guidance according to feedbacks about their own errors in English. As a result, the system feedback and artificial feedback can bring out the best in each other.

4.3 The Key Technical Supports of Formative Assessment System

The whole system is a multiple assessment form. It includes summative assessment and formative assessment, combining self-assessment with other person's assessment, which consist of practice subsystem, data processing subsystem, and feedback subsystem (see Fig. 2).

Practice subsystem. This subsystem, comprised of learning materials, interactive communication, practice homework, teacher assessment, peer response and so on, collects relevant information and transmits to the data processing subsystem. The information includes learning notes, knowledge sharing, learning courseware, the number of question, the situations of participating in community discussions, individual or group homework and examination, as well as the assessments of teachers and students and so on.

Data processing subsystem. This subsystem includes four steps: data collection, data preprocessing, data conversion and data analysis. At first, data acquisition, like collecting effective learning traces, records and other, should be carried out as

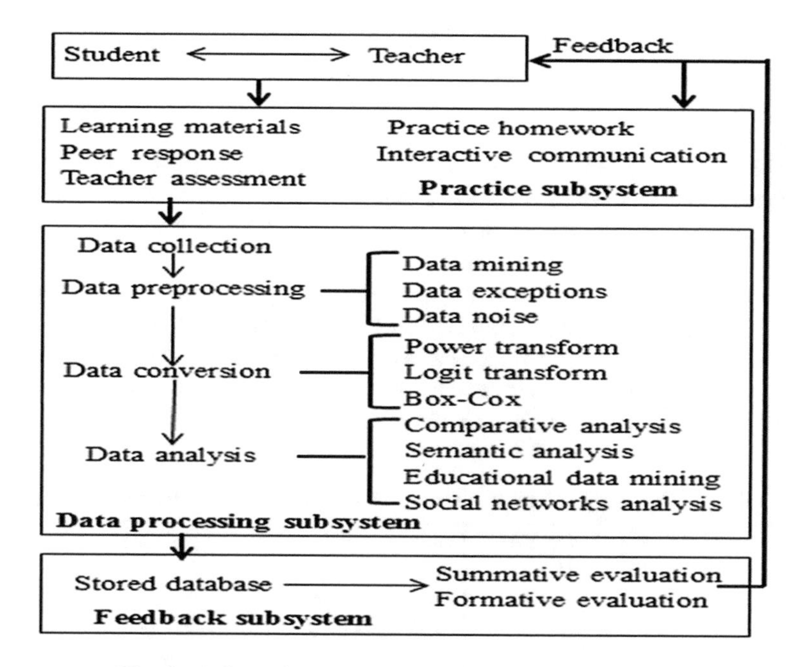

Fig. 2. A formative assessment system under the Big Data.

comprehensively as possible, according to assessment objectives and requirements. Specially, the network sniffer can collect preliminary information and filter it simultaneously. Next, data should be preprocessed because there may be missing, exceptions, noise, and redundancy which can be directly deleted, or replaced by new figures, or solved through estimation method. The mean, standard deviation and quantile can detect outliers. The data conversion serves the data mining, comparison and analysis. For example, the symbol can be transformed or reclassified into a number, various proportions and functions. The last is the further analysis of the data. Some data got on the above steps can be directly stored in the database, such as test scores. Some should compare with standard or coefficient, which is in the database, before passing to the database. Others should input database after text mining. Commonly, educational data mining, social networks analysis and semantic analysis are used to do that.

Feedback subsystem. This is mainly to send the timely feedback from the stored database students and teachers as well as information bank for update. The feedback can be divided into two types, summative assessment feedback and formative assessment feedback. The former refers to the various scores after the end of the course. The latter includes practice records, interaction, and tests of chapters during the course.

4.4 Experimental Results

Taking the English course, Comprehensive English I, in the accounting of Guangdong University of Foreign Studies as an example. 10 students in English Class B21 are chosen. From their learning situation of the second term in 2017 that is recorded in the system, their first chapter information into the charts as well as final results of the semester are selected in order to observe how Class B21 students accept knowledge. The system has appropriate practices and tests in each chapter and provides automatic scoring function.

These 10 students, in the first semester of 2017, who studied in the Summative assessment. Their first semester's academic performance is taken to compare with the second semester's score. The academic performance of the first semester consists of two parts, 20% of in-class score and 80% of score in the final exam. In-class score depends on the student attendance and quizzes. There is no doubt that evaluating the student's achievement through his attendance is unscientific which will increase the student's performance to a certain extent. And in one or two quizzes, teacher only give students a score without any suggestions or guidance. Most students have made progress in the second term, which is shown in Table 2.

Table 2. A comparison of final scores between the first and second semester.

Student ID	Attendance 100 *10%	Quizzes 100 *10%	Final exam 100 *80%	Final score of the first semester (100)	Final score of the second semester (100)	Difference
01	100	90	82	84.6	83.8	−0.8
02	80	80	75	76	77.6	1.6
03	90	85	90	89.5	92.4	2.9
04	90	89	87	87.5	87.6	0.1
05	100	100	98	98.4	98.8	0.4
06	90	90	89	89.2	94.2	5
07	100	100	90	92	94	2
08	90	82	60	65.2	68.8	3.6
09	100	74	75	77.4	81.6	4.2
10	100	88	86	87.6	88.8	1.2

In a survey of these 10 students, most students indicated that the formation evaluation system would allow them to study more seriously, even if they had to spend more time on study. Some students agreed that because every chapter in the system had exercises with appropriate specific feedback, they could understand clearly where should be corrected and be given reference. Another student indicated that the proportion of daily performance was larger, and every composition should be evaluated by others, so they were forced to pay more attention to daily practice and homework. One student believed students would make great progress and no long review knowledge only before the final exam.

Indeed, because the system can automatically correct and give feedback to students, as well as record evaluation from students and teachers, students can explicitly know the specific components of their achievements, but also get the criteria of score, understand what knowledge should be obtain and what problems they have. For example, composition practice includes self-evaluation and peer evaluation, so students can find out the gap between themselves and others which help students overcome shortcomings by learning from each other's strength. Moreover, the further evaluations by teachers provide individual guidance. After every chapter, students have to think about whether they get new knowledge in this period and adjust the learning method for next study. Teachers also should arrange the next teaching according to students' learning situation in this section. The formative assessment system will change the wrong view of students and teachers that is to only concern the final exam.

5 Summary

The technology of Big Data is the product under the rapid development of information technology, with the progress of cloud computing, distributed processing technology, storage technology and sensing technology. Big Data tells us "what" instead of "why" and liberates the human from exploring the causal relationship to the correlativity. Along with the development of informatization in education, experts think more about how to learn efficiently, how to realize individualizes teaching, and how to achieve students' individual development. Big Data is developing rapidly in recent years. Its educational value is discovered gradually and contributes to recording students' learning process, formative assessment, and improving educational business. In this new era of Big Data, we should keep pace with the times, actively learn to use Big Data, and then reform the education.

References

1. Qiuxian, Ch., Margaret, K., Val, K., Lyn, M.: Interpretations of formative assessment in the teaching of English at two Chinese universities: a sociocultural perspective. Assess. Eval. High. Educ. **38**(7), 831–846 (2013)
2. Ling, L., Xiaoliang, L.: Effective approaches to promote the implementation of developmental assessment among English teachers at Normal Universities. J. Jiangsu Teach. Univ. Technol. **16**(10), 100–104 (2010)
3. Jingyu, Z., Xi, W.: The impact of Big Data on English teaching: discussion on the application of Big Data in English teaching. Contemp. Teach. Educ. **9**(1), 43–47 (2016)
4. Macfadyen, L.P., Dawson, S., Pardo, A.: Embracing Big Data in complex educational systems: the learning analytics imperative and the policy challenge. Res. Pract. Assess. **9**, 17–28 (2014)
5. Zhenchao, L., Lin, C.H., Xudong, Z.H.: The research and design of developmental assessment system for E-learning supported by Big Data. Mod. Educ. Technol. **25**(6), 108–114 (2015)
6. Jing, Q.: Current investigation and analysis of formative assessment in oral English teaching of non-English majors. J. Chengdu Univ. Technol. **20**, 85–89 (2012)
7. Mohamadi, Z.: Comparative effect of online summative and formative assessment on EFL student writing ability. Stud. Educ. Eval. **59**, 29–40 (2018)

8. Scriven, M.: Beyond formative and summative evaluation. In: McLaughlin, M.W., Phillips, D.C. (eds.) Evaluation and Education: Atquarter Century, 5(11–12) 48–51. University of Chicago Press, Chicago (2005)
9. Havnes, A., Smith, K., Dysthe, O., Ludvigsen, K.: Formative assessment and feedback: making learning visible. Stud. Educ. Eval. **38**(1), 21–27 (2012)
10. Bloom, B.S.: Handbook on formative and summative evaluation of student learning. Stud. Art Educ. **14**(1), 932 (1971)
11. Wang, B.L., Wang, J.: A survey on the undergraduate English teaching and learning assessment in the East China Universities. Shandong Foreign Lang. Teach. J. **14**(1), 9–14 (2014)
12. Lee, I., Coniam, D.: Introducing assessment for learning for EFL writing in an assessment of learning examination-driven system in Hong Kong. J. Second Lang. Writ. **22**(1), 34–50 (2013)

Microplatform for Autonomous Experimenting on Journalism and Communication

Wei-Bo Huang(✉), Wen-Xin Liang, and Ge-Ling Lai

Guangdong University of Foreign Studies,
Guangzhou 510006, People's Republic of China
hwb@oamail.gdufs.edu.cn

Abstract. In order to meet the needs of media professionals in the new media era and make students majoring in journalism and communication have more opportunity to experiment and practice independently, this paper gives corresponding solutions based on the analysis of the current situation and problems in the experimental teaching of news communication at home and abroad. It combines development technology of Java Web and the development of WeChat public platform. On the interactive principle of the learner as the center, it is based on the functional requirements and emphasizes the autonomy of the experiment. This paper focuses on the scientific characteristic of the experiments and the diversification of the evaluation methods. It designs the corresponding functional modules and develops independent experimental microplatform of news communication, in order to improve the experimental enthusiasm and effectiveness of the news communication students. The experimental results show that the micro-platform can better assist students to carry out independent experiments and improve the effect of experimental learning.

Keywords: Journalism and communication · Autonomous experimenting
Micro learning · Microplatform

1 Preface

With the increasing demand for personalized and diversified talents, higher education in our country must conform to such a trend and pay more attention to the cultivation of individualized and multiple talents. In this new media era, especially because of the great leap forward development of the mobile Internet technology and big data technology, science communication presents a development process of continuous integration and innovation [1]. The journalism students' practice is currently relying on social media. However, the social media is limited by the system, equipment and venues, which means it lacks adequate capacity channels for accepting students' news practice. What's worse, the microplatform of comprehensive experiment has not received wide attention. In the rapid development of new media, the experimental center is required to create a micro-platform based on modern information communication technology and digital technology, with major media convergence and more complete and updated equipment, to provide students better opportunities to study and practice [2].

N. Xiong et al. (Eds.): ISCSC 2018, AISC 877, pp. 481–488, 2019.
https://doi.org/10.1007/978-3-030-02116-0_56

2 Current Situation of Experimental Teaching of Journalism and Communication

RENMIN UNIVERSITY OF CHINA has mixed broadcast media, flat newspaper, network multimedia and other media into an experimenting platform of journalism and communication, including four functional modules, radio and television production, design and editing of network multimedia, advertising design, photo editing and research of communication effect. The University has formed a set of relatively complete experimental teaching system, and actively provide students the extracurricular practice teaching platform, such as digital newspaper publishing practice platform, "New Weekly" platform, "Jieli Media" practice platform. The omnimedia has become the development direction of the media industry and put forward a higher demand of the quality of journalists [3]. Missouri School of Journalism began to set the media fusion curriculum in 2005, which aimed to cultivate comprehensive talent with various skills, because media convergence contributes students to fully grasping the technology of camera, recording, shooting and network. In Missouri School of Journalism, every student must enter the television station to participate in interviews, video recording, and program post-production. As the saying goes, "Genuine knowledge comes from practice". This study method makes students master new technology from pre-collection to post-production so that they can improve the professional skills faster through the practical operation [4]. Additionally, America Cromarte School of Journal has built a digital media lab and the Knight Digital Media Entrepreneurship Center, both of which are new media practice platforms. Digital media lab is an interdisciplinary project whose personnel are some students from School of Journalism, School of business and Computer Engineering and Design. They create new media multimedia products for the some media companies.

3 Problems and Solutions of Experimental Teaching of Journalism and Communication

Nowadays, journalism education requires qualified personnel in the field of new media. The journalism teaching platforms in colleges generally originated from the traditional media period. In current researches about microplatform for experimental study, experts mostly concentrated on the problems of teaching laboratory, less from the perspective of students to explore. Basically, this kind of microplatform is a single form with media form as the unit [5]. And the application of it to education is still in the initial stage [6].

Micro learning is a new learning way, oriented everyone's personalized learning needs. Learners can use intelligent communication tools to choose, obtain and process a variety of modular, loosely coupled and self-contained knowledge whenever and wherever possible, and also can interact with others with all kinds of approaches [7]. With the development of mobile network technology and mobile devices, micro learning will become the new trend of public learning, whose amount of users will furtherly increase. Chinese scholars have been thinking about and exploring micro learning from 2008. The professor, Xu Fuyin, from media characteristics, defined the micro learning to be a learning way using a micro media to carry micro knowledge. Professor Zhu Zhiting emphasized that micro learning is not only micro and small, but also contains the meaning and character of learning that can affect the learners' learning attitude [8]. Based on the concepts of micro learning abroad, Gu Xiaoqing and Gu Fengjia summarizes that the micro learning has some significant features, like occupying short time, fragmentation, individualization, multimedia and so on [9]. It is different to define the concept of micro learning among foreign researchers. Lindner mainly from the media features to define that micro learning is a new study of a micro-content and micro-media existing in the new media ecosystem. In Hug's opinion, micro learning is a learning activity that divides learning content into smaller learning modules and focuses on shorter time study [10]. Silvia Gabriellie believed that the implementation of micro-learning benefited from the development of small pieces of learning content and improved communication technology, so learners can easily have access to learning content in specific occasions, such as at recess or on the way [11].

4 Design of Microplatform for Independent Experiment of Journalism and Communication

According to the complete credit system and actual circumstance of Journalism and Communication students in Guangdong University of Foreign Studies, it is of great significance to put self-learning and practice as a whole to create a media comprehensive experimental platform, such as experimental television station, radio station, news website and social media.

4.1 Functional Requirements and System Characteristics

The microplatform of autonomous experimenting of Journalism and Communication with four original functional modules provides teachers and students exclusive experimental information and intelligently classifies them. The functional modules of this microplatform are shown in Fig. 1.

The platform combines the Java Web technology with WeChat public platform to achieve multi-adaptable. It can collect users' data, then constantly improve them through data analysis, and ultimately achieve intelligent goal. Users can get services from the platform through the computer and mobile. This system consists of four core functional modules, message notification module, task schedule module, experiment management module and experiment communication module.

Message Notification Module. The message notification module enhances the transmission efficiency of the message notification from three aspects. First of all, through the "Advanced Customer Service" interface of the microplatform, a message will be instantly twittered to the user's WeChat client as soon as it is issued. Because of the popularity of mobile devices and a huge amount of WeChat users, it can ensure that the notification can be transmitted timely. And then students can receive a notice through the function of "One Key to Confirm Receipt", which simplifies the process of

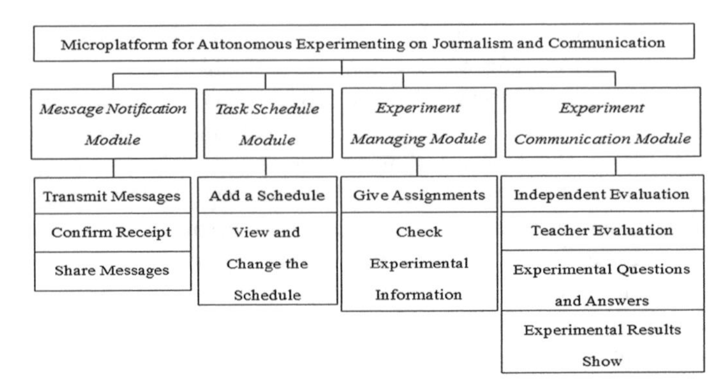

Fig. 1. The functional modules of the microplatform for autonomous experimenting on journalism and communication.

feedback. If the feedback is really needed, get it through the message. Thirdly, it is easier for students to continue to release and spread messages. According to the great structure of this microplatform, users can share information with other class or their own student organization through the function of "One Key to Share", which will improve the convenience.

Task Schedule Module. The microplatform can extract core information through experiment tasks that will be pre-filled to the shortcut schedule bar cleverly embedded in the notification page. Users can add important time points of messages or experiments to their schedule by the function of "Add".

Experiment Managing Module. Students' experiment information will be integrated into the management module, including traditional empty laboratory inquiry, performance inquiry, experiment schedule inquiry and other functions. At the same time, teachers can give assignments on their personal computer, which obviously improve the work efficiency. The experiment categories include professional basic autonomous experimenting, comprehensive experiments, and innovative experiments.

Professional Basic Autonomous Experimenting. Students should upload class and extracurricular news works, such as press releases, newspaper typesetting, broadcast news, television news. With laboratory of each major as a platform, professional basic experiments are to complete some necessary basic teaching task about experiments which enable students to systematically master basic experimental methods and skills on the various branches of journalism and communication and promote their autonomous experimenting ability and the enthusiasm of autonomic learning.

Comprehensive Experiments. Students use some high simulation platforms, like the experimental television station, the experimental broadcast station, the experimental news website, the university student news agency and so on, to carry out comprehensive or design experiments. In order to improve the ability of using modern media technology and cultivate the spirit of innovation, students are required to use high simulation experimental media platform independently. More than ten professional experimental courses that are divided into so detailed types that their experiments are crossed are converged at a comprehensive experimental teaching module.

Innovative Experiments. School of Journalism and Communication at Guangdong University of Foreign Studies, signed cooperation plans with People's network, Xin Hua Net and Nan Fang Metropolis Daily. Students can publish their own news works on above Media through the platform, really involved in the journalism practice and improving their practical ability. The platform provides information and opportunities for students to participate in media practice and learn actual skills rather than pure theoretical knowledge.

Experiment Communication Module. This model establishes a system of work quality evaluation divided into four parts, independent evaluation, teacher evaluation, experimental questions and answers, and experimental results show. This model can effectively change the previous situation that experiments lack of feedback, and it is difficult for students to make progress. It can put like-minded students together to carry out various experiments, at the same time, also facilitate teachers to manage and improve the experimental content, create a new evaluation system instead of the traditional unidirectional type, and increase communication between teachers and students.

4.2 Design of the Microplatform

According to the function and characteristics of the microplatform, it can be divided into five layers, data resource layer, application service layer, autonomous experimenting service layer, transmission layer and access layer, shown in Fig. 2.

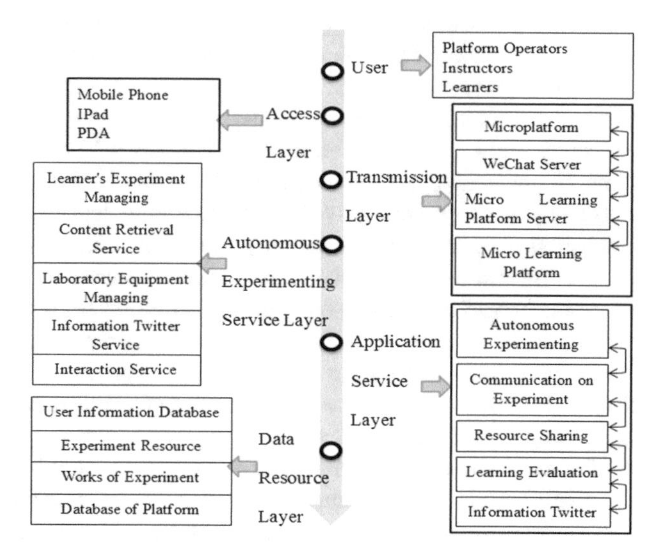

Fig. 2. The general framework of the microplatform for autonomous experimenting of journalism and communication.

From the functional analysis of WeChat, it can be found that its wide use and powerful function can reduce the difficulties of development and promotion. According to each element's characteristics of microplatform, the design of autonomous experimenting microplatform should follow the learner centered interactive principle and emphasize the autonomy of the experiment, focusing on the scientificity of experimental contents and diversified evaluation methods. The data resource layer contains user information database, experiment resource, works of experiment and database of platform. It is the basic part of this experimental platform, providing the related learning data and resources for the whole platform. The application service layer includes autonomous experimenting, communication on experiment, resource sharing, learning evaluation and information twitter. These modules can jump to each other. The autonomous experimenting service layer is the realization of the application service layer and to provide the corresponding services directly to users, including learner's experiment managing, content retrieval service, laboratory equipment managing, information twitter service, and interaction service. The transmission layer is responsible for data exchange. The transport layer provides transparent data transmission between end terminal users, achieving end-to-end delivery of data and a reliable data transmission. In the access layer, learners, instructors and platform operators use mobile phones, iPads, personal digital assistants or other mobile devices, through the WeChat client, to access the microplatform for study and communication.

5 Experimental Data and Analysis of the Microplatform

56 students in Journalism at Guangdong University of Foreign Studies who have been using the microplatform are chosen to analyze their study effects and the promotion of the platform during two months, from August 1st to September 30th in 2016.

The microplatform will regularly publish a new message, an article or a lecture about journalism and communication every Sunday. Figure 3 shows the number of readers after publishing a new article for three consecutive weeks. Demonstrably, it will reach the peak number of reading on Sunday, but will fall back rapidly in the subsequent one or two days. Thus, students pay attention to the content of microplatform quickly and spend short time in reading, which means that there is a high demand for the timeliness of making and publishing such information. Furthermore, appropriately shortening the time interval of each article can promote students' learning enthusiasm.

As shown in Fig. 4, 10 students have forwarded learning information, accounting for 18% of 56 students. Most users have forwarded less than four times, and one student has forwarded the learning information 5 times. Only four users have sent 5 messages in the microplatform that are some evaluation and views of the published information. In addition, it shows six articles were praised among 8 published articles. This can be seen that students rarely interact on the platform after reading. So at the next study stage, students should strengthen their awareness of communication and actively participate in the discussion in order to accept new knowledge better and create positive learning and experiment atmosphere.

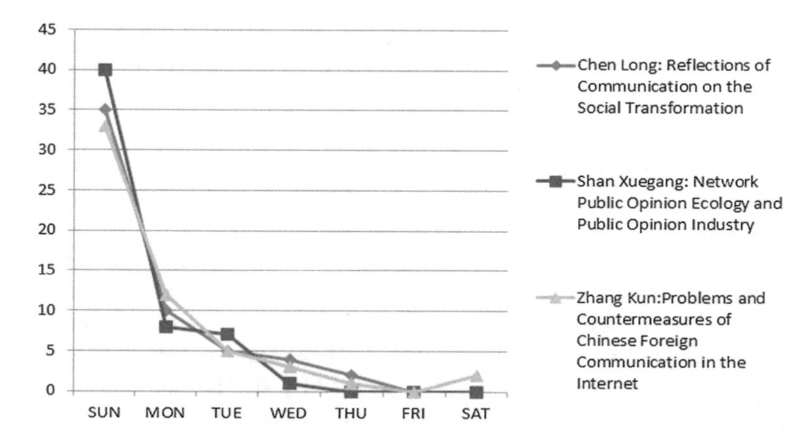

Fig. 3. The number of readers after publishing article for three weeks.

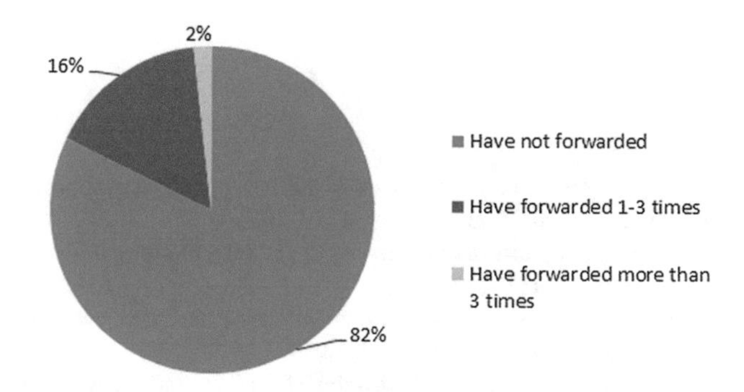

Fig. 4. The distribution of the number of students who have forwarded the information.

6 Conclusion

The microplatform for autonomous experimenting of journalism and communication not only exerts the flexibility of mobile devices, but also enhances the efficiency of teaching transmission and communication, combining reality with virtual space. It makes experimental teaching and independent practice more interactive and convenient, fully mobilizes students, and provides them a new way of autonomous experimenting, in line with modern educational requirements.

With the aim of better realizing this microplatform and improving students' professional knowledge and ability, what should be paid attention to is the quality of interaction in the microplatform for autonomous experimenting, which includes the breadth and depth of interaction. The breadth of interaction is the scope of interaction for users in microplatform, which means it need to has a variety of interactive ways for users to show or share something, like one to one exchange, interacting between teachers and students or between users and resources, and even cross platform. Students can use mobile devices to communicate with somebody across time and regional, to accept, find or share information and experimental results they need. The depth of interaction means the usefulness of interactive information. The microplatform for autonomous experimenting should enable students to have more opportunities to discuss, get effective experimental learning materials, show their experimental results, evaluate others' experiment results with self-reflection and promote mutual learning, breaking the barrier in traditional learning. At the same time, the microplatform would covert the shortage that is the limitation of teacher's energy.

Moreover, in order to strengthen the interaction of microplatform for autonomous experimenting, it should be set up two types of feedbacks between the platform and students. One is to set keywords in the background for automatically reply, or to use the high level development mode with robot, which allows users to input specific words to obtain message and reach back. There is no doubt that this form can develop the transmission efficiency, convenient for information inquiry. The other is that teachers respond artificially in the background, giving students one-on-one personalized response and study guidance, which is a supplement to the automatic response function.

References

1. Hongxia, H., Feng, L.: Research on the mode of cultivating innovative media talents based on cloud computing. Journal. Bimon. **6**, 116–121 (2014)
2. Yaqin, X.: Construction of journalism experimental teaching platform in media convergence era. Exp. Sci. Technol. **14**, 181–182 (2016)
3. Ning, S.: Analysis of the new media environment based on the construction of teaching resources. China Educ. Technol. **7**, 91–95 (2013)
4. Wen, C., Xinfeng, Z.: The 'Missouri Method' of news education – analysis of the educational model of Missouri school of journalism. Mod. Commun. **2**, 123–125 (2006)
5. Ping, W.: Analysis of support functions and design principles of mobile learning based on WeChat. J. Distance Educ. **6**, 34–41 (2013)
6. Hao, B., Jingjing, H.: Research on the application of WeChat public platform in higher education. Chin. J. ICT Educ. **4**, 78–81 (2013)
7. Xuewei, Z., Yu, Z., Xiaoli, X.: Research and design of mobile learning platform supported by WeChat. Distance Educ. China **4**, 77–83 (2014)
8. Zhiting, Z., Hao, Z., Xiaoqing, G.: Miniature learning – a practical model of informal learning. China Educ. Technol. **2**, 17–21 (2008)
9. Xiaoqing, G., Fengjia, G.: Miniature learning strategies: designing mobile learning. China Educ. Technol. **3**, 17–23 (2008)
10. Tscherteu, G.: The blogosphere map visualizing microcontent dissemination – inspired by maria montessori. In: Hug, T., Lindner, M., Bruck, P.A. (eds.) Proceedings of the Microlearning, pp. 109–120. Innsbruck University Press (2005)
11. Gabrielli, S., Kimani, S., Catarci, T.: The design of microlearning experiences: a research agenda. In: Proceedings of Microlearning 2005 Conference, pp. 45–49. Innsbruck University Press, Innsbruck (2005)

Multi-source Geospatial Vector Data Fusion Technology and Software Design

Huadong Yang[1,2(✉)] and Hongping Tuo[3(✉)]

[1] Naval Engineering University, Wuhan, China
control_beijing@sina.com
[2] Naval Research Academy, Beijing, China
[3] System Design Institute of Mechanical Electrical Engineering, Beijing, China
yhd_beijing@163.com

Abstract. Firstly, according to the problem consisting with the uniformity of multi-source geospatial data model and the coding of elements, the method of unified design of geospatial data model and consistent expression of feature code is put forward, and the problem of factor layer conversion and coding in the process of coding is analyzed and resolved. Then, the digital map space vector data fusion system is designed with digital topographic map and digital chart, and the fusion of multi-source geospatial data is realized. Finally, the multi-source fusion geographic information demonstration software is designed; the human-computer interaction interfaces are abundantly put forward to meet the application requirements of multi-source geospatial data at all level.

Keywords: Multi-source geospatial data · Data fusion
Geographic information system · Demonstration software

1 Introduction

With the rapid development of computer technology and database technology, digital terrain map database technology, which is the basic information data source of vector electronic map, has also been developed by leaps and bounds, and electronic topographic products have emerged. As the digital topographic map and digital chart are produced in different attribute coding system, the digital topographic map and digital chart are independently organized in accordance with their own characteristics in the map representation and map elements of the classification and classification methods. Thus, there is always no uniformity of spatial data model and data structure. This situation is feasible for the specific application areas of electronic topographic map and electronic chart products. However, when the unified joint digital map platform is required for both joint applications from the sea to land and from the land to sea, the malpractice will be undoubtedly exposed. Therefore, the digital terrain and digital chart joint problem is needed to be studied. The key to the joint is that: for the data with different origins, it is important to keep consistent with the spatial rules and data models and feature classification and grading methods in accordance with specific joint rules to ensure that vector map data from different sources are located in spatial data and attribute information. Then the unified digital joint database is needed to be obtained

N. Xiong et al. (Eds.): ISCSC 2018, AISC 877, pp. 489–496, 2019.
https://doi.org/10.1007/978-3-030-02116-0_57

effectively to support the joint use of the requirements. Thus, the study of multi-source geospatial vector data fusion technology has important and practical significance and significant application value.

2 Research Status of Digital Map Data Fusion

The vector map fusion technology belongs to the research category of geospatial vector data fusion. The research content related to the integration of geospatial vector data is the problem of digital map merging [1]. In the mid-1980s, the US Geological Survey (USGS) and the Census Bureau collaborated on the map merger technology [2], and successfully developed the world's first map mapping automatic merger system, which can improve the quality of data and eliminate the errors as well as the exchange of attributes and entity information. Subsequently, experts and scholars from all over the world have conducted various studies on this issue. In 1996, the map merging technology was listed as one of the top ten research topics by the American University Geographic Information Science Association (UCGIS) as one of the main research subtopics in Data Acquisition and Integration [3]. In 1998, Cobb studied the knowledge-based attribute data matching strategy, and proposed the American vector data standard format file fusion method [4]. In 2004, Hoseok gave the vector map fusion model [5]. In 2005, based on the rubber band transformation, Haunert proposed an entity element conversion method of topological consistency to support the map fusion of the different periods [6].

At the same time, the world's geographic information systems (GIS) software providers have launched their own vector map fusion software or functional modules. The more typical examples are as follows: Intergraph Company has developed a map to achieve physical matching and merger of the software package [7]. E-Sports Entertainment Association (ESEA) has introduced the map integration system and map Merger; GIS/Trans has developed Arc Info based map fusion system; Geography Data Technology has introduced commercialized map fusion software.

At this stage, the research on vector data fusion is still limited. Yin studied Arctic + Geodatabase's multisource data fusion [8], in which the Arc GIS's geospatial data model is utilized and multi-source data fusion techniques is used to incorporate data from different formats into a unified data mode. By this way, the purpose of the integration of different sources of sea chart is realized. In his academic papers, Li studied space vector data fusion problem using ontology-based semantic matching method to achieve the same name entity matching problem [3]. Hu gave the basic framework of spatial vector data fusion [9]. Lu studied the spatial data fusion problem and its application in the chart [10]. Li explored the problem of the integration of multiple heterogeneous data based on the electronic chart platform [11].

3 Digital Map Space Vector Data Fusion Technology

3.1 Map Projection Data Fusion

The law of describing the latitude and longitude lines of the surface of the earth's ellipsoid to the map (plane) is called the map projection. Map projection transformation is the basis of map data fusion from different sources. Most of the map data from different sources are different in terms of map projection and geographic coordinates, and different map data may be inconsistent between scales and the length units of scale and digitizer. Therefore, in order to ensure the identity of the map data, the first need for unified data is the benchmark of coordinate system.

Usually, topographic and sea measurements are mostly made by Gaussian-Kruger projection (also known as isometric cross-elliptical projection). As the name suggests, the projection has isometric nature, there is no angular deformation, and that is, all of the length ratio in the direction from a point are equal. And the mathematical derivation also shows that the length and area of the projection is not deformed, such as by 6 ° zonation, the length of deformation is about 0.2%, area deformation is not more than 0.4%. Just as this projection has the nature of smaller deformation. This projection has been widely used in the large-scale topographic map [12] to meet the needs of different applications. Therefore, the projection method used in the joint GIS in this paper is also Gauss-Kruger projection.

The most common projection method in the chart is isometric cylindrical projection [13, 14] (Mercator projection). Since the equidistant route after the projection is shown as a straight line on the chart, the ship can easily reach the destination in the direction of the straight line, and after the projection, the latitude and longitude lines on the map are parallel lines, the drawing is simple and clear. Due to this reason, for several centuries, Mercator projection has been commonly used in the world's charts. In this paper, the use of GIS in the chart is just the Mercator projection.

In this paper, the Gaussian-Kruger projection and the Mercator projection are used to describe the topographic data and the chart data in the joint GIS. As the projection method is different when the topographic map and the chart are projected, it is necessary to unify the projection mode before the data fusion. The way is described as follows: according to the different application focus, a projection method is selected as the benchmark, by using coordinate conversion formula, the other projection mode of the data can be converted into a selected baseline benchmark projection mode. For example, the Mercator projection method is selected as the reference projection, and the data under the Gaussian-Kruger projection used in the topographic map is converted into topographic map data under the Mercator projection. The converting method has steps described as follows:

1. From the original geodetic coordinate system Gauss- Kruger projection plane coordinates (x, y) directly into the WGS2000 national geodetic coordinate system Mercator projection plane coordinates (x, y);
2. The geographic coordinates (l, b) of the original geodetic coordinate system is firstly converted to the geographical coordinates (L, B) under the WGS2000 national geodetic coordinate system by using antisolution formula. and then, the

geographical coordinates (L, B) are converted to the coordinates (x, y) of the Mercator projection plane by the projection transformation formula of the Mercator projection;

3. The geometric coordinates (x, y) of the Gaussian- Kruger projection by the projection transformation and the transformation of the Gaussian-Kruger project are transformed into the space coordinates (l, b) by using anti-solution formula. And then converted into the WGS2000 country by Bursa spatial coordinates of the space coordinate system by using conversion formula. Finally, those are converted to the geographical coordinates under the Mercator projection using the Mercator projection transformation formula. By analyzing, there are a variety of spatial Cartesian coordinates into the WGS2000 national geodetic coordinate under the space coordinates of mature technology, and the conversion accuracy is very high. Therefore, this method is adopted in the map projection uniform. The concrete coordinate system transformation steps are shown in Fig. 1.

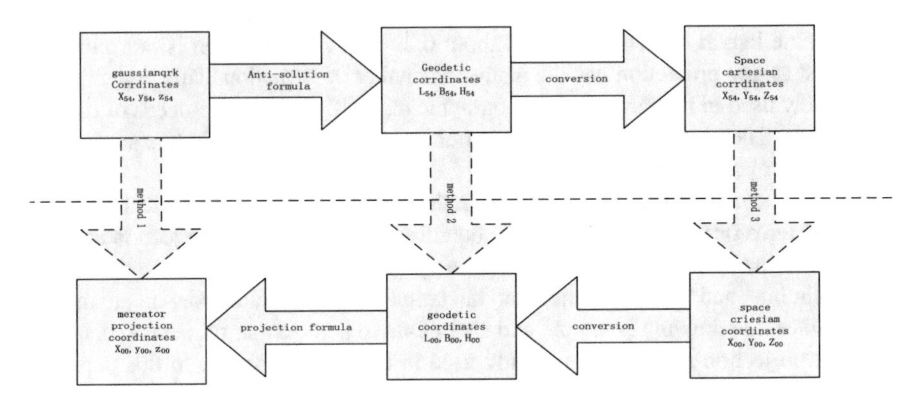

Fig. 1. Coordinate transformation steps diagram.

3.2 Unified Spatial Data Model Design

For the current vector map and vector chart, the no topological structure is adopted on the spatial data model; the geometric location information and attribute information are used to adopt the entity-oriented spatial data type. The unified spatial data model is still in the form of non-topological structure. At the same time, the organizing way of two layers data is inherited. The design of the entity-oriented unified spatial data model is shown in Fig. 2.

The model inherits the advantages of the vector description data in the vector map data, which can effectively avoid the errors caused by the overrun in the entity drawing, and also facilitate the operation of the map entity. This model also inherits the characteristics of determining symbols of vector chart data through the geographical elements of the attribute code. At the same time, the defects that caused by excessive space occupied by the amount of data due to storage of drawing information are overcome.

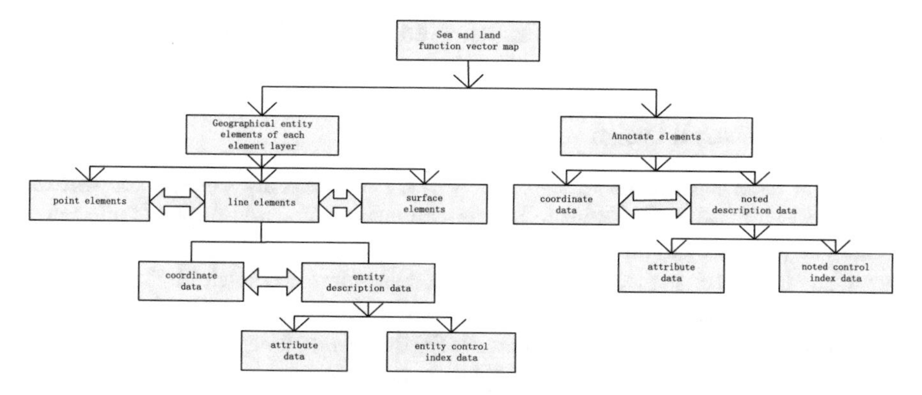

Fig. 2. The entity-oriented unified spatial data model.

3.3 The Realization of Fusion of Feature Classification

In the design of the unified map attribute coding table, a unified map attribute code table should be designed based on multi-source geospatial data elements with the principles of coding and fusion coding and attribute coding table rules. In addition, the advantages of existing map and chart attribute code table should be also fully absorbed.

Based on the above requirements, the coding method can be used to clearly show the classification of the object category, and the code structure has a strict subordinate relationship between the sub-code and the application which can be fully utilized relational database mature processing technology to achieve the length of coding methods effective management of data. Therefore, the integrated application level code and fixed length coding structure are used in this paper to achieve the integration of map element coding.

The element code consists of classification code, sub classification code and identification code, and three kinds of code from front to back each two digits, together constituting with the elements of the unique identification code. In the process of coding elements fusion, the coordination problem of different factor layers should be also considered. Elements are to be classified and encoded according to the essential attributes. And the basic information elements and navigation elements should be integrated into the corresponding categories to realize that all elements of the comprehensive classification are to be unified definition and unified coding. It is also necessary to pay attention to the descriptive information of the elements in the attribute items, which are no longer involved in the coding, so that the encoding object to form a complete entity, this encoding is more suitable for computers, geographic information systems, database technology of the data processing and application.

4 Design of Multi-source Fusion Geographic Information System Demonstration Software

4.1 System Overall Design

In order to meet the various needs of the users, it is necessary to make the software simple and intuitive and friendly in human-computer interaction for convenient operation. The joint geographic information system demonstration software uses a digital menu, and the screen is divided into five functional areas. As shown in Fig. 3, where: I area is for the map display area, which displays the contents of the data after the integration of the map information; II area is for the system function menu display area, including system management, map calculation, map query, thematic map, plotting, style settings and map display mode and other function menu; III area is for the current map of the basic information and measurement results information display area, showing the current map zoom scale, the location of the corresponding coordinates of the map and map calculation results of the operation information; IV area is convenient tools bar area, containing the commonly used map operation command buttons; V area is for the layer management and hawkeyed browsing function area.

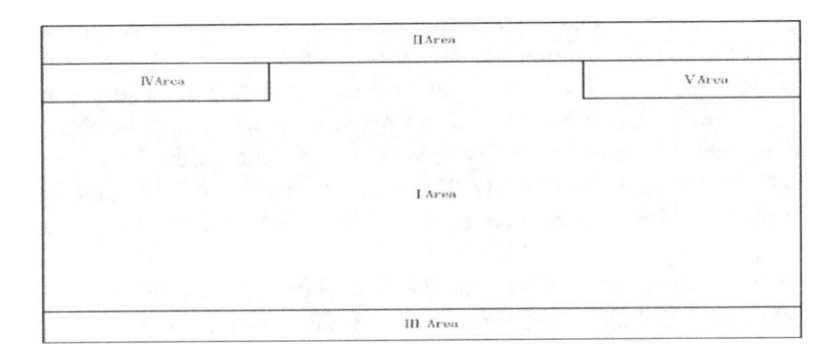

Fig. 3. System main interface design.

4.2 System Process Design

Digital map space vector data fusion system includes two modes of human-computer interaction and automatic operation. The basic processing process is divided into human-computer interaction process and format automatic conversion process two parts:

1. Human-computer interaction process. After the operation, the system will automatically display the electronic map on the current system main interface, which can support the map related to various types of operating functions, including the map zoom, zoom pan, full map display and map calculation, map feature information query, layer display control, latitude and longitude display, latitude and longitude grid line display, map screenshots and other functions. Human computer interaction process is shown in Fig. 4.

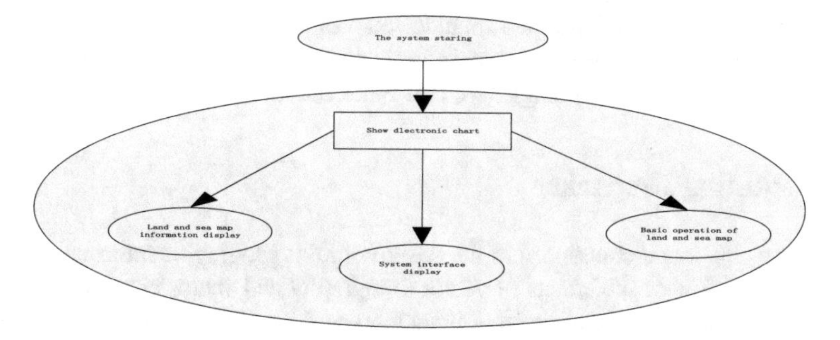

Fig. 4. Human-computer interaction process.

2. Automatic operation mode process. After the operation, the system automatically loads the electronic map data in the specified directory, and converts the data into the unified format that the system can recognize and operate, and then automatically displays the map data in the system map displaying area interface. The format automatic conversion process is shown in Fig. 5.

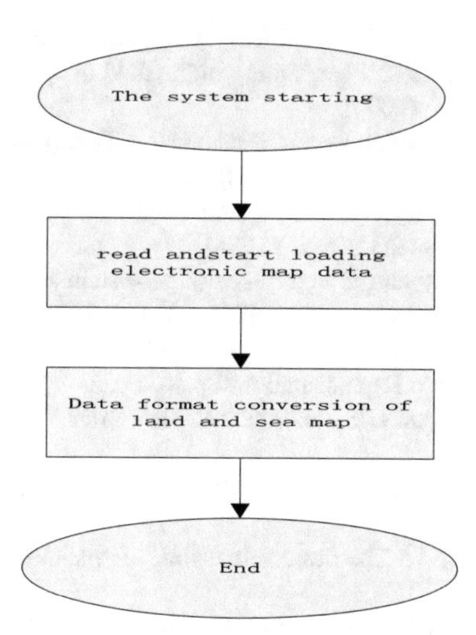

Fig. 5. A figure caption is always placed below the illustration. Short captions are centered, while long ones are justified. The macro button chooses the correct format automatically.

4.3 System Function Design

The map display section includes: ① map display. The software can provide two-dimensional vector electronic map display and digital topographic map display with digital elevation model (DEM) elevation information. ② basic information display. The software can display the current map of the scale, the location of the mouse latitude and longitude information.

The basic functions of the map include: map browsing, translation, roaming, element selection, feature information query, layer management, measurement, screenshot saving, magnifying glass and eagle eye and other functions.

5 Concluding Remarks

In this paper, the key technologies in the design of joint geographic information system are studied. In view of the effects of vector topography and vector chart on data fusion and the differences between projection method and spatial data model and feature classification and factor coding, the key problems of data fusion are studied deeply, and the concrete realization method of data fusion is given. On this basis, the design software of joint geographic information system is designed, and the design flow and the steps of the demonstration software are given. The demonstration software validates the correctness and validity of the multi - source geospatial vector data fusion method.

References

1. Wang, J.: Spatial Information System Principle. Science Press, Beijing (2001)
2. Lan, R.: US military science and technology in the field of spatial data integration research. Military Mapp. Inf. **2**, 5–6 (2006)
3. Li, G.: Multi-source Geospatial Vector Data Fusion Theory and Method Research. PLA Information Engineering University, Zhejiang (2008)
4. Cobb, M., Chung, M., Foley, H.: A rule-based approach for the conflation of attributed vector data. GeoInformatica **2**(1), 7–35 (1998)
5. Hoseok, K.: Geometrically and topographically consistent map conflation for federal and local government. J. Korean Geogr. Soc. **39**(5), 804–818 (2004)
6. Haunert, J.H.: Link based conflation of geographic datasets. In: 8th ICA Workshop on Generalization and Multiple Representation, Spain, pp. 1–7 (2005)
7. Open GIS Consortium, Open GIS Reference Model, Draft Version 0.1.3. Open Geospatial Consortium Inc. (2003)
8. Yin, X.: GeoDatabase-based multi-source map data fusion research. Beijing Surv. Mapp. **4**, 11–14 (2010)
9. Hu, S., Zhang, J., Wang, X.: The basic framework of spatial data fusion. Surv. Sci. **32**(3), 175–177 (2007)
10. Lu, P., Zhang, L., Wang, G.: Application of spatial data fusion in charting. Bull. Surv. Mapp. **11**, 43–45 (2007)
11. Li, H., Chen, Y.: Study on multivariate heterogeneous data fusion based on electronic chart platform. Inf. Technol. **5**, 156–158 (2006)
12. GIS Guangyi, the characteristics of several projection and zoning method, 23 October 2008/20 November 2013. http://blog.sina.com.cn/s/blog_5f0f65b90100v6d1.html
13. Tang, W.: Land and Sea Geospatial Space Vector Data Fusion Technology. Harbin Engineering University, Harbin (2009)
14. Zhang, Q.: Map Database Entity Matching and Merging Technology Research. Wuhan University Library, Wuhan (2002)

Numerical Simulation for Flow Dynamic Two-Phase Turbulent of ZhuangLi Reservoir Spillway

Jia-rong Wang, Xiu-ling Sun(✉), Rong Sun, and Yu-xiang Zhou

Shandong University, Jinan, Shandong, People's Republic of China
xiuling-sun@sdu.edu.cn

Abstract. According to three-opening overflow spillway complex flow problems in ZhuangLi Reservoir based on FLUENT software, Using RNG-κ-ε double-equation turbulence model and Volume of Fluid (VOF) method simulate the 2-D and 3-D Gas-liquid two-phase flow of spillway in different working conditions, using PISO algorithm to solve the discrete equations. Simulated the Gas-liquid two-phase flow velocity field, the pressure field distribution, the mixed aerated concentration and other characteristic parameters surface in the different conditions, analyses spillway capacity and comparing with the results of hydraulic model test, and combination with the results of hydraulic model test verify the reliability, rationality of the various hydraulic parameters calculated by the numerical model, more accurate and comprehensive reflect the spillway flow distribution, The hydraulic parameters are in good agreement, which indicates that the numerical model is practical in engineering practice. It can provide a reliable basis for the hydraulic design of the ZhuangLi Reservoir spillway.

Keywords: Pick flow spillway · RNG-κ-ε · VOF method · Non-orthogonal Finite volume method

1 Introduction

The spillway is one of the main drainage building in Hydraulic Engineering; discharge mode is adopted in the form of flip trajectory bucket. In the last 90's, domestic researchers did the experimental study and the 2-D numerical simulation for the spillway flip bucket flow turbulence characteristics [1, 2]. With the development of numerical simulation technology and method [3–7], the application of fluent software in hydraulic engineering is beginning [8]. For proposed the ZhuangLi Reservoir spillway using the pick flow spillway, discharge flow is influenced by many factors, there are strong local complex flow field which has the strong three-dimensional motion characteristics. In this paper, using FLUENT software to propose the three-opening overflow spillway of ZhuangLi Reservoir spillway, a 3-D numerical model is established by using the finite volume method according to the specific working conditions. Simulated the water velocity, water surface profile, pressure distribution on weir surface in the different conditions, and analyzed the aeration condition of the high-speed flow. The existing research results shows the numerical model has been able to

N. Xiong et al. (Eds.): ISCSC 2018, AISC 877, pp. 497–505, 2019.
https://doi.org/10.1007/978-3-030-02116-0_58

reflect the characteristics of water flow, and the size of the numerical model is consistent with the practical engineering project, the measured parameters are more comprehensive and the model can provide the more efficient and convenient way for the optimal design of the Reservoir spillway scheme.

2 Hydraulic Model

Proposed the hydraulic model of ZhuangLi Reservoir includes the water inlet section of sluice, the overflow channel, the downstream water cushion pool and spillway sets the 3 opening gate, the width of each hole 13 m, the thickness of the middle pier is 3 m, total width 39 m. Spillway is using the WES practical weir, crest elevation is 108.30 m, the maximum dam height of spillway is 43.90 m, the Weir surface curve equation of spillway is $y = 0.077x^{1.85}$. The origin of weir surface curve is 108.30 m, the last point elevation is 99.634 m, the end point of the curved surface is connected with the slope 1:0.7 dam slope, turn to the flip bucket and it's radius is 15 m. Flip bucket top elevation is 91.71 m, the degree of the flip shot is 25°; the height of pier top is 119.9 m, after the flip bucket set up the water cushion pool, set up the dam at the downstream 120 m of the flip bucket, the height of dam is 0.5 m, crest elevation is 86.0 m. The physical 3-D model designs sketch of spillway as Fig. 1.

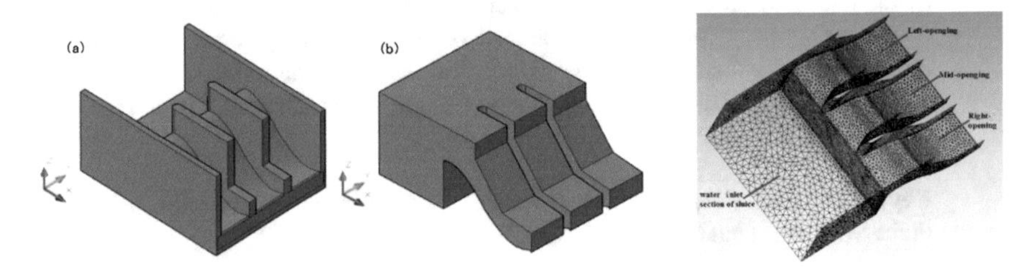

Fig. 1. The hydraulic model of spillway (a), numerical model (b) and Grid subdivision (right) for 3-D effect chart

3 Numerical Model

3.1 Governing Equations

The numerical model using RNG-κ-ε double-equation turbulence model, by using VOF method to deal with problem of the free surface of air-water two-phase flows, Tracking the Volume fraction of each fluid in each control volume. 3-D turbulent flow governing equation under the rectangular coordinate system [9] are shown as Eq. (1) and Eq. (2):

$$\text{Equation of continuity: } \frac{\partial \rho}{\partial t} + \frac{\partial}{\partial x_i}(\rho u_i) = 0 \tag{1}$$

Momentum equation: $$\frac{\partial(\rho u_i)}{\partial t} + \frac{\partial}{\partial x_j}(\rho u_i u_j) = -\frac{\partial p}{\partial x_i} + \frac{\partial}{\partial x_j}\left(\mu \frac{\partial u_i}{\partial x_j} - \rho \overline{u_i' u_j'}\right) + S_i \quad (2)$$

In the above equation, ρ is water density, (i = 1, 2, 3) is the velocity component of x, y, z direction, μ is molecular dynamic viscosity coefficient of water body, is Reynolds stress term, P is intensity of pressure, Si is generalized source terms in momentum equations.

3.2 Solution Method and Boundary Condition

Numerical model using unstructured mesh generation, the grid spacing used in the 3-D model is 0.9–2 m. The total number of grids is about 139 thousand. Encrypted local grid in the spillway weir surface near the wall, and the air area in the upper part of the model use the sparse grid. The specific grid is shown in Fig. 1(right).

Model of the finite volume method is adopted to control equation discretization method, pressure-velocity coupling used PISO algorithm based on pressure correction. PISO algorithm through forecast-amendment-re-amendment three steps in every time step, obtain the after two amendment pressure field P^* and velocity field (U^*, V^*), compared with the SIMPLE or SIMPLEC algorithm improve the accuracy of calculation, speed up convergence, and increase the mesh skew correction, more suitable for the calculation of unsteady flow.

The 3-D model is at the end of the spillway export. Water inlet of upstream set up the speed inlet boundary, flow velocity use mean velocity of cross section, downstream outlet boundary set up pressure-outlet boundary condition, and the outlet boundary is directly connected with the atmosphere. Outlet pressure is set to atmospheric pressure. Upstream air inlet and the upper part of the computational domain are air pressure inlet boundary, and use atmospheric pressure as its boundary pressure value. Other boundary is the solid wall boundary condition which meets the condition of no slip and can't penetrate. The numerical model simulation condition and velocity inlet boundary design parameters are shown in Table 1.

Table 1. Velocity inlet boundary parameters of different calculation conditions

Flood standard	Working condition	Inlet velocity (m/s)	Turbulent kinetic energy (m^2/s^2)	Dissipation rate (m^2/s^3)
2%	C	1.5	8.49e−3	7.45e−5
1%	B	1.607	9.68e−3	8.92e−5
0.05%	A	2.09	1.64e−2	1.42e−3

4 Calculation Result and Analysis

4.1 Velocity Field Analysis

Water flow through the entire calculation area of water inlet section of sluice and after the top of the weir, water flow through the change of slow flow-gradually varied flow-race. Figure 2 gives distribution of flow velocity in air-water two phase flows in vertical section of the spillway and 5.1 m range before the gate in A, B conditions. Combine with the water volume fraction distribution map in the Fig. 4 and air-water interface partition method of high speed water flow, taking water volume fraction is $\alpha q = 0.5$ as the Hydro-pneumatic interfaces, in the green part in Fig. 3, the thickness of the water layer above the weir surface is 2.7–3.2 m. It is observed that, boundary layer and trajectory bucket is not considered, when the flow is increased, flow velocity of air-water two phase is more uniformity in the vertical direction of the spillway, after entering the reflex arc segment, the water flow is obviously affected by the mixed air flow rate, mix into the air to cause the water energy consumption and partial water kinetic energy converted into internal energy. Water flow suffers the gravity and centrifugal force in the same time and flow velocity achieve the maximum value at the spillway flip bucket. The air-water two phase velocities radiating outward distribution centered with the flip-bucket.

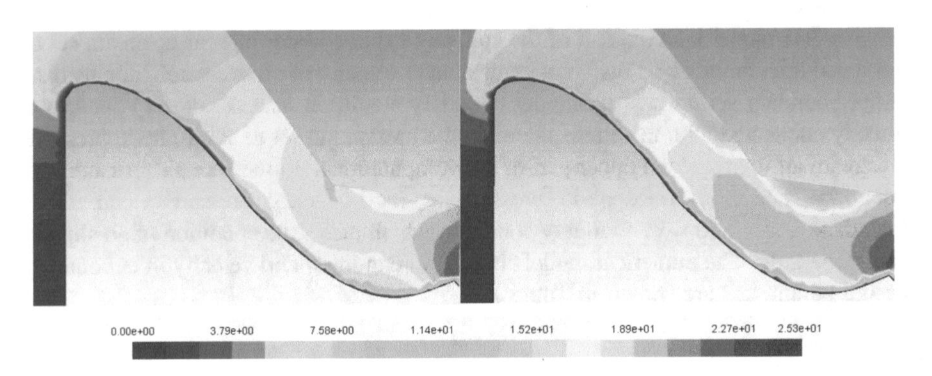

Fig. 2. The velocity contour map of spillway numerical calculation in working condition A (Up) and B (Middle) (Color indicates flow velocity: m/s)

Figure 3 shows that in working condition A and B, spillway 0+000, 0+010, 0+020, 0+027, 0+034 flow velocity vector distribution of five cross sections, the direction of flow velocity is shown by the arrow. From the vector of cross section, the flow velocity is gradually increased from the weir top to the flip bucket, get the maximum flow velocity at the end of the flip bucket, maximum flow velocity is about 24 m/s, the maximum flow velocity of weir top is about 10.12 m/s. Before the lowest point of the arc segment, high-speed flow area located the upper layer and after the lowest point, it located in the under layer. And the greater of the flow, the farther distance between the main stream velocity zone (high-speed fillet) and water surface, in other words, nearing the vertical flow central. Combined with the flow chart in working condition A and B,

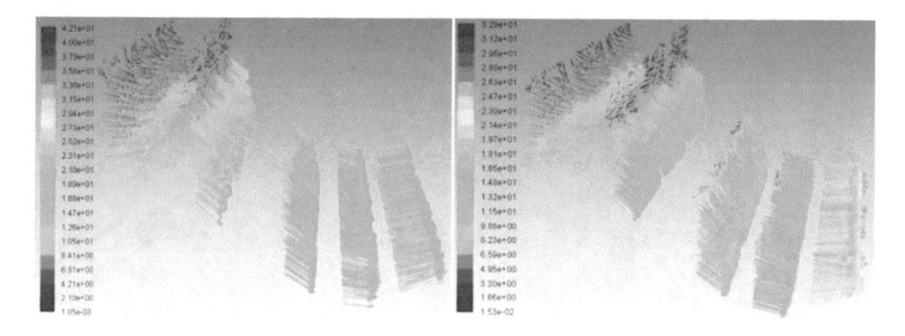

Fig. 3. The velocity vector distribution numerical calculation of typical section of spillway in working condition A (Up) and B (Down) (Flow rate unit: m/s)

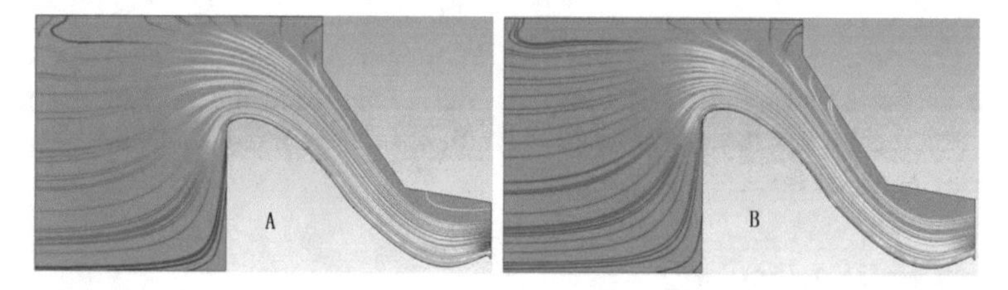

Fig. 4. The flow streamline distribution diagram in working condition A (Up) and B (Down)

shown in Fig. 4, can more clearly know the spillway vertical and longitudinal flow velocity distribution (color indicates flow velocity, the redder the color indicates the greater the flow rate).

4.2 Water Flow Mixed Air Analysis

Figure 5 shows that in working condition A and B, ZhuangLi Reservoir spillway, water-air flow volume fraction distribution of arc segment, the red part is water, the blue part is air, the green part is water-air interface, the figure shows that with the increase of discharge, thickness of the drainage groove water layer also increase, that is to say, water depth increases, (A:3.2 m, B:2.7 m), with the increase of flow, the turbulence of the water body mixed air is more obvious, water tongue forward on the weir surface (without air) as the flow rate increases, the closer to the arc segment lowest point, like Fig. 6, the red part of working condition A, but condition B do not have it, water volume fraction maximum value of the arc segment weir surface is 0.9, the minimum value is 0.05, in another word, aerated concentration is 10%, 5%. We can know from the existing research results, if the discharge flow in the discharge structure aerated concentration at the range of 3%–7%, the cavitation damage can be avoided, when aerated concentration up to 10% can be completely avoided. Cavitation damage can be avoided in the high-speed flow [12] which means velocity is 22 m/s of the arc segment. From the vertical direction of the weir surface, aerated concentration gradient has the great change in the range of up and down within 3 m taken water vapor

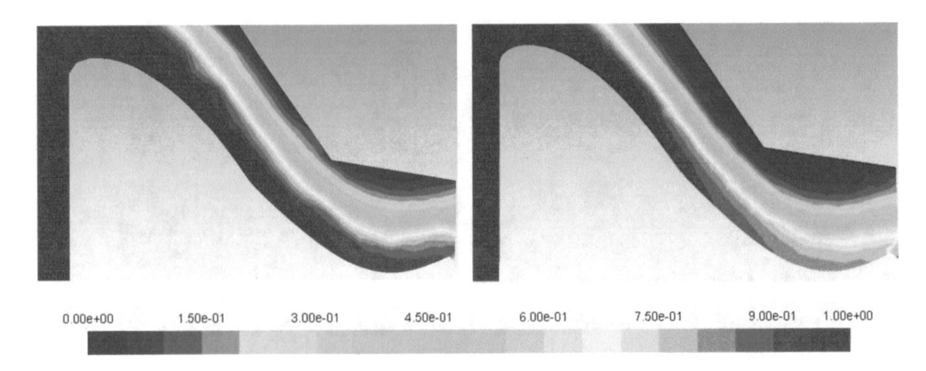

Fig. 5. The calculation of water volume fraction distribution of the numerical in working condition A (Up) and B (Down) (The volume fraction of water in a color strip: 0–1)

interface as the central line. The closer to the center line, the greater the gradient change, and the higher the concentration gradient of the mixed air in the same range with the increase of flow.

4.3 Pressure Distribution

As shown in Fig. 7 is the absolute pressure distribution of the longitudinal section of the spillway in working condition A, B. WES Curve section absolute pressure has the uniform distribution, after entering the arc segment, absolute pressure is reduced radiating outward centered with the lowest point of the arc, but the radiation radius is unequal of the upstream and downstream, the pressure gradient is uniform. It is shown from the figure that the whole weir at the arc segment lowest point and up and down about 1 m range achieve maximum. Influenced by water flow, the minimum value of working conditions A appears at the flip bucket (red circle in Figure), pressure value is positive (1.97×104 Pa), working condition B appears in the section 0+007 (red arrow in Figure), the pressure value is equal to the atmospheric pressure.

Figure 8 (left) shows the absolute pressure distribution of numerical calculated value in the spillway weir surface under the three working conditions. Weir surface pressure distribution conforms to the law of water flow in practical engineering and it can reflect the effect of weir surface convex and concave curvature on the pressure distribution. In the three working conditions, the pressure change of the weir surface is basically consistent, the pressure of working condition A is about two times of working condition B, the pressure distribution on the weir surface flip bucket of working condition C is basically the same as the working condition A, the pressure of the flip bucket has no obvious change. In all these three conditions, the maximum value of the weir surface pressure is at the lowest point (section 0+027) of the spillway flip bucket and then after the lowest point it will be reduced, the minimum pressure of the weir surface appears in the section 0+007. Comprehensive comparison the vertically and horizontally absolute pressure of the weir surface, the variation law is basically consistent.

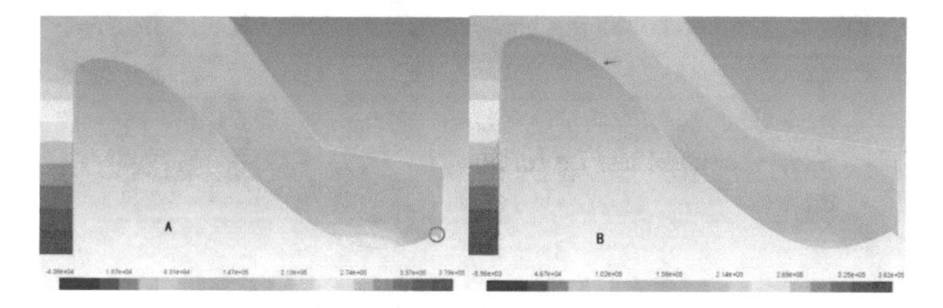

Fig. 6. The longitudinal profile of the absolute pressure distribution map (Color representation pressure: Pa) in working condition A and B

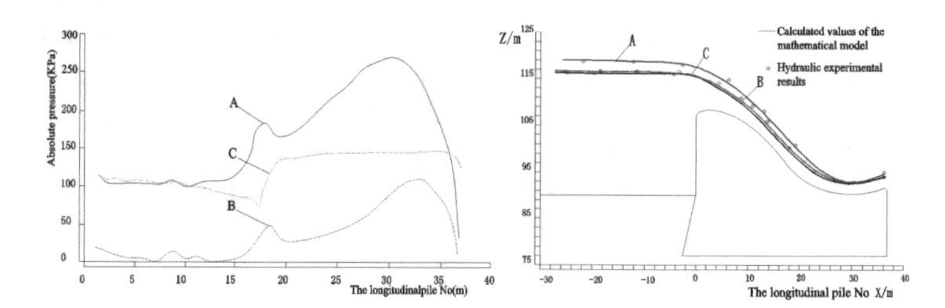

Fig. 7. Distribution of absolute pressure numerical calculated value in the spillway weir surface under various working conditions (left); Numerical calculation and water surface distribution of hydraulic model test under different working conditions (right)

4.4 Results Comparative Analysis of Numerical Simulation and Hydraulic Model Test

A hydraulic test model is established in the laboratory as shown in Fig. 1, divided into A, B, C three kinds of working conditions to carry out the water test, measured spillway discharge capacity, flow velocity in each section, water surface line distribution, compared and analyzed the calculation results with the numerical model.

Table 2 shows the spillway discharge capacity contrast the numerical simulation with the hydraulic model test, the wall surface roughness of spillway in numerical calculation is n = 0.014. In working condition A, B, numerical calculation value is higher than the hydraulic model test, with the increase of the upstream water level, the error also increases. The 2-D model considering the downstream water cushion pool, there is a large amount of gas in the water, and has the influence of backwater, so the value has a great computing statistical error, and outlet flow is small, the value of the error is $\varepsilon_{max} = -3.489\%$.

Figure 8(right) gives the waterline distribution of the water inlet section of sluice and spillway among the three working conditions, the curve is the value of the numerical model, the scattered points is the test values of hydraulic model. The calculated value of the characteristic section of the numerical model is 0.134 m higher than that of the physical model, the maximum difference value of the waterline between these two model is $\Delta h_{max} = 0.84$ m and the minimum difference value is $\Delta h_{min} = -0.01$ m, both the maximum difference value and the minimum difference value are all

Table 2. The flood discharge capacity of the spillway in each frequency

Working condition	A	B	C
Numerical model Value Q_T (m^3/s)	2720.98	1934.69	1717.56
Hydraulic model test Q_C (m^3/s)	2717.53	1932.83	1777.49
Error $\varepsilon = (Q_T - Q_C)/Q_T \times 100\%$	0.127	0.096	−3.489

appeared in the section 0+027.77, working condition C has the maximum difference, working condition A has the minimum difference. The variation of waterline between two models Δh has the error ε, the value of the error is within 0.78%. Three working conditions are all in the flow contraction section between the 0+022 section and the 0 +030 section. In the three working conditions, water surface line elevation difference is only 0.04–0.16 m, The results show that the numerical model is in good agreement with the experimental values of hydraulic model. In conclusion, numerical simulation results are verified that the spillway design scheme of the preliminary design provided for the ZhuangLi reservoir is reasonable, and it can be able to meet the requirements.

5 Conclusions

This paper established the air-water two-phase turbulence model based on finite volume method to do the 3-D numerical simulation for the ZhuangLi reservoir three-opening spillway over dam flow, numerical model calculation results accord with the basic theory.

(a) Using RNG k-ε turbulence Models simulated and calculated the 3-D velocity field and pressure field of spillway in different conditions more comprehensive. It can evaluate the design parameters of rationality; evaluate the spillway flow velocity distribution, flow characteristics and the possibility of cavitation damage to the building.
(b) Using water-air two phase VOF mixed model, Obtained the aeration distribution of vertical and horizontal flow in the spillway computational domain. Through $\alpha_q = 0.5$ determine the water vapor interface, the aeration concentration of calculation of water flow mixing part >5%, it can avoid cavitation damage.
(c) By comparing the numerical simulation results with the hydraulic model test results, the calculated results agree well with experimental values. That shows the model established can accurately simulate the flow field distribution and hydrodynamic characteristics of multi-opening spillway. The parameter information of the spillway flow can be more comprehensive and detailed reflection and it can be accurately applied to the prototype.

References

1. Zhang, Z., Zhou, T.: The numerical simulation of the flow dynamic spillway ogee section. J. Hydraul. Eng. **6**(6), 31–36 (1994)
2. Dong, Z., Chen, C., Zhao, J.: Turbulence characteristics of the spillway arc segment. J. Hydraul. Eng. **12**(12), 9–17 (1990)
3. Li, Y., Hua, G., Zhang, J.: Analysis on energy dissipation of the single-lever with multi-strand and multi-lever with multi-strand horizontal submerged jets. J. Hydrodyn.: Ser. A **21**(1), 26–31 (2006)
4. Xu, J., Chen, B., Wang, X.: Experimental research and numerical simulation of flow field near the narrow side in a rectangular narrow channel. J. Hydrodyn.: Ser. A **21**(6), 770–775 (2006)
5. Sha, H., Zhou, H., Wu, S.: 3-D numerical simulation on discharge of spillway with muti-opening. Water Resour. Hydro-Power Eng. **36**(10), 42–46 (2005)
6. Mu, L., Gao, X.: Numerical simulation of water-air two-phase flow of spillway flip bucket. J. Yangtze River Sci. Res. Inst. **29**(4), 35–39 (2012)
7. Wang, Y., Bao, Z.: Three-dimensional numerical simulation of flow in stilling basin based on flow-3D. Eng. J. Wuhan Univ. **45**(4), 454–476 (2012)
8. Liang, Z.: Engineering Turbulence. Wuhan Cent. China Univ. Sci. Technol. **21**(1), 26–31 (2006)
9. Ye, M., Wu, C., Chen, Y.: Application of FLUENT to hydraulic projects. Adv. Sci. Technol. Water Resour. **26**(3), 78–81 (2006)
10. Ju-jiang.: Numerical Simulation and Visualization of Engineering Hydraulics, vol. 10, pp. 139–143. China Water Conservancy and Hydropower Publishing House, Beijing (2010)
11. Ferziger, J.H., Peric, M.: Computational Methods for Fluid Dynamics, pp. 149–208. Springer, Berlin (1996)
12. Yang, H.: China Water Conservancy Encyclopedia. Water Conservancy and Electric Power Press, Beijing (1991)

Spatial Interpolation of Monthly Mean Temperatures Based on Cokriging Method

Qingyan Guo(✉), Min Zhang, Xiandong Yang, and Min Zheng

Fujian Provincial Meteorological Information Center, Fuzhou, China
876920285@qq.com

Abstract. Cokriging methods can simulate the spatial distribution characteristics of meteorology data. In this study, we proposed a cokriging method to interpolate monthly mean air temperatures. This cokriging method used elevation as a covariate. We used the cross validation method to select the optimal theoretical model, the optimal segmentation distance and other parameters of the semi variogram in the interpolation process. We collected monthly average air temperature data from 2000 to 2015 totally 16 years by 88 national meteorological observation sites in Fujian Province and its neighboring province. Experiments on this real dataset show that our method achieves superior performance to ordinary kriging method and inverse distance weighting method.

Keywords: Spatial interpolation · Cokriging · Ordinary kriging
Inverse distance weighting · Cross validation

1 Introduction

Air temperature is the most direct meteorological factor for high temperature monitoring, evaluation and early warning. The accurately acquisition of temperature data at any position is of great significance to improve the accuracy of Fujian high temperature prediction and prediction [1]. However, due to the uneven distribution of meteorological stations, the temperature data acquired is discrete in space, usually, in order to obtain the temperature data [2] at each position in global space range, it is to estimate based on the existing site. Many scholars at home and abroad have used kriging interpolation method to interpolate meteorological elements [3–10], We can see that the kriging method has certain advantages in the practical application compared with other traditional interpolation methods. However, the above researches are more inclined to compare with different interpolation methods, but the selection of parameter set and variogram model is seldom discussed. Therefore, in the support of ArcGIS software, we selected Fujian Province as the study area, and we used the monthly average air temperature data from 2000 to 2015 totally 16 years, which is provided by the national meteorological observation sites by Fujian province and its neighboring province, and then, we adopted cokriging method to interpolate the monthly mean air temperature, in this process we selected elevation as a covariate. In the course of the study, the cross validation method [11] was used to select the optimal theoretical model, the optimal segmentation distance and other parameters of the semi variogram in the interpolation process. Then we

N. Xiong et al. (Eds.): ISCSC 2018, AISC 877, pp. 506–513, 2019.
https://doi.org/10.1007/978-3-030-02116-0_59

validated the cokriging interpolation results with the ordinary kriging method and inverse distance weight method to obtain the more accurate interpolation results.

2 The Cokriging Interpolation Method

The method of cokriging interpolation is used to develop the optimal estimation method of regionalized variable from single attribute to two and more than two cooperative regionalized attributes, there was a correlation between these secondary attributes and the main attributes, it is assuming that the correlations between variables can be used to improve the precision of the predicted value. In the use of cokriging interpolation, we first assume that the regionalized variables in the region satisfy the two order stationarity hypothesis and the hypothesis of this hypothesis. The formula of cokriging is shown in Eq. (1):

$$Z(x,y) = \sum_{i=1}^{n} w_i z_{1i} + \sum_{j=1}^{n} q_j z_{2j} \tag{1}$$

where, z_{1i} and z_{2j} are two distinct attributes of an observation point, w_i and q_j are weight coefficients.

3 Results and Analysis

3.1 Distribution Characteristics of Data

In this paper, spatial data exploration and analysis tools are used to analyze the distribution characteristics of the average temperature of 86 sample stations. The results are shown in Table 1.

Table 1. Monthly mean temperature distribution characteristics

Months	Mean	sd	Min	Max	Skewness	Kurtosis
Jan	9.882	2.647	5.2	14.506	−0.0284	1.8172
Feb	11.963	2.157	7.325	15.9	−0.2063	2.2268
Mar	14.554	1.838	9.23	18.069	−0.4374	3.0557
Apr	19.059	1.607	12.813	21.863	−1.1585	5.2361
May	22.953	1.509	15.925	25.469	−1.9326	8.5149
June	25.882	1.421	17.863	27.75	−2.8236	14.397
July	28.26	1.463	19.462	30.265	−3.0684	17.262
Aug	27.709	1.487	19.163	29.5	−2.9415	15.181
Sept	25.663	1.734	16.9	27.956	−2.1424	10.041
Oct	21.523	2.074	13.25	24.962	−0.999	4.9072
Nov	16.843	2.497	10.144	21.1	−0.1897	2.1955
Dec	11.577	2.784	6.0125	16.369	−0.0034	1.7753

Where, sd stands for standard deviation.

As can be seen from Table 1, the overall trend of the monthly average temperature is consistent with the seasonal variation of the climate, the highest monthly average temperature occurred in July and August, and the lowest monthly average temperature occurred in January. From the value of skewness and kurtosis, the spatial distribution of air temperature elements varies during different periods of the year. Winter (from December to February) changed little, followed by spring (from March to May) and autumn (September to November), and the biggest change was in summer (June to August).

The cokriging interpolation method requires the data to obey normal distribution [12], the skewness and kurtosis, Karl kolmogoroff Smirno test (K-S) are used to check the data of normal distribution. As can be seen from Table 1, the skewness and kurtosis of the data show that the data distribution has strong negative skewness and high kurtosis, which will affect the interpolation accuracy. The K-S test results of the data are shown in Table 2, it can be seen that, before the conversion, the K-S P-value in January, February, March, October, November and December were greater than 0.05, the data is close to the normal distribution. But the K-S P-value in the remaining months are less than 0.05, therefore, the data needs to be converted. In this paper, the Box-Cox transform is used to make the data tend to be normal distribution. After the Box-Cox conversion, the data for all months passed the K-S test, as shown in Table 2.

Table 2. K-S test P-value

Months	Raw data	Box-Cox P-value
Jan	0.200	0.318
Feb	0.200	0.259
Mar	0.200	0.304
Apr	0.046	0.245
May	0.025	0.113
June	0.021	0.105
July	0.016	0.154
Aug	0.013	0.109
Sept	0.020	0.132
Oct	0.200	0.245
Nov	0.198	0.201
Dec	0.093	0.298

3.2 Selection of Semi Variogram Model and Parameter Setting

In this paper, three commonly models of Gaussian functions, exponential function and spherical function are used to interpolate, and the simulation results of each model are compared according to the sill coefficients of each model and the cross validation parameters.

With the support of ArcGIS software, when the three models are used to interpolate, the lag size and the number of lags determine the distance of segmentation, and the segmentation distance has an important influence on the variogram. For a specific set of data points, the small spacing means that the experimental variogram curve is not smooth, and the logarithm of the points falling in each distance segment is also small, which is not conducive to the analysis. While the spacing increases, means increasing the nugget effect, which indicates that semivariable function is affected by human factors to some extent, so when calculating and fitting, it is particularly important to the selection of distance. From the distribution of observation sites in Fujian Province, the distance between Zhaoan and Fuding is the longest, which is 505 km. Therefore, when the variogram parameters are set up, half of the distance (202 km) is selected as the maximum lag size. Accordingly, this paper uses the mean nearest neighbor analysis tool provided by ArcGIS to calculate the average distance between interpolation point and nearest neighbor observation sites, which is 30.448 km, which can provide a good lag size reference. On the basis of this good lag size reference, the optimum parameters are selected to interpolate by the method of trial optimization.

Table 3 shows the interpolation results of the Gaussian functions for the model. Table 4 shows the interpolation results of the exponential function. Table 5 shows the interpolation results of the spherical functions. The results of the nugget, partial sill, major range and cross validation results in the three tables show that the interpolation with the Gaussian functions for the model is relatively good, the Mean Error and the Mean Standardized Error are close to 0, the Root-Mean-Square Error and the Average Standard Error are closed. Although there are a few months' Root-Mean-Square Error of exponential and spherical function small than the Root-Mean-Square Error of Gaussian function, the difference between the Root-Mean-Square Error and the Average Standard Error is too big. In combination with Tables 3 and 5, it is found that the interpolation results of Gaussian function and spherical function in January, February, March, October, November and December are close and basically meet the requirements of precision parameters. From the nugget and partial sill, exponential function interpolation results are the worst.

Table 3. Interpolation results of Gaussian functions model

Months	Nugget (C0)	Partial sill (C)	Major range (km)	Mean error	RMS error	MS error	RMSS error	AS error
Jan	0.8112	13.46	487560	−0.0007	1.0525	−0.0038	1.0687	1.0011
Feb	0.8204	5.427	372000	−0.0050	1.0490	−0.0101	1.0584	1.0063
Mar	0.9572	3.343	372000	0.0026	1.0604	−0.0004	1.0089	1.0657
Apr	1.2322	2.692	465000	0.0046	1.1289	0.0032	0.9703	1.1772
May	1.661	0.7070	372000	−0.0034	1.1974	−0.0019	0.8957	1.3457
June	2.2403	0.0088	372000	−0.0149	1.2893	−0.0095	0.8385	1.5428
July	2.9419	0.0009	372000	−0.0267	1.4689	−0.0149	0.8330	1.7678
Aug	2.9027	0.00180	372000	−0.0315	1.4432	−0.0176	0.8241	1.7560
Sept	2.9275	0.0340	365860	−0.0219	1.4756	−0.0120	0.8389	1.7642
Oct	2.2301	3.0412	372000	0.0056	1.3927	0.0054	0.8734	1.6126
Nov	1.3058	7.8525	372000	0.0109	1.1978	0.0086	0.9619	1.2656
Dec	0.8433	10.951	372000	0.0006	1.0987	−0.0015	1.0715	1.0437

Table 4. Interpolation results of exponential function model

Months	Nugget (C0)	Partial sill (C)	Major range (km)	Mean error	RMS error	MS error	RMSS error	AS error
Jan	0	10.417	372000	0.0087	0.9785	0.0047	0.6267	1.6498
Feb	0	6.0242	372000	0.0101	0.9806	0.0073	0.8276	1.2546
Mar	0	3.8892	372000	0.0046	0.9943	0.0060	1.0495	1.0080
Apr	0.4263	2.403	372000	0.0010	1.0495	0.0013	1.0001	1.0807
May	1.0537	1.3321	372000	0.0048	1.1318	0.0029	0.9097	1.2584
June	2.1889	0.0446	372000	−0.0135	1.2845	−0.0087	0.8419	1.5310
July	3.19	0.0065	372000	−0.0262	1.4681	−0.0140	0.7983	1.8437
Aug	3.0682	0.0121	372000	−0.0308	1.4416	−0.167	0.8001	1.8067
Sept	2.9322	0.0304	372000	−0.0219	1.4758	−0.0120	0.8384	1.7655
Oct	0.5283	4.5946	372000	−0.0083	1.2497	−0.0040	0.9460	1.3718
Nov	0	8.7183	372000	0.0018	1.1041	0.0020	0.7788	1.5093
Dec	0	11.72	372000	0.0049	1.0202	0.0028	0.6153	1.7500

Table 5. Interpolation results of spherical function model

Months	Nugget (C0)	Partial sill (C)	Major range (km)	Mean error	RMS error	MS error	RMSS error	AS error
Jan	0.0345	9.2876	495000	0.0028	0.9888	0.0031	0.9334	1.1294
Feb	0.0462	5.3859	403000	0.0039	0.9832	0.0037	1.1642	0.8965
Mar	0.3609	3.5197	372000	0.0036	0.9929	0.0017	1.0518	0.9749
Apr	0.8224	1.9289	465000	0.0028	1.0732	0.0019	0.9594	1.1369
May	1.4394	0.8873	372000	−0.0037	1.1675	−0.0022	0.8934	1.3167
June	2.2758	0.0264	372000	−0.0144	1.2876	−0.0091	0.8298	1.5571
July	3.128	0.0039	372000	−0.0266	1.4686	−0.0144	0.0876	1.8231
Aug	3.051	0.0072	372000	−0.0313	1.4426	−0.0170	0.0833	1.8008
Sept	2.8134	0.16123	365860	−0.0200	1.4649	−0.0110	0.8445	1.7409
Oct	1.5254	3.558	372000	−0.0041	1.2989	−0.0012	0.8678	1.5209
Nov	0.0876	7.927	372000	0.0008	1.1088	0.0017	1.0734	1.0998
Dec	0.2657	9.517	372000	0.0010	1.0308	0.0022	0.9132	1.2019

Tag: Mean Error is Mean Error. RMS Error is Root-Mean-Square Error. MS Error is Mean Standardized Error. RMSS Error is Root-Mean-Square Standardized Error. AS Error is Average Standard Error.

Because of the limitation of space, in this paper, the Fig. 1 only lists the three semi variogram model fitting curves of the average temperature in December, and also can be seen from the fitting graph that the Gaussian function is better.

From the three tables, in terms of the lag size 30.488 km, the interpolation results of three models on April, May, June, July, August, September is not very good, and the Root-Mean-Square Error and the Average Standard Error of these months compared with other months is to larger and less similar, from the values of the sill and partial sill, the substrate effect of these months is too large. To the end, this paper chooses the Gaussian function model to interpolate the mean temperatures in April, May, June, July, August, and September by increasing or reducing the lag size, it is found that reducing the lag size can improve the interpolation results, as shown in Table 6, it is the

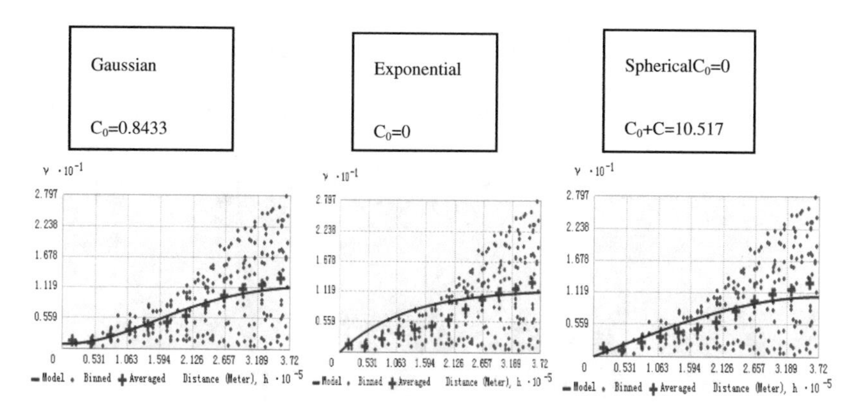

Fig. 1. The fitting curves of three semi variogram to monthly mean temperature in December

improved interpolation result. Compared with Tables 3, 4 and 5 in April, May, June, July, August and September, the interpolation accuracy has been improved.

From the sill efficient and accuracy evaluation parameters, compared to the interpolation results of January, February, March, October, November, December, the interpolation results of April, May, June, July, August, September was worse, The reason may be that the temperature change is great. This period is the rainy season and typhoon period in Fujian Province, which affect the temperature range. The other reasons need to be further studied.

Table 6. Improved interpolation results of Gauss function model

Months	Nugget (C0)	Partial sill (C)	Major range (km)	Mean error	RMS error	MS error	RMSS error	AS error
Apr	0.0558	1.7687	54192	−0.0000	1.1349	0.0089	1.0502	1.1867
May	0.1847	1.6394	59070	−0.0121	1.1888	0.0005	1.0989	1.1720
June	0.33508	1.7074	62106	−0.0188	1.2576	−0.0064	1.0805	1.2552
July	0.2894	2.1625	59855	−0.0179	1.3886	−0.0030	1.1006	1.3618
Aug	0.0258	2.612	55356	−0.0104	1.3671	0.0032	1.1006	1.3962
Sept	0.0745	2.9934	67625	−0.0450	1.4841	−0.0138	1.3521	1.3262

3.3 Analysis of Interpolation Results

When we use cokriging method, we selected Gaussian functions and selected parameters above as interpolation model, the monthly average temperature distribution map of Fujian Province in 2000–2015 years is obtained, and the interpolation results are compared with those of the inverse distance weighting method and the ordinary kriging method, the Mean Error and the Root-Mean-Square Standardized Error are selected as accuracy evaluation index. The interpolation accuracy of each method is shown in Table 7, and the interpolation results are shown in Fig. 2.

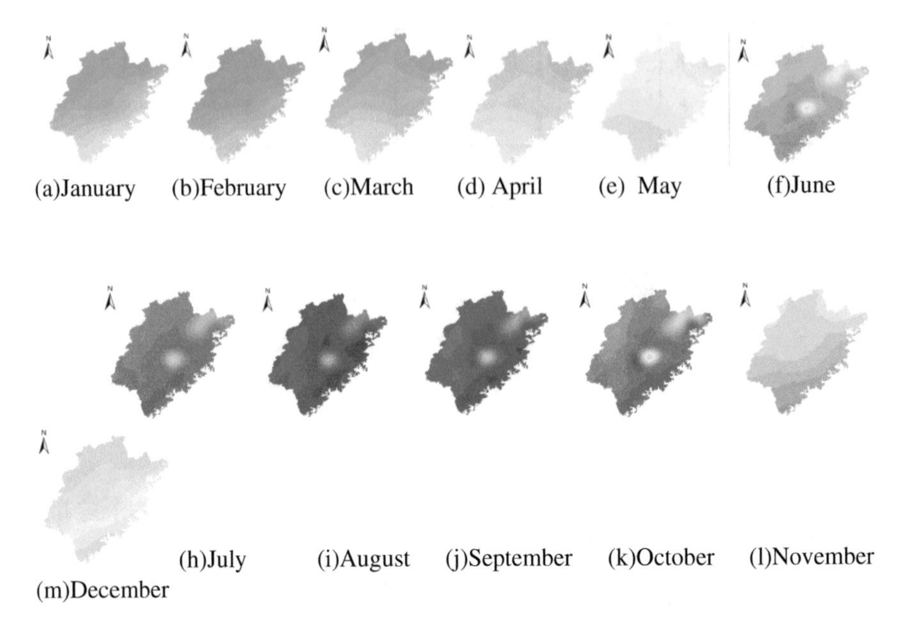

(a)January (b)February (c)March (d) April (e) May (f)June

(h)July (i)August (j)September (k)October (l)November

(m)December

Fig. 2. Results of cokriging for average temperature of each month

Table 7. Interpolation accuracy comparison table

	IDW		OK		CK	
Months	Mean error	RMSS error	Mean error	RMSS error	Mean error	RMSS error
Jan	0.0759	0.9993	0.0242	0.9367	−0.0007	1.0687
Feb	0.0664	1.0059	0.0231	0.9507	−0.0050	1.0584
Mar	0.0374	1.0068	0.0094	0.9695	0.0026	1.0089
Apr	0.0153	1.0671	0.0025	1.0019	−0.0000	1.0502
May	0.0002	1.1193	−0.0083	0.9967	−0.0121	1.0989
June	0.0051	1.1882	−0.0123	0.9934	−0.0188	1.0805
July	0.0016	1.3217	−0.0200	1.0107	−0.0179	1.1006
Aug	−0.0037	1.2752	−0.0247	1.0252	−0.0104	1.1006
Sept	0.0005	1.3107	−0.0239	1.0135	−0.0450	1.3521
Oct	0.0161	1.2781	−0.0162	1.0060	−0.0041	0.8734
Nov	0.0495	1.1316	0.0164	0.9134	0.0008	0.9619
Dec	0.0668	1.0522	0.0218	0.9209	0.0010	1.0715

As can be seen from Table 7, the Mean Error of the cokriging method is closest to 0 in the three interpolation methods, the minimum is −0.0007 in January, and the largest is t −0.0450 in September. While minimum value of the Mean Error of the ordinary kriging method is −0.0025, the maximum value is −0.0247. The Root-Mean-Square Standardized Error of the inverse distance weighting method and the Root-Mean-Square

Standardized Error of the ordinary kriging method is closed, therefore, the interpolation accuracy of the cokriging, which considering elevation values, is relatively good.

References

1. Wang, Y., Lin, X., Zou, Y.: Analysis of some characteristics of winter temperature anomalies in Fujian province. Fujian Meteorol. **1**, 7–11 (2012)
2. Liu, Z.-H., Li, L.-T., McVicar T.R., et al.: Introduction of the professional interpolation software for meteorology data: ANUSPLIN. Meteorol. Mon. **34**(2), 92–100 (2008)
3. Zheng, X.-B., Luo, Y.-X., Yu, F., et al.: Comparisons of spatial interpolation methods for agro-climate factors in complex mountain areas of southwest China. Chin. J. Agrometeorol. **29**(4), 458–462 (2008)
4. Xu, C., Wu, D.-Q., Zhang, Z.-G.: Comparative study of spatial interpolation methods on weather data in Shandong Province. J. Shandong Univ. (Nat. Sci.) **43**(3), 1–5 (2008)
5. Li, Y., He, Z., Liu, C.: Review of methods of temperature spatialization based on site observation data. Adv. Geogr. Sci. **33**(8), 1019–1028 (2014)
6. Ashraf, M., Loftis, J.C., Hubbard, K.G.: Application of geostatistics to evaluate partial weather station networks. Agric. For. Meteorol. **84**, 255–271 (1997)
7. Peng, B., Zhou, Y., Gao, P., et al. Applicability analysis of different spatial interpolation methods in temperature interpolation. J. Geogr. Inf. Sci. **13**(4), 539–548 (2011)
8. Wang, Z., Zhu, Y.: Study on spatial interpolation method of air temperature in Inner Mongolia area. Geospatial Inf. **12**(4), 1–6 (2014)
9. Chen, S., Li, X., Chen, W., et al.: Comparison of gridded methods of meteorological elements. J. Subtrop. Resour. Environ. **5**(4), 43–51 (2010)
10. Yang, F., Sun, Y., et al.: In Taiyi, nearly 10 years of temporal and spatial variation of temperature in Heilongjiang province and analysis. J. Geogr. Inf. Sci. **11**(5), 585–596 (2009)
11. Li, J., Huang, J.: Spatial interpolation method and spatial distribution of mean temperature in China in 1961–2000. Ecol. Environ. **15**(1), 109–114 (2006)
12. Shi, T., Yang, X., Zhang, T., et al.: BME spatiotemporal interpolation of precipitation in Fujian Province based on TRMM data. Geo Inf. Sci. **16**(3), 470–481 (2014)

The Quantitative Evaluation of a Translator's Translation Literacy

Mei Dong(✉) and Ying Xue

The School of Foreign Languages, Xi'an Shiyou University, Xi'an 710065, People's Republic of China
2548121615@qq.com

Abstract. The Evaluation of a translator's translation literacy is an essential part of translator training. In order to evaluate the progress made by each translator at various stages of the training, this study firstly carefully examines the existing definitions regarding translation literacy contributed by researchers both within and outside China from 1970s till 2015; and then devises an index system to objectively reflect the gradual progress of translation literacy of each translator. Analytic Hierarchy Process (AHP) is adopted when designing this index system for quantifying translation literacy. It is proved that the index system is student-oriented and able to indicate the individual differences of translation learners. The index system may not only be applied to personnel responsible for evaluation and management of translator training, but also to individual translation learner who expects to conduct self-evaluation on a regular basis.

Keywords: Translation literacy · Index system · AHP

1 Introduction

Nowadays, IT technology has exerted tremendous influence over translator training. In order to adapt to this trend, it is thought that transformation should be carried out accordingly in the aspect of translation learner evaluation. It is well-known that traditional outcome-oriented evaluation method cannot reflect the gradual progress made by translation learners and thus shall give way to process-based method which is believed to be more sensitive to the achievement of each translation learner at various stages of training. Besides, digital computing shall be adopted in this process in order to make the evaluation more reliable and objective. Under such a background, this study has devised a translation literacy evaluation index system with an aim to digitally represent and assess the gradual development of translation literacy of each translation learner. The system highlights the individual differences among translation learners and is able to quantitatively demonstrate their constantly evolutionary translation performance. It is hoped that such a system may be applied to evaluate the translation literacy of translation learners. It may also possibly be employed by translation learners themselves to conduct self-evaluation on a regular basis and meanwhile get a clear picture of their own translation capabilities in a timely manner.

N. Xiong et al. (Eds.): ISCSC 2018, AISC 877, pp. 514–518, 2019.
https://doi.org/10.1007/978-3-030-02116-0_60

2 An Approach to the Evaluation of Translation Literacy

So far, literature review indicates that people's understanding of translation competence has always been going towards two extremes. That is, it is either too objective or too subjective. Besides, researchers seem to have made things more and more complicated [8]. They have included those elements that belong to premises of translation, such as bilingualism, encyclopedic knowledge, and tool employment ability, to name just a few, into existing definitions of translation competence.

However, after careful examination of translation process re-collectedly and rationally, we have to admit that it is no wrong to say that translation competence is a cognitive concept that is related to a series of capabilities, including translation problem solving ability, translation strategy selection ability, linguistic aesthetic choosing ability and across-cultural communication ability as well.

Though translation competence is of a cognitive nature, we have to evaluate it in an objective way. A division have to be made between defining the concept and evaluating it. These are two distinctively different things and cannot be mixed together. The reason is quite clear. Since there is no scientific tools advanced enough to help us to assess something which is tactfully embedded in people's mind yet, we have no choice but to achieve this purpose by conducting the so-called backward infer. To put it much clearer, due to limitation in measuring instrument, we have to evaluate a translator's translation competence by referring to his translation products. In order to ensure the fairness of the conclusions, the whole evaluation process must be made objective and reasonable.

3 The Construction of Translator Attainment Evaluation Index System

Based on Professor Li's description of translation literacy, this study has designed a translation literacy evaluation index system. The system is hierarchical in structure. It is of three logically related levels with each level being represented by its particular indexes hereafter explained as follows.

3.1 The Hierarchical Structure of the Translation Literacy Evaluation Index System and Selection of Indexes

Since translation literacy is represented by language proficiency, knowledge literacy, aesthetic appreciation, translation strategy, digital literacy and communication skills as mentioned above, it is decided that they are selected as level-one indexes in the translation literacy evaluation index system.

As for the selection of Level-two and Level-three indexes, the SMART principle is strictly observed. The acronym of SMART stands for specific, measurable, attainable, realistic and time-bound. The hierarchical structure and indexes are shown in Table 1.

Table 1. The hierarchical structure and indexes of the Translation Literacy Evaluation Index System

Level one index	Level two index	Level three index
A. Language proficiency	A1. Pragmatics	A11. No negative transfer A12. No translationese A13. No source text comprehension mistakes A.14. No grammatical mistake A15. Conciseness A16. Accurate diction
	A2. Discourse	A21. Cohesion and coherence A22. Sign of restructuring in translated text
B. Knowledge literacy	B1. Successfully solve cultural conflict B2. Know about technologies of related discipline	
C. Aesthetic appreciation	C1. Reasonable when commenting on works translated by others C2. Reasonable when commenting on self-translated works	
D. Translation strategy	D1. Selection of appropriate translation strategies D2. The creativity of translation strategy	
E. Digital literacy	E1. Use of search engine E2. Use of translation software E3. Use of other digital technologies	
F. Communication skills	F1. Management skills	F11. Like to serve others F12. Sign of sound project management capabilities
	F2. Cooperation skills	F21. Sign of collaborative skills F22. Appropriately position oneself within a project team

3.2 Judgment Matrix and Consistency Test

A ten-member panel is set up. Five of them are expert translators who have more than 15 years of translation. The other five are professors whose research interest is translation. Firstly, each expert is given a printed copy of Table 2 (See Table 2). Then, they are explained about how to weigh each index according to Table 2. Secondly, the ten experts start to give weight to each index. The results are compared and discussed by the panel. Any dispute would lead to repeating discussion among experts, until consensus is reached.

Based on the weight given by the 10 experts, 25 judgment matrix were formed (see Appendix 1). With Matlab7.0, the consistency of each matrix Is tested. Due to the existence of subjective elements, the matrix consistency in this study is hard to achieve,

Table 2. The weighing standard based on the relative importance of each index

Implications	Weight
Xi is as important as *Xj*	1
Xi is a bit more important than *Xj*	3
Xi is more important than *Xj*	5
Xi is clearly more important than *Xj*	7
Xi is much more important than *Xj*	9
The importance of Xi over Xj is between every two levels mentioned above	2, 4, 6, 8

due to which the notion of C.R. is introduced. C. R. is the quotient of Consistency Index (CI) and Random Index (RI). If C.R. of a matrix is less than 0.1, it indicates that all the indexes within the matrix are consistent with each other in logic. In this case, the relative weight of importance of the matrix can be gained. If not, the matrix has to be reconstructed.

3.3 The Calculation of the Weight of Importance

After going through consistency test, the relative weight of importance of each index within the 25 matrix are calculated, based on which a complete translator attainment evaluation system is obtained (see Table 3).

Table 3. Translator attainment evaluation system

Level one index	Level two index	Level three index
A. 30.25	A1. 25.20	A11. 0.77616 A12. 6.85944 A13 8.4924 A.14 6.85944 A15 1.25244 A16 0.9576
	A2. 5.05	A21. 1.2625 A22. 3.7875
B. 28.18	B1. 14.0730 B2. 8.000302	
C. 26.11	C1. 13.055 C2. 13.055	
D. 6.73	D1. 1.6825 D2. 5.0475	
E. 6.73	E1. 2.2433 E2. 2.2433 E3. 2.2433	
F. 2.01	F1. 1.005	F11. 0.7537 F12. 0.25125
	F2. 1005	F21. 0.7537 F22. 0.25125

4 Conclusion

The translator attainment evaluation system is a multi-source evaluation system, which is thus relatively more objective and reliable. It may reflect the evolution of a translator's translation accomplishment at different stages. With analytic hierarchy process, the system may control subjective impact on translation evaluation process to a large extent, and be more fair and fine-grained. It is hoped that the system can be widely adopted and trialed in various translator training centers. The researcher is looking forward to constructive suggestions regarding improvement of the system.

References

1. Colina, S.: Translation Teaching: From Research to the Classroom. McG raw-H ill, Boston (2003)
2. Gutt, E.A.: Challenges of metarepresentation to translation competence. In: Fleischmann, E., Schmitt, P.A., Wotjak, G. (eds.) Translationskocompetenz, Taguangsberichte der LICTRA (Leipzig International Conference on Translation Studies 4), pp. 77–89. Stauffenberg Tubingen (2014)
3. Haris, B.: The importance of natural translation. Working Papers on Bilingualism, no. 12, pp. 96–114 (1977)
4. Lin, L., Lin, G.: The application of fuzzy mathematics and AHP in performance evaluation. China Manag. Inf. **9**(11), 13–16 (2006)
5. Neubert, A.: Competence in language, in languages and in translation. In: Schaffner, C., Adab, B. (eds.) Developing Translation Competence, pp. 3–18. John Benjamins, Amsterdam (2000)
6. PACTE Group, Building a translation competency model, In: Vlves, F. (ed.) Triangulating Translation: Perspectives in Process-Oriented Research, pp. 43–66. John Benjamins, Amsterdam (2003)
7. Pym, A.: Redefining translation competence in an electronic age: in defense of a minimalist approach. Meta (XLVIII-4), 481–497 (2003)
8. Ruilin, Li: From translation competency to translator's attainment. Chin. Transl. J. **33**(1), 46–50 (2011)
9. Li, R., He. Y.: Research on translation project learning mode from the perspective of learning sciences. Foreign Lang. Educ. **32**(1), 94–98 (2011)
10. Risku, H.A.: Cognitive scientific view on technical communication and translation: Doembodiment and situatedness really make a difference? Target **22**(1), 94–111 (2010)
11. Rothe-Neves, R.: Notes on the concept of translator competence. Quaderns Traduccio **4**, 125–138 (2007)
12. Shreve, G.M.: Cognition and the evolution of translation competence. In: Danks, J.H., Shreve, G.M., Fountain, S.B., Mcbeath, M.K. (eds.) Cognitive Processes in Translation and Interpreting, pp. 120–136. Sage, Thousand Oaks (1997)
13. Wilss, W.: Perspectives and limitations of a didactic framework for the teaching of translation. In: Brislin, R.W. (ed.) Translation, Application and Research, pp. 117–137. Gardner Press, New York (1976)

Author Index

N. Xiong et al. (Eds.): ISCSC 2018, AISC 877, pp. 519–521, 2019.
https://doi.org/10.1007/978-3-030-02116-0

Printed in the United States
By Bookmasters